线性系统理论与设计
(双语版)
Linear Systems Theory and Design
(Chinese & English Bilingual Edition)

江驹　史爽　彭聪　陈谋　编著

科学出版社

北京

内 容 简 介

本书内容涵盖控制相关学科各专业所必需的基础知识，以时域中的线性系统理论知识为主要内容，同时兼顾控制的频域知识。主要内容包括系统的数学描述、系统的动态响应、系统的能控性和能观性、系统的最小实现、系统的稳定性、系统的时域综合等。本书在内容论述上力求精练，在概念叙述上力求清晰，在理论分析上力求严谨，在系统设计方法和算法介绍上力求实用，在例证说明上力求简明。

本书可以作为控制科学与工程及相关学科各专业的高年级本科生和研究生的教材，也可供相关专业的科技工作者参考。

图书在版编目(CIP)数据

线性系统理论与设计：双语版：汉、英 / 江驹等编著.—北京：科学出版社，2024.5
ISBN 978-7-03-076632-8

Ⅰ.①线⋯ Ⅱ.①江⋯ Ⅲ.①线性系统理论–双语教学–高等学校–教材–汉、英 Ⅳ.①O231.1

中国国家版本馆 CIP 数据核字(2023)第 187301 号

责任编辑：惠 雪 曾佳佳 / 责任校对：郝璐璐
责任印制：张 伟 / 封面设计：许 瑞

科学出版社 出版
北京东黄城根北街 16 号
邮政编码：100717
http://www.sciencep.com

中煤（北京）印务有限公司印刷
科学出版社发行 各地新华书店经销

*

2024 年 5 月第 一 版　开本：787 × 1092　1/16
2024 年 5 月第一次印刷　印张：27 3/4
字数：655 000

定价：109.00 元
(如有印装质量问题，我社负责调换)

前　言

在控制科学与工程及相关领域，线性系统是最基本的研究对象。对于线性系统的研究已经形成一套完整且成熟的理论体系。线性系统理论中的许多概念和方法，对于非线性系统研究、鲁棒控制、最优控制、自适应控制、模糊控制、随机控制、系统辨识、数字滤波等都具有重要的支撑作用，成为深入研究控制科学各分支必备的基础知识。因此，线性系统理论是控制专业本科生和研究生以及许多与控制相关专业的学生应具备的基础知识。

本书内容涵盖控制相关学科各专业所必需的基础知识，以时域中的线性系统理论知识为主要内容，同时兼顾控制的频域知识。主要内容包括系统的数学描述、系统的动态响应、系统的能控性和能观性、系统的最小实现、系统的稳定性、系统的时域综合等。本书在内容论述上力求精练，在概念叙述上力求清晰，在理论分析上力求严谨，在系统设计方法和算法介绍上力求实用，在例证说明上力求简明。

本书是为了适应研究生国际化培养的需求，在姜长生教授主编的《线性系统理论与设计》的基础上重新编著的中英双语教材。特别感谢姜长生教授、吴庆宪教授对本书编写的大力支持。本书在编写过程中参考了国内外同行的相关著作和文献，并引用了他们的成果和论述，感谢书中所引文献的作者们。

由于作者水平有限，书中疏漏和不当之处在所难免，热忱欢迎来自各方面的批评、指教。衷心感谢每一位提出批评和指教的读者。

作　者
2023 年 12 月 7 日

Preface

The study of linear systems is the foundation for the control field, and has been formed as a complete and mature theoretical system. Many concepts and methods in the linear systems theory are prerequisites for nonlinear systems, robust control, optimal control, adaptive control, fuzzy control, stochastic control, system identification, digital filtering and some other disciplines. Therefore, mastering the linear systems theory is essential for students in control science and engineering and relevant majors.

This text focuses on the linear systems theory in both time and frequency domains, which is the cornerstone of the modern control theory. The main contents include mathematical description, response, controllability, observability, minimal realization, stability and synthesis in time domain of linear systems. Authors try to present refined contents, clear concepts, rigorous theoretical analysis, practical system design procedures and concise examples.

To satisfy the requirement of the international talent cultivation, this text is rewritten to be Chinese-English bilingual based on "Linear Systems Theory and Design" compiled by Professor Jiang Changsheng. We are indebted to Professor Jiang Changsheng and Professor Wu Qingxian for revising this text. Lots of pertinent literature and works are referred and cited to complete this text. Thank them all.

Owing to the limited level of editing, errors and defects are inevitable. Thanks a lot for guidance and suggestions.

Editors

December 7, 2023

目　录

第 1 章　线性系统的数学描述 ··· 1
  1.1　基本概念 ·· 1
    1.1.1　因果性 ··· 2
    1.1.2　线性 ·· 3
    1.1.3　时不变性 ·· 3
  1.2　系统的输入-输出描述 ·· 4
    1.2.1　线性系统的脉冲响应 ··· 5
    1.2.2　线性定常系统的传递函数矩阵 ································ 7
    1.2.3　传递函数矩阵的 Smith-McMillan 形 ························ 8
    1.2.4　传递函数矩阵的极点和零点 ································· 13
  1.3　线性系统的状态空间描述 ·· 14
    1.3.1　线性系统的动态方程 ·· 14
    1.3.2　线性定常系统的特征结构 ····································· 17
    1.3.3　根据状态空间描述求传递函数矩阵 ······················· 18
    1.3.4　动态系统的等价 ·· 22

第 2 章　线性系统的动态响应 ··· 31
  2.1　概述 ·· 31
  2.2　线性时变系统的动态响应 ·· 32
    2.2.1　线性时变系统的状态转移矩阵 ······························ 33
    2.2.2　线性时变系统的零输入响应 ································· 37
    2.2.3　线性时变系统的零状态响应 ································· 38
    2.2.4　线性时变系统的状态响应和输出响应 ···················· 39
  2.3　线性定常系统的动态响应 ·· 41
    2.3.1　线性定常系统的矩阵指数 ····································· 41
    2.3.2　线性定常系统的零输入响应 ································· 48
    2.3.3　线性定常系统的零状态响应 ································· 50
    2.3.4　线性定常系统的状态响应和输出响应 ···················· 51

第 3 章　线性系统的能控性和能观性 ······································ 54
  3.1　线性系统的能控性 ·· 54
    3.1.1　基本概念 ·· 54
    3.1.2　线性时变系统的能控性判据 ································· 56
    3.1.3　线性定常系统的能控性判据 ································· 60
    3.1.4　线性定常系统的能控性指数 ································· 66

3.2 线性系统的能观性 ·· 69
　　3.2.1 基本概念 ·· 69
　　3.2.2 线性时变系统的能观性判据 ·· 71
　　3.2.3 线性定常系统的能观性判据 ·· 73
　　3.2.4 线性定常系统的能观性指数 ·· 76
3.3 线性系统的对偶原理 ··· 78
　　3.3.1 对偶系统 ·· 78
　　3.3.2 对偶原理 ·· 79
3.4 线性定常系统的结构分解 ·· 80
　　3.4.1 线性系统在等价变换下的能控性和能观性 ································· 81
　　3.4.2 线性定常系统的能控性结构分解 ··· 81
　　3.4.3 线性定常系统的能观性结构分解 ··· 85
　　3.4.4 线性定常系统的Kalman分解 ·· 86
　　3.4.5 极点、零点和特征值间的关系 ··· 89

第4章 线性定常系统的标准形和实现 ··· 94
4.1 线性定常系统的标准形 ·· 94
　　4.1.1 线性定常SISO系统的标准形 ·· 94
　　4.1.2 线性定常MIMO系统的标准形 ·· 99
4.2 能控性和能观性的频域形式 ·· 107
4.3 线性定常系统的实现 ·· 108
　　4.3.1 基本概念 ··· 108
　　4.3.2 线性定常SISO系统的实现 ·· 109
　　4.3.3 线性定常MIMO系统的实现 ·· 111
4.4 线性定常系统的最小实现 ··· 114
　　4.4.1 基本概念 ··· 114
　　4.4.2 线性定常SISO系统的最小实现 ··· 118
　　4.4.3 线性定常MIMO系统的最小实现 ··· 121

第5章 线性系统的稳定性 ··· 125
5.1 Lyapunov稳定理论 ·· 125
　　5.1.1 基本概念 ··· 125
　　5.1.2 Lyapunov第二法的主要定理 ··· 129
5.2 线性系统的内部稳定性 ·· 134
　　5.2.1 线性时变系统的内部稳定性 ·· 134
　　5.2.2 线性定常系统的内部稳定性 ·· 138
5.3 线性系统的外部稳定性 ·· 142
　　5.3.1 线性系统的BIBO稳定性 ··· 142
　　5.3.2 线性系统的BIBS稳定性 ··· 144
5.4 外部稳定和内部稳定的关系 ··· 145

第 6 章 线性系统的时域综合与反馈控制 ··· 148
6.1 状态反馈控制 ··· 148
6.1.1 状态反馈系统的能控性和能观性 ··· 148
6.1.2 SISO 系统的状态反馈特征值配置 ··· 150
6.1.3 MIMO 系统的状态反馈特征值配置 ··· 154
6.1.4 状态反馈镇定 ··· 162
6.2 输出反馈控制 ··· 166
6.2.1 输出反馈系统的能控性和能观性 ··· 166
6.2.2 输出反馈特征值配置 ··· 167
6.2.3 动态输出反馈补偿器 ··· 174
6.3 解耦控制 ··· 177
6.3.1 基本概念 ··· 177
6.3.2 状态反馈解耦 ··· 180
6.3.3 稳态解耦 ··· 186
6.4 状态观测器 ··· 188
6.4.1 全维状态观测器 ··· 188
6.4.2 降维状态观测器 ··· 191
6.4.3 函数状态观测器 ··· 195
6.4.4 带状态观测器的动态系统 ··· 198

参考文献 ··· 203

Contents

Chapter 1 Mathematical Description of Linear Systems ·················· 205
  1.1 Basic Concepts ·· 205
      1.1.1 Causality ·· 206
      1.1.2 Linearity ·· 207
      1.1.3 Time-Invariance ·· 208
  1.2 Input-Output Description of Linear Systems ···················· 209
      1.2.1 Impulse Response of Linear Systems ······················ 209
      1.2.2 Transfer Function Matrix of Linear Time-Invariant Systems ·········· 212
      1.2.3 Smith-McMillan Form of Transfer Function Matrices ·················· 213
      1.2.4 Poles and Zeros of Transfer Function Matrices ·············· 218
  1.3 State Space Description of Linear Systems ······················ 219
      1.3.1 Dynamical Equations of Linear Systems ·················· 220
      1.3.2 Characteristic Structures of Linear Time-Invariant Systems ············ 223
      1.3.3 Deriving Transfer Function Matrix from State Space Description ·········· 224
      1.3.4 Equivalence of Dynamical Systems ························ 228
  Assignments ·· 236

Chapter 2 Response of Linear Dynamical Systems ·················· 238
  2.1 Introduction ·· 238
  2.2 Response of Linear Time-Varying Dynamical Systems ·············· 240
      2.2.1 State Transition Matrix of Linear Time-Varying Systems ·············· 240
      2.2.2 Zero-Input Response of Linear Time-Varying Systems ············ 245
      2.2.3 Zero-State Response of Linear Time-Varying Systems ············ 246
      2.2.4 State Response and Output Response of Linear Time-Varying Systems ········ 247
  2.3 Response of Linear Time-Invariant Dynamical Systems ·················· 249
      2.3.1 Matrix Exponential of Linear Time-Invariant Systems ·················· 250
      2.3.2 Zero-Input Response of Linear Time-Invariant Systems ················ 257
      2.3.3 Zero-State Response of Linear Time-Invariant Systems ·················· 259
      2.3.4 State Response and Output Response of Linear Time-Invariant Systems ········ 260
  Assignments ·· 261

Chapter 3 Controllability and Observability of Linear Systems ·············· 264
  3.1 Controllability of Linear Systems ································ 264
      3.1.1 Basic Concepts ·· 264
      3.1.2 Controllability Criteria of Linear Time-Varying Systems ·············· 267

    3.1.3 Controllability Criteria of Linear Time-Invariant Systems ············ 270

    3.1.4 Controllability Index of Linear Time-Invariant Systems ············· 277

  3.2 Observability of Linear Systems ·· 279

    3.2.1 Basic Concepts ··· 279

    3.2.2 Observability Criteria of Linear Time-Varying Systems ············ 282

    3.2.3 Observability Criteria of Linear Time-Invariant Systems ·········· 284

    3.2.4 Observability Index of Linear Time-Invariant Systems ············· 287

  3.3 Duality Principle of Linear Systems ······································ 289

    3.3.1 Dual Systems ·· 290

    3.3.2 Duality Principle ·· 291

  3.4 Structure Decomposition of Linear Time-Invariant Systems ············· 292

    3.4.1 Controllability and Observability of Linear Systems with Equivalent

      Transformation ·· 292

    3.4.2 Controllability Structure Decomposition of Linear Time-Invariant Systems ····· 293

    3.4.3 Observability Structure Decomposition of Linear Time-Invariant Systems ······ 296

    3.4.4 Kalman's Decomposition of Linear Time-Invariant Systems ········· 298

    3.4.5 Relationships among Poles, Zeros and Eigenvalues ················· 301

  Assignments ··· 305

Chapter 4 Canonical Form and Realization of Linear Time-Invariant Systems ········· 308

  4.1 Canonical Form of Linear Time-Invariant Systems ······················· 308

    4.1.1 Canonical Form of Linear Time-Invariant SISO Systems ············ 308

    4.1.2 Canonical Form of Linear Time-Invariant MIMO Systems ·········· 313

  4.2 Controllability and Observability in Frequency Domain ················· 321

  4.3 Realization of Linear Time-Invariant Systems ···························· 323

    4.3.1 Basic Concepts ··· 323

    4.3.2 Realization of Linear Time-Invariant SISO Systems ················ 324

    4.3.3 Realization of Linear Time-Invariant MIMO Systems ·············· 326

  4.4 Minimal Realization of Linear Time-Invariant Systems ·················· 329

    4.4.1 Basic Concepts ··· 329

    4.4.2 Minimal Realization of Linear Time-Invariant SISO Systems ······· 334

    4.4.3 Minimal Realization of Linear Time-Invariant MIMO Systems ····· 337

  Assignments ··· 340

Chapter 5 Stability of Linear Systems ·· 342

  5.1 Lyapunov Stability Theory ·· 342

    5.1.1 Basic Concepts ··· 342

    5.1.2 Main Theorems of Lyapunov's Second Method ···················· 346

  5.2 Internal Stability of Linear Systems ······································· 351

		5.2.1	Internal Stability of Linear Time-Varying Systems ························· 352
		5.2.2	Internal Stability of Linear Time-Invariant Systems ······················· 356
	5.3	External Stability of Linear Systems ·· 360	
		5.3.1	BIBO Stability of Linear Systems ··· 361
		5.3.2	BIBS Stability of Linear Systems ·· 363
	5.4	Relationships between External Stability and Internal Stability ············ 364	
	Assignments ··· 366		

Chapter 6 Synthesis and Feedback Control of Linear Systems in Time Domain ··· 369

- 6.1 State Feedback Control ··· 369
  - 6.1.1 Controllability and Observability of System with State Feedback ············· 369
  - 6.1.2 Eigenvalue Assignment with State Feedback for SISO Systems ··············· 371
  - 6.1.3 Eigenvalue Assignment with State Feedback for MIMO Systems ············· 375
  - 6.1.4 Stabilization with State Feedback ·· 384
- 6.2 Output Feedback Control ·· 388
  - 6.2.1 Controllability and Observability of System with Output Feedback ············ 388
  - 6.2.2 Eigenvalue Assignment with Output Feedback ································ 389
  - 6.2.3 Dynamic Output Feedback Compensator ····································· 396
- 6.3 Decoupling Control ··· 399
  - 6.3.1 Basic Concepts ·· 399
  - 6.3.2 Decoupling with State Feedback ·· 403
  - 6.3.3 Steady State Decoupling ··· 408
- 6.4 State Observer ··· 410
  - 6.4.1 Full-Order State Observer ·· 410
  - 6.4.2 Reduced-Order State Observer ·· 414
  - 6.4.3 Functional State Observer ·· 418
  - 6.4.4 Dynamical Systems with State Observer Feedback ·························· 421
- Assignments ·· 425

第 1 章

线性系统的数学描述

根据系统运动过程的物理或化学规律所写出的、描述系统动态或静态行为的数学表达式，称为系统的数学描述或数学模型。实际系统的数学模型通常是非线性的，为了便于研究，在系统的工作点附近可以将其看作是线性的。也就是说，可以在工作点附近对系统进行线性化处理，这对于解决多数实际的工程问题是足够精确的。这一线性化处理后的数学模型被称为线性模型，由线性模型描述的系统被称为线性系统。线性系统理论不研究具体的物理或化学系统，而是在线性数学模型的基础上研究系统的结构性质以及分析与设计系统的方法。

描述系统的方法可以分为实变量法和复变量法两大类。最常用的实变量法是状态空间描述，它以时间 t 作为实变量并直接研究系统行为随时间的变化。最常用的复变量法则是传递函数矩阵描述，它以和频域相关的复变量 s 作为变元，并通过研究系统随输入频率的变化所表现出的行为来研究系统特性，这一方法也通常被称为频域法。

1.1 基本概念

如果以连续信号作为系统的输入，同时系统的输出也是连续信号，那么这一系统被称作连续系统。本书只考虑连续系统。如果系统仅有一个输入端和一个输出端，那么称之为单变量系统或单输入单输出（single-input single-output，SISO）系统，经典控制理论主要研究的就是这类系统。如果系统具有多个输入端和（或）输出端，则称之为多变量系统。特别地，系统具有多个输入端和多个输出端时，称之为多输入多输出（multi-input multi-output，MIMO）系统。

考虑一个系统具有 p 个输入和 q 个输出，见图 1.1，输入向量记为 $u = [u_1 \ u_2 \ \cdots u_p]^{\mathrm{T}} \in \mathbb{R}^{p \times 1}$，输出向量记为 $y = [y_1 \ y_2 \ \cdots \ y_q]^{\mathrm{T}} \in \mathbb{R}^{q \times 1}$。

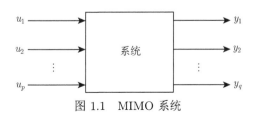

图 1.1　MIMO 系统

1.1.1 因果性

如果系统在 t_0 时刻的输出 $y(t_0)$ 仅取决于 t_0 时刻的输入,那么称该系统为无记忆系统。此时 $y(t_0)$ 与 t_0 时刻之前和之后的输入都无关。仅由电阻构成的电路网络就是一类无记忆系统。然而,大多数系统都不是无记忆系统,也就是说,系统在 t_0 时刻的输出不仅取决于 t_0 时刻的输入,也取决于 t_0 时刻之前和(或)之后的输入,即系统当前的输出可能和过去、现在及未来的输入都相关,这类系统被称为记忆系统。

如果系统当前的输出仅取决于过去和现在的输入,且不受未来的输入影响,则称该系统为因果系统。非因果系统当前时刻的输出也会受未来输入的影响,也就是说,非因果系统可以预测未来的输入。事实上,实际的物理系统都不具有这种特性。任何实际的物理过程,其结果不可能在引起这种结果的原因之前产生,所以实际的物理系统都具有因果性。本书只考虑因果系统。

一般地,对于一个因果系统,从 $-\infty$ 到当前时刻 t 的输入都会影响当前的输出 $y(t)$,即因果系统当前时刻的输出会受到过去所有输入的影响。但是,从 $t=-\infty$ 开始追踪系统的输入是非常不方便的,因此,状态的概念被提出用以解决这一问题。系统在 t_0 时刻的信息被称为系统在 t_0 时刻的状态 $x(t_0)$。由系统在 t_0 时刻的状态 $x(t_0)$ 和系统的输入 $u(t)$,$t \geqslant t_0$,可以唯一确定系统的输出 $y(t)$,$t \geqslant t_0$。也就是说,在已知系统的状态 $x(t_0)$ 的情况下,无须知道 t_0 时刻之前的输入,即可确定系统在 t_0 时刻之后的输出。例如,考虑如图 1.2 所示的电路网络,如果已知在 t_0 时刻两个电容两端的电压 $x_1(t_0)$ 和 $x_2(t_0)$ 以及通过电感的电流 $x_3(t_0)$,那么对于任意的在 t_0 时刻后施加的输入,系统的输出都可以唯一确定。这一电路网络在 t_0 时刻的状态是 $x(t_0) = [\ x_1(t_0)\ \ x_2(t_0)\ \ x_3(t_0)\ ]^{\mathrm{T}} \in \mathbb{R}^3$。状态 x 中的每个元被称为状态变量。一般情况下,初始状态可以被简单地看作是初始条件的集合。

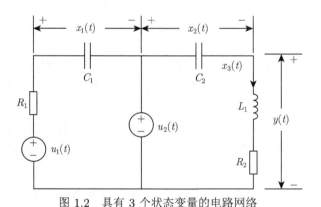

图 1.2 具有 3 个状态变量的电路网络

基于 t_0 时刻的状态,系统的输入和输出关系可以被描述为

$$\left.\begin{array}{r} x(t_0) \\ u(t), t \geqslant t_0 \end{array}\right\} \to y(t),\ t \geqslant t_0 \tag{1.1}$$

它表明系统的输出同时受到初始状态 $x(t_0)$ 和 t_0 时刻之后的输入影响。这种情况下,不再

1.1 基本概念

需要获取从 $-\infty$ 到 t_0 时刻的输入,可以更加方便地研究系统的行为。一般地,式 (1.1) 被称作一个状态-输入-输出对。

1.1.2 线性

如果一个系统的任意两个状态-输入-输出对

$$\left.\begin{aligned} x_1(t_0) \\ u_1(t), t \geqslant t_0 \end{aligned}\right\} \to y_1(t),\ t \geqslant t_0$$

$$\left.\begin{aligned} x_2(t_0) \\ u_2(t), t \geqslant t_0 \end{aligned}\right\} \to y_2(t),\ t \geqslant t_0$$

对于任意时刻 t_0 和任意实数 α,

$$\left.\begin{aligned} x_1(t_0) + x_2(t_0) \\ u_1(t) + u_2(t), t \geqslant t_0 \end{aligned}\right\} \to y_1(t) + y_2(t),\ t \geqslant t_0$$

$$\left.\begin{aligned} \alpha x_1(t_0) \\ \alpha u_1(t), t \geqslant t_0 \end{aligned}\right\} \to \alpha y_1(t),\ t \geqslant t_0$$

总是成立,该系统被称为线性系统。这里,前一条件所描述的性质被称为可加性,后一条件所描述的性质被称为齐次性。两者合称为叠加性,可以被描述为

$$\left.\begin{aligned} \alpha_1 x_1(t_0) + \alpha_2 x_2(t_0) \\ \alpha_1 u_1(t) + \alpha_2 u_2(t), t \geqslant t_0 \end{aligned}\right\} \to \alpha_1 y_1(t) + \alpha_2 y_2(t),\ t \geqslant t_0$$

其中,α_1 和 α_2 是任意两个常数。不满足叠加性的系统被称为非线性系统。

需要注意的是,可加性和齐次性是两个不可替代的概念。具有齐次性的系统不一定具有可加性,反之亦然。例如,考虑一个 SISO 系统,对于任意时刻 t,它的输入和输出满足

$$y(t) = \begin{cases} \dfrac{u^2(t)}{u(t-1)}, & u(t-1) \neq 0 \\ 0, & u(t-1) = 0 \end{cases}$$

可以很容易验证这一系统具有齐次性,但不具有可加性。

1.1.3 时不变性

如果系统的任意状态-输入-输出对

$$\left.\begin{aligned} x(t_0) \\ u(t), t \geqslant t_0 \end{aligned}\right\} \to y(t),\ t \geqslant t_0$$

对任意的常值 T，总是有

$$\left.\begin{array}{r}x(t_0+T)\\u(t-T), t \geqslant t_0+T\end{array}\right\} \to y(t-T),\ t \geqslant t_0+T$$

那么这一系统具有时不变性，称其为定常系统。

上述定义表明，如果初始状态位移到 t_0+T，同时从 t_0+T 而非 t_0 时刻为系统施加具有相同波形的输入，则其输出信号从 t_0+T 时刻开始出现，波形保持不变。也就是说，对于定常系统，不论何时施加输入信号，输出信号的波形总是相同的，如图 1.3 所示。所以不失一般性地，对于定常系统，总是可以假设初始时刻 $t_0=0$。不满足时不变性的系统被称为时变系统。

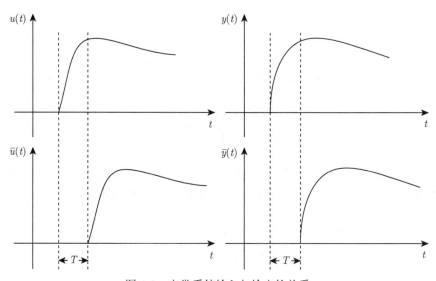

图 1.3　定常系统输入与输出的关系

例如，图 1.2 中所示的电路网络，如果 R_1，R_2，C_1，C_2 和 L_1 都是常值，那么这一系统是定常的，否则，该系统是时变的。

值得一提的是，许多实际的物理系统在有限的时间区间内都可以被建模为定常系统。但是一些物理系统必须被建模为时变系统，一个典型的例子是飞行中的火箭，它的质量会随着时间迅速下降。

1.2　系统的输入-输出描述

系统的输入-输出描述揭示了系统的输入与输出之间的数学关系。当建立系统的输入-输出描述时，可以认为系统的内部特性是未知的，即系统可以被看作是一个黑箱。对一个系统施加不同的输入信号，可以测量到相应的输出信号，根据这些输入-输出数据，可以确定系统的输入-输出描述。

1.2 系统的输入-输出描述

当线性系统采用输入-输出描述时，通常假设在给定时刻 t_0，系统的输出 $y(t)$，$t \geqslant t_0$ 可以由输入 $u(t)$，$t \geqslant t_0$ 唯一确定，此时称系统在 t_0 时刻松弛。实际应用中，总是可以假设系统在 $t = -\infty$ 时松弛。这种情况下，如果对系统施加输入 $u(t)$，$t \in (-\infty, \infty)$，系统的输出 $y(t)$ 可以由这一输入 $u(t)$ 唯一确定，$t \in (-\infty, \infty)$。事实上，系统的松弛性可以被看作零初始条件。

1.2.1 线性系统的脉冲响应

系统的脉冲响应是一类重要的输入-输出描述。

考虑图 1.4 所示的脉冲函数 $\delta_\Delta(t - t_1)$，它可以被描述为

$$\delta_\Delta(t - t_1) = \begin{cases} 0, & t < t_1 \\ \dfrac{1}{\Delta}, & t_1 \leqslant t \leqslant t_1 + \Delta \\ 0, & t > t_1 + \Delta \end{cases}$$

对于所有的 Δ，$\delta_\Delta(t - t_1)$ 的面积始终为 1。

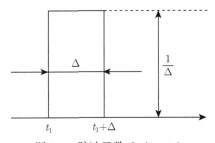

图 1.4 脉冲函数 $\delta_\Delta(t - t_1)$

如图 1.5 所示，每个分段连续的输入信号 $u(t)$ 均可以用一系列脉冲函数来逼近，由于 $\delta_\Delta(t - t_i)\Delta$ 的值为 1，输入 $u(t)$ 可以表示为

$$u(t) \approx \sum_i u(t_i) \delta_\Delta(t - t_i) \Delta \tag{1.2}$$

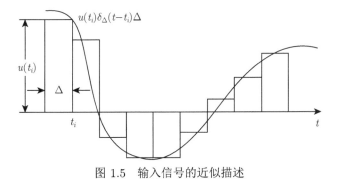

图 1.5 输入信号的近似描述

把线性 SISO 系统在 t 时刻由脉冲 $\delta_\Delta(t-t_i)$ 引起的输出记作 $g_\Delta(t,t_i)$。由于线性系统具有齐次性，这一系统由 $\delta_\Delta(t-t_i)u(t_i)\Delta$ 引起的输出可以表示为 $g_\Delta(t,t_i)u(t_i)\Delta$。同时，由于线性系统具有可加性，由 $\sum_i \delta_\Delta(t-t_i)u(t_i)\Delta$ 引起的输出则可以表示为 $\sum_i g_\Delta(t,t_i)u(t_i)\Delta$。所以，由输入 $u(t)$（式 (1.2)）引起的输出可以被表示为

$$y(t) \approx \sum_i g_\Delta(t,t_i)u(t_i)\Delta \tag{1.3}$$

如果 $\Delta \to 0$，脉冲 $\delta_\Delta(t-t_i)$ 可以被看作是 t_i 时刻的理想脉冲，用 $\delta(t-t_i)$ 表示，相应的输出记作 $g(t,t_i)$。随着 $\Delta \to 0$，式 (1.3) 中的近似关系转变为等式关系，求和运算变为积分运算，离散时刻 t_i 变为连续时间，用 τ 表示，Δ 可以被写作 $\mathrm{d}\tau$。因此，式 (1.3) 可以被表示为

$$y(t) = \int_{-\infty}^{\infty} g(t,\tau)u(\tau)\mathrm{d}\tau \tag{1.4}$$

注意到 $g(t,\tau)$ 是一个具有两个变量的函数，变量 t 代表了观测输出的时刻，变量 τ 则代表理想脉冲作用到系统的时刻。由于 $g(t,\tau)$ 是由脉冲信号引起的响应，所以通常称之为脉冲响应。

式 (1.4) 可以被进一步表示为

$$y(t) = \int_{-\infty}^{t_0} g(t,\tau)u(\tau)\mathrm{d}\tau + \int_{t_0}^{t} g(t,\tau)u(\tau)\mathrm{d}\tau + \int_{t}^{\infty} g(t,\tau)u(\tau)\mathrm{d}\tau$$

如果系统具有因果性，输出不会在输入信号施加之前出现。因此，当 $t < \tau$ 时，$g(t,\tau) = 0$。进一步可知 $\int_{t}^{\infty} g(t,\tau)u(\tau)\mathrm{d}\tau = 0$。同时，如果系统在 t_0 时刻松弛，输出 $y(t)$，$t \geqslant t_0$ 由输入 $u(t)$，$t \geqslant t_0$ 唯一地确定。所以 $\int_{-\infty}^{t_0} g(t,\tau)u(\tau)\mathrm{d}\tau = 0$。综上所述，一个因果的、在 t_0 时刻松弛的线性 SISO 系统可以被描述为

$$y(t) = \int_{t_0}^{t} g(t,\tau)u(\tau)\mathrm{d}\tau \tag{1.5}$$

如果一个因果的、在 t_0 时刻松弛的线性系统具有 p 个输入和 q 个输出，可以拓展式 (1.5) 来描述这一系统，

$$y(t) = \int_{t_0}^{t} G(t,\tau)u(\tau)\mathrm{d}\tau \tag{1.6}$$

其中，

$$G(t,\tau) = \begin{bmatrix} g_{11}(t,\tau) & \cdots & g_{1p}(t,\tau) \\ \vdots & & \vdots \\ g_{q1}(t,\tau) & \cdots & g_{qp}(t,\tau) \end{bmatrix}$$

1.2 系统的输入-输出描述

$G(t,\tau)$ 被称作系统的脉冲响应矩阵。$g_{ij}(t,\tau)$ 是 $G(t,\tau)$ 的元，它的物理意义是仅对系统的第 j 个输入端在 τ 时刻施加一个理想脉冲，在 t 时刻系统的第 i 个输出端的响应。简言之，$g_{ij}(t,\tau)$ 是第 i 个输出端对第 j 个输入端的脉冲响应。

考虑一个因果的、在 t_0 时刻松弛的线性定常 SISO 系统，在 τ 时刻对系统施加理想脉冲信号，系统在 t 时刻观测到的输出是 $g(t,\tau)$。如果施加脉冲信号的时间是 $\tau+T$（T 是一个实数），系统在 $t+T$ 时刻观测到的输出将和 $g(t,\tau)$ 相等，即

$$g(t+T,\tau+T) = g(t,\tau)$$

这一等式对任意的 t, τ 和 T 均成立。如果 $T = -\tau$，这一等式可以被写作

$$g(t,\tau) = g(t-\tau,0)$$

方便起见，把 $g(t-\tau,0)$ 记作 $g(t-\tau)$。这一结果也可以拓展到线性定常 MIMO 系统中，即对任意的 t 和 τ，

$$G(t,\tau) = G(t-\tau,0) = G(t-\tau)$$

对于定常情形，总是可以假设在 $t_0 = 0$ 时才施加输入信号。所以，对于一个因果的、在 t_0 时刻松弛的线性定常系统，它的输入-输出描述可以表示为

$$y(t) = \int_0^t G(t-\tau)u(\tau)\mathrm{d}\tau = \int_0^t G(\tau)u(t-\tau)\mathrm{d}\tau \tag{1.7}$$

1.2.2 线性定常系统的传递函数矩阵

在经典控制理论中，传递函数是系统的一类重要的数学描述。一个因果的、在 t_0 时刻松弛的线性定常 SISO 系统的输入-输出描述是

$$y(t) = \int_0^t g(t-\tau)u(\tau)\mathrm{d}\tau$$

取 Laplace（拉普拉斯）变换，可以得到[①]

$$y(s) = \mathcal{L}[y(t)] = \int_0^\infty y(t)\mathrm{e}^{-st}\mathrm{d}t$$

根据 Laplace 变换的卷积定理，可以进一步得到

$$y(s) = g(s)u(s)$$

其中，$g(s)$ 被称作系统的传递函数。可以看出，传递函数的 Laplace 反变换是脉冲响应，反之，脉冲响应的 Laplace 变换是传递函数。

[①] 注意到 $y(t)$ 和 $y(s)$ 是具有不同参数的不同向量函数。但是方便起见，这里使用相同的符号 y 表示。

经典控制理论中传递函数的概念可以拓展应用到 MIMO 系统中。一个因果的、在 t_0 时刻松弛的线性定常 MIMO 系统，它的输入-输出描述（式 (1.7)）的 Laplace 变换是

$$y(s) = \mathcal{L}[y(t)] = \int_0^\infty y(t)\mathrm{e}^{-st}\mathrm{d}t = G(s)u(s)$$

其中，

$$G(s) = \int_0^\infty G(t)\mathrm{e}^{-st}\mathrm{d}t$$

被称作系统的传递函数矩阵，它是脉冲响应矩阵的 Laplace 变换。

一般地，$G(s)$ 可以表示为

$$G(s) = \begin{bmatrix} g_{11}(s) & \cdots & g_{1p}(s) \\ \vdots & & \vdots \\ g_{q1}(s) & \cdots & g_{qp}(s) \end{bmatrix}$$

其中，$g_{ij}(s)$ 是 $G(s)$ 的元，它是系统第 i 个输出与第 j 个输入之间的传递函数，等价于在零初始条件下，系统第 i 个输出与第 j 个输入的 Laplace 变换之比。传递函数矩阵是系统输入-输出描述的另一种形式。

1.2.3 传递函数矩阵的 Smith-McMillan 形

Smith-McMillan（史密斯-麦克米伦）形是有理分式矩阵的一种重要的标准形，基于这一标准形，可以定义和分析传递函数矩阵的极点与零点。有理分式矩阵的 Smith-McMillan 形是多项式矩阵的 Smith 形的拓展。

考虑 $N(s)$ 是一个实多项式矩阵，它的元是关于 s 的多项式。其非零子式的最大阶次被称为 $N(s)$ 的秩。如果 $N(s)$ 是一个方阵并且 $\det N(s) \neq 0$，则称 $N(s)$ 是正则的或非奇异的，且它的逆矩阵 $N^{-1}(s)$ 存在。$N^{-1}(s)$ 是多项式矩阵的充要条件是 $\det N(s)$ 是一个非零常数。行列式为非零常数的多项式方阵被称为单位模阵，单位模阵的积仍是单位模阵。

对于一个多项式矩阵 $N(s)$，$\operatorname{rank} N(s) = r$，它的行列式因子是一簇多项式 $d_i(s)$，$i \in \{0, \cdots, r\}$，其中 $d_0(s) = 1$，$d_i(s)$ 是 $N(s)$ 所有的 i 阶非零子式的首一最大公因子。

应用多项式矩阵 $N(s)$ 的行列式因子，可以定义它的 Smith 形。

定义 1.1（多项式矩阵的 Smith 形） 对于一个 $q \times p$ 的实多项式矩阵 $N(s)$，$\operatorname{rank} N(s) = r$，它的 Smith 形是具有

$$N_\mathrm{S}(s) = \left[\begin{array}{cccc|c} s_1(s) & & & & \\ & s_2(s) & & & 0 \\ & & \ddots & & \\ & & & s_r(s) & \\ \hline & & 0 & & 0 \end{array}\right] \tag{1.8}$$

1.2 系统的输入-输出描述

形式的对角实多项式矩阵，其中

$$s_i(s) = \frac{d_i(s)}{d_{i-1}(s)}, \quad i \in \{1, \cdots, r\}$$

$s_i(s)$ 被称作 $N(s)$ 的不变因子。

例 1.1 实多项式矩阵

$$N(s) = \begin{bmatrix} s(s+3) & 0 \\ 0 & (s+2)^2 \\ s+3 & (s+2)(s+3) \\ 0 & s(s+2) \end{bmatrix}$$

的非零子式、行列式因子和不变因子如表 1.1 所示。

表 1.1 $N(s)$ 的非零子式、行列式因子和不变因子

阶次	非零子式	行列式因子	不变因子
$i=0$	无	$d_0(s)=1$	
$i=1$	$s(s+3), (s+2)^2,$ $s+3, (s+2)(s+3),$ $s(s+2)$	$d_1(s)=1$	$s_1(s)=1$
$i=2$	$s(s+2)^2(s+3),$ $s(s+2)(s+3)^2,$ $s^2(s+2)(s+3),$ $-(s+2)^2(s+3),$ $s(s+2)(s+3)$	$d_2(s)=(s+2)(s+3)$	$s_2(s)=(s+2)(s+3)$

由此可得，$N(s)$ 的 Smith 形是

$$N_S(s) = \begin{bmatrix} 1 & 0 \\ 0 & (s+2)(s+3) \\ 0 & 0 \\ 0 & 0 \end{bmatrix}$$

实际上，一个实多项式矩阵总是可以通过初等变换得到它的 Smith 形。多项式矩阵的加、减和乘运算与常数矩阵相同，它的初等变换有以下三类。

(1) 行（列）交换的变换矩阵为

$$U_1(s) = \begin{bmatrix} 1 & & & & & & \\ & \ddots & & & & & \\ & & 0 & \cdots & 1 & & \\ & & \vdots & & \vdots & & \\ & & 1 & \cdots & 0 & & \\ & & & & & \ddots & \\ & & & & & & 1 \end{bmatrix} \begin{matrix} \\ \\ i\text{行} \\ \\ j\text{行} \\ \\ \end{matrix}$$

$$\;\;i\text{列}\;\;\;\;j\text{列}$$

用 $U_1(s)$ 左乘 $N(s)$ 可以实现行交换，用 $U_1(s)$ 右乘 $N(s)$ 可以实现列交换。

(2) 用 $f(s)$ 乘以 $N(s)$ 的第 j 行 (第 i 列)，并加到第 i 行 (第 j 列) 上，相应的变换矩阵为

$$U_2(s) = \begin{bmatrix} 1 & & & & & & \\ & \ddots & & & & & \\ & & 1 & \cdots & f(s) & & \\ & & & \ddots & \vdots & & \\ & & & & 1 & & \\ & & & & & \ddots & \\ & & & & & & 1 \end{bmatrix} \begin{matrix} \\ \\ i\text{行} \\ \\ j\text{行} \\ \\ \end{matrix}$$

$$\;\;i\text{列}\;\;\;\;j\text{列}$$

用 $U_2(s)$ 左乘 $N(s)$ 表示行变换，用 $U_2(s)$ 右乘 $N(s)$ 表示列变换。

(3) 将 $N(s)$ 的第 i 行（列）乘以一个非零的常数 α，相应的变换矩阵为

$$U_3(s) = \begin{bmatrix} 1 & & & & & & \\ & \ddots & & & & & \\ & & 1 & & & & \\ & & & \alpha & & & \\ & & & & 1 & & \\ & & & & & \ddots & \\ & & & & & & 1 \end{bmatrix} \begin{matrix} \\ \\ \\ i\text{行} \\ \\ \\ \end{matrix}$$

$$\;\;\;\;i\text{列}$$

用 $U_3(s)$ 左乘（右乘）$N(s)$，表示将 $N(s)$ 的第 i 行（列）的所有元素均乘以 α。

定义 1.2（多项式矩阵的等价） 对于两个多项式矩阵 $N(s)$ 和 $M(s)$，如果存在一个单位模阵 $U_L(s)$ 使得

$$N(s) = U_L(s)M(s)$$

那么 $N(s)$ 和 $M(s)$ 是行等价的；如果存在一个单位模阵 $U_R(s)$ 使得

$$N(s) = M(s)U_R(s)$$

那么 $N(s)$ 和 $M(s)$ 是列等价的；如果存在两个单位模阵 $U_L(s)$ 和 $U_R(s)$ 使得

$$N(s) = U_L(s)M(s)U_R(s)$$

那么 $N(s)$ 和 $M(s)$ 是等价的。

应用初等变换，可以将多项式矩阵转化为它的 Smith 形。

定理 1.1 对于 $q \times p$ 的实多项式矩阵 $N(s)$，$\text{rank}\, N(s) = r$，总是存在 $q \times q$ 和 $p \times p$ 的单位模阵 $U(s)$ 和 $V(s)$ 使得

$$U(s)N(s)V(s) = N_S(s)$$

其中 $N_S(s)$ 由式 (1.8) 定义，是 $N(s)$ 的 Smith 形。

应用初等变换可以证明这一结论，这里不详细证明。

不失一般性地，一个 $q \times p$ 的传递函数矩阵 $G(s)$ 可以表示为

$$G(s) = \frac{1}{g(s)}N(s) \tag{1.9}$$

其中，$g(s)$ 是 $G(s)$ 所有元的最小公分母，$N(s)$ 是 $q \times p$ 的多项式矩阵，$\text{rank}\, N(s) = r$。

定义 1.3（有理分式矩阵的 Smith-McMillan 形） 对于式 (1.9) 中的 $q \times p$ 实真有理分式矩阵 $G(s)$，它的 Smith-McMillan 形是

$$G_M(s) = \frac{1}{g(s)}N_S(s) = \begin{bmatrix} \frac{\varepsilon_1(s)}{\psi_1(s)} & & & & \\ & \frac{\varepsilon_2(s)}{\psi_2(s)} & & & 0 \\ & & \ddots & & \\ & & & \frac{\varepsilon_r(s)}{\psi_r(s)} & \\ \hdashline & & 0 & & 0 \end{bmatrix} \tag{1.10}$$

形式的对角实有理分式矩阵，其中，$N_S(s)$ 是 $N(s)$ 的 Smith 形，式 (1.10) 中所有元的分子和分母的公因子应被消去，即多项式对 $\{\varepsilon_i(s), \psi_i(s)\}$ 是互质的，$i \in \{1, \cdots, r\}$。

根据定理 1.1 中给出的多项式矩阵的 Smith 形分解，可以很容易得出有理分式矩阵的 Smith-McMillan 形分解。

定理 1.2 对于式 (1.9) 中的 $q \times p$ 实真有理分式矩阵 $G(s)$，总是存在 $q \times q$ 和 $p \times p$ 的单位模阵 $U(s)$ 和 $V(s)$ 使得

$$U(s)G(s)V(s) = G_M(s)$$

其中 $G_M(s)$ 是由式 (1.10) 定义的 $G(s)$ 的 Smith-McMillan 形。

例 1.2 求下述传递函数矩阵的 Smith-McMillan 形:

$$G(s) = \begin{bmatrix} \dfrac{s}{(s+1)^2(s+2)^2} & \dfrac{s}{(s+2)^2} \\ -\dfrac{s}{(s+2)^2} & -\dfrac{s}{(s+2)^2} \end{bmatrix}$$

$G(s)$ 所有元的最小公分母为 $g(s) = (s+1)^2(s+2)^2$。由此，$G(s)$ 可以表示为 $G(s) = N(s)/g(s)$，其中，

$$N(s) = \begin{bmatrix} s & s(s+1)^2 \\ -s(s+1)^2 & -s(s+1)^2 \end{bmatrix}$$

$N(s)$ 的一阶子式是它的每个元，相应的行列式因子为 $d_1(s) = s$。$N(s)$ 的二阶子式是 $s^3(s+1)^2(s+2)$，相应的行列式因子是 $d_2(s) = s^3(s+1)^2(s+2)$。进一步，可以求出 $N(s)$ 的不变因子是 $s_1(s) = s$ 和 $s_2(s) = s^2(s+1)^2(s+2)$。所以，$N(s)$ 的 Smith 形是

$$\begin{bmatrix} s & 0 \\ 0 & s^2(s+1)^2(s+2) \end{bmatrix}$$

$N(s)$ 的 Smith 形也可以通过初等变换得到。

将 $N(s)$ 的第二行加上第一行的 $(s+1)^2$ 倍，得到

$$\begin{bmatrix} s & s(s+1)^2 \\ 0 & s(s+1)^4 - s(s+1)^2 \end{bmatrix}$$

进一步，将上式的第二列加上式第一列的 $-(s+1)^2$ 倍，可以得到

$$\begin{bmatrix} s & 0 \\ 0 & s^2(s+1)^2(s+2) \end{bmatrix}$$

用最小公分母 $g(s)$ 除上式中的每个元，可以得到 $G(s)$ 的 Smith-McMillan 形,

$$G_M(s) = U(s)G(s)V(s) = \begin{bmatrix} \dfrac{s}{(s+1)^2(s+2)^2} & \\ & \dfrac{s^2}{s+2} \end{bmatrix}$$

其中，

$$U(s) = \begin{bmatrix} 1 & 0 \\ (s+1)^2 & 1 \end{bmatrix}, \quad V(s) = \begin{bmatrix} 1 & -(s+1)^2 \\ 0 & 1 \end{bmatrix}$$

1.2.4 传递函数矩阵的极点和零点

在线性定常 SISO 系统中，传递函数的极点和零点具有重要作用，它们决定了系统的稳定性、动态特性和频率特性。在 SISO 系统中，零点被定义为输入 u 为有限值时使输出 y 为 0 的 s 值，极点被定义为使 y 为 ∞ 的 s 值。为了便于计算，通常取传递函数分子多项式的根作为零点，取传递函数分母多项式的根作为极点。但是，线性定常 MIMO 系统的零点和极点要复杂很多，需要从概念、计算和物理意义等多方面加以理解。

基于 Smith-McMillan 形可以定义传递函数矩阵 $G(s)$ 的极点和零点。

定义 1.4（传递函数矩阵的极点和零点） 对于 $q \times p$ 的实有理分式矩阵 $G(s)$，它的秩满足 $\text{rank}\, G(s) = r \leqslant \min\{q, p\}$。它的 Smith-McMillan 形是式 (1.10)，则

$$p(s) = \prod_{i=1}^{r} \psi_i(s)$$

被称为 $G(s)$ 的极点（或特征）多项式，$p(s)$ 的阶被称为 $G(s)$ 的 McMillan 阶，$p(s) = 0$ 的根被称为 $G(s)$ 的极点。

$$z(s) = \prod_{i=1}^{r} \varepsilon_i(s)$$

被称作 $G(s)$ 的零点多项式，$z(s) = 0$ 的根被称作 $G(s)$ 的零点。

例 1.3 给定一个 $q \times p$ 的有理分式矩阵，

$$G(s) = \begin{bmatrix} \dfrac{s}{(s+1)^2(s+2)^2} & \dfrac{s}{(s+2)^2} \\ -\dfrac{s}{(s+2)^2} & -\dfrac{s}{(s+2)^2} \end{bmatrix}$$

例 1.2 中已经给出 $G(s)$ 的 Smith-McMillan 形，

$$G_{\text{M}}(s) = \begin{bmatrix} \dfrac{s}{(s+1)^2(s+2)^2} & \\ & \dfrac{s^2}{s+2} \end{bmatrix}$$

根据定义 1.4，$G(s)$ 在 $s = 0$ 有 3 个零点，在 $s = -1$ 有 2 个极点，在 $s = -2$ 有 3 个极点。$G(s)$ 的零点多项式是 $z(s) = s^3$，极点多项式是 $p(s) = (s+1)^2(s+2)^3$。

值得注意的是，定义 1.4 仅适用于确定 $G(s)$ 在有限复平面内的极点和零点。$G_{\text{M}}(s)$ 在 ∞ 处的极点和零点可能不是 $G(s)$ 的极点和零点。

与线性定常 SISO 系统的传递函数不同，传递函数矩阵在复平面内的同一点同时出现零点和极点时，可以不形成对消。这是由于式 (1.10) 所描述的 Smith-McMillan 形 $G_{\text{M}}(s)$ 中，$\varepsilon_i(s)$ 和 $\psi_j(s)$ 可能具有公因子，$i \in \{1, \cdots, r\}$，$j \in \{1, \cdots, r\}$，$i \neq j$。

传递函数矩阵 $G(s)$ 的极点和零点有许多其他定义的方法。下述定义在工程应用中十分常用。

定义 1.5（传递函数矩阵的极点和零点） 对于 $q \times p$ 的实有理分式矩阵 $G(s)$, rank $G(s) = r$, 它所有非零子式的最小公分母是 $G(s)$ 的极点多项式 $p(s)$。$p(s) = 0$ 的根是 $G(s)$ 的极点。当 $G(s)$ 的所有 r 阶子式以 $p(s)$ 作为公分母时，它们分子的首一最大公因子是 $G(s)$ 的零点多项式 $z(s)$。$z(s) = 0$ 的根是 $G(s)$ 的零点。

例 1.4 给定一个 $q \times p$ 的有理分式传递函数矩阵 $G(s)$,

$$G(s) = \begin{bmatrix} \dfrac{1}{s+1} & 0 & \dfrac{s-1}{(s+1)(s+2)} \\ \dfrac{-1}{s-1} & \dfrac{1}{s+2} & \dfrac{1}{s+2} \end{bmatrix}$$

应用定义 1.5, rank $G(s) = r = 2 \leqslant \min\{p, q\}$。$G(s)$ 的一阶子式是 $G(s)$ 的各元，二阶子式是

$$\frac{1}{(s+1)(s+2)}, \quad \frac{-(s-1)}{(s+1)(s+2)^2}, \quad \frac{2}{(s+1)(s+2)}$$

$G(s)$ 所有子式的最小公分母是 $(s+1)(s+2)^2(s-1)$, 它是 $G(s)$ 的极点多项式。因此, $G(s)$ 的极点是 -1, -2, -2 和 1。

当上述三个二阶子式应用极点多项式 $p(s)$ 作为分母时，它们转化为

$$\frac{(s+2)(s-1)}{(s+1)(s+2)^2(s-1)}, \quad \frac{-(s-1)^2}{(s+1)(s+2)^2(s-1)}, \quad \frac{2(s+2)(s-1)}{(s+1)(s+2)^2(s-1)}$$

它们分子的首一最大公因子是 $s-1$。所以, $G(s)$ 的零点是 1。

1.3 线性系统的状态空间描述

系统的输入-输出描述值只适用于具有松弛特性的系统，且不能研究系统的内部信息。20 世纪 60 年代初，Kalman(卡尔曼) 等将状态空间的概念引入系统和控制理论的研究中，极大地推动了现代控制的发展。

1.3.1 线性系统的动态方程

一般地，对于具有 p 个输入、q 个输出的连续线性系统，它的状态空间描述可以表示为

$$\dot{x}(t) = A(t)x(t) + B(t)u(t) \tag{1.11}$$

$$y(t) = C(t)x(t) + D(t)u(t) \tag{1.12}$$

其中, $x(t) \in \mathbb{R}^n$ 是状态向量; $u(t) \in \mathbb{R}^p$ 是输入向量; $y(t) \in \mathbb{R}^q$ 是输出向量; $A(t) \in \mathbb{R}^{n \times n}$, $B(t) \in \mathbb{R}^{n \times p}$, $C(t) \in \mathbb{R}^{q \times n}$ 和 $D(t) \in \mathbb{R}^{q \times p}$ 是定义在 $(-\infty, \infty)$ 上的时间 t 的连续函数矩阵，它们分别被称作系统的状态系数矩阵、控制系数矩阵、输出系数矩阵和前馈系数矩阵。一阶微分方程 (1.11) 被称作状态方程，方程 (1.12) 则被称作输出方程，两者合称为系统的动态方程。

图 1.6 描述了动态方程所表示的系统的内部结构和信号之间的传递关系。

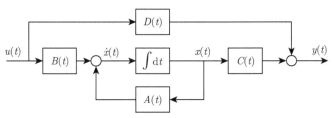

图 1.6　线性系统的状态空间结构图

在方程 (1.11) 和 (1.12) 中，系数矩阵随时间变化，因此由上述方程描述的系统是线性时变系统。如果上述方程中所有的系数矩阵都是常值矩阵，动态方程可以表示为

$$\dot{x}(t) = Ax(t) + Bu(t) \tag{1.13}$$

$$y(t) = Cx(t) + Du(t) \tag{1.14}$$

其中，A，B，C 和 D 是常值系数矩阵，由方程 (1.13) 和 (1.14) 所描述的系统是线性定常系统。

例 1.5　考虑图 1.7 所示的电路网络，输入 $u(t)$ 是电压 $u_\mathrm{s}(t)$，满足

$$LC\frac{\mathrm{d}^2 u_\mathrm{C}(t)}{\mathrm{d}t^2} + RC\frac{\mathrm{d}u_\mathrm{C}(t)}{\mathrm{d}t} + u_\mathrm{C}(t) = u(t)$$

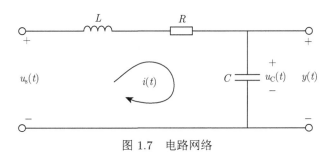

图 1.7　电路网络

选取状态变量 $x_1(t) = u_\mathrm{C}(t)$，$x_2(t) = i(t)$，可以得到

$$\dot{x}_1(t) = \frac{1}{C}x_2(t)$$

$$\dot{x}_2(t) = -\frac{1}{L}x_1(t) - \frac{R}{L}x_2(t) + \frac{1}{L}u(t)$$

它们可以表示为式 (1.13) 形式的状态方程，其中，$x(t) = [\ x_1(t)\ \ x_2(t)\ ]^\mathrm{T}$，

$$A = \begin{bmatrix} 0 & \dfrac{1}{C} \\ -\dfrac{1}{L} & -\dfrac{R}{L} \end{bmatrix},\ B = \begin{bmatrix} 0 \\ \dfrac{1}{L} \end{bmatrix}$$

把电容两端的电压作为输出，即 $y(t) = u_C(t)$，可以进一步写作式 (1.14) 形式的输出方程，其中，

$$C = \begin{bmatrix} 1 & 0 \end{bmatrix}, \quad D = 0$$

例 1.6 考虑图 1.8 所示的电枢控制的直流电动机。

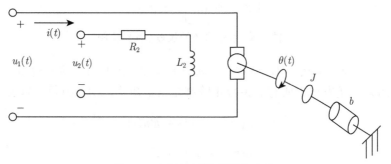

图 1.8 电枢控制的直流电动机

它的激励电压为恒值，控制电压 $u_1(t)$ 施加在电枢回路中，电动机轴的转角为 $\theta(t)$，电枢回路的电压平衡方程为

$$u_1(t) = Ri(t) + L\frac{\mathrm{d}i(t)}{\mathrm{d}t} + \varepsilon(t)$$

其中，R 和 L 分别为电枢回路的等效电阻和电感；$\varepsilon(t)$ 是电动机的反电势。由电动机学可知，

$$\varepsilon(t) = k_\mathrm{e}\frac{\mathrm{d}\theta(t)}{\mathrm{d}t}$$

其中，k_e 是电动机的电势系数，电动机的转动力矩平衡方程为

$$J\frac{\mathrm{d}^2\theta(t)}{\mathrm{d}t^2} + b\frac{\mathrm{d}\theta(t)}{\mathrm{d}t} = M = k_\mathrm{m}i(t)$$

其中，J 是电动机轴的等效转动惯量；b 是轴上的等效阻尼系数；k_m 是电动机的力矩系数。根据上述方程，可以写出

$$\frac{\mathrm{d}i(t)}{\mathrm{d}t} = -\frac{R}{L}i(t) - \frac{k_\mathrm{e}}{L}\frac{\mathrm{d}\theta(t)}{\mathrm{d}t} + \frac{1}{L}u_1(t)$$

$$\frac{\mathrm{d}^2\theta(t)}{\mathrm{d}t^2} = \frac{k_\mathrm{m}}{J}i(t) - \frac{b}{J}\frac{\mathrm{d}\theta(t)}{\mathrm{d}t}$$

选取状态变量 $x_1(t) = \theta(t)$，$x_2(t) = \dot{\theta}(t)$，$x_3(t) = i(t)$，选取 $u_1(t)$ 作为控制输入，选取 $y_1(t) = x_1(t)$ 和 $y_2(t) = x_3(t)$ 作为输出变量，则电枢控制的直流电动机系统的状态方程和输出方程可以写作

$$\dot{x}_1(t) = x_2(t)$$

$$\dot{x}_2(t) = -\frac{b}{J}x_2(t) + \frac{k_{\mathrm{m}}}{J}x_3(t)$$

$$\dot{x}_3(t) = -\frac{k_{\mathrm{e}}}{L}x_2(t) - \frac{R}{L}x_3(t) + \frac{1}{L}u_1(t)$$

$$y_1(t) = x_1(t)$$

$$y_2(t) = x_3(t)$$

进一步，上述方程可以被写作式 (1.13) 和式 (1.14) 形式的动态方程，其中，$x(t) = [\ x_1(t)\ \ x_2(t)\ \ x_3(t)\ ]^{\mathrm{T}}$ 是状态向量，$u(t) = u_1(t)$ 是控制输入，$y(t) = [\ y_1(t)\ \ y_2(t)\ ]^{\mathrm{T}}$ 是输出向量，且相应的系数矩阵为

$$A = \begin{bmatrix} 0 & 1 & 0 \\ 0 & -\dfrac{b}{J} & \dfrac{k_{\mathrm{m}}}{J} \\ 0 & -\dfrac{k_{\mathrm{e}}}{L} & -\dfrac{R}{L} \end{bmatrix}, \quad B = \begin{bmatrix} 0 \\ 0 \\ \dfrac{1}{L} \end{bmatrix}, \quad C = \begin{bmatrix} 1 & 0 & 0 \\ 0 & 0 & 1 \end{bmatrix}, \quad D = 0$$

需要指出，状态变量的选择不是唯一的，对同一个系统，状态变量不同，动态方程也可以不同。

1.3.2 线性定常系统的特征结构

线性定常系统的特征结构主要通过特征值和特征向量表征，它们对系统的运动特性具有重要影响。

考虑一个线性定常系统，它的状态方程为式 (1.13)。$sI - A$ 被称作特征矩阵，其中，s 是复变量，I 是和 A 具有相同维数的单位矩阵，$sI - A$ 是非奇异的多项式矩阵，它的逆 $(sI - A)^{-1}$ 被称作预解矩阵。$\det(sI - A)$ 被称作特征多项式，它是一个 n 阶多项式，可以写作

$$\alpha(s) = \det(sI - A) = s^n + a_1 s^{n-1} + \cdots + a_{n-1} s + a_n \tag{1.15}$$

其中，a_1，$\cdots$，a_n 是由矩阵 A 确定的实常数。$\det(sI - A) = 0$ 被称作特征方程。根据 Cayley-Hamilton（凯莱-哈密顿）定理，A 是特征方程的"矩阵解"，即

$$A^n + a_1 A^{n-1} + \cdots + a_{n-1} A + a_n I = 0 \tag{1.16}$$

这表明对于 $A \in \mathbb{R}^{n \times n}$，所有 A^i，$i \geqslant n$，都可以表示为 $\{I, A, \cdots, A^{n-1}\}$ 的线性组合。

预解矩阵 $(sI - A)^{-1}$ 满足

$$(sI - A)^{-1} = \frac{\mathrm{adj}(sI - A)}{\alpha(s)} = \frac{P(s)}{\phi(s)} \tag{1.17}$$

其中，$\mathrm{adj}(sI - A)$ 是 $sI - A$ 的伴随矩阵，它是一个多项式矩阵。$\mathrm{adj}(sI - A)$ 中多项式 s 的阶次不会大于 $\alpha(s)$ 中 s 的阶次。如果进一步消除 $\alpha(s)$ 和 $\mathrm{adj}(sI - A)$ 中每个元的公因

子，可以得到式 (1.17) 中最右端的形式，其中 $P(s)$ 是多项式矩阵，$\phi(s)$ 是阶次不会大于 $\alpha(s)$ 的多项式，且 $\phi(s)$ 和 $P(s)$ 中的每个多项式都是互质的。基于此，$\phi(s)$ 被称作 A 的最小多项式。$\phi(s)$ 也满足 Cayley-Hamilton 定理，即 $\phi(A) = 0$。由此，最小多项式 $\phi(s)$ 也可以被定义为使得 $\phi(A) = 0$ 的最小阶的首一多项式。当且仅当矩阵 A 的特征多项式等于最小多项式，矩阵 A 被称作是循环的。

特征方程 $\det(sI - A) = 0$ 的根被称作 A 的特征值。$n \times n$ 的矩阵 A 具有 n 个特征值，考虑 λ_i 是 A 的一个特征值，那么

$$\det(\lambda I - A) = (\lambda - \lambda_i)^{\sigma_i} \beta_i(\lambda), \ \beta_i(\lambda) \neq 0$$

其中，σ_i 被称作 λ_i 的代数重数。代数重数为 1 的特征值被称作单特征值，代数重数大于或等于 2 的特征值被称作重特征值。同时，

$$\alpha_i = n - \mathrm{rank}\,(\lambda_i I - A)$$

被称作 λ_i 的几何重数。如果 λ_i 是单特征值，那么 $\sigma_i = \alpha_i = 1$，否则 $\sigma_i \geqslant \alpha_i$。

满足 $A v_i = \lambda_i v_i$ 的非零向量 v_i 被称作 A 的与 λ_i 相关的（右）特征向量。特征向量的定义通常被写作

$$(\lambda_i I - A) v_i = 0$$

特征向量是不唯一的，如果 A 的特征值全部互异，那么与 $\{\lambda_1, \cdots, \lambda_n\}$ 相关的特征向量 $\{v_1, \cdots, v_n\}$ 是线性无关的。

如果非零向量 v_i 满足

$$(\lambda_i I - A)^k v_i = 0, \ (\lambda_i I - A)^{k-1} v_i \neq 0$$

那么 v_i 被称作与特征值 λ_i 相关的 k 级广义特征向量。当 $k = 1$ 时，它就是普通的特征向量。如果 v_i 是与特征值 λ_i 相关的 k 级广义特征向量，那么下列 k 个向量是线性无关的，

$$v_i^{(k)} = v_i, \ v_i^{(k-1)} = (\lambda_i I - A) v_i, \ \cdots, \ v_i^{(1)} = (\lambda_i I - A)^{k-1} v_i$$

其中，向量组 $v_i^{(1)}, \cdots, v_i^{(k)}$ 被称作广义特征向量链。如果 λ_i 是代数重数为 σ_i 的特征向量，那么与 λ_i 相关的广义特征向量组具有 σ_i 个线性无关的非零向量。与 A 的不同特征值相关的广义特征向量是线性无关的。

这一小节介绍了线性定常系统特征结构相关的一些定义和性质，它们在后续章节中具有重要应用。

1.3.3 根据状态空间描述求传递函数矩阵

考虑由状态空间模型 (1.13) 和 (1.14) 描述的线性定常系统。假设初始条件满足 $x(0) = 0$ 和 $u(0) = 0$。对方程 (1.13) 和 (1.14) 两端同时做 Laplace 变换，可以得到

$$s x(s) = A x(s) + B u(s)$$

1.3 线性系统的状态空间描述

$$y(s) = Cx(s) + Du(s)$$

消除 $x(s)$，可以得到

$$G(s) = C(sI - A)^{-1}B + D \tag{1.18}$$

式 (1.18) 表明当 $D = 0$ 时，$G(s)$ 是严格真有理分式矩阵。当 $D \neq 0$ 时，$G(s)$ 是真有理分式矩阵，同时

$$\lim_{s \to \infty} G(s) = D$$

$G(s)$ 的所有极点的集合表示为 Λ_G，A 的所有特征值的集合表示为 Λ。当且仅当式 (1.13) 和式 (1.14) 描述的系统完全能控能观时，$\Lambda_G = \Lambda$，否则 $\Lambda_G \subset \Lambda$。能控性和能观性的定义将在第 3 章介绍。

因为式 (1.18) 中含有逆运算，所以很难用来直接计算 $G(s)$。Faddeev-Leverrier（法捷耶夫-勒韦里耶）算法给出了计算式 (1.18) 中预解矩阵的方法。

算法 1.1（Faddeev-Leverrier 算法） 根据式 (1.17)，$(sI - A)^{-1}$ 可以被写作

$$(sI - A)^{-1} = \frac{R_{n-1}s^{n-1} + R_{n-2}s^{n-2} + \cdots + R_1 s + R_0}{s^n + a_1 s^{n-1} + \cdots + a_{n-1} s + a_n} \tag{1.19}$$

其中，$R_0, \cdots, R_{n-1}$ 是 $n \times n$ 的常值矩阵。在式 (1.19) 两端左乘 $\alpha(s)(sI - A)$，其中，$\alpha(s)$ 是 A 的特征多项式，可以得到

$$\left(s^n + a_1 s^{n-1} + \cdots + a_{n-1} s + a_n\right) I = (sI - A)\left(R_{n-1}s^{n-1} + R_{n-2}s^{n-2} + \cdots + R_1 s + R_0\right)$$

比较上述等式两端具有相同阶次的 s 的矩阵系数，可以得到

$$\begin{cases} R_{n-1} = I \\ R_{n-2} = AR_{n-1} + a_1 I = A + a_1 I \\ R_{n-3} = AR_{n-2} + a_2 I = A^2 + a_1 A + a_2 I \\ \vdots \\ R_0 = AR_1 + a_{n-1} I = A^{n-1} + a_1 A^{n-2} + \cdots + a_{n-1} I \\ 0 = AR_0 + a_n I \end{cases} \tag{1.20}$$

计算特征多项式 (1.15) 的系数的递推公式为

$$\begin{cases} a_1 = -\operatorname{tr}(R_{n-1}A) \\ a_2 = -\dfrac{1}{2}\operatorname{tr}(R_{n-2}A) \\ \vdots \\ a_n = -\dfrac{1}{n}\operatorname{tr}(R_0 A) \end{cases} \tag{1.21}$$

其中，$\mathrm{tr}(X)$ 是方阵 X 的迹。使用式 (1.20) 和式 (1.21) 可以交替计算 R_{n-i} 和 a_i，$i=1$, $2,\cdots,n$，进一步可以求得预解矩阵。式 (1.20) 中的最后一个等式可以用来验证计算的准确性。

例 1.7 给定矩阵 $A = \begin{bmatrix} 1 & 1 & 2 \\ 0 & 1 & 3 \\ 0 & 0 & 2 \end{bmatrix}$，使用式 (1.17) 计算 $(sI-A)^{-1}$。

A 的特征多项式是

$$\alpha(s) = \det(sI-A) = \det \begin{bmatrix} s-1 & -1 & -2 \\ 0 & s-1 & -3 \\ 0 & 0 & s-2 \end{bmatrix}$$

$$= (s-1)^2(s-2) = s^3 - 4s^2 + 5s - 2$$

所以，$a_1 = -4$，$a_2 = 5$，$a_3 = -2$。使用式 (1.20) 来计算 $R(s) = R_2 s^2 + R_1 s + R_0$，

$$R_2 = I, R_1 = AR_2 + a_1 I = \begin{bmatrix} -3 & 1 & 2 \\ 0 & -3 & 3 \\ 0 & 0 & -2 \end{bmatrix}, R_0 = AR_1 + a_2 I = \begin{bmatrix} 2 & -2 & 1 \\ 0 & 2 & -3 \\ 0 & 0 & 1 \end{bmatrix}$$

根据式 (1.19)，可以得到

$$(sI-A)^{-1} = \frac{R(s)}{\alpha(s)} = \frac{1}{(s-1)^2(s-2)} \begin{bmatrix} s^2-3s+2 & s-2 & 2s+1 \\ 0 & s^2-3s+2 & 3s-3 \\ 0 & 0 & s^2-2s+1 \end{bmatrix}$$

$$= \begin{bmatrix} \dfrac{1}{s-1} & \dfrac{1}{(s-1)^2} & \dfrac{2s+1}{(s-1)^2(s-2)} \\ 0 & \dfrac{1}{s-1} & \dfrac{3}{(s-1)(s-2)} \\ 0 & 0 & \dfrac{1}{s-2} \end{bmatrix}$$

根据 Cayley-Hamilton 定理，也可以导出一个求解预解矩阵的算法。

算法 1.2（Cayley-Hamilton 算法） $(sI-A)^{-1}$ 可以表示为幂级数，

$$(sI-A)^{-1} = \sum_{k=0}^{\infty} \frac{1}{s^{k+1}} A^k$$

根据 Cayley-Hamilton 定理，这一幂级数可以被表示为

$$(sI-A)^{-1} = q_0(s)I + q_1(s)A + \cdots + q_{n-1}(s)A^{n-1} \tag{1.22}$$

其中，$q_i(s)$，$i \in \{0,1,\cdots,n-1\}$ 是关于 s 的有理多项式。在式 (1.22) 两端左乘 $sI-A$，可以得到

$$I = sq_0(s)I + sq_1(s)A + \cdots + sq_{n-1}(s)A^{n-1} - \left[q_0(s)A + q_1(s)A^2 + \cdots + q_{n-1}(s)A^n\right]$$

1.3 线性系统的状态空间描述

再次使用 Cayley-Hamilton 定理，经整理可以得到

$$I = [sq_0(s) + a_n q_{n-1}(s)] I + [sq_1(s) - q_0(s) + a_{n-1} q_{n-1}(s)] A + \cdots$$
$$+ [sq_{n-1}(s) - q_{n-2}(s) + a_1 q_{n-1}(s)] A^{n-1}$$

比较上述等式两端 A 的同次幂的系数，可以得到

$$\begin{cases} sq_0(s) + a_n q_{n-1}(s) = 1 \\ sq_1(s) - q_0(s) + a_{n-1} q_{n-1}(s) = 0 \\ \vdots \\ (s + a_1) q_{n-1}(s) - q_{n-2}(s) = 0 \end{cases} \quad (1.23)$$

依次在上式两端分别乘以 1, s, s^2, $\cdots$, s^{n-1} 后，再将上式两边分别相加，可以得到

$$\left(s^n + a_1 s^{n-1} + \cdots + a_{n-1} s + a_n\right) q_{n-1}(s) = \alpha(s) q_{n-1}(s) = 1$$

由此可知

$$q_{n-1}(s) = \frac{1}{\alpha(s)}$$

其中，$\alpha(s)$ 是 A 的特征多项式。把 $q_{n-1}(s)$ 代入式 (1.23) 中的最后一个等式中，可以得到

$$q_{n-2}(s) = \frac{1}{\alpha(s)} (s + a_1)$$

把 $q_{n-1}(s)$ 和 $q_{n-2}(s)$ 代入式 (1.23) 中的倒数第二个等式中，并依次递推，可以得到

$$q_{n-1}(s) = \frac{1}{\alpha(s)} = \frac{1}{\alpha(s)} p_{n-1}(s)$$

$$q_{n-2}(s) = \frac{1}{\alpha(s)} (s + a_1) = \frac{1}{\alpha(s)} p_{n-2}(s)$$

$$\vdots$$

$$q_1(s) = \frac{1}{\alpha(s)} \left(s^{n-2} + a_1 s^{n-3} + \cdots + a_{n-3} s + a_{n-2}\right) = \frac{1}{\alpha(s)} p_1(s)$$

$$q_0(s) = \frac{1}{\alpha(s)} \left(s^{n-1} + a_1 s^{n-2} + \cdots + a_{n-2} s + a_{n-1}\right) = \frac{1}{\alpha(s)} p_0(s)$$

其中，

$$\begin{bmatrix} p_0(s) \\ p_1(s) \\ \vdots \\ p_{n-2}(s) \\ p_{n-1}(s) \end{bmatrix} = \begin{bmatrix} 1 & a_1 & a_2 & \cdots & a_{n-1} \\ & 1 & a_1 & \cdots & a_{n-2} \\ & & 1 & \ddots & \vdots \\ & & & \ddots & a_1 \\ & & & & 1 \end{bmatrix} \begin{bmatrix} s^{n-1} \\ s^{n-2} \\ \vdots \\ s \\ 1 \end{bmatrix}$$

根据式 (1.22)，可以得到

$$(sI - A)^{-1} = \sum_{k=0}^{n-1} q_k(s) A^k = \sum_{k=0}^{n-1} \frac{1}{\alpha(s)} p_k(s) A^k$$

1.3.4 动态系统的等价

例 1.8 在例 1.5 中，如果选择 $\bar{x}(t) = [\ u_C(t)\quad \dot{u}_C(t)\ ]^T$ 作为状态向量，相应的状态方程为

$$\dot{\bar{x}}(t) = \begin{bmatrix} 0 & 1 \\ -\dfrac{1}{LC} & -\dfrac{R}{L} \end{bmatrix} \bar{x}(t) + \begin{bmatrix} 0 \\ \dfrac{1}{LC} \end{bmatrix} u(t)$$

可以看出，状态变量的选择是不唯一的。因此，存在许多不同的动态方程可以描述同一个动态系统。这些动态方程之间的关系非常值得研究。

定义 1.6（线性定常动态系统的等价） 考虑非奇异变换 $\bar{x}(t) = Tx(t)$，T 是非奇异的常值矩阵。线性定常动态系统

$$\dot{\bar{x}}(t) = \bar{A}\bar{x}(t) + \bar{B}u(t) \tag{1.24}$$

$$y(t) = \bar{C}\bar{x}(t) + \bar{D}u(t) \tag{1.25}$$

被称作与式 (1.13) 和式 (1.14) 所描述的系统（代数）等价，其中，

$$\bar{A} = TAT^{-1},\ \bar{B} = TB,\ \bar{C} = CT^{-1},\ \bar{D} = D \tag{1.26}$$

$\bar{x}(t) = Tx(t)$ 被称作等价变换。

根据定义 1.6，可以得到 $x(t) = T^{-1}\bar{x}(t)$，进一步可知

$$\dot{\bar{x}}(t) = T\dot{x}(t) = T(Ax(t) + Bu(t)) = TAT^{-1}\bar{x}(t) + TBu(t) = \bar{A}\bar{x}(t) + \bar{B}u(t)$$

$$y(t) = Cx(t) + Du(t) = CT^{-1}\bar{x}(t) + Du(t) = \bar{C}\bar{x}(t) + \bar{D}u(t)$$

由此，可以得到式 (1.26) 成立。

考虑式 (1.24) 和式 (1.25) 所描述的线性定常系统的传递函数矩阵是 $\bar{G}(s)$，结合式 (1.18) 和式 (1.26)，可以得到

$$\bar{G}(s) = \bar{C}(sI - \bar{A})^{-1}\bar{B} + \bar{D} = CT^{-1}\left(sI - TAT^{-1}\right)^{-1}TB + D$$

$$= CT^{-1}\left(TsT^{-1} - TAT^{-1}\right)^{-1}TB + D = CT^{-1}\left[T(sI - A)T^{-1}\right]^{-1}TB + D$$

$$= CT^{-1}\left[T(sI - A)^{-1}T^{-1}\right]TB + D = C(sI - A)^{-1}B + D = G(s)$$

1.3 线性系统的状态空间描述

这表明等价的动态系统具有相同的传递函数矩阵。事实上，任何等价变换都不会改变式 (1.13) 和式 (1.14) 所描述的系统的全部特性。考虑 $\bar{A}$ 的特征多项式是 $\bar{\alpha}(s)$，结合 $\det T \det T^{-1} = 1$，可以得到

$$\bar{\alpha}(s) = \det(sI - \bar{A}) = \det\left(TsT^{-1} - TAT^{-1}\right) = \det\left[T(sI-A)T^{-1}\right]$$
$$= \det T \det(sI-A) \det T^{-1} = \det(sI-A) = \alpha(s)$$

上式表明等价的动态系统具有相同的特征多项式。相应地，在任何等价变换下，特征值也不会改变。同时，传递函数矩阵的极点也不会改变。

等价变换也可以拓展到线性时变系统中。这种情况下，变换矩阵应该是时变的，同时满足可微性要求。

定义 1.7（线性时变动态系统的等价） 考虑非奇异变换 $\bar{x}(t) = T(t)x(t)$，$T(t)$ 是对于所有 $t \in (-\infty, \infty)$ 都非奇异且连续可微的矩阵，线性时变动态系统

$$\dot{\bar{x}}(t) = \bar{A}(t)\bar{x}(t) + \bar{B}(t)u(t) \tag{1.27}$$

$$y(t) = \bar{C}(t)\bar{x}(t) + \bar{D}(t)u(t) \tag{1.28}$$

被称作与式 (1.11) 和式 (1.12) 所描述的系统（代数）等价，其中，

$$\bar{A}(t) = T(t)A(t)T^{-1}(t) + \dot{T}(t)T^{-1}(t), \ \bar{B}(t) = T(t)B(t)$$
$$\bar{C}(t) = C(t)T^{-1}(t), \ \bar{D}(t) = D(t) \tag{1.29}$$

$\bar{x}(t) = T(t)x(t)$ 被称作等价变换。

根据定义 1.7，可以得到

$$\dot{\bar{x}}(t) = T(t)\dot{x}(t) + \dot{T}(t)x(t) = T(t)[A(t)x(t) + B(t)u(t)] + \dot{T}(t)x(t)$$
$$= \left[T(t)A(t)T^{-1}(t) + \dot{T}(t)T^{-1}(t)\right]\bar{x}(t) + T(t)B(t)u(t) = \bar{A}(t)\bar{x}(t) + \bar{B}(t)u(t)$$
$$y(t) = C(t)x(t) + D(t)u(t) = C(t)T^{-1}(t)\bar{x}(t) + D(t)u(t) = \bar{C}(t)\bar{x}(t) + \bar{D}(t)u(t)$$

由此可知式 (1.29) 成立。

根据等价变换，线性定常系统的状态空间描述可以转化为 Jordan（若尔当）标准形，它相应的状态系数矩阵是对角的或块对角的。

定义 1.8（Jordan 矩阵） 如果矩阵 J 具有下述形式：

$$J = \begin{bmatrix} J_1 & & & & \\ & \ddots & & & \\ & & J_i & & \\ & & & \ddots & \\ & & & & J_l \end{bmatrix} \tag{1.30}$$

其中，

$$J_i = \begin{bmatrix} J_{i1} & & \\ & \ddots & \\ & & J_{i\alpha_i} \end{bmatrix} \tag{1.31}$$

$$J_{ik} = \begin{bmatrix} \lambda_i & 1 & & & \\ & \lambda_i & 1 & & \\ & & \ddots & \ddots & \\ & & & \lambda_i & 1 \\ & & & & \lambda_i \end{bmatrix} \tag{1.32}$$

则称 J 为 Jordan 矩阵，这里，λ_i 是代数重数为 σ_i 的 J 的特征值，$i \in \{1, \cdots, l\}$，J_i 被称作与 λ_i 相关的 Jordan 块矩阵，J_{ik}，$k \in \{1, \cdots, \alpha_i\}$ 被称作 Jordan 块，α_i 是 λ_i 的几何重数。

定理 1.3 考虑线性定常系统 (1.13) 和 (1.14)，λ_i 是 A 的特征值，代数重数为 σ_i，$i \in \{1, \cdots, l\}$，且 $\sigma_1 + \cdots + \sigma_i + \cdots + \sigma_l = n$。存在等价变换 $\bar{x}(t) = Tx(t)$ 使得 A 转化为 Jordan 标准形

$$J = TAT^{-1} \tag{1.33}$$

其中，J 具有式 (1.30) 的形式。

根据 Jordan 标准形的构造方法可以证明这一定理，后续进行详细介绍。

1.3.2 节中介绍了代数重数和几何重数的概念。λ_i 的几何重数等于与 λ_i 相关的 Jordan 块的个数，即与 λ_i 相关的线性无关的特征向量的个数。也就是说，对于每一个 Jordan 块，仅存在一个线性无关的特征向量。与 λ_i 相关的所有 Jordan 块的阶次之和等于 λ_i 的代数重数。进一步，式 (1.30) 中 Jordan 块的个数等于所有特征值的几何重数之和，即 A 的线性无关的特征向量的个数。对于 A 的每一个特征值，如果它的代数重数总是等于几何重数，那么 A 的 Jordan 标准形是对角的。

用 σ_{ik} 表示与 λ_i 相关的第 k 个 Jordan 块的阶次 $i \in \{1, \cdots, l\}$，$k \in \{1, \cdots, \alpha_i\}$，且 $\sigma_{i\max} = \max_{1 \leqslant k \leqslant \alpha_i} \{\sigma_{ik}\}$。那么，$A$ 的最小多项式也可以被定义为

$$\phi(s) = \prod_{i=1}^{l} (s - \lambda_i)^{\sigma_{i\max}}$$

为了构造 Jordan 标准形，分为以下情况讨论。

首先，考虑 A 的特征值全部互异。用 λ_i，$i \in \{1, \cdots, n\}$ 表示 A 的特征值，v_i 表示相应的特征向量，那么特征向量 $\{v_1, \cdots, v_n\}$ 是线性无关的并且可以被用来作为一组基。因此，$Q = \begin{bmatrix} v_1 & \cdots & v_n \end{bmatrix}$ 是非奇异的，选择 $T = Q^{-1}$。注意到 $Av_i = \lambda_i v_i$，进行等价变换 $\bar{x}(t) = Tx(t)$，根据式 (1.26) 可得

$$\bar{A} = TAT^{-1} = TA \begin{bmatrix} v_1 & \cdots & v_n \end{bmatrix}$$

1.3 线性系统的状态空间描述

$$\begin{aligned}
&= T \begin{bmatrix} Av_1 & \cdots & Av_n \end{bmatrix} = T \begin{bmatrix} \lambda_1 v_1 & \cdots & \lambda_n v_n \end{bmatrix} \\
&= T \begin{bmatrix} v_1 & \cdots & v_n \end{bmatrix} \begin{bmatrix} \lambda_1 & & \\ & \ddots & \\ & & \lambda_n \end{bmatrix} = \begin{bmatrix} \lambda_1 & & \\ & \ddots & \\ & & \lambda_n \end{bmatrix}
\end{aligned} \tag{1.34}$$

可以看出，所有特征值都互异的方阵可以转化为一个对角阵，且对角阵的元素是它的特征值。这里选择了 $\begin{bmatrix} v_1 & \cdots & v_n \end{bmatrix}^{-1}$ 作为变换矩阵 T，特征向量不同的排列顺序会使得同一个 A 转化为不同的对角矩阵。

例 1.9 考虑矩阵 $A = \begin{bmatrix} 0 & 0 & 0 \\ 1 & 0 & 2 \\ 0 & 1 & 1 \end{bmatrix}$，求它的 Jordan 标准形。

它的特征多项式是

$$\alpha(s) = \det(sI - A) = \det \begin{bmatrix} s & 0 & 0 \\ -1 & s & -2 \\ 0 & -1 & s-1 \end{bmatrix} = (s-2)(s+1)s$$

A 的特征值是 2，-1 和 0。与 $\lambda_1 = 2$ 相关的特征向量是

$$(\lambda_1 I - A) v_1 = \begin{bmatrix} 2 & 0 & 0 \\ -1 & 2 & -2 \\ 0 & -1 & 1 \end{bmatrix} v_1 = 0$$

的任意非零解。因此，$v_1 = \begin{bmatrix} 0 & 1 & 1 \end{bmatrix}^\mathrm{T}$ 是与 $\lambda_1 = 2$ 相关的一个特征向量。注意到特征向量是不唯一的，$\begin{bmatrix} 0 & \alpha & \alpha \end{bmatrix}^\mathrm{T}$ 都可以被选为它的特征向量，α 是任意非零实数。与 $\lambda_2 = -1$ 相关的特征向量则是

$$(\lambda_2 I - A) v_2 = \begin{bmatrix} -1 & 0 & 0 \\ -1 & -1 & -2 \\ 0 & 1 & -2 \end{bmatrix} v_2 = 0$$

的解，可以解得 $v_2 = [0 \ -2 \ 1]^\mathrm{T}$。类似地，与 $\lambda_3 = 0$ 相关的特征向量是 $v_3 = [2 \ 1 \ -1]^\mathrm{T}$。

由此，可以写出 A 的 Jordan 标准形是

$$\bar{A} = \begin{bmatrix} 2 & 0 & 0 \\ 0 & -1 & 0 \\ 0 & 0 & 0 \end{bmatrix}$$

它是一个对角线上的元素是特征值的对角矩阵。这一矩阵也可以通过

$$\bar{A} = TAT^{-1}$$

计算得到，其中

$$T = \begin{bmatrix} v_1 & v_2 & v_3 \end{bmatrix}^{-1} = \begin{bmatrix} \dfrac{1}{6} & \dfrac{1}{3} & \dfrac{2}{3} \\ \dfrac{1}{3} & -\dfrac{1}{3} & \dfrac{1}{3} \\ \dfrac{1}{2} & 0 & 0 \end{bmatrix}$$

值得一提的是，计算得到的特征值和特征向量中可能出现复数。这种情况下，对角 Jordan 标准形（式 (1.30)）中会包含复数元。相应地，实线性空间应被拓展为复线性空间，且用共轭转置替换转置。尽管在实际应用中这一形式没有物理意义，但它有助于分析系统的结构。实际上，含有复特征值的对角矩阵可以转化为一个非常实用的实矩阵。如果一个矩阵含有一对共轭复特征值 $\alpha \pm \mathrm{j}\beta$，那么式 (1.34) 可以进一步被写作

$$\bar{A} = \begin{bmatrix} \lambda_1 & & & & \\ & \ddots & & & \\ & & \lambda_{n-2} & & \\ & & & \alpha+\mathrm{j}\beta & \\ & & & & \alpha-\mathrm{j}\beta \end{bmatrix}$$

通过等价变换，可得

$$\tilde{A} = \bar{T}\bar{A}\bar{T}^{-1}$$

$$= \begin{bmatrix} I & & \\ & 1 & 1 \\ & \mathrm{j} & -\mathrm{j} \end{bmatrix} \begin{bmatrix} \lambda_1 & & & & \\ & \ddots & & & \\ & & \lambda_{n-2} & & \\ & & & \alpha+\mathrm{j}\beta & \\ & & & & \alpha-\mathrm{j}\beta \end{bmatrix} \begin{bmatrix} I & & \\ & 0.5 & -0.5\mathrm{j} \\ & 0.5 & 0.5\mathrm{j} \end{bmatrix}$$

$$= \begin{bmatrix} \lambda_1 & & & & \\ & \ddots & & & \\ & & \lambda_{n-2} & & \\ & & & \alpha & \beta \\ & & & -\beta & \alpha \end{bmatrix} \tag{1.35}$$

可以看出，上述等价变换将对角线上的复特征值转化为块矩阵，这一块矩阵对角线上的元为特征值实部，反对角线上的元为特征值虚部。如果存在两对或更多对共轭复特征值，可以应用与式 (1.35) 类似的变换。事实上，式 (1.34) 和式 (1.35) 两个等价变换可以合并为一个变换，变换矩阵满足

$$\tilde{T}^{-1} = T^{-1}\bar{T}^{-1}$$

$$= \begin{bmatrix} v_1 & \cdots & v_{n-2m} & \operatorname{Re}(v_{n-2m+1}) & \operatorname{Im}(v_{n-2m+1}) & \cdots & \operatorname{Re}(v_{n+m}) & \operatorname{Im}(v_{n+m}) \end{bmatrix}$$

其中，m 是共轭复特征值的对数。这一形式在用状态空间描述的系统的设计中非常实用。

接下来，讨论 A 的特征值不全互异的情况。此时，如果仍然存在 n 个线性无关的特征向量，那么仍可以选择这 n 个特征向量作为一组基。与特征值全部互异的情况类似，相应的 Jordan 标准形是对角的。另一种情况是线性无关的特征向量的个数小于 n，那么相应的 Jordan 标准形是块对角的。考虑 λ_i 是 A 的特征值，它的代数重数为 σ_i，$i \in \{1, \cdots, l\}$，分以下两种情况来讨论这一问题。

第一种情况是每个特征值都仅存在一个线性无关的特征向量。此时，对于所有的 $i \in \{1, \cdots, l\}$，$\alpha_i = 1$。也就是说，每个特征值都仅对应一个 Jordan 块，且 Jordan 块的阶次等于相应特征值的代数重数。这种情况下，所有特征值求出后即可确定相应的 Jordan 标准形，

$$\bar{A} = \begin{bmatrix} \bar{A}_1 & & \\ & \ddots & \\ & & \bar{A}_l \end{bmatrix}$$

其中，$\bar{A}_i$ 是对应于 λ_i 的 $\sigma_i \times \sigma_i$ 的 Jordan 块。

通过求解下述矩阵方程，可以得到变换矩阵 $T = Q^{-1}$，

$$Q\bar{A} = AQ \tag{1.36}$$

记 $Q = \begin{bmatrix} Q_1 & \cdots & Q_l \end{bmatrix}$，式 (1.36) 可以被写作下述矩阵方程组：

$$\begin{cases} Q_1 \bar{A}_1 = AQ_1 \\ \vdots \\ Q_l \bar{A}_l = AQ_l \end{cases} \tag{1.37}$$

记 $Q_i = \begin{bmatrix} v_{i1} & \cdots & v_{i\sigma_i} \end{bmatrix}$，并代入式 (1.37) 中，可以得到

$$\begin{bmatrix} v_{i1} & \cdots & v_{i\sigma_i} \end{bmatrix} \begin{bmatrix} \lambda_i & 1 & & & \\ & \lambda_i & 1 & & \\ & & \ddots & \ddots & \\ & & & \lambda_i & 1 \\ & & & & \lambda_i \end{bmatrix} = A \begin{bmatrix} v_{i1} & \cdots & v_{i\sigma_i} \end{bmatrix}$$

上式等价于

$$\begin{cases} \lambda_i v_{i1} = Av_{i1} \\ v_{i1} + \lambda_i v_{i2} = Av_{i2} \\ \vdots \\ v_{i(\sigma_i-1)} + \lambda_i v_{i\sigma_i} = Av_{i\sigma_i} \end{cases} \tag{1.38}$$

求解式 (1.38)，可以确定 $Q_i = \begin{bmatrix} v_{i1} & \cdots & v_{i\sigma_i} \end{bmatrix}$。式 (1.38) 可以被写作

$$\begin{cases} (\lambda_i I - A) v_{i1} = 0 \\ (\lambda_i I - A) v_{i2} = -v_{i1} \\ \vdots \\ (\lambda_i I - A) v_{i\sigma_i} = -v_{i(\sigma_i-1)} \end{cases} \tag{1.39}$$

根据式 (1.39)，可以看出 v_{i1} 是与 λ_i 相关的特征向量，$v_{i2}, \cdots, v_{i\sigma_i}$ 是与 λ_i 相关的广义特征向量。由此，可以得到变换矩阵

$$T = Q^{-1} = \begin{bmatrix} Q_1 & \cdots & Q_l \end{bmatrix}^{-1} = \begin{bmatrix} v_{11} & \cdots & v_{1\sigma_1} & \cdots & v_{l1} & \cdots & v_{l\sigma_l} \end{bmatrix}^{-1}$$

第二种情况是一些特征值存在不止一个线性无关的特征向量。这里，假设 λ_i 是一个重特征值，且相应的 $\alpha_i > 1$。这种情况下，存在 α_i 个线性无关的特征向量 $v_{i1}, \cdots, v_{i\alpha_i}$，它们可以通过

$$(\lambda_i I - A) v_{ik} = 0$$

来确定，$k \in \{1, \cdots, \alpha_i\}$。进一步，可以利用式 (1.39)，根据 $\{v_{i1}, \cdots, v_{i\alpha_i}\}$ 中的任何一个特征向量来计算 $\sigma_i - \alpha_i$ 个广义特征向量 $v_{i\alpha_i+1}, \cdots, v_{i\sigma_i}$。构造

$$Q_i = \begin{bmatrix} v_{i1} & \cdots & v_{i\alpha_i} & v_{i\alpha_i+1} & \cdots & v_{i\sigma_i} \end{bmatrix}$$

最终，可以得到相应的变换矩阵 $T = Q^{-1} = \begin{bmatrix} Q_1 & \cdots & Q_l \end{bmatrix}^{-1}$。

例 1.10 计算下述动态方程的 Jordan 标准形。

$$\dot{x}(t) = \begin{bmatrix} 0 & 1 & 0 \\ 0 & 0 & 1 \\ 2 & 3 & 0 \end{bmatrix} x(t) + \begin{bmatrix} 0 \\ 0 \\ 1 \end{bmatrix} u(t), \; y(t) = \begin{bmatrix} 1 & 0 & 0 \end{bmatrix} x(t)$$

由于 $\det(sI-A) = s^3 - 3s - 2 = (s+1)^2(s-2) = 0$，所以，$A$ 的特征值为 $\lambda_1 = \lambda_2 = -1$ 和 $\lambda_3 = 2$。

接下来计算变换矩阵 T。

第一步，计算当 $\lambda_1 = \lambda_2 = -1$ 时，特征矩阵的秩满足

$$\text{rank}\,(\lambda_1 I - A) = \text{rank} \begin{bmatrix} -1 & -1 & 0 \\ 0 & -1 & -1 \\ -2 & -3 & -1 \end{bmatrix} = 2 = 3 - 1 = n - \alpha_1$$

由于 $\alpha_1 = 1$，仅存在一个与 $\lambda_1 = \lambda_2 = -1$ 相关的特征向量，即仅对应一个 Jordan 块。

与 $\lambda_1 = -1$ 相关的特征向量可以根据下式计算：

$$(\lambda_1 I - A)\, v_1 = \begin{bmatrix} -1 & -1 & 0 \\ 0 & -1 & -1 \\ -2 & -3 & -1 \end{bmatrix} v_1 = 0$$

上式的一个解是 $v_1 = \begin{bmatrix} 1 & -1 & 1 \end{bmatrix}^{\text{T}}$。

进一步，计算与 $\lambda_1 = -1$ 相关的另一个广义特征向量，可以通过求解下式得到：

$$(\lambda_1 I - A)\, v_2 = -v_1$$

即

$$\begin{bmatrix} -1 & -1 & 0 \\ 0 & -1 & -1 \\ -2 & -3 & -1 \end{bmatrix} v_2 = \begin{bmatrix} -1 \\ 1 \\ -1 \end{bmatrix}$$

上式的解为 $v_2 = \begin{bmatrix} 1 & 0 & -1 \end{bmatrix}^{\text{T}}$。

最后可以通过下式计算与 $\lambda_3 = 2$ 相关的特征向量 v_3，

$$(\lambda_3 I - A)\, v_3 = \begin{bmatrix} 2 & -1 & 0 \\ 0 & 2 & -1 \\ -2 & -3 & 2 \end{bmatrix} v_3 = 0$$

上式的解是 $v_3 = \begin{bmatrix} 1 & 2 & 4 \end{bmatrix}^{\text{T}}$。

所以，

$$Q = \begin{bmatrix} v_1 & v_2 & v_3 \end{bmatrix} = \begin{bmatrix} 1 & 1 & 1 \\ -1 & 0 & 2 \\ 1 & -1 & 4 \end{bmatrix}$$

且

$$T = Q^{-1} = \frac{1}{9}\begin{bmatrix} 2 & -5 & 2 \\ 6 & 3 & -3 \\ 1 & 2 & 1 \end{bmatrix}$$

进一步，可以计算得到

$$\bar{A} = TAT^{-1} = \begin{bmatrix} -1 & 1 & 0 \\ 0 & -1 & 0 \\ 0 & 0 & 2 \end{bmatrix}, \quad \bar{B} = TB = \frac{1}{9}\begin{bmatrix} 2 \\ -3 \\ 1 \end{bmatrix}, \quad \bar{C} = CT^{-1} = \begin{bmatrix} 1 & 1 & 1 \end{bmatrix}$$

第 2 章

线性系统的动态响应

状态空间描述为定量分析系统的运动行为提供了基础。在定量分析中，主要关注的问题是输入和初始条件如何影响系统的响应。这在控制理论中具有重要意义，例如，确定合适的输入使得系统的输出满足稳定性、能控性、能观性等要求。系统的动态响应通常包括状态响应和输出响应，其中状态响应的数学意义实际上就是状态方程的解析解。本章将详细讨论线性系统的动态响应。

2.1 概述

对于连续线性系统，运动分析问题可以归结为在初始条件和外部输入确定的情况下，求解状态方程。第 1 章中指出，线性时变系统的状态方程为

$$\dot{x}(t) = A(t)x(t) + B(t)u(t),\ x(t_0) = x_0,\ t \in [t_0, t_\text{f}] \tag{2.1}$$

线性定常系统的状态方程为

$$\dot{x}(t) = Ax(t) + Bu(t),\ x(0) = x_0,\ t \in [0, \infty) \tag{2.2}$$

求解向量微分方程 (2.1) 和 (2.2)，可以得到解析解 $x(t) \in \mathbb{R}^n$，把它代入到输出方程中，可以进一步获得系统的输出响应 $y(t)$。尽管系统的响应由初始状态和外部输入激励，但是系统的运动模态主要由系统的结构和参数决定，即由参数矩阵 $(A(t), B(t))$ 或 (A, B) 决定。

显然，只有在状态方程具有唯一解时，分析系统的运动才有意义。所以，有必要对系统的参数矩阵和输入引入额外的约束条件，以确保解的存在性和唯一性。

对于线性时变系统方程 (2.1)，如果 $A(t)$ 和 $B(t)$ 的所有元在 $[t_0, t_\text{f}]$ 上都是关于 t 的实连续函数，$u(t)$ 的所有元在 $[t_0, t_\text{f}]$ 上也都是关于 t 的实连续函数，那么方程 (2.1) 的解存在且唯一。对于实际的物理系统，这些条件通常都是满足的，但是，从数学分析的角度，这些条件的限制性过强，它们可以退化为下述三个条件：

(1) 状态系数矩阵 $A(t)$ 的每个元，$a_{ij}(t)$ 在 $[t_0, t_\text{f}]$ 上是绝对可积的，即

$$\int_{t_0}^{t_\text{f}} |a_{ij}(t)|\,\mathrm{d}t < \infty,\ i,j = 1, \cdots, n$$

(2) 控制系数矩阵 $B(t)$ 的每个元，$b_{ik}(t)$ 在 $[t_0, t_\text{f}]$ 上是平方可积的，即

$$\int_{t_0}^{t_\text{f}} (b_{ik}(t))^2\,\mathrm{d}t < \infty,\ i = 1, \cdots, n,\ k = 1, \cdots, p$$

(3) $u(t)$ 的每个元, $u_k(t)$ 在 $[t_0, t_\mathrm{f}]$ 上是平方可积的, 即

$$\int_{t_0}^{t_\mathrm{f}} (u_k(t))^2 \, \mathrm{d}t < \infty, \ k = 1, \cdots, p$$

其中, n 是状态 $x(t)$ 的维数, p 是输入 $u(t)$ 的维数。应用 Schwarz(施瓦茨)不等式, 可得

$$\sum_{k=1}^{p} \int_{t_0}^{t_\mathrm{f}} |b_{ik}(t) u_k(t)| \, \mathrm{d}t \leqslant \sum_{k=1}^{p} \left(\int_{t_0}^{t_\mathrm{f}} (b_{ik}(t))^2 \, \mathrm{d}t \int_{t_0}^{t_\mathrm{f}} (u_k(t))^2 \, \mathrm{d}t \right)^{\frac{1}{2}}$$

这表明条件 (2) 和 (3) 等价于 $B(t)u(t)$ 的每个元在 $[t_0, t_\mathrm{f}]$ 上绝对可积。

对于线性定常系统方程 (2.2), 由于 A 和 B 是常值矩阵并且它们的元都是有限值, 所以, 条件 (1) 和 (2) 总是成立的。

后续章节中, 假设考虑的线性系统总是满足上述条件。

注意到线性系统满足叠加原理, 因此, 由初始状态 x_0 和输入 $u(t)$ 同时激励的状态响应 $x(t)$ 可以被分解为由 x_0 和 $u(t)$ 分别单独激励的状态响应 $x(t)$。

仅由初始状态 x_0 激励的状态响应 $x(t)$, 即 $x_0 \neq 0$, $u(t) = 0$ 时的响应, 被称为零输入响应。它代表状态的自由运动。一般地, 没有外部输入的系统被称为自治系统, 时变情形下可以表示为

$$\dot{x}(t) = A(t)x(t), \ x(t_0) = x_0, \ t \in [t_0, t_\mathrm{f}] \tag{2.3}$$

定常情形下可以表示为

$$\dot{x}(t) = Ax(t), \ x(0) = x_0, \ t \in [0, \infty) \tag{2.4}$$

式 (2.3) 和式 (2.4) 的解是相应系统的零输入响应, 一般被记作 $\phi(t; t_0, x_0, 0)$。

仅由输入 $u(t)$ 激励的状态响应 $x(t)$, 即 $x_0 = 0$, $u(t) \neq 0$ 时的响应, 被称为零状态响应。它表示系统状态的受迫运动。这种情况下, 连续线性时变系统可以表示为

$$\dot{x}(t) = A(t)x(t) + B(t)u(t), \ x(t_0) = 0, \ t \in [t_0, t_\mathrm{f}] \tag{2.5}$$

连续线性定常系统则可以表示为

$$\dot{x}(t) = Ax(t) + Bu(t), \ x(t_0) = 0, \ t \in [0, \infty) \tag{2.6}$$

式 (2.5) 和式 (2.6) 的解是相应系统的零状态响应, 通常被记作 $\phi(t; t_0, 0, u)$。

系统由初始状态 x_0 和输入 $u(t)$ 同时激励的状态响应 $x(t)$, 通常被记作 $\phi(t; t_0, x_0, u)$, 是零输入响应和零状态响应的叠加, 即

$$\phi(t; t_0, x_0, u) = \phi(t; t_0, x_0, 0) + \phi(t; t_0, 0, u) \tag{2.7}$$

2.2 线性时变系统的动态响应

本节讨论线性时变系统的动态响应的特性。

2.2.1 线性时变系统的状态转移矩阵

首先,讨论齐次状态方程的解空间。

考虑以下线性时变齐次状态方程,

$$\dot{x}(t) = A(t)x(t), \ x(t_0) = x_0, \ t \in [t_0, t_{\mathrm{f}}] \tag{2.8}$$

其中,$x(t) \in \mathbb{R}^n$ 是状态向量;$A(t)$ 是时变矩阵,且它的每个元对于 $t \in [t_0, t_{\mathrm{f}}]$ 都绝对可积。2.1节中指出,这一方程具有唯一解。

定理 2.1 n 维线性时变齐次方程 (2.8) 的解集构成实域内的 n 维线性空间。

证明 首先,证明 n 维线性时变齐次方程 (2.8) 的解集构成了实域内的线性空间。

考虑 $\psi_1(t)$ 和 $\psi_2(t)$ 是 (2.8) 的任意两个解,那么对于任意实数 α_1 和 α_2,$\alpha_1\psi_1(t) + \alpha_2\psi_2(t)$ 也是 (2.8) 的解,这是由于

$$\frac{\mathrm{d}}{\mathrm{d}t}(\alpha_1\psi_1(t) + \alpha_2\psi_2(t)) = \alpha_1 \frac{\mathrm{d}}{\mathrm{d}t}\psi_1(t) + \alpha_2 \frac{\mathrm{d}}{\mathrm{d}t}\psi_2(t)$$
$$= \alpha_1 A(t)\psi_1(t) + \alpha_2 A(t)\psi_2(t)$$
$$= A(t)(\alpha_1\psi_1(t) + \alpha_2\psi_2(t))$$

所以,它的解集构成了实域内的线性空间。

接下来,证明这一线性空间是 n 维的。

假设 $e_1, \cdots, e_n$ 是 n 个线性无关的向量,$\psi_i(t)$ 是式 (2.8) 在 $\psi_i(t_0) = e_i$ 时的解,$i \in \{1, \cdots, n\}$。由反证法,假设 $\psi_1(t), \cdots, \psi_n(t)$ 是线性相关的,那么一定存在实向量 $\alpha \neq 0 \in \mathbb{R}^n$,使得

$$\begin{bmatrix} \psi_1(t) & \cdots & \psi_n(t) \end{bmatrix} \alpha = 0, \ \forall t \in [t_0, \infty)$$

当 $t = t_0$ 时,上述等式仍然成立,即

$$\begin{bmatrix} \psi_1(t_0) & \cdots & \psi_n(t_0) \end{bmatrix} \alpha = \begin{bmatrix} e_1 & \cdots & e_n \end{bmatrix} \alpha = 0$$

这表明向量组 $e_1, \cdots, e_n$ 是线性相关的,与假设矛盾。由此表明 $\psi_1(t), \cdots, \psi_n(t)$ 对于 $t \in [t_0, \infty)$ 都是线性无关的,进一步可知解空间是 n 维的。□

假设 $\psi(t)$ 是式 (2.8) 的解,且 $\psi(t_0) = e$。显然,e 可以被表示为

$$e = \alpha_1 e_1 + \alpha_2 e_2 + \cdots + \alpha_n e_n$$

其中,$\alpha_1, \cdots, \alpha_n$ 是实常数。考虑 $\sum_{i=1}^{n} \alpha_i \psi_i(t)$ 也是式 (2.8) 的一个解,且满足初始条件 $e = \sum_{i=1}^{n} \alpha_i e_i$。由微分方程具有唯一解,可知

$$\psi(t) = \sum_{i=1}^{n} \alpha_i \psi_i(t) \tag{2.9}$$

这表明 $\psi(t)$ 可以被描述为 $\psi_i(t)$ 的线性组合，$i \in \{1, \cdots, n\}$。

根据定理 2.1，一个 n 维的线性时变齐次方程 (2.8) 有且仅有 n 个线性无关的解。基于这一事实，引出下述定义。

定义 2.1（线性时变系统的基本解矩阵） 考虑 $\psi_1(t), \cdots, \psi_n(t)$ 是 n 维线性时变齐次方程 (2.8) 的 n 个线性无关解，$t \in [t_0, t_f]$，则矩阵

$$\Psi(t) = \begin{bmatrix} \psi_1(t) & \cdots & \psi_n(t) \end{bmatrix}$$

被称作线性时变系统的基本解矩阵。

定理 2.2 线性时变系统的基本解矩阵 $\Psi(t)$ 满足矩阵微分方程

$$\dot{\Psi}(t) = A(t)\Psi(t),\ \Psi(t_0) = E,\ t \in [t_0, t_f] \tag{2.10}$$

其中，E 是非奇异的实常值矩阵。

证明 可以推得

$$\dot{\Psi}(t) = \begin{bmatrix} \dot{\psi}_1(t) & \cdots & \dot{\psi}_n(t) \end{bmatrix} = \begin{bmatrix} A(t)\psi_1(t) & \cdots & A(t)\psi_n(t) \end{bmatrix} = A(t)\Psi(t)$$

在此基础上，根据 $\psi_1(t), \cdots, \psi_n(t)$ 的线性无关性，很容易推得 E 是非奇异的。 □

事实上，线性时变系统的基本解矩阵 $\Psi(t)$ 也可以基于定理 2.2 来定义，即对于任意非奇异的实常值矩阵 $E \in \mathbb{R}^{n \times n}$，满足式 (2.10) 的 $n \times n$ 的矩阵 $\Psi(t)$ 被称作线性时变系统的基本解矩阵。显然，由于矩阵 E 的选择是不唯一的，线性时变系统的基本解矩阵 $\Psi(t)$ 也是不唯一的。

定理 2.3 如果 $\Psi_1(t)$ 是线性时变系统方程 (2.8) 的基本解矩阵，那么对于任意的非奇异矩阵 $C \in \mathbb{R}^{n \times n}$，$\Psi_2(t) = \Psi_1(t)C$ 也是这一系统的基本解矩阵。另一方面，如果 $\Psi_1(t)$ 和 $\Psi_2(t)$ 都是线性时变系统方程 (2.8) 的基本解矩阵，那么一定存在一个非奇异矩阵 $C \in \mathbb{R}^{n \times n}$ 使得 $\Psi_2(t) = \Psi_1(t)C$。

证明 可以证得

$$\dot{\Psi}_2(t) = \dot{\Psi}_1(t)C = A(t)\Psi_1(t)C = A(t)\Psi_2(t)$$

与此同时，由于 $\Psi_1(t_0)$ 和 C 都是非奇异的实常值矩阵，所以 $\Psi_2(t_0) = \Psi_1(t_0)C$ 也是非奇异的实常值矩阵。因此，$\psi_2(t)$ 也是这一系统的基本解矩阵。

接下来，注意到 $\Psi_1(t)$ 对于 $\forall t \in [t_0, t_f]$ 都是非奇异的，所以 $\Psi_1^{-1}(t)$ 存在。考虑到 $\Psi_1(t)\Psi_1^{-1}(t) = I$，对其两端取微分，可以得到

$$\left(\frac{\mathrm{d}}{\mathrm{d}t}\Psi_1(t)\right)\Psi_1^{-1}(t) + \Psi_1(t)\frac{\mathrm{d}}{\mathrm{d}t}\Psi_1^{-1}(t) = 0$$

在此基础上，可以计算出

$$\frac{\mathrm{d}}{\mathrm{d}t}\left(\Psi_1^{-1}(t)\Psi_2(t)\right) = \left(\frac{\mathrm{d}}{\mathrm{d}t}\Psi_1^{-1}(t)\right)\Psi_2(t) + \Psi_1^{-1}(t)\frac{\mathrm{d}}{\mathrm{d}t}\Psi_2(t)$$

2.2 线性时变系统的动态响应

$$= - \Psi_1^{-1}(t) \left(\frac{\mathrm{d}}{\mathrm{d}t} \Psi_1(t) \right) \Psi_1^{-1}(t) \Psi_2(t) + \Psi_1^{-1}(t) A(t) \Psi_2(t)$$

$$= - \Psi_1^{-1}(t) A(t) \Psi_1(t) \Psi_1^{-1}(t) \Psi_2(t) + \Psi_1^{-1}(t) A(t) \Psi_2(t)$$

$$= 0$$

这表明 $\Psi_1^{-1}(t)\Psi_2(t)$ 是一个常值矩阵。 □

下面，介绍线性时变系统的状态转移矩阵。

齐次状态方程 (2.8) 可以被写作

$$\mathrm{d}x(t) = A(t)x(t)\mathrm{d}t$$

对上式两端从 t_0 到 t 进行积分，可以得到

$$x(t) - x(t_0) = \int_{t_0}^{t} A(\tau_1)x(\tau_1)\mathrm{d}\tau_1 \tag{2.11}$$

对于给定的初始条件 $x(t_0) = x_0$，选取第零次近似 $x(t) \approx x_0$。把它代入

$$x(t) = x_0 + \int_{t_0}^{t} A(\tau_1)x_0 \mathrm{d}\tau_1 = x_0 + \int_{t_0}^{t} A(\tau_1)\mathrm{d}\tau_1 x_0$$

选取第一次近似 $x(t) \approx x_0 + \int_{t_0}^{t} A(\tau_1)\mathrm{d}\tau_1 x_0$，并代入方程 (2.11) 右端，可以得到

$$x(t) = x_0 + \int_{t_0}^{t} A(\tau_1) \left(x_0 + \int_{t_0}^{\tau_1} A(\tau_2)\mathrm{d}\tau_2 x_0 \right) \mathrm{d}\tau_1$$

$$= x_0 + \int_{t_0}^{t} A(\tau_1)\mathrm{d}\tau_1 x_0 + \int_{t_0}^{t} A(\tau_1) \int_{t_0}^{\tau_1} A(\tau_2)\mathrm{d}\tau_2 \mathrm{d}\tau_1 x_0$$

$$= \left(I + \int_{t_0}^{t} A(\tau_1)\mathrm{d}\tau_1 + \int_{t_0}^{t} A(\tau_1) \int_{t_0}^{\tau_1} A(\tau_2)\mathrm{d}\tau_2 \mathrm{d}\tau_1 \right) x_0$$

重复这一过程，可以无穷次近似逼近 $x(t)$，从而有

$$x(t) = \left(I + \int_{t_0}^{t} A(\tau_1)\mathrm{d}\tau_1 + \int_{t_0}^{t} A(\tau_1) \int_{t_0}^{\tau_1} A(\tau_2)\mathrm{d}\tau_2 \mathrm{d}\tau_1 \right.$$

$$\left. + \int_{t_0}^{t} A(\tau_1) \int_{t_0}^{\tau_1} A(\tau_2) \int_{t_0}^{\tau_2} A(\tau_3)\mathrm{d}\tau_3 \mathrm{d}\tau_2 \mathrm{d}\tau_1 + \cdots \right) x_0$$

定义

$$\Phi(t, t_0) \triangleq I + \int_{t_0}^{t} A(\tau_1)\mathrm{d}\tau_1 + \int_{t_0}^{t} A(\tau_1) \int_{t_0}^{\tau_1} A(\tau_2)\mathrm{d}\tau_2 \mathrm{d}\tau_1$$

$$+ \int_{t_0}^{t} A(\tau_1) \int_{t_0}^{\tau_1} A(\tau_2) \int_{t_0}^{\tau_2} A(\tau_3)\mathrm{d}\tau_3 \mathrm{d}\tau_2 \mathrm{d}\tau_1 + \cdots \tag{2.12}$$

式 (2.12) 中的级数被称为 Peano-Baker（佩亚诺-贝克）级数，$\Phi(t,t_0)$ 通常被称为线性时变系统的状态转移矩阵。

注意到

$$\frac{\partial}{\partial t}\Phi(t,t_0) = A(t) + A(t)\int_{t_0}^{t} A(\tau_1)\mathrm{d}\tau_1 + A(t)\int_{t_0}^{t} A(\tau_1)\int_{t_0}^{\tau_1} A(\tau_2)\mathrm{d}\tau_2\mathrm{d}\tau_1 + \cdots$$

$$= A(t)\left(I + \int_{t_0}^{t} A(\tau_1)\mathrm{d}\tau_1 + \int_{t_0}^{t} A(\tau_1)\int_{t_0}^{\tau_1} A(\tau_2)\mathrm{d}\tau_2\mathrm{d}\tau_1 + \cdots\right)$$

且

$$\Phi(t_0,t_0) = I + \int_{t_0}^{t_0} A(\tau_1)\mathrm{d}\tau_1 + \int_{t_0}^{t_0} A(\tau_1)\int_{t_0}^{\tau_1} A(\tau_2)\mathrm{d}\tau_2\mathrm{d}\tau_1 + \cdots = I$$

也就是说，线性时变系统的状态转移矩阵满足

$$\frac{\partial}{\partial t}\Phi(t,t_0) = A(t)\Phi(t,t_0),\ \Phi(t_0,t_0) = I,\ t \in [t_0, t_\mathrm{f}] \tag{2.13}$$

由此，线性时变系统的状态转移矩阵通常以下述方式定义。

定义 2.2（线性时变系统的状态转移矩阵） 对于线性时变系统方程 (2.8)，满足方程 (2.13) 的 $n \times n$ 矩阵 $\Phi(t,t_0)$ 被称为线性时变系统的状态转移矩阵。

定理 2.4 如果 $\Psi(t)$ 是线性时变系统方程 (2.8) 的基本解矩阵，那么状态转移矩阵 $\Phi(t,t_0)$ 满足

$$\Phi(t,t_0) = \Psi(t)\Psi^{-1}(t_0),\ t \in [t_0, t_\mathrm{f}] \tag{2.14}$$

证明 可以证得

$$\frac{\partial}{\partial t}\Phi(t,t_0) = \frac{\mathrm{d}}{\mathrm{d}t}\Psi(t)\Psi^{-1}(t_0) = A(t)\Psi(t)\Psi^{-1}(t_0) = A(t)\Phi(t,t_0)$$

同时，当 $t = t_0$ 时，$\Phi(t_0,t_0) = \Psi(t_0)\Psi^{-1}(t_0) = I$。根据定义 2.2，$\Phi(t,t_0)$ 是状态转移矩阵。 □

定理 2.5 线性时变系统的状态转移矩阵 $\Phi(t,t_0)$ 是唯一的。

证明 考虑 $\Phi_1(t,t_0)$ 和 $\Phi_2(t,t_0)$ 是系统方程 (2.8) 的两个状态转移矩阵，它们分别由两个基本解矩阵 $\Psi_1(t)$ 和 $\Psi_2(t)$ 生成。结合定理 2.3 和定理 2.4，可以得到

$$\Phi_2(t,t_0) = \Psi_2(t)\Psi_2^{-1}(t_0) = \Psi_1(t)CC^{-1}\Psi_1^{-1}(t_0) = \Psi_1(t)\Psi_1^{-1}(t_0) = \Phi_1(t,t_0)$$

这表明对于同一个线性时变系统，$\Phi(t,t_0)$ 是唯一的，且与 $\Psi(t)$ 的选择无关。 □

接下来，讨论状态转移矩阵 $\Phi(t,t_0)$ 的性质。

(1) $\Phi(t_0,t_0) = I$。

(2) $\Phi^{-1}(t,t_0) = \Phi(t_0,t)$。

(3) $\Phi(t_2,t_1)\Phi(t_1,t_0) = \Phi(t_2,t_0)$。

根据式 (2.14)，很容易导出上述性质。

(4) $\dfrac{\partial}{\partial t}\Phi^{-1}(t,t_0) = \dfrac{\partial}{\partial t}\Phi(t_0,t) = -\Phi(t_0,t)A(t)$。

由于 $\Phi(t_0,t_0) = \Phi(t_0,t)\Phi(t,t_0) = I$，进一步可得

$$\left(\dfrac{\partial}{\partial t}\Phi(t_0,t)\right)\Phi(t,t_0) + \Phi(t_0,t)\dfrac{\partial}{\partial t}\Phi(t,t_0) = 0$$

这表明

$$\begin{aligned}\dfrac{\partial}{\partial t}\Phi(t_0,t) &= -\Phi(t_0,t)\left(\dfrac{\partial}{\partial t}\Phi(t,t_0)\right)\Phi^{-1}(t,t_0)\\ &= -\Phi(t_0,t)A(t)\Phi(t,t_0)\Phi^{-1}(t,t_0)\\ &= -\Phi(t_0,t)A(t)\end{aligned}$$

2.2.2 线性时变系统的零输入响应

这一小节中，讨论线性时变系统方程 (2.8) 的零输入响应。

定理 2.6 线性时变系统的零输入响应，即齐次状态方程 (2.8) 的解具有以下形式，

$$x(t) = \phi(t;t_0,x_0,0) = \Phi(t,t_0)x_0, \ t \in [t_0,t_{\mathrm{f}}] \tag{2.15}$$

其中，$\Phi(t,t_0)$ 是线性时变系统方程 (2.8) 的状态转移矩阵。

根据前一小节的讨论，可以很容易证明这一定理。

为了得到齐次状态方程 (2.8) 封闭形式的解，需要确定封闭形式的 $\Phi(t,t_0)$，但是对于一般的线性时变系统这很难做到。特别地，有以下结论。

定理 2.7 对于线性时变系统方程 (2.8)，如果 $A(t)$ 和 $\int_{t_0}^{t}A(\tau)\mathrm{d}\tau$ 是可交换的，即

$$A(t)\int_{t_0}^{t}A(\tau)\mathrm{d}\tau = \int_{t_0}^{t}A(\tau)\mathrm{d}\tau A(t) \tag{2.16}$$

那么相应的状态转移矩阵为

$$\Phi(t,t_0) = \exp\left\{\int_{t_0}^{t}A(\tau)\mathrm{d}\tau\right\} \tag{2.17}$$

证明 可知下式成立

$$\exp\left\{\int_{t_0}^{t}A(\tau)\mathrm{d}\tau\right\} = I + \int_{t_0}^{t}A(\tau)\mathrm{d}\tau + \dfrac{1}{2!}\left(\int_{t_0}^{t}A(\tau)\mathrm{d}\tau\right)^2 + \cdots$$

在此基础上，可以得到

$$\dfrac{\mathrm{d}}{\mathrm{d}t}\exp\left\{\int_{t_0}^{t}A(\tau)\mathrm{d}\tau\right\} = A(t) + \dfrac{1}{2!}\left(A(t)\int_{t_0}^{t}A(\tau)\mathrm{d}\tau + \int_{t_0}^{t}A(\tau)\mathrm{d}\tau A(t)\right) + \cdots$$

$$= A(t) + A(t)\int_{t_0}^t A(\tau)\mathrm{d}\tau + \cdots = A(t)\exp\left\{\int_{t_0}^t A(\tau)\mathrm{d}\tau\right\}$$

且

$$\exp\left\{\int_{t_0}^t A(\tau)\mathrm{d}\tau\right\}\bigg|_{t=t_0} = I$$

所以，根据定义 2.2，可知式 (2.17) 成立。□

如果 $A(t)$ 是对角阵，式 (2.16) 成立，进一步可以使用式 (2.17) 来计算 $\Phi(t,t_0)$。对于一般的线性时变系统，式 (2.16) 不成立，$\Phi(t,t_0)$ 很难表示为封闭形式。

2.2.3 线性时变系统的零状态响应

这一小节讨论线性时变系统的零状态响应。

考虑线性时变系统

$$\dot{x}(t) = A(t)x(t) + B(t)u(t),\ x(t_0) = 0,\ t \in [t_0, t_\mathrm{f}] \tag{2.18}$$

其中，$x(t) \in \mathbb{R}^n$ 是状态向量，$u(t) \in \mathbb{R}^p$ 是输入向量，$A(t)$ 和 $B(t)$ 是时变矩阵。假设式 (2.18) 满足唯一解条件，它的零状态响应满足下述结论。

定理 2.8 线性时变系统的零状态响应，即非齐次状态方程 (2.18) 的解具有下述形式，

$$x(t) = \phi(t;t_0,0,u) = \int_{t_0}^t \Phi(t,\tau)B(\tau)u(\tau)\mathrm{d}\tau,\ t \in [t_0, t_\mathrm{f}] \tag{2.19}$$

证明 根据 $\Phi(t,t_0)$ 的性质，可得

$$\frac{\partial}{\partial t}\Phi(t_0,t) = -\Phi(t_0,t)A(t) \tag{2.20}$$

在式 (2.18) 的两端分别乘 $\Phi(t_0,t)$ 可以得到

$$\Phi(t_0,t)\dot{x}(t) - \Phi(t_0,t)A(t)x(t) = \Phi(t_0,t)B(t)u(t)$$

结合式 (2.20)，上式可以被写作

$$\frac{\partial}{\partial t}\left(\Phi(t_0,t)x(t)\right) = \Phi(t_0,t)B(t)u(t)$$

对上式两端从 t_0 到 t 取积分，在零初始条件下，可以得到

$$\Phi(t_0,t)x(t) = \int_{t_0}^t \Phi(t_0,\tau)B(\tau)u(\tau)\mathrm{d}\tau$$

注意到 $\Phi(t_0,t)\Phi(t,t_0) = I$，$\Phi(t,t_0)\Phi(t_0,\tau) = \Phi(t,\tau)$，进一步能够得到

$$x(t) = \Phi(t,t_0)\int_{t_0}^t \Phi(t_0,\tau)B(\tau)u(\tau)\mathrm{d}\tau = \int_{t_0}^t \Phi(t,\tau)B(\tau)u(\tau)\mathrm{d}\tau, t \in [t_0, t_\mathrm{f}] \quad □$$

2.2.4 线性时变系统的状态响应和输出响应

结合式 (2.15) 与式 (2.19)，线性时变系统的状态响应，即状态方程同时由初始状态 x_0 和输入 $u(t)$ 激励的解具有以下形式，

$$x(t) = \phi(t; t_0, x_0, u) = \Phi(t, t_0) x_0 + \int_{t_0}^{t} \Phi(t, \tau) B(\tau) u(\tau) \mathrm{d}\tau, \ t \in [t_0, t_\mathrm{f}] \tag{2.21}$$

线性系统同时由初始状态 x_0 和输入 $u(t)$ 激励的状态响应可以记作 $\phi(t; t_0, x_0, u)$，它是零输入响应和零状态响应的叠加。

可以直观地看出，状态响应由两部分组成，一部分是与 x_0 相关的初始状态转移项，另一部分是输入 $u(t)$ 激励的受迫项。受迫项的存在使通过设计合适的输入 $u(t)$ 以改善系统的性能变得可能。

如果已知输出方程，根据式 (2.21) 可以进一步导出系统的输出响应为

$$y(t) = C(t)\Phi(t, t_0) x_0 + \int_{t_0}^{t} C(t)\Phi(t, \tau) B(\tau) u(\tau) \mathrm{d}\tau + D(t)u(t), \ t \in [t_0, t_\mathrm{f}] \tag{2.22}$$

在式 (2.22) 中，若考虑 $x_0 = 0$，同时 $u(t)$ 可以被表示为

$$u(t) = \int_{t_0}^{t} \delta(t - \tau) u(\tau) \mathrm{d}\tau$$

其中，$\delta(t)$ 是脉冲函数。由此，可以得到

$$y(t) = \int_{t_0}^{t} (C(t)\Phi(t, \tau) B(\tau) + D(t)\delta(t - \tau)) u(\tau) \mathrm{d}\tau, \ t \in [t_0, t_\mathrm{f}] \tag{2.23}$$

1.2.1 节中指出，在 t_0 时刻松弛的线性时变系统可以被描述为

$$y(t) = \int_{t_0}^{t} G(t, \tau) u(\tau) \mathrm{d}\tau \tag{2.24}$$

其中，$G(t, \tau)$ 是脉冲响应矩阵。比较式 (2.23) 和式 (2.24)，可以得到在 t_0 时刻松弛的线性时变系统的脉冲响应矩阵为

$$G(t, \tau) = C(t)\Phi(t, \tau) B(\tau) u(\tau) + D(t)\delta(t - \tau) \tag{2.25}$$

例 2.1 计算下述动态方程的输出响应和脉冲响应矩阵，

$$\dot{x}(t) = \begin{bmatrix} 0 & \cos t \\ 0 & 0 \end{bmatrix} x(t) + \begin{bmatrix} 0 \\ 1 \end{bmatrix} u(t), \ y(t) = \begin{bmatrix} 1 & 0 \\ 0 & 1 \end{bmatrix} x(t) + \begin{bmatrix} 1 \\ 1 \end{bmatrix} u(t)$$

首先，计算基本解矩阵 $\Psi(t)$。由于 $\Psi(t)$ 满足式 (2.9)，这里，选择 $E = I$，可以得到

$$\dot{\psi}_{11}(t) = \cos t \psi_{21}(t), \ \psi_{11}(t_0) = 1, \ \dot{\psi}_{21}(t) = 0, \ \psi_{21}(t_0) = 0$$

$$\dot{\psi}_{12}(t) = \cos t\, \psi_{22}(t),\ \psi_{12}(t_0) = 0,\ \dot{\psi}_{22}(t) = 0,\ \psi_{22}(t_0) = 1$$

分别求解上述方程，可以得到

$$\psi_1(t) = \begin{bmatrix} \psi_{11}(t) \\ \psi_{21}(t) \end{bmatrix} = \begin{bmatrix} 1 \\ 0 \end{bmatrix},\quad \psi_2(t) = \begin{bmatrix} \psi_{12}(t) \\ \psi_{22}(t) \end{bmatrix} = \begin{bmatrix} \sin t - \sin t_0 \\ 1 \end{bmatrix}$$

所以，基本解矩阵 $\Psi(t)$ 为

$$\Psi(t) = \begin{bmatrix} \psi_1(t) & \psi_2(t) \end{bmatrix} = \begin{bmatrix} 1 & \sin t - \sin t_0 \\ 0 & 1 \end{bmatrix}$$

状态转移矩阵 $\Phi(t, t_0)$ 为

$$\Phi(t, t_0) = \Psi(t)\Psi^{-1}(t_0) = \begin{bmatrix} 1 & \sin t - \sin t_0 \\ 0 & 1 \end{bmatrix}$$

根据式 (2.22)，输出响应可以表示为

$$\begin{aligned} y(t) &= C(t)\Phi(t, t_0)x_0 + \int_{t_0}^{t} C(t)\Phi(t, \tau)B(\tau)u(\tau)\mathrm{d}\tau + D(t)u(t) \\ &= \begin{bmatrix} 1 & \sin t - \sin t_0 \\ 0 & 1 \end{bmatrix} x_0 + \int_{t_0}^{t} \begin{bmatrix} \sin t - \sin \tau \\ 1 \end{bmatrix} u(\tau)\mathrm{d}\tau + \begin{bmatrix} 1 \\ 1 \end{bmatrix} u(t) \end{aligned}$$

根据式 (2.25)，本例中系统的脉冲响应矩阵为

$$G(t, \tau) = C(t)\Phi(t, \tau)B(\tau) + D(t)\delta(t - \tau) = \begin{bmatrix} \sin t - \sin \tau \\ 1 \end{bmatrix} + \begin{bmatrix} 1 \\ 1 \end{bmatrix}\delta(t - \tau)$$

在 1.3.4 节中已经介绍了线性动态的等价性。接下来，讨论具有等价性的线性时变系统的响应特性。

如果两个线性时变系统 $(A(t), B(t), C(t), D(t))$ 和 $(\bar{A}(t), \bar{B}(t), \bar{C}(t), \bar{D}(t))$ 是等价的[①]，那么总是存在非奇异的、连续可微的变换矩阵 $T(t)$ 使得 $\bar{A}(t) = T(t)A(t)T^{-1}(t) + \dot{T}(t)T^{-1}(t)$，$\bar{B}(t) = T(t)B(t)$，$\bar{C}(t) = C(t)T^{-1}(t)$，$\bar{D}(t) = D(t)$，且 $\bar{x}(t) = T(t)x(t)$。令 $\Psi(t)$ 是 $(A(t), B(t), C(t), D(t))$ 的一个基本解矩阵，那么结合式 (2.10) 可知

$$\begin{aligned} \frac{\mathrm{d}}{\mathrm{d}t}(T(t)\Psi(t)) &= \frac{\mathrm{d}}{\mathrm{d}t}(T(t))\Psi(t) + T(t)\frac{\mathrm{d}}{\mathrm{d}t}\Psi(t) = \left(\frac{\mathrm{d}}{\mathrm{d}t}T(t) + T(t)A(t)\right)\Psi(t) \\ &= \left(\frac{\mathrm{d}}{\mathrm{d}t}T(t) + T(t)A(t)\right)T^{-1}(t)T(t)\Psi(t) = \bar{A}(t)T(t)\Psi(t) \end{aligned}$$

---

① 方便起见，用 $(A(t), B(t), C(t), D(t))$ 表示动态方程 (1.11) 和 (1.12)，用 $(\bar{A}(t), \bar{B}(t), \bar{C}(t), \bar{D}(t))$ 表示动态方程 (1.24) 和 (1.25)。

同时可知，$T(t_0)\Psi(t_0) = T(t_0)E$ 是非奇异的常值矩阵。所以，$\bar{\Psi}(t) = T(t)\Psi(t)$ 是 $(\bar{A}(t), \bar{B}(t), \bar{C}(t), \bar{D}(t))$ 的一个基本解矩阵。进一步，用 $\Phi(t,t_0)$ 和 $\bar{\Phi}(t,t_0)$ 分别表示两个系统的状态转移矩阵，结合式 (2.14) 可知，

$$\bar{\Phi}(t,t_0) = \bar{\Psi}(t)\bar{\Psi}^{-1}(t_0) = T(t)\Phi(t,t_0)T^{-1}(t_0) \tag{2.26}$$

考虑零输入的情况，在初始状态为 x_0 的条件下，系统 $(A(t), B(t), C(t), D(t))$ 的输出为

$$y(t) = C(t)\Phi(t,t_0)x_0$$

而系统 $(\bar{A}(t), \bar{B}(t), \bar{C}(t), \bar{D}(t))$ 在初始状态 $\bar{x}_0 = T(t_0)x_0$ 作用下的输出为

$$\bar{y}(t) = \bar{C}(t)\bar{\Phi}(t,t_0)\bar{x}_0 = C(t)T^{-1}(t)T(t)\Phi(t,t_0)T^{-1}(t_0)T(t_0)x_0 = C(t)\Phi(t,t_0)x_0$$

也就是说，对于两个等价的线性时变系统，当一个系统在初始状态 x_0 的作用下的输出为 $y(t)$ 时，另一个系统总是能够找到一个相应的初始状态，使得在零输入情况下，两个系统具有相同的输出。

用 $G(t,\tau)$ 和 $\bar{G}(t,\tau)$ 表示系统 $(A(t), B(t), C(t), D(t))$ 和 $(\bar{A}(t), \bar{B}(t), \bar{C}(t), \bar{D}(t))$ 的脉冲响应矩阵，结合式 (2.25) 和式 (2.26)，可以得到

$$\begin{aligned}\bar{G}(t,\tau) &= \bar{C}(t)\bar{\Phi}(t,\tau)\bar{B}(\tau) + \bar{D}(t)\delta(t-\tau) \\ &= C(t)T^{-1}(t)T(t)\Phi(t,\tau)T^{-1}(\tau)T(\tau)B(\tau) + D(t)\delta(t-\tau) \\ &= C(t)\Phi(t,\tau)B(\tau) + D\delta(t-\tau) = G(t,\tau)\end{aligned} \tag{2.27}$$

这表明两个等价的线性时变系统具有相同的脉冲响应矩阵。

进一步，考虑零状态的情况，对于相同的输入 $u(t)$，两个等价的线性时变系统的输出总是相同的，这一结论可以根据式 (2.24) 和式 (2.27) 很容易推得。

2.3 线性定常系统的动态响应

本节将讨论线性定常系统的动态响应的性质。

2.3.1 线性定常系统的矩阵指数

在讨论线性定常系统的响应之前，首先介绍状态系数矩阵 A 的矩阵指数。

定义 2.3（矩阵指数） 矩阵 $A \in \mathbb{R}^{n \times n}$ 的矩阵指数定义为

$$e^{At} \triangleq I + At + \frac{1}{2!}A^2t^2 + \cdots = \sum_{k=0}^{\infty}\frac{1}{k!}A^kt^k \tag{2.28}$$

矩阵指数在分析线性定常系统的响应时具有重要作用。下面，首先讨论 e^{At} 的性质。

(1) e^{At} 在 $t=0$ 的值为

$$e^{At}|_{t=0} = I \tag{2.29}$$

(2) 考虑两个无关的时间变量 t 和 τ,
$$e^{A(t+\tau)} = e^{At}e^{A\tau} = e^{A\tau}e^{At} \tag{2.30}$$

(3) e^{At} 的转置满足
$$\left(e^{At}\right)^{\mathrm{T}} = e^{A^{\mathrm{T}}t} \tag{2.31}$$

根据式 (2.28) 很容易推得以上性质。

(4) e^{At} 的逆满足
$$(e^{At})^{-1} = e^{-At} \tag{2.32}$$

对于任意的状态系数矩阵 A, e^{At} 是非奇异的。由此,在式 (2.30) 中选择 $\tau = -t$ 并结合式 (2.28),可以推导出 $e^{At}e^{-At} = I$。所以式 (2.32) 成立。

(5) 对于非负整数 k,
$$(e^{At})^k = e^{A(kt)} \tag{2.33}$$

根据式 (2.30) 很容易得到式 (2.33)。

(6) 如果矩阵 A 和 F 可交换,即 $AF = FA$,那么
$$e^{(A+F)t} = e^{At}e^{Ft} = e^{Ft}e^{At} \tag{2.34}$$

根据式 (2.28),可以得到

$$\begin{aligned}
e^{At}e^{Ft} &= \left(I + At + \frac{1}{2!}A^2t^2 + \frac{1}{3!}A^3t^3 + \cdots\right)\left(I + Ft + \frac{1}{2!}F^2t^2 + \frac{1}{3!}F^3t^3 + \cdots\right) \\
&= I + (A+F)t + \frac{1}{2!}(A^2 + 2AF + F^2)t^2 + \frac{1}{3!}(A^3 + 3A^2F + 3F^2A + F^3)t^3 + \cdots
\end{aligned}$$

且

$$\begin{aligned}
e^{(A+F)t} &= I + (A+F)t + \frac{1}{2!}(A+F)^2t^2 + \frac{1}{3!}(A+F)^3t^3 + \cdots \\
&= I + (A+F)t + \frac{1}{2!}(A^2 + AF + FA + F^2)t^2 \\
&\quad + \frac{1}{3!}(A^3 + A^2F + AFA + AF^2 + FA^2 + FAF + F^2A + F^3)t^3 + \cdots
\end{aligned}$$

由此可以归纳出如果 $AF = FA$,$e^{(A+F)t} = e^{At}e^{Ft}$。采用类似的方法,可以证明 $e^{(A+F)t} = e^{Ft}e^{At}$。

(7) e^{At} 的导数满足
$$\frac{\mathrm{d}}{\mathrm{d}t}e^{At} = Ae^{At} = e^{At}A \tag{2.35}$$

2.3 线性定常系统的动态响应

根据式 (2.28)，可得

$$\frac{\mathrm{d}}{\mathrm{d}t}\mathrm{e}^{At} = A + A^2 t + \frac{1}{2!}A^3 t^2 + \cdots = A\sum_{k=0}^{\infty}\frac{1}{k!}A^k t^k = \left(\sum_{i=0}^{\infty}\frac{1}{k!}A^k t^k\right)A = A\mathrm{e}^{At} = \mathrm{e}^{At}A$$

这表明式 (2.35) 成立。

(8) e^{-At} 的导数满足

$$\frac{\mathrm{d}}{\mathrm{d}t}\mathrm{e}^{-At} = -A\mathrm{e}^{-At} = -\mathrm{e}^{-At}A \tag{2.36}$$

式 (2.36) 的证明与式 (2.35) 类似。

(9) 如果存在等价变换矩阵 T 使得 $\bar{A} = TAT^{-1}$，那么

$$\mathrm{e}^{At} = T^{-1}\mathrm{e}^{\bar{A}t}T \tag{2.37}$$

根据式 (2.28)，可以得到

$$\mathrm{e}^{TAT^{-1}t} = I + TAT^{-1}t + \frac{1}{2!}(TAT^{-1})^2 t^2 + \frac{1}{3!}(TAT^{-1})^3 t^3 + \cdots$$
$$= T\left(I + At + \frac{1}{2!}A^2 t^2 + \frac{1}{3!}A^3 t^3 + \cdots\right)T^{-1}$$

这表明式 (2.37) 成立。

(10) e^{At} 等于预解矩阵的 Laplace 反变换，即

$$\mathrm{e}^{At} = \mathcal{L}^{-1}\left[(sI - A)^{-1}\right] \tag{2.38}$$

这一结论的证明将在下一小节中给出。

接下来，讨论如何计算 e^{At}。

(1) 使用无穷级数求和的方法来计算 e^{At}。

利用矩阵指数式 (2.28) 的定义可以直接计算 e^{At}，但是这一方法很难获得具有封闭形式的解析解。

(2) 使用 Laplace 反变换来计算 e^{At}。

在 1.3.3 小节中，已经介绍了求解预解矩阵 $(sI - A)^{-1}$ 的方法。根据式 (2.38)，可以对 $(sI - A)^{-1}$ 进行 Laplace 反变换来计算 e^{At}。

(3) 使用 Jordan 标准形来计算 e^{At}。

在讨论这一方法前，首先介绍当 A 具有特殊形式时，一些典型的 e^{At}。

如果 A 是对角阵，

$$A = \begin{bmatrix} \lambda_1 & & \\ & \ddots & \\ & & \lambda_n \end{bmatrix}$$

那么 e^{At} 也是对角阵，

$$e^{At} = \begin{bmatrix} e^{\lambda_1 t} & & \\ & \ddots & \\ & & e^{\lambda_n t} \end{bmatrix} \qquad (2.39)$$

可以根据式 (2.28) 来证明这一点，

$$e^{At} = \begin{bmatrix} 1 & & \\ & \ddots & \\ & & 1 \end{bmatrix} + \begin{bmatrix} \lambda_1 & & \\ & \ddots & \\ & & \lambda_n \end{bmatrix} t + \frac{1}{2!} \begin{bmatrix} \lambda_1^2 & & \\ & \ddots & \\ & & \lambda_n^2 \end{bmatrix} t^2 + \cdots$$

$$= \begin{bmatrix} \sum_{k=0}^{\infty} \frac{1}{k!} \lambda_1^k t^k & & \\ & \ddots & \\ & & \sum_{k=0}^{\infty} \frac{1}{k!} \lambda_n^k t^k \end{bmatrix} = \begin{bmatrix} e^{\lambda_1 t} & & \\ & \ddots & \\ & & e^{\lambda_n t} \end{bmatrix}$$

如果 A 是块对角矩阵，

$$A = \begin{bmatrix} A_1 & & \\ & \ddots & \\ & & A_l \end{bmatrix}$$

那么 e^{At} 也是块对角矩阵，

$$e^{At} = \begin{bmatrix} e^{A_1 t} & & \\ & \ddots & \\ & & e^{A_l t} \end{bmatrix} \qquad (2.40)$$

可以根据式 (2.28) 来证明，这里省略详细过程。

如果 $n \times n$ 的矩阵 A 具有以下形式，

$$A = \begin{bmatrix} 0 & 1 & & & \\ & 0 & 1 & & \\ & & \ddots & \ddots & \\ & & & 0 & 1 \\ & & & & 0 \end{bmatrix}$$

2.3 线性定常系统的动态响应

那么可以验证当 $k = n,\ n+1,\ \cdots$ 时，$A^k = 0$。这种情况下，根据式 (2.28) 可以推导出

$$e^{At} = \begin{bmatrix} 1 & t & \dfrac{t^2}{2!} & \dfrac{t^3}{3!} & \cdots & \dfrac{t^{n-1}}{(n-1)!} \\ 0 & 1 & t & \dfrac{t^2}{2!} & \cdots & \dfrac{t^{n-2}}{(n-2)!} \\ & 0 & 1 & t & \ddots & \vdots \\ \vdots & & \ddots & 1 & \ddots & \dfrac{t^2}{2!} \\ & & & & \ddots & t \\ 0 & & \cdots & & 0 & 1 \end{bmatrix} \tag{2.41}$$

如果 A 是一个 Jordan 块，那么它可以被写作

$$A = \begin{bmatrix} \lambda & 1 & & & \\ & \lambda & 1 & & \\ & & \ddots & \ddots & \\ & & & \lambda & 1 \\ & & & & \lambda \end{bmatrix} = \begin{bmatrix} \lambda & & & & \\ & \lambda & & & \\ & & \ddots & & \\ & & & \lambda & \\ & & & & \lambda \end{bmatrix} + \begin{bmatrix} 0 & 1 & & & \\ & 0 & 1 & & \\ & & \ddots & \ddots & \\ & & & 0 & 1 \\ & & & & 0 \end{bmatrix}$$

$$\triangleq A_1 + A_2$$

可以看出 $A_1 A_2 = A_2 A_1$，根据式 (2.34) 可以得到 $e^{At} = e^{(A_1+A_2)t} = e^{A_1 t} e^{A_2 t}$。结合式 (2.39) 和式 (2.41)，可以得到

$$e^{At} = \begin{bmatrix} e^{\lambda t} & t e^{\lambda t} & \dfrac{t^2 e^{\lambda t}}{2!} & \dfrac{t^3 e^{\lambda t}}{3!} & \cdots & \dfrac{t^{n-1} e^{\lambda t}}{(n-1)!} \\ 0 & e^{\lambda t} & t e^{\lambda t} & \dfrac{t^2 e^{\lambda t}}{2!} & \cdots & \dfrac{t^{n-2} e^{\lambda t}}{(n-2)!} \\ & 0 & e^{\lambda t} & t e^{\lambda t} & \ddots & \vdots \\ \vdots & & \ddots & e^{\lambda t} & \ddots & \dfrac{t^2 e^{\lambda t}}{2!} \\ & & & & \ddots & t e^{\lambda t} \\ 0 & & \cdots & & 0 & e^{\lambda t} \end{bmatrix} \tag{2.42}$$

如果 A 具有以下形式，

$$A = \begin{bmatrix} & \beta \\ -\beta & \end{bmatrix}$$

那么

$$e^{At} = \begin{bmatrix} 1 & \\ & 1 \end{bmatrix} + \begin{bmatrix} & \beta \\ -\beta & \end{bmatrix} t + \frac{1}{2!} \begin{bmatrix} -\beta^2 & \\ & -\beta^2 \end{bmatrix} t^2 + \frac{1}{3!} \begin{bmatrix} & -\beta^3 \\ -\beta^3 & \end{bmatrix} t^3 + \cdots$$

$$= \begin{bmatrix} \cos\beta t & \sin\beta t \\ -\sin\beta t & \cos\beta t \end{bmatrix} \tag{2.43}$$

如果 A 具有以下形式，

$$A = \begin{bmatrix} \alpha & \beta \\ -\beta & \alpha \end{bmatrix} = \begin{bmatrix} \alpha & \\ & \alpha \end{bmatrix} + \begin{bmatrix} & \beta \\ -\beta & \end{bmatrix}$$

与获得式 (2.42) 的方法类似，可以得到

$$\mathrm{e}^{At} = \begin{bmatrix} \mathrm{e}^{\alpha t}\cos\beta t & \mathrm{e}^{\alpha t}\sin\beta t \\ -\mathrm{e}^{\alpha t}\sin\beta t & \mathrm{e}^{\alpha t}\cos\beta t \end{bmatrix} \tag{2.44}$$

1.3.4 节中指出，状态系数矩阵 A 可以转化为 Jordan 标准形 $\bar{A} = TAT^{-1}$。变换矩阵 T 的求取方法可以参考 1.3.4 节。接下来，分两种情况来讨论。

如果 A 的特征值都不相同，结合式 (2.37) 和式 (2.39)，可以得到

$$\mathrm{e}^{At} = T^{-1}\mathrm{e}^{\bar{A}t}T = T^{-1}\begin{bmatrix} \mathrm{e}^{\lambda_1 t} & & \\ & \ddots & \\ & & \mathrm{e}^{\lambda_n t} \end{bmatrix} T \tag{2.45}$$

如果式 (2.45) 中的一些特征值为复数，1.3.4 节中指出 Jordan 标准形可以进一步变换。结合式 (2.40) 和式 (2.44)，采用类似于获得式 (2.45) 的方法，可以进一步获得相应的 e^{At}。

如果 A 存在重特征值，相应的 Jordan 标准形可以被分为一些 Jordan 块，那么结合式 (2.40) 和式 (2.42)，可以得到 e^{At}。

(4) 使用 Cayley-Hamilton 定理来计算 e^{At}。

1.3.3 节中指出，预解矩阵可以通过求解

$$(sI - A)^{-1} = \sum_{k=0}^{n-1} \frac{1}{\alpha(s)} p_k(s) A^k$$

得到，其中，$\alpha(s) = s^n + a_1 s^{n-1} + \cdots + a_{n-1}s + a_n$ 是 A 的特征多项式，且

$$\begin{bmatrix} p_0(s) \\ p_1(s) \\ \vdots \\ p_{n-2}(s) \\ p_{n-1}(s) \end{bmatrix} = \begin{bmatrix} 1 & a_1 & a_2 & \cdots & a_{n-1} \\ & 1 & a_1 & \cdots & a_{n-2} \\ & & 1 & \ddots & \vdots \\ & & & \ddots & a_1 \\ & & & & 1 \end{bmatrix} \begin{bmatrix} s^{n-1} \\ s^{n-2} \\ \vdots \\ s \\ 1 \end{bmatrix}$$

记

$$f_k(t) \triangleq \mathcal{L}^{-1}\left[\frac{p_k(s)}{\alpha(s)}\right], \ k = 0, 1, \cdots, n-1$$

2.3 线性定常系统的动态响应

可以得到

$$e^{At} = \sum_{k=0}^{n-1} f_k(t) A^k$$

例 2.2 考虑 $A = \begin{bmatrix} -1 & 2 \\ 0 & -3 \end{bmatrix}$,求 e^{At}。

使用无穷级数求和的方法来计算 e^{At}。根据式 (2.28),可以得到

$$\begin{aligned} e^{At} &= I + At + \frac{1}{2!}A^2 t^2 + \cdots \\ &= \begin{bmatrix} 1 & 0 \\ 0 & 1 \end{bmatrix} + \begin{bmatrix} -t & 2t \\ 0 & -3t \end{bmatrix} + \begin{bmatrix} -\frac{1}{2}t^2 & -4t^2 \\ 0 & \frac{9}{2}t^2 \end{bmatrix} + \cdots \\ &= \begin{bmatrix} 1 - t - \frac{1}{2}t^2 + \cdots & 2t - 4t^2 + \cdots \\ 0 & 1 - 3t + \frac{9}{2}t^2 + \cdots \end{bmatrix} \end{aligned}$$

使用 Laplace 反变换来计算 e^{At}。状态系数矩阵 A 的预解矩阵为

$$\begin{aligned} (sI - A)^{-1} &= \begin{bmatrix} s+1 & -2 \\ 0 & s+3 \end{bmatrix}^{-1} = \begin{bmatrix} \dfrac{1}{s+1} & \dfrac{2}{(s+1)(s+3)} \\ 0 & \dfrac{1}{s+3} \end{bmatrix} \\ &= \begin{bmatrix} \dfrac{1}{s+1} & \dfrac{1}{s+1} - \dfrac{1}{s+3} \\ 0 & \dfrac{1}{s+3} \end{bmatrix} \end{aligned}$$

进行 Laplace 反变换,可以得到

$$e^{At} = \begin{bmatrix} e^{-t} & e^{-t} - e^{-3t} \\ 0 & e^{-3t} \end{bmatrix}$$

使用 Jordan 标准形来计算 e^{At}。A 的特征值为 $\lambda_1 = -1$ 和 $\lambda_2 = -3$。把 A 变换为 Jordan 标准形的变换矩阵 T 满足

$$T = \begin{bmatrix} 1 & 1 \\ 0 & -1 \end{bmatrix}, \; T^{-1} = \begin{bmatrix} 1 & 1 \\ 0 & -1 \end{bmatrix}$$

根据式 (2.45) 可以得到

$$e^{At} = T^{-1} e^{\bar{A}t} T = \begin{bmatrix} 1 & 1 \\ 0 & -1 \end{bmatrix} \begin{bmatrix} e^{\lambda_1 t} & \\ & e^{\lambda_2 t} \end{bmatrix} \begin{bmatrix} 1 & 1 \\ 0 & -1 \end{bmatrix} = \begin{bmatrix} e^{-t} & e^{-t} - e^{-3t} \\ 0 & e^{-3t} \end{bmatrix}$$

使用 Cayley-Hamilton 定理来计算 e^{At}。A 的特征多项式为 $\alpha(s) = s^2 + 4s + 3$，进一步可以计算出

$$\begin{bmatrix} p_0(s) \\ p_1(s) \end{bmatrix} = \begin{bmatrix} 1 & 4 \\ 0 & 1 \end{bmatrix} \begin{bmatrix} s \\ 1 \end{bmatrix} = \begin{bmatrix} s+4 \\ 1 \end{bmatrix}$$

由此，可以得到

$$f_0(t) = \mathcal{L}^{-1}\left[\frac{p_0(s)}{\alpha(s)}\right] = \mathcal{L}^{-1}\left[\frac{s+4}{s^2+4s+3}\right] = \frac{3}{2}e^{-t} - \frac{1}{2}e^{-3t}$$

$$f_1(t) = \mathcal{L}^{-1}\left[\frac{p_1(s)}{\alpha(s)}\right] = \mathcal{L}^{-1}\left[\frac{1}{s^2+4s+3}\right] = \frac{1}{2}e^{-t} - \frac{1}{2}e^{-3t}$$

进一步可得

$$\begin{aligned}
e^{At} &= f_0(t)I + f_1(t)A \\
&= \begin{bmatrix} \frac{3}{2}e^{-t} - \frac{1}{2}e^{-3t} & 0 \\ 0 & \frac{3}{2}e^{-t} - \frac{1}{2}e^{-3t} \end{bmatrix} + \begin{bmatrix} -\frac{1}{2}e^{-t} + \frac{1}{2}e^{-3t} & e^{-t} - e^{-3t} \\ 0 & -\frac{3}{2}e^{-t} + \frac{3}{2}e^{-3t} \end{bmatrix} \\
&= \begin{bmatrix} e^{-t} & e^{-t} - e^{-3t} \\ 0 & e^{-3t} \end{bmatrix}
\end{aligned}$$

2.3.2 线性定常系统的零输入响应

考虑以下线性定常自治系统，

$$\dot{x}(t) = Ax(t), \ x(0) = x_0, \ t \in [0, \infty) \tag{2.46}$$

其中，$x(t) \in \mathbb{R}^n$ 是状态向量，A 是常值矩阵。

定理 2.9 线性定常系统的零输入响应，即齐次状态方程 (2.46) 的解具有下述形式

$$x(t) = \phi(t; 0, x_0, 0) = e^{At}x_0, \ t \geqslant 0 \tag{2.47}$$

证明 式 (2.46) 的解可以表示为幂级数的形式

$$x(t) = b_0 + b_1 t + b_2 t^2 + \cdots = \sum_{k=0}^{\infty} b_k t^k, \ t \geqslant 0 \tag{2.48}$$

其中，$b_0, b_1, \cdots$ 是待定的系数向量。把式 (2.48) 代入式 (2.46) 中，可以得到

$$b_1 + 2b_2 t + 3b_3 t^2 + \cdots = Ab_0 + Ab_1 t + Ab_2 t^2 + \cdots$$

2.3 线性定常系统的动态响应

这表明

$$\begin{cases} b_1 = Ab_0 \\ b_2 = \dfrac{1}{2}Ab_1 = \dfrac{1}{2!}A^2 b_0 \\ b_3 = \dfrac{1}{3}Ab_2 = \dfrac{1}{3!}A^3 b_0 \\ \vdots \\ b_k = \dfrac{1}{k}Ab_{k-1} = \dfrac{1}{k!}A^k b_0 \\ \vdots \end{cases} \quad (2.49)$$

把式 (2.49) 代入式 (2.48) 中，可以得到

$$x(t) = \left(I + At + \dfrac{1}{2!}A^2 t^2 + \cdots\right) b_0, \ t \geqslant 0 \quad (2.50)$$

在式 (2.50) 中考虑 $t = 0$ 且 $x(0) = x_0$，可以进一步计算得到 $b_0 = x_0$。根据矩阵指数的定义式 (2.28)，可以推导出式 (2.47) 成立。□

值得一提的是对于线性定常系统，一般选择 $t_0 = 0$。当 $t_0 \neq 0$ 时，式 (2.47) 可以被写作

$$x(t) = \phi(t; t_0, x_0, 0) = e^{A(t-t_0)} x_0, \ t \geqslant t_0 \quad (2.51)$$

根据定理 2.9 可以证明式 (2.38) 成立。

在状态方程 (2.46) 的两端取 Laplace 变换，可以得到

$$sx(s) - x_0 = Ax(s)$$

这表明

$$x(s) = (sI - A)^{-1} x_0$$

进行 Laplace 反变换，可以进一步得到

$$x(t) = \mathcal{L}^{-1}\left[(sI - A)^{-1}\right] x_0$$

与式 (2.47) 对比，即可知式 (2.38) 成立。

接下来，定义线性定常系统方程 (2.46) 的基本解矩阵和状态转移矩阵。

定义 2.4（线性定常系统的基本解矩阵） 考虑 $\psi_1(t), \cdots, \psi_n(t)$ 是 n 维线性时变齐次方程 (2.46) 的 n 个线性无关的解，$t \geqslant t_0$，则矩阵

$$\Psi(t) = \begin{bmatrix} \psi_1(t) & \cdots & \psi_n(t) \end{bmatrix}$$

被称为线性定常系统方程 (2.46) 的基本解矩阵。

可以很容易验证对于 $\forall t \geqslant t_0$，$\Psi(t)$ 是非奇异的。与定理 2.2 类似，$\Psi(t)$ 也满足

$$\dot{\Psi}(t) = A\Psi(t), \ \Psi(t_0) = E, \ t \geqslant t_0 \tag{2.52}$$

其中，E 是非奇异的实常值矩阵。

线性定常系统的基本解矩阵也可以根据式 (2.52) 来定义。同时，如果 $\Psi_1(t)$ 是线性定常系统的一个基本解矩阵，那么对于任意非奇异矩阵 $C \in \mathbb{R}^{n \times n}$，$\Psi_2(t) = \Psi_1(t)C$ 也是这一系统的一个基本解矩阵。同时，如果 $\Psi_1(t)$ 和 $\Psi_2(t)$ 是同一个线性定常系统的两个基本解矩阵，那么总是存在非奇异矩阵 $C \in \mathbb{R}^{n \times n}$ 使得 $\Psi_2(t) = \Psi_1(t)C$。

注意到 e^{At} 满足式 (2.52)。所以，e^{At} 是相应线性定常系统的一个基本解矩阵。

定义 2.5（线性定常系统的状态转移矩阵） 对于线性定常系统方程 (2.46)，满足

$$\frac{\partial}{\partial t}\Phi(t-t_0) = A\Phi(t-t_0), \ \Phi(0) = I, \ t \geqslant t_0 \tag{2.53}$$

的 $n \times n$ 矩阵 $\Phi(t-t_0)$ 被称为线性定常系统的状态转移矩阵。

与定理 2.4 类似，线性定常系统的状态转移矩阵满足

$$\Phi(t-t_0) = \Psi(t)\Psi^{-1}(t_0), \ t \geqslant t_0 \tag{2.54}$$

其中，$\Psi(t)$ 是线性定常系统的一个基本解矩阵。进一步可知，线性定常系统的状态转移矩阵 $\Phi(t-t_0)$ 也是唯一的。根据定理 2.9，可以推导出 $\Phi(t-t_0) = \mathrm{e}^{A(t-t_0)}$，$t \geqslant t_0$。同时，当 $t_0 = 0$ 时，$\Phi(t) = \mathrm{e}^{At}$，$t \geqslant 0$。这表明线性定常系统的状态转移矩阵由状态系数矩阵 A 唯一确定。

结合式 (2.51)，线性定常系统的零输入响应也可以表示为

$$x(t) = \phi(t; t_0, x_0, 0) = \Phi(t-t_0)x_0, \ t \geqslant t_0$$

这与线性时变系统的零输入响应的形式一致。

2.3.3 线性定常系统的零状态响应

考虑以下线性定常系统，

$$\dot{x}(t) = Ax(t) + Bu(t), \ x(t_0) = 0, \ t \in [0, \infty) \tag{2.55}$$

其中，$x(t) \in \mathbb{R}^n$ 是状态向量，$u(t) \in \mathbb{R}^p$ 是输入向量，A 和 B 是常值矩阵。假设方程 (2.55) 满足唯一解条件，它的零状态响应满足下述结论。

定理 2.10 线性定常系统的零状态响应，即非齐次状态方程 (2.55) 的解具有下述形式，

$$x(t) = \phi(t; 0, 0, u) = \int_0^t \mathrm{e}^{A(t-\tau)}Bu(\tau)\mathrm{d}\tau, \ t \geqslant 0 \tag{2.56}$$

证明 在方程 (2.55) 两端左乘 e^{-At}，可以得到

$$\mathrm{e}^{-At}(\dot{x}(t) - Ax(t)) = \mathrm{e}^{-At}Bu(t) \tag{2.57}$$

2.3 线性定常系统的动态响应

结合式 (2.36)，可以得到

$$\frac{\mathrm{d}}{\mathrm{d}t}\mathrm{e}^{-At}x(t) = -\mathrm{e}^{-At}Ax(t) + \mathrm{e}^{-At}\dot{x}(t) = \mathrm{e}^{-At}\left(\dot{x}(t) - Ax(t)\right) \tag{2.58}$$

结合式 (2.57) 与式 (2.58)，可以获得

$$\frac{\mathrm{d}}{\mathrm{d}t}\mathrm{e}^{-At}x(t) = \mathrm{e}^{-At}Bu(t) \tag{2.59}$$

对式 (2.59) 两端从 0 到 t 进行积分，可以得到

$$\mathrm{e}^{-At}x(t) - x_0 = \int_0^t \mathrm{e}^{-A\tau}Bu(\tau)\mathrm{d}\tau \tag{2.60}$$

注意到 $x_0 = 0$。同时，在方程 (2.55) 两端左乘 e^{At}，可以推导出式 (2.56)。 □

在定理 2.10 中，假设了 $t_0 = 0$。当 $t_0 \neq 0$ 时，式 (2.56) 可以被写作

$$x(t) = \phi(t; t_0, 0, u) = \int_{t_0}^t \mathrm{e}^{A(t-\tau)}Bu(\tau)\mathrm{d}\tau,\ t \geqslant 0 \tag{2.61}$$

例 2.3 考虑以下线性定常系统，

$$\dot{x}(t) = \begin{bmatrix} -1 & 2 \\ 0 & -3 \end{bmatrix} x(t) + \begin{bmatrix} 0 \\ 1 \end{bmatrix} u(t),\ t \geqslant 0$$

初始状态为 $x_0 = [\ 0\ \ 0\ ]^\mathrm{T}$，输入为 $u(t) = 1(t)$，即单位阶跃输入。求它的零状态响应。

例 2.1 中已求出，A 的矩阵指数为

$$\mathrm{e}^{At} = \begin{bmatrix} \mathrm{e}^{-t} & \mathrm{e}^{-t} - \mathrm{e}^{-3t} \\ 0 & \mathrm{e}^{-3t} \end{bmatrix}$$

根据式 (2.61)，可以计算出系统的零状态响应为

$$x(t) = \int_0^t \mathrm{e}^{A(t-\tau)}Bu(\tau)\mathrm{d}\tau = \int_0^\tau \begin{bmatrix} \mathrm{e}^{-(t-\tau)} & \mathrm{e}^{-(t-\tau)} - \mathrm{e}^{-3(t-\tau)} \\ 0 & \mathrm{e}^{-3(t-\tau)} \end{bmatrix} \begin{bmatrix} 0 \\ 1 \end{bmatrix} \mathrm{d}\tau$$

$$= \int_0^t \begin{bmatrix} \mathrm{e}^{-(t-\tau)} - \mathrm{e}^{-3(t-\tau)} \\ \mathrm{e}^{-3(t-\tau)} \end{bmatrix} \mathrm{d}\tau = \begin{bmatrix} \dfrac{2}{3} - \mathrm{e}^{-t} + \dfrac{1}{3}\mathrm{e}^{-3t} \\ \dfrac{1}{3} - \dfrac{1}{3}\mathrm{e}^{-3t} \end{bmatrix},\ t \geqslant 0$$

2.3.4 线性定常系统的状态响应和输出响应

2.1节中指出，线性系统的状态响应 $x(t)$ 是零输入响应和零状态响应的叠加。所以，对于线性定常系统，可以总结出它的状态响应，即状态方程同时由初始状态 x_0 和输入 $u(t)$ 激励的解具有下述形式，

$$x(t) = \phi(t;0,x_0,u) = \mathrm{e}^{At}x_0 + \int_0^t \mathrm{e}^{A(t-\tau)}Bu(\tau)\mathrm{d}\tau, \ t \geqslant 0 \qquad (2.62)$$

当 $t_0 \neq 0$ 时，方程 (2.62) 可以写作

$$x(t) = \phi(t;t_0,x_0,u) = \mathrm{e}^{A(t-t_0)}x_0 + \int_{t_0}^t \mathrm{e}^{A(t-\tau)}Bu(\tau)\mathrm{d}\tau, \ t \geqslant 0 \qquad (2.63)$$

在解得线性定常系统的状态响应后，结合它的输出方程，可以得到输出响应为

$$y(t) = Cx(t) + Du(t) = C\mathrm{e}^{At}x_0 + \int_0^t C\mathrm{e}^{A(t-\tau)}Bu(\tau)\mathrm{d}\tau + Du(t), \ t \geqslant 0 \qquad (2.64)$$

当 $t_0 \neq 0$ 时，式 (2.64) 可以写作

$$y(t) = Cx(t) + Du(t) = C\mathrm{e}^{A(t-t_0)}x_0 + \int_{t_0}^t C\mathrm{e}^{A(t-\tau)}Bu(\tau)\mathrm{d}\tau + Du(t), \ t \geqslant 0 \qquad (2.65)$$

进一步能够得到，对于在 $t_0 = 0$ 松弛的线性定常系统，它的脉冲响应矩阵可以表示为

$$G(t-\tau) = C\mathrm{e}^{A(t-\tau)}B + D\delta(t-\tau) \qquad (2.66)$$

或

$$G(t) = C\mathrm{e}^{At}B + D\delta(t) \qquad (2.67)$$

因为 $\Phi(t-t_0) = \mathrm{e}^{A(t-t_0)}$, $t \geqslant t_0$，式 (2.63)，式 (2.65) 和式 (2.67) 可以表示为

$$x(t) = \phi(t;t_0,x_0,u) = \Phi(t-t_0)x_0 + \int_{t_0}^t \Phi(t-\tau)Bu(\tau)\mathrm{d}\tau, \ t \geqslant 0$$

$$y(t) = C\Phi(t-t_0)x_0 + \int_{t_0}^t C\Phi(t-\tau)Bu(\tau)\mathrm{d}\tau + Du(t), \ t \geqslant 0$$

和

$$G(t) = C\Phi(t)B + D\delta(t)$$

这与线性时变系统的结果一致。

1.3.4 节中指出，如果两个线性定常系统 (A,B,C,D) 和 $(\bar{A},\bar{B},\bar{C},\bar{D})$ 等价[①]，那么总是存在等价变换矩阵 T 使得 $\bar{A} = TAT^{-1}$, $\bar{B} = TB$, $\bar{C} = CT^{-1}$, $\bar{D} = D$，且 $\bar{x}(t) = Tx(t)$。用 $\Phi(t-t_0)$ 和 $\bar{\Phi}(t-t_0)$ 分别表示两个系统的状态转移矩阵，可以得到

$$\bar{\Phi}(t-t_0) = \mathrm{e}^{\bar{A}(t-t_0)} = \mathrm{e}^{TAT^{-1}(t-t_0)} = T\mathrm{e}^{A(t-t_0)}T^{-1} = T\Phi(t-t_0)T^{-1}$$

---

① 方便起见，用 (A,B,C,D) 表示动态方程 (1.11) 和 (1.12)，用 $(\bar{A},\bar{B},\bar{C},\bar{D})$ 表示动态方程 (1.24) 和 (1.25)。

不难证明如果两个线性定常系统等价，那么一个系统在初始状态 x_0 作用下的输出确定后，另一个系统总是能找到一个相应的初始状态使得在零输入情况下两者具有相同的输出。同时，(A, B, C, D) 和 $(\bar{A}, \bar{B}, \bar{C}, \bar{D})$ 具有相同的脉冲响应矩阵，也具有相同的传递函数矩阵。换句话说，在零状态情况下，对于相同的输入 $u(t)$，两个等价的线性定常系统的输出总是相同的。

第 3 章

线性系统的能控性和能观性

能控性和能观性是线性系统的两个基本结构性质。在 20 世纪 60 年代初，Kalman 率先研究了这两个概念。控制科学的发展表明，能控性和能观性的概念对控制理论研究和工程应用都具有重要意义。本章的主要目的是深入研究能控性和能观性的性质。

3.1 线性系统的能控性

3.1.1 基本概念

在系统的状态空间描述中，输入和输出构成系统的外部变量，而状态是系统的内部变量。能控性主要研究系统的内部状态是否可以由外部输入加以控制和影响。如果系统的每一个状态变量都可以被输入变量影响和控制，并且可以由状态空间的任意初值回到原点，那么系统就被称为状态能控的，否则被称为状态不完全能控的。

例 3.1 给定系统的状态方程如下，

$$\begin{bmatrix} \dot{x}_1(t) \\ \dot{x}_2(t) \end{bmatrix} = \begin{bmatrix} 1 & 0 \\ 0 & -3 \end{bmatrix} \begin{bmatrix} x_1(t) \\ x_2(t) \end{bmatrix} + \begin{bmatrix} 5 \\ 1 \end{bmatrix} u(t)$$

显然，系统的状态方程由两个标量子方程并联而成，

$$\dot{x}_1(t) = x_1(t) + 5u(t)$$

$$\dot{x}_2(t) = -3x_2(t) + u(t)$$

上述两个方程表明，状态变量 $x_1(t)$ 和 $x_2(t)$ 都可以通过选择合适的输入 $u(t)$ 而从状态空间内的任意初始状态回到原点。

例 3.2 考虑图 3.1 所示的运算放大电路，系统的状态变量为运算放大器的输出 $x_1(t)$ 和 $x_2(t)$，运算放大器的输入为 $u(t)$。

图 3.1表明如果 $x_1(t_0) = x_2(t_0)$，输入 $u(t)$ 会使得 $x_1(t)$ 和 $x_2(t)$ 转移到任意相同的目标值，或者使它们同时回到 $x_1(t) = x_2(t) = 0$，$t \geqslant t_0$。然而，$u(t)$ 不能使 $x_1(t)$ 和 $x_2(t)$ 转移到不同的目标值。如果 $x_1(t_0) \neq x_2(t_0)$，$u(t)$ 也不能使它们同时回到 $x_1(t) = x_2(t) = 0$，$t \geqslant t_0$。

3.1 线性系统的能控性

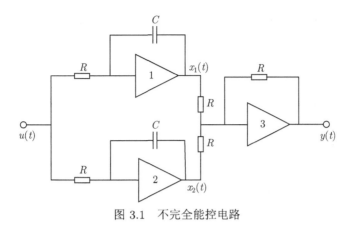

图 3.1 不完全能控电路

例 3.3 考虑图 3.2 所示的桥式电路，系统的状态变量 $x(t)$ 为电容的端电压，输入为电源电压 $u(t)$。

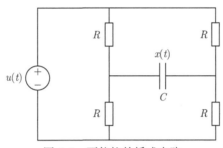

图 3.2 不能控的桥式电路

从这一电路可以看出，如果初始状态 $x(t_0) = 0$，那么无论 $u(t)$ 为何值，对于 $t \geqslant t_0$，$x(t) = 0$，即 $u(t)$ 不能影响和控制 $x(t)$。

上述三个例子给出了能控性的直观描述，而未给出严格的定义。为了深入讨论能控性的特性，下面给出它的严格定义。

线性时变系统的状态方程为

$$\dot{x}(t) = A(t)x(t) + B(t)u(t), \ x(t_0) = x_0, \ t \in [t_0, t_{\mathrm{f}}] \tag{3.1}$$

其中，$x(t) \in \mathbb{R}^n$ 是系统的状态向量，$u(t) \in \mathbb{R}^p$ 是输入向量，$A(t)$ 和 $B(t)$ 是时变矩阵，且它们的元在 $[t_0, t_{\mathrm{f}}]$ 上分别是绝对可积和平方可积的。方便起见，把状态方程 (3.1) 记为 $(A(t), B(t))$。

对于线性定常系统，相应的状态方程为

$$\dot{x}(t) = Ax(t) + Bu(t), \ x(t_0) = x_0, \ t \geqslant t_0 \tag{3.2}$$

其中，A 和 B 是常值矩阵。这里把方程 (3.2) 简记为 (A, B)。

定义 3.1（状态的能控性） 对于有限时刻 $t_{\mathrm{f}} > t_0$，如果存在输入 $u(t)$，$t \in [t_0, t_{\mathrm{f}}]$，能使 t_0 时刻的状态 x_0 在 t_{f} 时刻转移到原点，即从 $x(t_0) = x_0$ 转移到 $x(t_{\mathrm{f}}) = 0$，那么称状态 x_0 在 t_0 时刻是能控的。

定义 3.2（系统的能控性） 如果状态空间内系统的所有非零状态 x_0 在 t_0 时刻都是能控的，那么称系统 $(A(t), B(t))$ 或 (A, B) 在 t_0 时刻是（完全状态）能控的。如果状态空间内系统至少有一个非零状态在 t_0 时刻是不能控的，那么称系统 $(A(t), B(t))$ 或 (A, B) 在 t_0 时刻是不完全（状态）能控的。如果状态空间内系统的所有非零状态在 t_0 时刻都是不能控的，那么称系统 $(A(t), B(t))$ 或 (A, B) 在 t_0 时刻是（完全状态）不能控的。

下面对上述定义进行一些说明。

定义 3.1 中，$u(t)$ 的每个元应满足在 $[t_0, t_f]$ 区间内是平方可积的，并且它们的幅值是有限的。但是 $u(t)$ 不一定要求物理可实现。另一方面，$u(t)$ 应能够把 t_0 时刻的非零状态 x_0 在有限时间内转移到零，但状态转移的轨迹可以是任意的。也就是说，能控性仅是反映系统状态转移的定性性质。

能控性是针对给定时刻 t_0 定义的，这对于时变系统是必要的。如果线性时变系统对于 $[t_0, t_f]$ 区间内的任意时刻都是能控的，那么称之为一致（完全状态）能控的。对于线性定常系统，能控性与初始时刻 t_0 的选取无关，即 (A, B) 能控表明这一系统就是一致能控的。

能控性是系统的结构性质，仅取决于系统的结构和参数，而与系统的任何输入和初始状态无关。

实际上，系统不完全能控是一种"奇异"情况。系统参数值的微小变化都会使系统变得完全能控。例如，图 3.2 中的电路，任何一个电阻 R 的值出现微小变动都会使得电路的对称性遭到破坏，进而电容上的电压，即状态 $x(t)$ 会从不能控变得能控。

能控性强调把非零的状态转移到零状态。如果把零状态转移到非零状态，则涉及能达性的概念。

定义 3.3（状态的能达性） 对于有限时刻 $t_0 < t_f$，如果存在输入 $u(t)$，$t \in [t_0, t_f]$，能使系统的状态 $x(t)$ 在 t_0 时刻从原点转移到 t_f 时刻的非零状态 x_f，即从 $x(t_0) = 0$ 转移到 $x(t_f) = x_f$，那么称状态 x_f 在 t_f 时刻是能达的。

定义 3.4（系统的能达性） 如果状态空间内系统的所有非零状态 x_f 在 t_f 时刻都是能达的，那么称系统 $(A(t), B(t))$ 或 (A, B) 在 t_f 时刻是（完全状态）能达的。如果状态空间内系统至少有一个非零状态在 t_f 时刻是不能达的，那么称系统 $(A(t), B(t))$ 或 (A, B) 在 t_f 时刻是不完全（状态）能达的。如果状态空间内系统的所有非零状态在 t_f 时刻是不能达的，那么称系统 $(A(t), B(t))$ 或 (A, B) 在 t_f 时刻是（完全状态）不能达的。

对于连续线性定常系统，能达性和能控性是等价的。但是对于时变系统，二者不等价。一个不完全能控的时变系统，可能是完全能达的。

3.1.2 线性时变系统的能控性判据

本小节主要介绍线性时变系统 $(A(t), B(t))$ 的能控性判据。

定理 3.1（Gram 判据） 线性时变系统 $(A(t), B(t))$ 在 t_0 时刻是完全能控的，当且仅当存在 $t_f > t_0$ 使得

$$W_c(t_0, t_f) \triangleq \int_{t_0}^{t_f} \Phi(t_0, \tau) B(\tau) B^T(\tau) \Phi^T(t_0, \tau) d\tau \tag{3.3}$$

是非奇异的，其中，$\Phi(t_0, \tau)$ 是 $(A(t), B(t))$ 的状态转移矩阵。

3.1 线性系统的能控性

$n \times n$ 的矩阵 $W_c(t_0, t_f)$ 被称为线性时变系统的能控性矩阵。

证明 首先，证明充分性，即证明当 $W_c(t_0, t_f)$ 是非奇异的，$(A(t), B(t))$ 在 t_0 时刻完全能控。

因为 $W_c(t_0, t_f)$ 非奇异，所以 $W_c^{-1}(t_0, t_f)$ 存在。在此基础上，对于任意的 $x(t_0) = x_0$，都可以构造输入

$$u(t) = -B^{\mathrm{T}}(t)\Phi^{\mathrm{T}}(t_0, t)W_c^{-1}(t_0, t_f)x_0, \ t \in [t_0, t_f]$$

系统被上述输入 $u(t)$ 激励时，在 t_f 时刻的状态响应为

$$\begin{aligned} x(t_f) &= \Phi(t_f, t_0)x_0 + \int_{t_0}^{t_f} \Phi(t_f, \tau)B(\tau)u(\tau)\mathrm{d}\tau \\ &= \Phi(t_f, t_0)x_0 - \int_{t_0}^{t_f} \Phi(t_f, t_0)\Phi(t_0, \tau)B(\tau)B^{\mathrm{T}}(\tau)\Phi^{\mathrm{T}}(t_0, \tau)W_c^{-1}(t_0, t_f)x_0 \mathrm{d}\tau \\ &= \Phi(t_f, t_0)\left[x_0 - \int_{t_0}^{t_f} \Phi(t_0, \tau)B(\tau)B^{\mathrm{T}}(\tau)\Phi^{\mathrm{T}}(t_0, \tau)\mathrm{d}\tau W_c^{-1}(t_0, t_f)x_0\right] \\ &= \Phi(t_f, t_0)\left[x_0 - W_c(t_0, t_f)W_c^{-1}(t_0, t_f)x_0\right] = 0 \end{aligned}$$

这表明存在有限时刻 $t_f > 0$ 和控制输入 $u(t)$，能够使得任意初始状态 x_0 在时刻 t_f 转移到 $x(t_f) = 0$。根据定义 3.2，$(A(t), B(t))$ 在 t_0 时刻是完全能控的。

接下来，证明必要性。使用反证法，假设 $(A(t), B(t))$ 在 t_0 时刻完全能控，但是 $W_c(t_0, t_f)$ 是奇异的。这种情况下，必然存在非零向量 $\alpha \in \mathbb{R}^n$ 使得

$$\alpha^{\mathrm{T}} W_c(t_0, t_f)\alpha = 0$$

结合式 (3.3)，可以得到

$$\begin{aligned} 0 &= \int_{t_0}^{t_f} \alpha^{\mathrm{T}}\Phi(t_0, \tau)B(\tau)B^{\mathrm{T}}(\tau)\Phi^{\mathrm{T}}(t_0, \tau)\alpha \mathrm{d}\tau \\ &= \int_{t_0}^{t_f} \alpha^{\mathrm{T}}\Phi(t_0, \tau)B(\tau)\left(\alpha^{\mathrm{T}}\Phi(t_0, \tau)B(\tau)\right)^{\mathrm{T}} \mathrm{d}\tau \end{aligned}$$

上式仅在满足下述条件时成立。

$$\alpha^{\mathrm{T}}\Phi(t_0, \tau)B(\tau) = 0, \ \forall t \in [t_0, t_f] \tag{3.4}$$

考虑 $x(t_0) = \alpha$ 是系统 $(A(t), B(t))$ 的一个初始状态，因为 $(A(t), B(t))$ 在 t_0 时刻是完全能控的，那么必然存在 $t_f > t_0$ 和输入 $u(t)$ 使得 $x(t_f)=0$，即

$$0 = x(t_f) = \Phi(t_f, t_0)\alpha + \int_{t_0}^{t_f} \Phi(t_f, \tau)B(\tau)u(\tau)\mathrm{d}\tau$$

这表明

$$\alpha = -\Phi^{-1}(t_f,t_0)\int_{t_0}^{t_f}\Phi(t_f,\tau)B(\tau)u(\tau)\mathrm{d}\tau$$

$$= -\Phi(t_0,t_f)\int_{t_0}^{t_f}\Phi(t_f,\tau)B(\tau)u(\tau)\mathrm{d}\tau$$

$$= -\int_{t_0}^{t_f}\Phi(t_0,\tau)B(\tau)u(\tau)\mathrm{d}\tau$$

进一步可得

$$\alpha^\mathrm{T}\alpha = -\int_{t_0}^{t_f}\alpha^\mathrm{T}\Phi(t_0,\tau)B(\tau)u(\tau)\mathrm{d}\tau$$

结合式 (3.4)，可以推导出 $\alpha^\mathrm{T}\alpha = 0$，进一步可得 $\alpha = 0$，这与假设矛盾。所以 $W_c(t_0,t_f)$ 是非奇异的。 □

可以验证如果 $(A(t),B(t))$ 在 t_0 时刻完全能控，那么 $W_c(t_0,t_f)$ 具有以下性质。

(1) $W_c(t_0,t_f)$ 是一个正定的常值矩阵。
(2) $W_c(t_0,t_f) = W_c(t_0,t) + \Phi(t_0,t)W_c(t,t_f)\Phi^\mathrm{T}(t_0,t)$。
(3) $W_c(t_f,t_f) = 0$。
(4) $\dfrac{\partial}{\partial t}W_c(t,t_f) = A(t)W_c(t,t_f) + W_c(t,t_f)A^\mathrm{T}(t) - B(t)B^\mathrm{T}(t)$。

定理 3.2 线性时变系统 $(A(t),B(t))$ 在 t_0 时刻是完全能控的，当且仅当存在 $t_f > t_0$ 使得 $\Phi(t_0,t)B(t)$ 的行对于 $t \in [t_0,t_f]$ 是线性无关的。

定理 3.2 可以根据定理 3.1 推得，这里省略证明。

值得注意的是，对于时间向量 $f_i(t)$，$i = 1, 2, \cdots, n$，它的元都是定义在区间 $[t_0,t_f]$ 内的连续实函数，如果存在不全为零的常值 c_i，$i = 1, 2, \cdots, n$，使得对于任意 $t \in [t_0,t_f]$，

$$c_1 f_1(t) + c_2 f_2(t) + \cdots + c_n f_n(t) = 0$$

那么，$f_1(t)$，$f_2(t)$，$\cdots$，$f_n(t)$ 是线性相关的。否则，它们是线性无关的。当向量组的元和这些常值属于不同数域时，不能用一般方法来检验它们的线性无关性。

定理 3.3（秩判据） 假设 $A(t)$ 是 $n-2$ 阶连续可微的，$B(t)$ 是 $n-1$ 阶连续可微的。线性时变系统 $(A(t),B(t))$ 在 t_0 时刻是完全能控的，当存在 $t_f > t_0$ 使得

$$\mathrm{rank}\begin{bmatrix} M_0(t_f) & M_1(t_f) & \cdots & M_{n-1}(t_f) \end{bmatrix} = n \tag{3.5}$$

其中，

$$M_0(t) = B(t)$$

$$M_1(t) = -A(t)M_0(t) + \dfrac{\mathrm{d}}{\mathrm{d}t}M_0(t)$$

$$\vdots$$
$$M_{n-1}(t) = -A(t)M_{n-2}(t) + \frac{\mathrm{d}}{\mathrm{d}t}M_{n-2}(t)$$

证明 使用反证法。假设 $(A(t), B(t))$ 在 t_0 时刻是不完全能控的,那么根据定理 3.2,对于任意的 $t_\mathrm{f} > t_0$, $\Phi(t_0,t)B(t)$ 的行总是线性相关的, $t \in [t_0, t_\mathrm{f}]$。也就是说,存在非零向量 α 使得

$$\alpha^\mathrm{T} \Phi(t_0, t) B(t) = 0, \ \forall t \in [t_0, t_\mathrm{f}]$$

对上式取关于 t 的偏导,

$$\frac{\partial}{\partial t}\left(\alpha^\mathrm{T} \Phi(t_0,t) M_0(t)\right) = \alpha^\mathrm{T} \left(\frac{\partial}{\partial t}\Phi(t_0,t)\right) M_0(t) + \alpha^\mathrm{T}\Phi(t_0,t)\frac{\mathrm{d}}{\mathrm{d}t}M_0(t)$$
$$= -\alpha^\mathrm{T}\Phi(t_0,t)A(t)M_0(t) + \alpha^\mathrm{T}\Phi(t_0,t)\frac{\mathrm{d}}{\mathrm{d}t}M_0(t)$$
$$= \alpha^\mathrm{T}\Phi(t_0,t)M_1(t) = 0$$

继续对 $\alpha^\mathrm{T}\Phi(t_0,t)M_1(t) = 0$ 取偏导并重复上述过程,最终可以得到对于 $\forall t \in [t_0, t_\mathrm{f}]$,

$$\alpha^\mathrm{T}\Phi(t_0,t)\left[\begin{array}{cccc} M_0(t) & M_1(t) & \cdots & M_{n-1}(t) \end{array}\right] = 0$$

考虑 $t = t_\mathrm{f}$,由于 $\Phi(t_0, t_\mathrm{f})$ 是非奇异的且 α 是非零的向量,所以下式成立,

$$\mathrm{rank}\left[\begin{array}{cccc} M_0(t_\mathrm{f}) & M_1(t_\mathrm{f}) & \cdots & M_{n-1}(t_\mathrm{f}) \end{array}\right] < n$$

这与式 (3.5) 矛盾。所以 $(A(t), B(t))$ 在 t_0 时刻是完全能控的。 □

值得说明的是,定理 3.3 给出的能控性判据是充分的但不是必要的。也就是说,当式 (3.5) 不满足时,不能得到 $(A(t), B(t))$ 在 t_0 时刻不完全能控的结论。

例 3.4 考虑系统

$$\left[\begin{array}{c} \dot{x}_1(t) \\ \dot{x}_2(t) \end{array}\right] = \left[\begin{array}{cc} t & 1 \\ 0 & t \end{array}\right]\left[\begin{array}{c} x_1(t) \\ x_2(t) \end{array}\right] + \left[\begin{array}{c} 0 \\ 1 \end{array}\right]u(t)$$

可以计算出

$$M_0(t) = \left[\begin{array}{c} 0 \\ 1 \end{array}\right], \ M_1(t) = -A(t)M_0(t) + \frac{\mathrm{d}}{\mathrm{d}t}M_0(t) = \left[\begin{array}{c} -1 \\ -t \end{array}\right]$$

进一步可得

$$\left[\begin{array}{cc} M_0(t) & M_1(t) \end{array}\right] = \left[\begin{array}{cc} 0 & -1 \\ 1 & -t \end{array}\right]$$

对于任意 t,它都是满秩的。所以,这一系统在任意时刻都是完全能控的。

当考虑线性时变系统 $(A(t), B(t))$ 的能达性时，可以引入能达性 Gram 矩阵

$$W_{\mathrm{r}}(t_0, t_{\mathrm{f}}) \triangleq \int_{t_0}^{t_{\mathrm{f}}} \Phi(t_{\mathrm{f}}, \tau) B(\tau) B^{\mathrm{T}}(\tau) \Phi^{\mathrm{T}}(t_{\mathrm{f}}, \tau) \mathrm{d}\tau \tag{3.6}$$

$(A(t), B(t))$ 在 t_{f} 时刻是完全能达的，当且仅当存在 $t_0 < t_{\mathrm{f}}$，使得 $W_{\mathrm{r}}(t_0, t_{\mathrm{f}})$ 是非奇异的。参考定理 3.1，可以推得这一结论。同时，结合式 (3.3) 和式 (3.6)，可以得到

$$W_{\mathrm{r}}(t_0, t_{\mathrm{f}}) = \Phi(t_{\mathrm{f}}, t_0) W_{\mathrm{c}}(t_0, t_{\mathrm{f}}) \Phi^{\mathrm{T}}(t_{\mathrm{f}}, t_0)$$

对于任意的 t_{f} 和 t_0，$\Phi(t_{\mathrm{f}}, t_0)$ 是非奇异的，所以 $\operatorname{rank} W_{\mathrm{r}}(t_0, t_{\mathrm{f}}) = \operatorname{rank} W_{\mathrm{c}}(t_0, t_{\mathrm{f}})$。也就是说，当 $(A(t), B(t))$ 在 t_{f} 时刻完全能达时，它在 t_0 时刻完全能控，反之亦然。需要注意的是，$(A(t), B(t))$ 在 t_{f} 时刻完全能达并不意味着它在 t_{f} 时刻完全能控，反之亦然。

3.1.3 线性定常系统的能控性判据

线性定常系统的能控性与初始时刻 t_0 无关。不失一般性地，本小节中后续总是假设 $t_0 = 0$。

下面介绍一些线性定常系统 (A, B) 常用的能控性判据。

定理 3.4（Gram 判据） 线性定常系统 (A, B) 是完全能控的，当且仅当存在 $t_{\mathrm{f}} > t_0$ 使得

$$W_{\mathrm{c}}(0, t_{\mathrm{f}}) \triangleq \int_0^{t_{\mathrm{f}}} \mathrm{e}^{-A\tau} B B^{\mathrm{T}} \mathrm{e}^{-A^{\mathrm{T}}\tau} \mathrm{d}\tau \tag{3.7}$$

是非奇异的。

$n \times n$ 的矩阵 $W_{\mathrm{c}}(0, t_{\mathrm{f}})$ 被称作线性定常系统的能控性 Gram 矩阵。

证明 首先，证明充分性，即证明当 $W_{\mathrm{c}}(0, t_{\mathrm{f}})$ 是非奇异的，(A, B) 完全能控。

因为 $W_{\mathrm{c}}(0, t_{\mathrm{f}})$ 是非奇异的，所以 $W_{\mathrm{c}}^{-1}(0, t_{\mathrm{f}})$ 存在。在此基础上，对于任意的初始状态 $x(t_0) = x_0$，可以构造以下输入，

$$u(t) = -B^{\mathrm{T}} \mathrm{e}^{-A^{\mathrm{T}} t} W_{\mathrm{c}}^{-1}(0, t_{\mathrm{f}}) x_0, \ t \in [0, t_{\mathrm{f}}]$$

被上述输入 $u(t)$ 激励时，在 t_{f} 时刻，系统的状态响应为

$$\begin{aligned} x(t_{\mathrm{f}}) &= \mathrm{e}^{A t_{\mathrm{f}}} x_0 + \int_0^{t_{\mathrm{f}}} \mathrm{e}^{A(t_{\mathrm{f}} - \tau)} B u(\tau) \mathrm{d}\tau \\ &= \mathrm{e}^{A t_{\mathrm{f}}} x_0 - \int_0^{t_{\mathrm{f}}} \mathrm{e}^{A t_{\mathrm{f}}} \mathrm{e}^{-A\tau} B B^{\mathrm{T}} \mathrm{e}^{-A^{\mathrm{T}}\tau} W_{\mathrm{c}}^{-1}(0, t_{\mathrm{f}}) x_0 \mathrm{d}\tau \\ &= \mathrm{e}^{A t_{\mathrm{f}}} x_0 - \mathrm{e}^{A t_{\mathrm{f}}} \int_0^{t_{\mathrm{f}}} \mathrm{e}^{-A\tau} B B^{\mathrm{T}} \mathrm{e}^{-A^{\mathrm{T}}\tau} \mathrm{d}\tau \ W_{\mathrm{c}}^{-1}(0, t_{\mathrm{f}}) x_0 \\ &= \mathrm{e}^{A t_{\mathrm{f}}} x_0 - \mathrm{e}^{A t_{\mathrm{f}}} x_0 = 0 \end{aligned}$$

这表明总是存在有限的时刻 $t_f > 0$ 和控制输入 $u(t)$ 使得任意的初始状态 x_0 都可以在 t_f 时刻转移到 $x(t_f) = 0$。根据定义 3.2，(A, B) 是完全能控的。

接下来，证明必要性。使用反证法，假设 (A, B) 完全能控，但是 $W_c(0, t_f)$ 是奇异的。这种情况下，必然存在非零向量 $\alpha \in \mathbb{R}^n$，使得

$$\alpha^T W_c(0, t_f) \alpha = 0$$

结合式 (3.7)，可以得到

$$0 = \int_0^{t_f} \alpha^T e^{-A\tau} B B^T e^{-A^T \tau} \alpha d\tau = \int_0^{t_f} \alpha^T e^{-A\tau} B \left(\alpha^T e^{-A\tau} B \right)^T d\tau$$

上式仅在以下条件满足时成立

$$\alpha^T e^{-At} B = 0, \ \forall t \in [0, t_f] \tag{3.8}$$

考虑 $x(t_0) = \alpha$ 是 (A, B) 的一个初始状态。因为 (A, B) 完全能控，那么必然存在 t_f 和输入 $u(t)$ 使得 $x(t_f) = 0$，即

$$0 = x(t_f) = e^{At_f} \alpha + \int_0^{t_f} e^{A(t_f - \tau)} B u(\tau) d\tau$$

这表明

$$\alpha = -\int_0^{t_f} e^{-A\tau} B u(\tau) d\tau$$

进一步，可得

$$\alpha^T \alpha = -\int_0^{t_f} \alpha^T e^{-A\tau} B u(\tau) d\tau$$

结合式 (3.8)，可以推导出 $\alpha^T \alpha = 0$，进一步可知 $\alpha = 0$，这与假设矛盾。所以 (A, B) 完全能控时，$W_c(0, t_f)$ 是非奇异的。 $\square$

事实上，线性定常系统的能控性 Gram 矩阵方程 (3.7) 是线性时变系统的能控性 Gram 矩阵方程 (3.6) 的特殊形式。所以，这一判据也可以根据定理 3.1 直接得到。

定理 3.5（秩判据） 线性定常系统 (A, B) 是完全能控的，当且仅当 $\text{rank } U = n$，其中

$$U \triangleq \begin{bmatrix} B & AB & A^2B & \cdots & A^{n-1}B \end{bmatrix} \tag{3.9}$$

U 被称为线性定常系统的能控性矩阵。

证明 首先，证明充分性。使用反证法，假设 (A, B) 是不完全能控的，但是 $\text{rank} U = n$。这种情况下，根据定理 3.4，$W_c(0, t_f)$ 对于 $\forall t_f > 0$ 总是奇异的。也就是说，总是存在非零向量 $\alpha \in \mathbb{R}^n$ 使得

$$\alpha^T W_c(0, t_f) \alpha = \int_0^{t_f} \alpha^T e^{-A\tau} B \left(\alpha^T e^{-A\tau} B \right)^T d\tau = 0$$

这表明
$$\alpha^\mathrm{T}\mathrm{e}^{-At}B = 0, \ \forall t \in [0, t_\mathrm{f}]$$

对上式求导，然后考虑 $t = 0$，可以得到
$$\alpha^\mathrm{T}B = 0, \ \alpha^\mathrm{T}AB = 0, \ \alpha^\mathrm{T}A^2B = 0, \ \cdots, \ \alpha^\mathrm{T}A^{n-1}B = 0$$

这表明
$$\alpha^\mathrm{T} \begin{bmatrix} B & AB & A^2B & \cdots & A^{n-1}B \end{bmatrix} = \alpha^\mathrm{T}U = 0$$

因为 $\alpha \neq 0$，所以 U 的行是线性相关的，即 $\mathrm{rank}\, U < n$，这与假设 $\mathrm{rank}\, U = n$ 矛盾。由此，充分性得证。

接下来，证明必要性。仍然采用反证法，假设 $\mathrm{rank}\, U < n$，那么存在非零向量 $\alpha \in \mathbb{R}^n$ 使得
$$\alpha^\mathrm{T}U = \alpha^\mathrm{T} \begin{bmatrix} B & AB & A^2B & \cdots & A^{n-1}B \end{bmatrix} = 0$$

由上式，进一步可以得到
$$\alpha^\mathrm{T}A^iB = 0, \ i = 0, 1, 2, \cdots, n-1$$

根据 Cayley-Hamilton 定理，A^k，$k = n, n+1, \cdots$，可以被表示为 $I, A, \cdots, A^{n-1}$ 的线性组合。由此可知
$$\alpha^\mathrm{T}A^iB = 0, \ i = 0, 1, 2, \cdots$$

根据矩阵指数的定义，可知
$$\alpha^\mathrm{T}\mathrm{e}^{-At}B = \alpha^\mathrm{T}\left(I - At + \frac{1}{2!}A^2t^2 - \cdots\right)B = 0, \ \forall t \in [0, t_\mathrm{f}]$$

在此基础上，可以推导出
$$\alpha^\mathrm{T}W_\mathrm{c}(0, t_\mathrm{f})\alpha = \int_0^{t_\mathrm{f}} \alpha^\mathrm{T}\mathrm{e}^{-A\tau}BB^\mathrm{T}\mathrm{e}^{-A^\mathrm{T}\tau}\alpha \mathrm{d}\tau = 0$$

因为 $\alpha \neq 0$，所以 $W_\mathrm{c}(0, t_\mathrm{f})$ 是奇异的。根据定理 3.4，(A, B) 是不完全能控的，这与假设矛盾。由此，必要性得证。□

例 3.5 系统的状态方程如下，
$$\dot{x}(t) = \begin{bmatrix} 1 & 2 & 3 \\ 0 & 4 & 0 \\ 5 & 0 & 6 \end{bmatrix} x(t) + \begin{bmatrix} 1 \\ 0 \\ 0 \end{bmatrix} u(t)$$

3.1 线性系统的能控性

因为

$$\operatorname{rank} \begin{bmatrix} B & AB & A^2B \end{bmatrix} = \operatorname{rank} \begin{bmatrix} 1 & 3 & 26 \\ 1 & 4 & 16 \\ 0 & 5 & 45 \end{bmatrix} = 3$$

所以系统是完全能控的。

接下来，介绍 Popov-Belevitch-Hautus 判据（PBH 判据），包括 PBH 特征向量判据和 PBH 秩判据。

定理 3.6（PBH 特征向量判据） 线性定常系统 (A, B) 是完全能控的，当且仅当 A 没有与 B 所有列都正交的非零左特征向量。或者说，对于 A 的任意特征值 $\lambda_i, i \in \{1, \cdots, n\}$，仅有 $\alpha = 0$ 能够满足

$$\alpha^\mathrm{T} A = \lambda_i \alpha^\mathrm{T}, \ \alpha^\mathrm{T} B = 0 \tag{3.10}$$

证明 首先，证明充分性，即证明不存在 $\alpha \neq 0$ 使得式 (3.10) 成立，那么 (A, B) 是完全能控的。

使用反证法，假设 (A, B) 不完全能控，根据定理 3.5，$\operatorname{rank} U < n$。所以，总是存在非零的 $\alpha \in \mathbb{R}^n$ 使得

$$\alpha^\mathrm{T} \begin{bmatrix} B & AB & A^2B & \cdots & A^{n-1}B \end{bmatrix} = 0$$

上式表明 $\alpha^\mathrm{T} B = 0$ 且 $\alpha^\mathrm{T} AB = 0$。根据这两个等式，可以进一步得到 A 至少存在一个特征值 λ_i，能够使得 $\alpha^\mathrm{T} A = \lambda_i \alpha^\mathrm{T}$。也就是说，总是存在非零的 α 使得式 (3.10) 成立，这与假设矛盾。

接下来，证明必要性。仍然采用反证法，假设存在非零的 $\alpha \in \mathbb{R}^n$，使得式 (3.10) 成立，但是 (A, B) 能控。由此可以得到

$$\alpha^\mathrm{T} B = 0, \ \alpha^\mathrm{T} AB = \lambda_i \alpha^\mathrm{T} B = 0, \ \cdots, \ \alpha^\mathrm{T} A^{n-1}B = \lambda_i^{n-1} \alpha^\mathrm{T} B = 0$$

这表明

$$\alpha^\mathrm{T} \begin{bmatrix} B & AB & A^2B & \cdots & A^{n-1}B \end{bmatrix} = 0$$

由此可知，$\operatorname{rank} U < n$。根据定理 3.5，$(A, B)$ 不完全能控的，与假设矛盾。 □

定理 3.7（PBH 秩判据） 线性定常系统 (A, B) 是完全能控的，当且仅当

$$\operatorname{rank} \begin{bmatrix} \lambda_i I - A & B \end{bmatrix} = n \tag{3.11}$$

对于 A 的所有特征值 λ_i 都成立，$i \in \{1, \cdots, n\}$。式 (3.11) 也可以等价地表示为在复域 $\mathbb{C}$ 中，

$$\operatorname{rank} \begin{bmatrix} sI - A & B \end{bmatrix} = n, \ \forall s \in \mathbb{C} \tag{3.12}$$

即 $sI - A$ 和 B 必须是左互质的。

证明 在式 (3.12) 中，如果 s 不是 A 的特征值，那么特征多项式 $\det(sI - A) \neq 0$。所以，$\operatorname{rank}(sI - A) = n$。进一步，可以得到 $\operatorname{rank}[\begin{array}{cc} sI - A & B \end{array}] = n$。因此，只需证明式 (3.11) 成立。

首先，证明充分性。因为式 (3.11) 成立，$[\begin{array}{cc} \lambda_i I - A & B \end{array}]$ 行是线性无关的。这种情况下，如果 $\alpha^{\mathrm{T}}[\begin{array}{cc} \lambda_i I - A & B \end{array}] = 0$，必有 $\alpha = 0$。或者说，仅有 $\alpha = 0$ 可以同时满足 $\alpha^{\mathrm{T}} A = \lambda_i \alpha^{\mathrm{T}}$ 和 $\alpha^{\mathrm{T}} B = 0$。根据定理 3.6，$(A, B)$ 是完全能控的。

接下来，证明必要性。使用反证法，假设存在特征值 λ_i 使得 $\operatorname{rank}[\begin{array}{cc} \lambda_i I - A & B \end{array}] < n$，这表明 $[\begin{array}{cc} \lambda_i I - A & B \end{array}]$ 的行是线性相关的。也就是说，存在非零的 $\alpha \in \mathbb{R}^n$，使得 $\alpha^{\mathrm{T}}[\begin{array}{cc} \lambda_i I - A & B \end{array}] = 0$。所以，存在 $\alpha \neq 0$，使得 $\alpha^{\mathrm{T}} A = \lambda_i \alpha^{\mathrm{T}}$ 和 $\alpha^{\mathrm{T}} B = 0$ 成立。根据定理 3.6，(A, B) 是不完全能控的，与假设矛盾。□

例 3.6 线性定常系统的状态方程如下，

$$\dot{x}(t) = \begin{bmatrix} 0 & 1 & 0 \\ 0 & 0 & -1 \\ 0 & 0 & 0 \end{bmatrix} x(t) + \begin{bmatrix} 0 & 1 \\ 1 & 0 \\ 0 & 1 \end{bmatrix} u(t)$$

注意到

$$\begin{bmatrix} sI - A & B \end{bmatrix} = \begin{bmatrix} s & -1 & 0 & 0 & 1 \\ 0 & s & 1 & 1 & 0 \\ 0 & 0 & s & 0 & 1 \end{bmatrix}$$

A 的特征值为 $\lambda_1 = \lambda_2 = \lambda_3 = 0$。只需要验证当 $s = \lambda_1 = \lambda_2 = \lambda_3 = 0$ 时上述矩阵的秩，

$$\operatorname{rank}[sI - A \quad B] = \operatorname{rank} \begin{bmatrix} 0 & -1 & 0 & 0 & 1 \\ 0 & 0 & 1 & 1 & 0 \\ 0 & 0 & 0 & 0 & 1 \end{bmatrix} = 3$$

所以，系统是完全能控的。

最后，介绍 Jordan 标准形判据。

定理 3.8（Jordan 标准形判据） 考虑线性定常系统 (A, B)，λ_i 是 A 的 σ_i 重特征值，$i \in \{1, \cdots, l\}$，且 $\sigma_1 + \cdots + \sigma_i + \cdots + \sigma_l = n$。存在等价变换 $\bar{x}(t) = Tx(t)$ 使得 (A, B) 转化为 Jordan 标准形 $(\bar{A}, \bar{B})$，其中，

$$\bar{A} = \begin{bmatrix} J_1 & & & \\ & J_2 & & \\ & & \ddots & \\ & & & J_l \end{bmatrix}, \quad \bar{B} = \begin{bmatrix} \bar{B}_1 \\ \bar{B}_2 \\ \vdots \\ \bar{B}_l \end{bmatrix}$$

3.1 线性系统的能控性

$$J_i = \begin{bmatrix} J_{i1} & & & \\ & J_{i2} & & \\ & & \ddots & \\ & & & J_{i\alpha_i} \end{bmatrix}, \bar{B}_i = \begin{bmatrix} \bar{B}_{i1} \\ \bar{B}_{i2} \\ \vdots \\ \bar{B}_{i\alpha_i} \end{bmatrix}$$

$$J_{ik} = \begin{bmatrix} \lambda_i & 1 & & \\ & \lambda_i & \ddots & \\ & & \ddots & 1 \\ & & & \lambda_i \end{bmatrix}, \bar{B}_{ik} = \begin{bmatrix} \bar{b}_{ik1} \\ \bar{b}_{ik2} \\ \vdots \\ \bar{b}_{ik\sigma_{ik}} \end{bmatrix}$$

J_i 是与 λ_i 相关的 $\sigma_i \times \sigma_i$ 的 Jordan 块矩阵，J_{ik}，$k \in \{1, \cdots, \alpha_i\}$，是 $\sigma_{ik} \times \sigma_{ik}$ 的 Jordan 块，α_i 是 λ_i 的几何重数，$\sigma_{i1} + \cdots + \sigma_{i\alpha_i} = \sigma_i$。

(A, B) 是完全能控的，当且仅当对于 $i \in \{1, \cdots, l\}$，矩阵

$$\tilde{B}_i = \begin{bmatrix} \bar{b}_{i1\sigma_{i1}} \\ \bar{b}_{i2\sigma_{i2}} \\ \vdots \\ \bar{b}_{i\alpha_i\sigma_{i\alpha_i}} \end{bmatrix}$$

都行线性无关。

首先，考虑 A 的所有特征值都互异。这种情况下，$J_1 = \lambda_1$，$J_2 = \lambda_2$，$\cdots$，$J_n = \lambda_n$，且 $\bar{B}_1$，$\bar{B}_2$，$\cdots$，$\bar{B}_n$ 都仅有一行。构造

$$\begin{bmatrix} sI - \bar{A} & \bar{B} \end{bmatrix} = \begin{bmatrix} s - \lambda_1 & & & & \bar{B}_1 \\ & s - \lambda_2 & & & \bar{B}_2 \\ & & \ddots & & \vdots \\ & & & s - \lambda_n & \bar{B}_l \end{bmatrix}$$

可以看出，当且仅当 $\bar{B}_i \neq 0$ 对于 $i \in \{1, \cdots, n\}$ 都成立时，$\text{rank}[\ sI - \bar{A} \quad \bar{B}\ ] = n$，根据定理 3.7，$(A, B)$ 是完全能控的。

当存在重特征值时，采用类似的方法可以证明上述定理，这里不详细介绍。

例 3.7 一个线性定常系统的 Jordan 标准形如下，判断它的能控性。

$$\dot{\bar{x}}(t) = \begin{bmatrix} -1 & 1 & & & & & & & \\ 0 & -1 & & & & & & & \\ & & -1 & & & & & & \\ & & & -1 & & & & & \\ & & & & 2 & 1 & & & \\ & & & & 0 & 2 & & & \\ & & & & & & 2 & & \\ & & & & & & & 4 \end{bmatrix} \bar{x}(t) + \begin{bmatrix} 0 & 0 & 0 \\ 1 & 0 & 0 \\ 0 & 2 & 0 \\ 0 & 0 & 4 \\ 0 & 0 & 3 \\ 1 & 2 & 0 \\ 0 & 3 & 6 \\ 4 & 0 & 0 \end{bmatrix} u(t)$$

系统的状态参数矩阵有 3 个不同的特征值，分别为 $\lambda_1 = -1$，$\lambda_2 = 2$ 和 $\lambda_3 = 4$。与 $\lambda_1 = -1$ 相关的由 $\bar{B}_{11}$，$\bar{B}_{12}$ 和 $\bar{B}_{13}$ 最后一行构成的矩阵是

$$\begin{bmatrix} \bar{b}_{112} \\ \bar{b}_{121} \\ \bar{b}_{131} \end{bmatrix} = \begin{bmatrix} 1 & 0 & 0 \\ 0 & 2 & 0 \\ 0 & 0 & 4 \end{bmatrix}$$

它是行线性无关的。

与 $\lambda_2 = 2$ 相关的由 $\bar{B}_{21}$ 和 $\bar{B}_{22}$ 最后一行构成的矩阵是

$$\begin{bmatrix} \bar{b}_{212} \\ \bar{b}_{221} \end{bmatrix} = \begin{bmatrix} 1 & 2 & 0 \\ 0 & 3 & 6 \end{bmatrix}$$

它的行是线性无关的。

与 $\lambda_3 = 4$ 相关的 $\bar{B}_3$ 的最后一行为 $\begin{bmatrix} 4 & 0 & 0 \end{bmatrix}$，它的元不全为零。

所以，这一系统是完全能控的。

应用 Jordan 标准形判据时，如果系统的状态方程不是 Jordan 标准形，可以通过等价变换将其转化为 Jordan 标准形。等价变换不会改变系统的固有特性。

当考虑线性定常系统 (A, B) 的能达性时，可以引入能达性 Gram 矩阵，

$$W_r(0, t_f) \triangleq \int_0^{t_f} e^{(t_f-\tau)A} BB^T e^{(t_f-\tau)A^T} d\tau \tag{3.13}$$

(A, B) 是完全能达的，当且仅当存在 $t_f > 0$ 使得 $W_r(0, t_f)$ 是非奇异的。参考定理 3.4，可以证明这一结论。结合式 (3.7) 和式 (3.13)，可以得到

$$W_r(0, t_f) = e^{At_f} W_c(0, t_f) \left(e^{At_f}\right)^T$$

因为 e^{At_f} 是非奇异的，$\operatorname{rank} W_r(0, t_f) = \operatorname{rank} W_c(0, t_f)$，也就是说，当且仅当 (A, B) 完全能控时，它也完全能达。

3.1.4 线性定常系统的能控性指数

对于线性定常系统 (A, B)，定义

$$U_\mu = \begin{bmatrix} B & AB & \cdots & A^{\mu-1}B \end{bmatrix} \tag{3.14}$$

其中，U_μ 是 $n \times \mu p$ 的常值矩阵，μ 是正整数。当 $\mu = n$ 时，U_μ 是系统的能控性矩阵，且如果系统是完全能控的，$\operatorname{rank} U_\mu = n$。考虑 b_i 是 B 的第 i 列，注意到如果 $A^k b_i$ 与它左边的各列线性相关，$k = 1, \cdots, \mu - 1$，$i = 1, \cdots, p$，那么 $A^l b_i$ 都和它左边的各列线性相关，$l > k$。由此可知，在式 (3.14) 中，当一个子块 $A^k B$ 引入到 U_μ 中后，U_μ 的秩至少增加 1，直到使 $\operatorname{rank} U_\mu = n$ 为止。

定义 3.5（线性定常系统的能控性指数） 当式 (3.14) 中的 μ 是使得

$$\operatorname{rank} U_\mu = \operatorname{rank} U_{\mu+1} = n \tag{3.15}$$

成立的最小正整数时，μ 被称作 (A, B) 的能控性指数。

定理 3.9 能控性指数 μ 满足

$$\frac{n}{p} \leqslant \mu \leqslant \min\{\bar{n}, n-r+1\} \tag{3.16}$$

其中，r 是 B 的秩，$\bar{n}$ 是 A 的最小多项式的阶次。

证明 考虑 $\phi(s) = s^{\bar{n}} + a_1 s^{\bar{n}-1} + \cdots + a_{\bar{n}-1} s + a_{\bar{n}}$ 是 A 的最小多项式，它满足 $\phi(A) = 0$，即

$$A^{\bar{n}} = -a_1 A^{\bar{n}-1} - \cdots - a_{\bar{n}-1} A - a_{\bar{n}} I$$

在上式两端左乘 B，可以得到

$$A^{\bar{n}} B = -a_1 A^{\bar{n}-1} B - \cdots - a_{\bar{n}-1} AB - a_{\bar{n}} B$$

这表明 $A^{\bar{n}} B$ 与 $B, AB, \cdots, A^{\bar{n}-1} B$ 是线性相关的。进一步，可以得到在式 (3.14) 中，$A^{\bar{n}} B$ 的每一列都和它左边的各列线性相关。由此可知，$\mu \leqslant \bar{n}$。

另一方面，随着 μ 增大 1，U_μ 的秩至少增大 1，直到 $\operatorname{rank} U_\mu = n$。这种情况下，$\mu - 1 \leqslant n - r$，即 $\mu \leqslant n - r + 1$。

U_μ 有 μp 列，为了保证 $\operatorname{rank} U_\mu = n$，$\mu p$ 应不小于 n，即 $n \leqslant \mu p$。

综上所述，式 (3.16) 成立。 □

根据定理 3.9，对于单输入系统，总是有 $\mu = n$。同时，定理 3.5 可以修改为下述结论。

推论 3.1 线性定常系统 (A, B) 是完全能控的，当且仅当

$$\operatorname{rank} U_{n-r+1} = \operatorname{rank} \begin{bmatrix} B & AB & A^2 B & \cdots & A^{n-r} B \end{bmatrix} = n \tag{3.17}$$

或

$$\operatorname{rank} U_{\bar{n}} = \operatorname{rank} \begin{bmatrix} B & AB & A^2 B & \cdots & A^{\bar{n}-1} B \end{bmatrix} = n \tag{3.18}$$

由于计算 $\bar{n}$ 较为复杂，所以与式 (3.18) 相比，式 (3.17) 更常用。

考虑 $b_1, \cdots, b_p$ 是 B 的列，U_μ 可以表示为

$$U_\mu = \begin{bmatrix} b_1 & \cdots & b_p & Ab_1 & \cdots & Ab_p & \cdots & A^{\mu-1} b_1 & \cdots & A^{\mu-1} b_p \end{bmatrix} \tag{3.19}$$

在式 (3.19) 中从左至右依次选择 n 个线性无关的列向量。如果一列不能被表示为已选择的它左边各列的线性组合，那么该列是这 n 个线性无关的列中的一列。如果一列和已选

择的各列线性相关，那么这一列不会被选择。假设 B 的秩是 r，已选择的 n 个线性无关的列可以被重新排列为

$$\begin{bmatrix} b_1 & Ab_1 & \cdots & A^{\mu_1-1}b_1 & b_2 & Ab_2 & \cdots & A^{\mu_2-1}b_2 & \cdots & b_r & Ab_r & \cdots & A^{\mu_r-1}b_r \end{bmatrix} \tag{3.20}$$

对于完全能控的系统，总是有

$$\mu_1 + \mu_2 + \cdots + \mu_r = n$$

能控性指数满足

$$\mu = \max\{\mu_1, \mu_2, \cdots, \mu_r\}$$

一般地，$\{\mu_1, \mu_2, \cdots, \mu_r\}$ 被称作 (A, B) 的能控性指数集或 Kronecker(克罗内克) 不变量。

定理 3.10 等价变换不会改变 (A, B) 的 μ 和 $\{\mu_1, \mu_2, \cdots, \mu_r\}$。

证明 考虑 T 是非奇异矩阵，且 $\bar{A} = TAT^{-1}$，$\bar{B} = TB$，那么

$$\bar{U}_\mu = \begin{bmatrix} \bar{B} & \bar{A}\bar{B} & \cdots & \bar{A}^{\mu-1}\bar{B} \end{bmatrix} = \begin{bmatrix} TB & TAB & \cdots & TA^{\mu-1}B \end{bmatrix}$$

因为 $\operatorname{rank} T = n$，由此可知 $\operatorname{rank} \bar{B} = \operatorname{rank} B$，$\cdots$，$\operatorname{rank} \bar{A}^{\mu-1}\bar{B} = \operatorname{rank} A^{\mu-1}B$。所以，$\mu$ 和 $\{\mu_1, \mu_2, \cdots, \mu_r\}$ 都不会变化。 □

例 3.8 给定线性定常系统的状态方程，

$$\dot{x}(t) = \begin{bmatrix} 0 & 1 & 0 & 0 \\ 3 & 0 & 0 & 0 \\ 0 & 1 & 1 & 0 \\ 0 & 0 & 0 & 0 \end{bmatrix} x(t) + \begin{bmatrix} 0 & 0 \\ 1 & 0 \\ 0 & 0 \\ 0 & 1 \end{bmatrix} u(t)$$

计算 U_μ，它满足 $\operatorname{rank} U_\mu = \operatorname{rank} U_{\mu+1} = n$，可得

$$\operatorname{rank} \begin{bmatrix} B & AB & A^2B \end{bmatrix} = \operatorname{rank} \begin{bmatrix} 0 & 0 & 1 & 0 & 0 & 0 \\ 1 & 0 & 0 & 0 & 3 & 0 \\ 0 & 0 & 1 & 0 & 1 & 0 \\ 0 & 1 & 0 & 0 & 0 & 0 \end{bmatrix} = 4 = n$$

所以，系统的能控性指数 μ 是 3。

在上面的 U_μ 中，依次选择 4 个线性无关的列向量并根据式 (3.20) 重新排列，可得

$$\begin{bmatrix} b_1 & Ab_1 & A^2b_1 & b_2 \end{bmatrix} = \begin{bmatrix} 0 & 1 & 0 & 0 \\ 1 & 0 & 3 & 0 \\ 0 & 1 & 1 & 0 \\ 0 & 0 & 0 & 1 \end{bmatrix}$$

显然，系统的能控性指数集为 $\{\mu_1, \mu_2\} = \{3, 1\}$，并且 $\mu = \max\{\mu_1, \mu_2\} = \mu_1 = 3$。

3.2 线性系统的能观性

3.2.1 基本概念

实际应用中，经常要求能够知晓系统的状态信息，但系统的状态信息可能不能获取。这就带来了一个问题，是否可以通过观测系统在某些时间区间内一些输入激励的输出响应来确定系统的状态信息，由此引出了系统的能观性的概念。能观性指的是系统所有状态变量的任意形式的运动是否可由输出完全反映，即系统的输出包含它内部状态运动的全部信息。

例 3.9 考虑例 3.1中的系统，它的输出方程为

$$y(t) = \begin{bmatrix} 0 & -2 \end{bmatrix} \begin{bmatrix} x_1(t) \\ x_2(t) \end{bmatrix}$$

显然，输出 $y(t)$ 只包含状态变量 $x_2(t)$ 的信息，状态变量 $x_1(t)$ 与输出 $y(t)$ 没有直接或间接的联系，所以系统是不完全能观的。

例 3.10 考虑图 3.3 所示的由运算放大器构成的积分电路，状态变量为两个积分器的输出 $x_1(t)$ 和 $x_2(t)$，系统的输出是 $x_1(t)$ 和 $x_2(t)$ 的和。

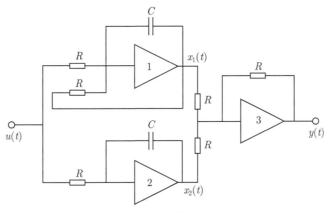

图 3.3 完全能观的运算电路

显然，输出 $y(t)$ 包含状态 $x_1(t)$ 和 $x_2(t)$ 的全部信息，并且可以通过测量 $y(t)$ 来获得 $x_1(t)$ 和 $x_2(t)$ 的初值 $x_1(t_0)$ 和 $x_2(t_0)$。因此，状态 $x_1(t)$ 和 $x_2(t)$ 是能观的。

例 3.11 考虑图 3.4 所示的桥式电路，如果 $u(t) = 0$，那么对于任意的 $t \geqslant t_0$，无论电容上的初始电压，即初始状态 $x(t)$ 为何值，$y(t) = 0$。状态 $x(t)$ 的信息完全不能反映在 $y(t)$ 中。这表明，这一电路系统的状态是完全不能观的。

上述三个例子仅仅是能观性的直观描述，为了深入讨论能观性的特性，下面介绍能观性的定义。

能观性表征系统的状态可以由输出反映的特性。为了讨论系统的能观性，需要同时考虑系统状态方程和输出方程，

$$\dot{x}(t) = A(t)x(t) + B(t)u(t), \; y(t) = C(t)x(t) + D(t)u(t), \; x(t_0) = x_0, \; t \in [t_0, t_\mathrm{f}] \quad (3.21)$$

其中，$x(t) \in \mathbb{R}^n$ 是状态向量，$u(t) \in \mathbb{R}^p$ 是输入向量，$y(t) \in \mathbb{R}^q$ 是输出向量。$A(t)$，$B(t)$，$C(t)$ 和 $D(t)$ 是满足唯一解条件的时变矩阵。2.2.4 节中给出了式 (3.21) 的解，可以表示为

$$x(t) = \Phi(t,t_0)x_0 + \int_{t_0}^{t} \Phi(t,\tau)B(\tau)u(\tau)\mathrm{d}\tau$$

$$y(t) = C(t)\Phi(t,t_0)x_0 + \int_{t_0}^{t} C(t)\Phi(t,\tau)B(\tau)u(\tau)\mathrm{d}\tau + D(t)u(t)$$

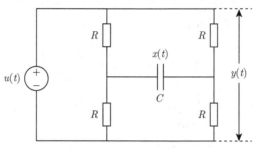

图 3.4 不能观的桥式电路

当研究能观性时，总是假设 $y(t)$ 和 $u(t)$ 是已知的，而初始状态 x_0 是未知的，记

$$\bar{y}(t) \triangleq y(t) - \left[\int_{t_0}^{t} C(t)\Phi(t,\tau)B(\tau)u(\tau)\mathrm{d}\tau + D(t)u(t) \right]$$

那么可以得到

$$\bar{y}(t) = C(t)\Phi(t,t_0)x_0$$

这表明，能观性就是研究 x_0 可由 $\bar{y}(t)$ 的可估计性。由于 $\bar{y}(t)$ 和 x_0 的任意性，能观性实质上就是研究当 $u(t) = 0$ 时，由 $y(t)$ 来估计 x_0 的可能性，即研究以下零输入系统的能观性，

$$\dot{x}(t) = A(t)x(t),\ y(t) = C(t)x(t),\ x(t_0) = x_0,\ t \in [t_0, t_\mathrm{f}] \tag{3.22}$$

方便起见，把动态方程 (3.22) 记作 $(A(t), C(t))$。

对于线性定常的情况，则考虑以下系统，

$$\dot{x}(t) = Ax(t),\ y(t) = Cx(t),\ x(t_0) = x_0,\ t \geqslant t_0 \tag{3.23}$$

并把式 (3.23) 简记为 (A, C)。

下面给出系统能观性的定义。

定义 3.6（状态的不能观性） 如果存在有限时刻 $t_\mathrm{f} > t_0$，对于给定的初始状态 x_0，和任意 $t \in [t_0, t_\mathrm{f}]$，$y(t) = 0$，那么称状态 x_0 在 t_0 时刻是不能观的。

定义 3.7（系统的能观性） 对于状态 $x_0 \in \mathbb{R}^n$，如果仅有零状态，即 $x_0 = 0$，在 t_0 时刻不能观，那么系统 $(A(t), C(t))$（或 (A, C)）在 t_0 时刻是（完全状态）能观的。如果存在非零状态 $x_0 \in \mathbb{R}^n$ 在 t_0 时刻不能观，那么系统 $(A(t), C(t))$（或 (A, C)）在 t_0 时刻是不完全（状态）能观的。

3.2 线性系统的能观性

能观性是系统的结构特性，仅取决于系统的结构和参数，和系统的输入与输出值无关。

能观性是基于给定时间 t_0 定义的，这对于时变系统是必要的。如果对于 $[t_0, t_\mathrm{f}]$ 区间内的任意的初始时刻，线性时变系统总是能观的，那么称这一系统是一致（完全状态）能观的。对于线性定常系统，它的能观性与 t_0 的选择无关，即如果 (A, C) 是能观的，那么它也是一致能观的。

实际应用中，系统不完全能观是一种"奇异"的情况，这是由于系统参数值的微小变化都会使不完全能观的系统变得完全能观。

能观性可以看作是通过测量未来的输出来确定当前时刻的状态。如果利用过去的输出的测量值来重构当前的状态，就涉及了能构性的概念。

定义 3.8（状态的不能构性） 如果存在有限时刻 $t_0 < t_\mathrm{f}$ 使得对于任意 $t \in [t_0, t_\mathrm{f}]$，零输入响应 $y(t) = 0$，那么称状态 $x(t_\mathrm{f})$ 在 t_f 时刻是不能构的。

定义 3.9（系统的能构性） 对于状态 $x(t_\mathrm{f}) \in \mathbb{R}^n$，如果仅有零状态，即 $x(t_\mathrm{f}) = 0$，在 t_f 时刻不能构，那么系统 $(A(t), C(t))$（或 (A, C)）在 t_f 时刻是（完全状态）能构的。如果存在非零状态 $x(t_\mathrm{f}) \in \mathbb{R}^n$ 在 t_f 时刻不能构，那么系统 $(A(t), C(t))$（或 (A, C)）在 t_f 时刻是不完全（状态）能构的。

3.2.2 线性时变系统的能观性判据

定理 3.11（Gram 判据） 线性时变系统 $(A(t), C(t))$ 在 t_0 时刻是完全能观的，当且仅当存在 $t_\mathrm{f} > t_0$ 使得

$$W_\mathrm{o}(t_0, t_\mathrm{f}) \triangleq \int_{t_0}^{t_\mathrm{f}} \Phi^\mathrm{T}(t, t_0) C^\mathrm{T}(t) C(t) \Phi(t, t_0) \mathrm{d}t \tag{3.24}$$

是非奇异的，其中，$\Phi(t, t_0)$ 是 $(A(t), C(t))$ 的状态转移矩阵。

$n \times n$ 的矩阵 $W_\mathrm{o}(t_0, t_\mathrm{f})$ 被称作线性时变系统的能观性 Gram 矩阵。

证明 首先，证明充分性。注意到对于 $t \in [t_0, t_\mathrm{f}]$，$y(t) = C(t)\Phi(t, t_0) x_0$。在上式两端左乘 $\Phi^\mathrm{T}(t, t_0) C^\mathrm{T}(t)$，并从 t_0 到 t_f 进行积分，可以得到

$$\int_{t_0}^{t_\mathrm{f}} \Phi^\mathrm{T}(t, t_0) C^\mathrm{T}(t) y(t) \mathrm{d}t = W_\mathrm{o}(t_0, t_\mathrm{f}) x_0$$

因为 $W_\mathrm{o}(t_0, t_\mathrm{f})$ 是非奇异的，所以 $W_\mathrm{o}^{-1}(t_0, t_\mathrm{f})$ 存在。可以进一步导出

$$x_0 = W_\mathrm{o}^{-1}(t_0, t_\mathrm{f}) \int_{t_0}^{t_\mathrm{f}} \Phi^\mathrm{T}(t, t_0) C^\mathrm{T}(t) y(t) \mathrm{d}t$$

这表明，当 $t \in [t_0, t_\mathrm{f}]$ 时，非零的状态 x_0 可以由 $y(t)$ 唯一确定。所以 $(A(t), C(t))$ 在 t_0 时刻是完全能观的。

接下来，证明必要性。使用反证法，假设 $(A(t), C(t))$ 在 t_0 时刻是完全能观的，但是 $W_\mathrm{o}(t_0, t_\mathrm{f})$ 是奇异的。此时，一定存在非零向量 $\alpha \in \mathbb{R}^n$ 使得

$$\alpha^\mathrm{T} W_\mathrm{o}(t_0, t_\mathrm{f}) \alpha = 0$$

结合式 (3.24)，可以得到

$$0 = \int_{t_0}^{t_f} \alpha^T \Phi^T(t,t_0) C^T(t) C(t) \Phi(t,t_0) \alpha \mathrm{d}t$$
$$= \int_{t_0}^{t_f} (C(t)\Phi(t,t_0)\alpha)^T C(t)\Phi(t,t_0)\alpha \mathrm{d}t$$

只有当

$$C(t)\Phi(t,t_0)\alpha = 0, \ \forall t \in [t_0, t_f]$$

时，上式成立。

考虑 $x(t_0) = \alpha$ 是 $(A(t), C(t))$ 的初始状态，相应的输出 $y(t) = C(t)\Phi(t,t_0)\alpha = 0$，这表明非零状态 $x(t_0) = \alpha$ 在 t_0 时刻是不能观的，与假设矛盾。必要性得证。□

可以验证，如果 $(A(t), C(t))$ 在 t_0 时刻是完全能观的，那么 $W_o(t_0, t_f)$ 具有以下性质。
(1) $W_o(t_0, t_f)$ 是正定的常值矩阵。
(2) $W_o(t_0, t_f) = W_o(t_0, t) + \Phi(t_0, t) W_o(t, t_f) \Phi^T(t_0, t)$。
(3) $W_o(t_f, t_f) = 0$。
(4) $\dfrac{\partial}{\partial t} W_o(t, t_f) = A(t) W_o(t, t_f) + W_o(t, t_f) A^T(t) - C^T(t) C(t)$。

定理 3.12　线性时变系统 $(A(t), C(t))$ 在 t_0 时刻是完全能观的，当且仅当存在 $t_f > t_0$ 使得 $C(t)\Phi(t, t_0)$ 的列对于 $t \in [t_0, t_f]$ 是线性无关的。

定理 3.13（秩判据）　假设 $A(t)$ 是 $n-2$ 阶连续可微的，$C(t)$ 是 $n-1$ 阶连续可微的。线性时变系统 $(A(t), C(t))$ 在 t_0 时刻是完全能观的，当存在 $t_f > t_0$ 使得

$$\mathrm{rank} \begin{bmatrix} N_0(t_f) \\ N_1(t_f) \\ \vdots \\ N_{n-1}(t_f) \end{bmatrix} = n \tag{3.25}$$

其中，

$$N_0(t) = C(t)$$
$$N_1(t) = N_0(t)A(t) + \dfrac{\mathrm{d}}{\mathrm{d}t} N_0(t)$$
$$\vdots$$
$$N_{n-1}(t) = N_{n-2}(t)A(t) + \dfrac{\mathrm{d}}{\mathrm{d}t} N_{n-2}(t)$$

这一判据的证明与定理 3.3 的证明类似。

例 3.12 考虑以下系统,

$$\dot{x} = \begin{bmatrix} 0 & \mathrm{e}^{-t} \\ 6 & 0 \end{bmatrix} x(t), \ y(t) = \begin{bmatrix} 1 & 2 \end{bmatrix} x(t)$$

可以计算得到

$$N_0(t) = \begin{bmatrix} 1 & 2 \end{bmatrix}, \ N_1(t) = N_0(t)A(t) + \frac{\mathrm{d}}{\mathrm{d}t}N_0(t) = \begin{bmatrix} 12 & \mathrm{e}^{-t} \end{bmatrix}$$

那么,

$$\begin{bmatrix} N_0(t) \\ N_1(t) \end{bmatrix} = \begin{bmatrix} 1 & 2 \\ 12 & \mathrm{e}^{-t} \end{bmatrix}$$

它对于任意的时刻 t 都是列满秩的,所以系统在任何时刻都是完全能观的。

研究线性时变系统 $(A(t), C(t))$ 的能构性时,引入以下能构性 Gram 矩阵,

$$W_{\mathrm{cn}}(t_0, t_{\mathrm{f}}) \triangleq \int_{t_0}^{t_{\mathrm{f}}} \Phi^{\mathrm{T}}(\tau, t_{\mathrm{f}}) C^{\mathrm{T}}(\tau) C(\tau) \Phi(\tau, t_{\mathrm{f}}) \mathrm{d}\tau \tag{3.26}$$

当且仅当存在 $t_0 < t_{\mathrm{f}}$ 使得 $W_{\mathrm{cn}}(t_0, t_{\mathrm{f}})$ 非奇异时,$(A(t), C(t))$ 在 t_{f} 时刻是完全能构的。同时,结合式 (3.24) 和式 (3.26),可以得到

$$W_{\mathrm{o}}(t_0, t_{\mathrm{f}}) = \Phi^{\mathrm{T}}(t_{\mathrm{f}}, t_0) W_{\mathrm{cn}}(t_0, t_{\mathrm{f}}) \Phi(t_{\mathrm{f}}, t_0)$$

这表明 $\mathrm{rank}\, W_{\mathrm{o}}(t_0, t_{\mathrm{f}}) = \mathrm{rank}\, W_{\mathrm{cn}}(t_0, t_{\mathrm{f}})$。

3.2.3 线性定常系统的能观性判据

不失一般性地,在这一小节中假设 $t_0 = 0$。下面给出一些线性定常系统 (A, C) 常用的能观性判据。

定理 3.14(Gram 判据) 线性定常系统 (A, C) 是完全能观的,当且仅当存在 $t_{\mathrm{f}} > 0$ 使得

$$W_{\mathrm{o}}(0, t_{\mathrm{f}}) \triangleq \int_0^{t_{\mathrm{f}}} \mathrm{e}^{A^{\mathrm{T}}\tau} C^{\mathrm{T}} C \mathrm{e}^{A\tau} \mathrm{d}\tau \tag{3.27}$$

是非奇异的。

$n \times n$ 的矩阵 $W_{\mathrm{o}}(0, t_{\mathrm{f}})$ 被称作是线性定常系统的能观性 Gram 矩阵。

根据定理 3.3可以直接导出这一判据,证明略。

以下能观性判据和能控性判据存在对应关系,且它们的证明与相应的能控性判据的证明类似。

定理 3.15（秩判据） 线性定常系统 (A,C) 是完全能观的，当且仅当 $\operatorname{rank} V = n$，其中，

$$V \triangleq \begin{bmatrix} C \\ CA \\ \vdots \\ CA^{n-1} \end{bmatrix} \tag{3.28}$$

V 被称作线性定常系统的能观性矩阵。

例 3.13 考虑以下系统，

$$\dot{x} = \begin{bmatrix} 1 & 0 \\ 0 & -3 \end{bmatrix} x(t) + \begin{bmatrix} 5 \\ 1 \end{bmatrix} u(t),\ y(t) = \begin{bmatrix} 0 & -2 \end{bmatrix} x(t)$$

因为

$$\operatorname{rank} \begin{bmatrix} C \\ CA \end{bmatrix} = \operatorname{rank} \begin{bmatrix} 0 & -2 \\ 0 & 6 \end{bmatrix} < 2$$

所以这一系统不完全能观。

定理 3.16（PBH 特征向量判据） 线性定常系统 (A,C) 是完全能观的，当且仅当 A 没有与 C 的所有行都正交的非零右特征向量。或者说，对于 A 的任意特征值 λ_i，$i \in \{1,\cdots,n\}$，仅有 $\alpha = 0$ 能够满足

$$A\alpha = \lambda_i \alpha,\ C\alpha = 0 \tag{3.29}$$

定理 3.17（PBH 秩判据） 线性定常系统 (A,C) 是完全能观的，当且仅当

$$\operatorname{rank} \begin{bmatrix} C \\ \lambda_i I - A \end{bmatrix} = n \tag{3.30}$$

对于 A 的所有特征值 λ_i 都成立，$i \in \{1,\cdots,n\}$。式 (3.30) 也可以等价地表示为在复域 $\mathbb{C}$ 中，

$$\operatorname{rank} \begin{bmatrix} C \\ sI - A \end{bmatrix} = n,\ \forall s \in \mathbb{C} \tag{3.31}$$

即 $sI - A$ 和 C 必须是右互质的。

定理 3.18（Jordan 标准形判据） 考虑线性定常系统 (A,C)，λ_i 是 A 的 σ_i 重特征值，$i \in \{1,\cdots,l\}$，且 $\sigma_1 + \cdots + \sigma_i + \cdots + \sigma_l = n$。存在等价变换 $\bar{x}(t) = Tx(t)$ 使得 (A,C) 转化为 Jordan 标准形 $(\bar{A},\bar{C})$，其中，

$$\bar{A} = \begin{bmatrix} J_1 & & & \\ & J_2 & & \\ & & \ddots & \\ & & & J_l \end{bmatrix},\ \bar{C} = \begin{bmatrix} \bar{C}_1 & \bar{C}_2 & \cdots & \bar{C}_l \end{bmatrix}$$

3.2 线性系统的能观性

$$J_i = \begin{bmatrix} J_{i1} & & & \\ & J_{i2} & & \\ & & \ddots & \\ & & & J_{i\alpha_i} \end{bmatrix}, \quad \bar{C}_i = \begin{bmatrix} \bar{C}_{i1} & \bar{C}_{i2} & \cdots & \bar{C}_{i\alpha_i} \end{bmatrix}$$

$$J_{ik} = \begin{bmatrix} \lambda_i & 1 & & \\ & \lambda_i & \ddots & \\ & & \ddots & 1 \\ & & & \lambda_i \end{bmatrix}, \quad \bar{C}_{ik} = \begin{bmatrix} \bar{C}_{ik1} & \bar{C}_{ik2} & \cdots & \bar{C}_{ik\sigma_{ik}} \end{bmatrix}$$

J_i 是与 λ_i 相关的 $\sigma_i \times \sigma_i$ 的 Jordan 块矩阵，J_{ik}, $k \in \{1, \cdots, \alpha_i\}$，是 $\sigma_{ik} \times \sigma_{ik}$ 的 Jordan 块，α_i 是 λ_i 的几何重数，$\sigma_{i1} + \cdots + \sigma_{i\alpha_i} = \sigma_i$。

(A, C) 是完全能观的，当且仅当对于 $i \in \{1, \cdots, l\}$，矩阵

$$\tilde{C}_i = \begin{bmatrix} \bar{c}_{i11} & \bar{c}_{i21} & \cdots & \bar{c}_{i\alpha_i 1} \end{bmatrix}$$

都列线性无关。

例 3.14 系统的 Jordan 标准形如下，判断这一系统的能观性，

$$\dot{\bar{x}}(t) = \begin{bmatrix} -1 & 1 & & & & & & \\ 0 & -1 & & & & & & \\ & & -1 & & & & & \\ & & & -1 & & & & \\ & & & & 2 & 1 & & \\ & & & & 0 & 2 & & \\ & & & & & & 2 & \\ & & & & & & & 4 \end{bmatrix} \bar{x}(t)$$

$$y(t) = \begin{bmatrix} 2 & 0 & 0 & 0 & 1 & 0 & 0 & 0 \\ 0 & 0 & 1 & 0 & 2 & 4 & 0 & 6 \\ 0 & 0 & 0 & 3 & 3 & 0 & 1 & 0 \end{bmatrix} \bar{x}(t)$$

根据 Jordan 标准形判据，写出分别与特征值 $\lambda_1 = -1$, $\lambda_2 = 2$ 和 $\lambda_3 = 4$ 对应的 Jordan 块的第一列构成的矩阵，

$$\begin{bmatrix} \bar{c}_{111} & \bar{c}_{121} & \bar{c}_{131} \end{bmatrix} = \begin{bmatrix} 2 & 0 & 0 \\ 0 & 1 & 0 \\ 0 & 0 & 3 \end{bmatrix}, \quad \begin{bmatrix} \bar{c}_{211} & \bar{c}_{221} \end{bmatrix} = \begin{bmatrix} 1 & 0 \\ 2 & 0 \\ 3 & 1 \end{bmatrix}, \quad \bar{c}_{311} = \begin{bmatrix} 0 \\ 6 \\ 0 \end{bmatrix}$$

上述三个矩阵的列都是线性无关的，所以，这一系统完全能观。

当考虑线性定常系统 (A,C) 的能构性时，引入能构性 Gram 矩阵

$$W_{\mathrm{cn}}(0,t_{\mathrm{f}}) \triangleq \int_0^{t_{\mathrm{f}}} \mathrm{e}^{A^{\mathrm{T}}(\tau-t_{\mathrm{f}})}C^{\mathrm{T}}C\mathrm{e}^{A(\tau-t_{\mathrm{f}})}\mathrm{d}\tau \tag{3.32}$$

当且仅当存在 $t_{\mathrm{f}}>0$ 使得 $W_{\mathrm{cn}}(0,t_{\mathrm{f}})$ 是非奇异的，(A,C) 是完全能构的。同时，结合式 (3.27) 和式 (3.32)，可以得到

$$W_{\mathrm{o}}(0,t_{\mathrm{f}}) = \left(\mathrm{e}^{At_{\mathrm{f}}}\right)^{\mathrm{T}} W_{\mathrm{cn}}(0,t_{\mathrm{f}})\mathrm{e}^{At_{\mathrm{f}}}$$

这表明 $\operatorname{rank} W_{\mathrm{o}}(0,t_{\mathrm{f}}) = \operatorname{rank} W_{\mathrm{cn}}(0,t_{\mathrm{f}})$，也就是说当且仅当 (A,C) 完全能观时，它完全能构。

3.2.4 线性定常系统的能观性指数

对于线性定常系统 (A,C)，定义

$$V_\nu = \begin{bmatrix} C^{\mathrm{T}} & A^{\mathrm{T}}C^{\mathrm{T}} & \cdots & (A^{\nu-1})^{\mathrm{T}}C^{\mathrm{T}} \end{bmatrix}^{\mathrm{T}} \tag{3.33}$$

其中，V_ν 是 $\nu q \times n$ 的常值矩阵，ν 是正整数。当 $\nu = n$ 时，V_ν 是系统的能观性矩阵。进一步，如果系统完全能观，那么 $\operatorname{rank} V_\nu = n$。注意到在 V_ν 中添加一个子块 CA^k 后，$k=1,\cdots,\nu-1$，V_ν 的秩至少增加 1，直到 $\operatorname{rank} V_\nu = n$。

定义 3.10（线性定常系统的能观性指数） 当式 (3.33) 中的 ν 是使得

$$\operatorname{rank} V_\nu = \operatorname{rank} V_{\nu+1} = n \tag{3.34}$$

成立的最小正整数时，ν 被称作 (A,C) 的能观性指数。

定理 3.19 能观性指数 ν 满足

$$\frac{n}{q} \leqslant \nu \leqslant \max\{\bar{n}, n-m+1\} \tag{3.35}$$

其中，m 是 C 的秩，$\bar{n}$ 是 A 的最小多项式的阶次。

证明与定理 3.9 类似，略。

根据定理 3.19，对于 SISO 系统，$\nu = n$。同时，定理 3.15 可以修改为以下形式。

推论 3.2 线性定常系统 (A,C) 是完全能观的，当且仅当

$$\operatorname{rank} V_{n-m+1} = \operatorname{rank} \begin{bmatrix} C^{\mathrm{T}} & A^{\mathrm{T}}C^{\mathrm{T}} & \cdots & (A^{n-m})^{\mathrm{T}}C^{\mathrm{T}} \end{bmatrix}^{\mathrm{T}} = n \tag{3.36}$$

或

$$\operatorname{rank} V_{\bar{n}} = \operatorname{rank} \begin{bmatrix} C^{\mathrm{T}} & A^{\mathrm{T}}C^{\mathrm{T}} & \cdots & (A^{\bar{n}-1})^{\mathrm{T}}C^{\mathrm{T}} \end{bmatrix}^{\mathrm{T}} = n. \tag{3.37}$$

3.2 线性系统的能观性

考虑 c_1, $\cdots$, c_q 是 C 的行, V_ν 可以表示为

$$V_\nu = \begin{bmatrix} c_1 \\ \vdots \\ c_q \\ \hdashline c_1 A \\ \vdots \\ c_q A \\ \hdashline \vdots \\ \hdashline c_1 A^{\nu-1} \\ \vdots \\ c_q A^{\nu-1} \end{bmatrix} \quad (3.38)$$

在 V_ν 中,按照从上至下的顺序搜寻 n 个线性无关的行。由于 C 的秩为 m,这 n 个线性无关的行可以重新排列为

$$\begin{bmatrix} c_1 \\ c_1 A \\ \vdots \\ c_1 A^{\nu_1 - 1} \\ \hdashline c_2 \\ c_2 A \\ \vdots \\ c_2 A^{\nu_2 - 1} \\ \hdashline \vdots \\ \hdashline c_m \\ c_m A \\ \vdots \\ c_m A^{\nu_m - 1} \end{bmatrix}$$

对于完全能观的系统,可以得到

$$\nu_1 + \nu_2 + \cdots + \nu_m = n$$

能观性指数满足

$$\nu = \max\{\nu_1, \nu_2, \cdots, \nu_m\}$$

一般地,$\{\nu_1, \nu_2, \cdots, \nu_m\}$ 被称作 (A, C) 的能观性指数集或 Kronecker 不变量。

定理 3.20 等价变换不会改变 (A, C) 的 ν 和 $\{\nu_1, \nu_2, \cdots, \nu_m\}$。

证明略。

例 3.15 考虑以下线性定常系统，

$$\dot{x}(t) = \begin{bmatrix} 0 & 1 & 0 & 0 \\ 3 & 0 & 0 & 2 \\ 0 & 0 & 0 & 1 \\ 0 & -2 & 0 & 0 \end{bmatrix} x(t), \quad y(t) = \begin{bmatrix} 1 & 0 & 0 & 0 \\ 0 & 0 & 1 & 0 \end{bmatrix} x(t)$$

因为

$$\operatorname{rank} \begin{bmatrix} C \\ CA \end{bmatrix} = \begin{bmatrix} 1 & 0 & 0 & 0 \\ 0 & 0 & 1 & 0 \\ 0 & 1 & 0 & 0 \\ 0 & 0 & 0 & 1 \end{bmatrix} = 4$$

这一系统的能观性指数为 $\nu = 2$。

选择 4 个线性无关的行向量并重新排列为

$$\begin{bmatrix} c_1 \\ c_1 A \\ c_2 \\ c_2 A \end{bmatrix} = \begin{bmatrix} 1 & 0 & 0 & 0 \\ 0 & 1 & 0 & 0 \\ 0 & 0 & 1 & 0 \\ 0 & 0 & 0 & 1 \end{bmatrix}$$

可以看出，能观性指数集为 $\{\nu_1, \nu_2\} = \{2, 2\}$，且 $\nu = \max\{\nu_1, \nu_2\} = 2$。

3.3 线性系统的对偶原理

从前边的讨论可以看出，能控性和能观性在概念和判据的形式上都具有对偶性。这种内在的对偶关系反映了系统控制问题和估计问题之间的对偶性。这一能够表征系统对偶关系的规律性结果，被称为对偶原理。

3.3.1 对偶系统

考虑以下线性时变系统，

$$\dot{x} = A(t)x(t) + B(t)u(t), \quad y(t) = C(t)x(t) + D(t)u(t) \tag{3.39}$$

其中，$x(t) \in \mathbb{R}^n$ 是状态向量，$u(t) \in \mathbb{R}^p$ 是输入向量，$y(t) \in \mathbb{R}^q$ 是输出向量，$A(t)$，$B(t)$，$C(t)$ 和 $D(t)$ 是时变矩阵。

定义 3.11（对偶系统） 根据线性时变系统方程 (3.39)，构造

$$\dot{z}(t) = -A^{\mathrm{T}}(t)z(t) + C^{\mathrm{T}}(t)v(t), \quad w(t) = B^{\mathrm{T}}(t)z(t) + D^{\mathrm{T}}(t)v(t) \tag{3.40}$$

其中，$z(t) \in \mathbb{R}^n$ 是协状态向量，$v(t) \in \mathbb{R}^q$ 和 $w(t) \in \mathbb{R}^p$ 分别是这一系统的输入向量和输出向量。系统方程 (3.40) 被称作是系统 (3.39) 的对偶系统。

线性时变系统和它的对偶系统的结构图如图 3.5 所示。

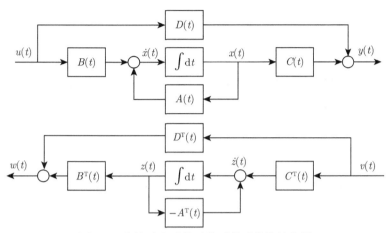

图 3.5 线性时变系统及其对偶系统的结构图

假设 $\Phi(t,t_0)$ 是系统方程 (3.39) 的状态转移矩阵，根据 $\Phi(t,t_0)$ 的性质，可知 $\Phi(t,t_0)\Phi^{-1}(t,t_0) = I$，且 $\Phi^{-1}(t,t_0) = \Phi(t_0,t)$。由此可以推导出

$$0 = \frac{\partial}{\partial t}\left(\Phi(t,t_0)\Phi^{-1}(t,t_0)\right) = \frac{\partial}{\partial t}\left(\Phi(t,t_0)\right)\Phi^{-1}(t,t_0) + \Phi(t,t_0)\frac{\partial}{\partial t}\Phi^{-1}(t,t_0)$$

$$= A(t)\Phi(t,t_0)\Phi^{-1}(t,t_0) + \Phi(t,t_0)\frac{\partial}{\partial t}\Phi(t_0,t)$$

这表明

$$\frac{\partial}{\partial t}\Phi(t_0,t) = -\Phi^{-1}(t,t_0)A(t) = -\Phi(t_0,t)A(t)$$

对上式进行转置，进一步可得

$$\frac{\partial}{\partial t}\Phi^{\mathrm{T}}(t_0,t) = -A^{\mathrm{T}}(t)\Phi^{\mathrm{T}}(t_0,t)$$

同时，$\Phi^{\mathrm{T}}(t_0,t_0) = I$。所以 $\Phi^{\mathrm{T}}(t_0,t)$ 是对偶系统方程 (3.40) 的状态转移矩阵。可以看出系统方程 (3.39) 的状态运动是从 t_0 到 t 正时向转移，而它的对偶系统方程 (3.40) 协状态的运动则是从 t 到 t_0 的反时向转移。

3.3.2 对偶原理

定理 3.21（对偶原理） 线性时变系统在 t_0 时刻是完全能控的，当且仅当它的对偶系统在 t_0 时刻是完全能观的。线性时变系统在 t_0 时刻是完全能观的，当且仅当它的对偶系统在 t_0 时刻是完全能控的。

证明 假设线性时变系统方程 (3.39) 在 t_0 时刻是完全能控的，那么存在 $t_f > t_0$ 使得能控性 Gram 矩阵 $W_c(t_0, t_f)$ 是非奇异的。注意到

$$W_c(t_0, t_f) \triangleq \int_{t_0}^{t_f} \Phi(t_0, \tau) B(\tau) B^T(\tau) \Phi^T(t_0, \tau) d\tau$$

$$= \int_{t_0}^{t_f} \left[\Phi^T(t_0, \tau)\right]^T \left[B^T(\tau)\right]^T \left[B^T(\tau)\right] \left[\Phi^T(t_0, \tau)\right] d\tau$$

这一能控性 Gram 矩阵等价于它的对偶系统方程 (3.40) 的能观性 Gram 矩阵。也就是说，系统方程 (3.39) 的能控性等价于它的对偶系统方程 (3.40) 的能观性。

采用类似的方法，可以证明系统方程 (3.39) 的能观性等价于它对偶系统方程 (3.40) 的能控性。 □

定理 3.21 不仅提出了基于对偶系统的结构特性来确定原系统结构特性的方法，也给出了系统的控制问题和估计问题之间的内在联系。对偶原理在后续章节中具有非常重要的应用。

如果系统方程 (3.39) 和方程 (3.40) 中的系数矩阵是常值矩阵，那么类似地可以定义线性定常系统的对偶系统，在线性定常的情况下，对偶原理同样成立。用 (A, B, C, D) 表示一个线性定常系统，而用 $(-A^T, C^T, B^T, D^T)$ 表示它的对偶系统。实际上，系统 (A, B, C, D) 和 (A^T, C^T, B^T, D^T) 之间也存在对偶关系。(A, B, C, D) 能控性矩阵和能观性矩阵分别为

$$U = \begin{bmatrix} B & AB & \cdots & A^{n-1}B \end{bmatrix}, \quad V = \begin{bmatrix} C^T & A^T C^T & \cdots & (A^{n-1})^T C^T \end{bmatrix}^T$$

同时，(A^T, C^T, B^T, D^T) 的能控性矩阵和能观性矩阵分别为

$$U_d = \begin{bmatrix} C^T & A^T C^T & \cdots & (A^{n-1})^T C^T \end{bmatrix}, \quad V_d = \begin{bmatrix} B & AB & \cdots & A^{n-1}B \end{bmatrix}^T$$

显然可得 $U = V_d^T$ 且 $V = U_d^T$。根据秩判据可知，(A, B, C, D) 是能控的，当且仅当 (A^T, C^T, B^T, D^T) 是能观的，(A, B, C, D) 是能观的，当且仅当 (A^T, C^T, B^T, D^T) 是能控的。后续章节中，考虑线性定常系统 (A, B, C, D) 的对偶系统是 (A^T, C^T, B^T, D^T)。

3.4 线性定常系统的结构分解

对于不完全能控和不完全能观的系统，一个重要的问题是如何找到一种方法把系统根据能控性和能观性进行结构分解。通过分解，系统的结构可以被区分为能控且能观的部分、能控但不能观的部分、不能控但能观的部分和不能控也不能观的部分。基于这一分解，可以更深刻地揭示系统的结构特性，更深入地理解状态空间描述和输入-输出描述之间的本质区别，更好地解决系统的分析与设计问题。

3.4.1 线性系统在等价变换下的能控性和能观性

考虑线性时变系统 $(A(t), B(t), C(t), D(t))$，对其进行等价变换 $\bar{x}(t) = T(t)x(t)$，其中，$T(t)$ 是非奇异且连续可微的变换矩阵。由此，$(A(t), B(t), C(t), D(t))$ 可以转换为 $(\bar{A}(t), \bar{B}(t),$

$\bar{C}(t), \bar{D}(t))$,其中,$\bar{A}(t) = T(t)A(t)T^{-1}(t)+\dot{T}(t)T^{-1}(t)$,$\bar{B}(t) = T(t)B(t)$,$\bar{C}(t) = C(t)T^{-1}(t)$,$\bar{D}(t) = D(t)$。用 $\Phi(t,t_0)$ 和 $\bar{\Phi}(t,t_0)$ 分别表示两个系统的状态转移矩阵,2.2.4 节中指出,

$$\bar{\Phi}(t,t_0) = T(t)\Phi(t,t_0)T^{-1}(t_0)$$

同时,系统 $(A(t), B(t), C(t), D(t))$ 的能控性和能观性 Gram 矩阵分别记作 $W_c(t_0, t_f)$ 和 $W_o(t_0, t_f)$,系统 $(\bar{A}(t), \bar{B}(t), \bar{C}(t), \bar{D}(t))$ 的能控性和能观性 Gram 矩阵分别记作 $\bar{W}_c(t_0, t_f)$ 和 $\bar{W}_o(t_0, t_f)$。进一步可以推导出

$$\begin{aligned}
\bar{W}_c(t_0, t_f) &= \int_{t_0}^{t_f} \bar{\Phi}(t_0, \tau)\bar{B}(\tau)\bar{B}^T(\tau)\bar{\Phi}^T(t_0, \tau)\mathrm{d}\tau \\
&= \int_{t_0}^{t_f} T(t_0)\Phi(t_0, \tau)B(\tau)B^T(\tau)\Phi^T(t_0, \tau)T^T(t_0)\mathrm{d}\tau \\
&= T(t_0) \int_{t_0}^{t_f} \Phi(t_0, \tau)B(\tau)B^T(\tau)\Phi^T(t_0, \tau)\mathrm{d}\tau \, T^T(t_0) \\
&= T(t_0)W_c(t_0, t_f)T^T(t_0)
\end{aligned}$$

因为 $T(t_0)$ 是非奇异的,由此可以证明

$$\operatorname{rank} \bar{W}_c(t_0, t_f) = \operatorname{rank} W_c(t_0, t_f) \tag{3.41}$$

类似地,可以得到

$$\operatorname{rank} W_o(t_0, t_f) = \operatorname{rank} \bar{W}_o(t_0, t_f) \tag{3.42}$$

式 (3.41) 和式 (3.42) 表明线性时变系统的等价变换不会改变系统的能控性和能观性,也不会改变系统不完全能控和不完全能观的程度。这一结论对线性定常系统也成立。这一事实为通过等价变换实现系统的结构分解提供了理论依据。

3.4.2 线性定常系统的能控性结构分解

考虑线性定常系统

$$\dot{x} = Ax(t) + Bu(t), \quad y(t) = Cx(t) \tag{3.43}$$

其中,$x(t) \in \mathbb{R}^n$ 是状态向量,$u(t) \in \mathbb{R}^p$ 是输入向量,$y(t) \in \mathbb{R}^q$ 是输出向量,A,B 和 C 是具有相应维数的常值矩阵。

假设系统方程 (3.43) 是不完全能控的。这种情况下,它的能控性矩阵 U 满足 $\operatorname{rank} U = k_1 < n$。从 U 中任意选取 k_1 个线性无关的列向量并记作 $q_1, \cdots, q_{k_1}$,在此基础上,进一步在 $\mathbb{R}^n$ 中选取 $n - k_1$ 个列向量并记作 $q_{k_1+1}, \cdots, q_n$,使得 $q_1, \cdots, q_{k_1}, q_{k_1+1}, \cdots, q_n$ 是线性无关的。由此,构造变换矩阵 T_1,它满足

$$T_1^{-1} = \begin{bmatrix} q_1 & \cdots & q_{k_1} & q_{k_1+1} & \cdots & q_n \end{bmatrix} \tag{3.44}$$

定理 3.22 对于不完全能控的线性定常系统方程 (3.43)，$\operatorname{rank} U = k_1 < n$，引入满足式 (3.44) 的等价变换矩阵 T_1，则 $\bar{x}(t) = T_1 x(t)$ 能够把系统方程 (3.43) 转化为

$$\begin{cases} \begin{bmatrix} \dot{\bar{x}}_1(t) \\ \dot{\bar{x}}_2(t) \end{bmatrix} = \begin{bmatrix} \bar{A}_{11} & \bar{A}_{12} \\ 0 & \bar{A}_{22} \end{bmatrix} \begin{bmatrix} \bar{x}_1(t) \\ \bar{x}_2(t) \end{bmatrix} + \begin{bmatrix} \bar{B}_1 \\ 0 \end{bmatrix} u(t) \\ y(t) = \begin{bmatrix} \bar{C}_1 & \bar{C}_2 \end{bmatrix} \begin{bmatrix} \bar{x}_1(t) \\ \bar{x}_2(t) \end{bmatrix} \end{cases} \quad (3.45)$$

其中，$\bar{x}_1(t) \in \mathbb{R}^{k_1}$ 是能控状态，$\bar{x}_2(t) \in \mathbb{R}^{n-k_1}$ 是不能控状态。

证明 记 $T_1 = [\, p_1 \; \cdots \; p_n \,]^{\mathrm{T}}$。因为 $T_1 T_1^{-1} = I$，可以得到

$$p_i^{\mathrm{T}} q_j = \begin{cases} 1 & i = j \\ 0 & i \neq j \end{cases} \quad (3.46)$$

其中，$i \in \{1, \cdots, n\}$ 且 $j \in \{1, \cdots, n\}$。

同时，考虑到系统的能控子空间是 A 的不变子空间。因为 $q_1, \cdots, q_{k_1}$ 是线性独立的，所以 Aq_j 是 $\{q_1, \cdots, q_{k_1}\}$ 的线性组合，$j = 1, \cdots, k_1$。结合式 (3.46)，可以推导出当 $i = k_1 + 1, \cdots, n$ 且 $j = 1, \cdots, k_1$ 时，$p_i^{\mathrm{T}} A q_j = 0$。所以，

$$\bar{A} = T_1 A T_1^{-1} = \begin{bmatrix} p_1^{\mathrm{T}} A q_1 & \cdots & p_1^{\mathrm{T}} A q_{k_1} & p_1^{\mathrm{T}} A q_{k_1+1} & \cdots & p_1^{\mathrm{T}} A q_n \\ \vdots & \ddots & \vdots & \vdots & \ddots & \vdots \\ p_{k_1}^{\mathrm{T}} A q_1 & \cdots & p_{k_1}^{\mathrm{T}} A q_{k_1} & p_{k_1}^{\mathrm{T}} A q_{k_1+1} & \cdots & p_{k_1}^{\mathrm{T}} A q_n \\ \hdashline p_{k_1+1}^{\mathrm{T}} A q_1 & \cdots & p_{k_1+1}^{\mathrm{T}} A q_{k_1} & p_{k_1+1}^{\mathrm{T}} A q_{k_1+1} & \cdots & p_{k_1+1}^{\mathrm{T}} A q_n \\ \vdots & \ddots & \vdots & \vdots & \ddots & \vdots \\ p_n^{\mathrm{T}} A q_1 & \cdots & p_n^{\mathrm{T}} A q_{k_1} & p_n^{\mathrm{T}} A q_{k_1+1} & \cdots & p_n^{\mathrm{T}} A q_n \end{bmatrix}$$

$$= \begin{bmatrix} \bar{A}_{11} & \bar{A}_{12} \\ 0 & \bar{A}_{22} \end{bmatrix}$$

另一方面，B 的列也都可以表示为 $\{q_1, \cdots, q_{k_1}\}$ 的线性组合。这种情况下，当 $i = k_1 + 1, \cdots, n$ 时，$p_i^{\mathrm{T}} B = 0$。由此可得

$$\bar{B} = T_1 B = \begin{bmatrix} p_1^{\mathrm{T}} B \\ \vdots \\ p_{k_1}^{\mathrm{T}} B \\ \hdashline p_{k_1+1}^{\mathrm{T}} B \\ \vdots \\ p_n^{\mathrm{T}} B \end{bmatrix}$$

3.4 线性定常系统的结构分解

$[\ \bar{C}_1\ \ \bar{C}_2\ ]$ 没有特殊形式。注意到 $\bar{x}(t) = T_1 x(t)$ 可以把系统方程 (3.43) 转化为式 (3.45) 的形式。进一步，可以推导出

$$k_1 = \operatorname{rank} \bar{U} = \operatorname{rank} [\ \bar{B}\ \ \bar{A}\bar{B}\ \ \cdots\ \ \bar{A}^{n-1}\bar{B}\ ]$$

$$= \operatorname{rank} \begin{bmatrix} \bar{B}_1 & \bar{A}_{11}\bar{B}_1 & \cdots & \bar{A}_{11}^{n-1}\bar{B}_1 \\ 0 & 0 & \cdots & 0 \end{bmatrix}$$

$$= \operatorname{rank} [\ \bar{B}_1\ \ \bar{A}_{11}\bar{B}\ \ \cdots\ \ \bar{A}_{11}^{n-1}\bar{B}_1\ ]$$

$\bar{A}_{11}$ 是 $k_1 \times k_1$ 维的矩阵。根据 Cayley-Hamilton 定理，$\bar{A}_{11}^{k_1}\bar{B}_1, \cdots, \bar{A}_{11}^{n-1}\bar{B}_1$ 可以表示为 $\bar{B}_1, \bar{A}_{11}\bar{B}, \cdots, \bar{A}_{11}^{k_1-1}\bar{B}_1$ 的线性组合，所以

$$\operatorname{rank} [\ \bar{B}_1\ \ \bar{A}_{11}\bar{B}\ \ \cdots\ \ \bar{A}_{11}^{n-1}\bar{B}_1\ ] = \operatorname{rank} [\ \bar{B}_1\ \ \bar{A}_{11}\bar{B}\ \ \cdots\ \ \bar{A}_{11}^{k_1-1}\bar{B}_1\ ] = k_1$$

这表明 $(\bar{A}_{11}, \bar{B}_1)$ 是完全能控的，且 $\bar{x}_1(t)$ 是能控的状态。 □

例 3.16 考虑线性定常系统

$$\dot{x}(t) = \begin{bmatrix} -1 & 0 & 0 \\ 1 & 0 & 5 \\ 0 & 1 & 3 \end{bmatrix} x(t) + \begin{bmatrix} 0 & 0 \\ 1 & 0 \\ 0 & 1 \end{bmatrix} u(t)$$

可以计算出

$$\operatorname{rank} [\ B\ \ AB\ ] = \operatorname{rank} \begin{bmatrix} 0 & 0 & 0 & 0 \\ 1 & 0 & 0 & 5 \\ 0 & 1 & 1 & 3 \end{bmatrix} = 2$$

$n = 3$，且 $\operatorname{rank} B = 2$。由此可知系统是不完全能控的。在能控性矩阵里选择线性无关的列 $q_1 = [\ 0\ \ 1\ \ 0\ ]^T$ 和 $q_2 = [\ 0\ \ 0\ \ 1\ ]^T$，同时选择能够保证 q_1，q_2 和 q_3 线性无关的列 $q_3 = [\ 1\ \ 0\ \ 0\ ]^T$。构造

$$T_1^{-1} = \begin{bmatrix} 0 & 0 & 1 \\ 1 & 0 & 0 \\ 0 & 1 & 0 \end{bmatrix}, T_1 = \begin{bmatrix} 0 & 1 & 0 \\ 0 & 0 & 1 \\ 1 & 0 & 0 \end{bmatrix}$$

可以得到

$$\bar{A} = T_1 A T_1^{-1} = \begin{bmatrix} 0 & 5 & 1 \\ 1 & 3 & 0 \\ 0 & 0 & -1 \end{bmatrix}, \bar{B} = T_1 B = \begin{bmatrix} 1 & 0 \\ 0 & 1 \\ 0 & 0 \end{bmatrix}, \bar{C} = C T_1^{-1} = [\ 0\ \ 1\ \ 1\ ]$$

根据定理 3.22，不完全能控的系统方程 (3.43) 被分解为两部分，一部分是能控的子系统，

$$\dot{\bar{x}}_1(t) = \bar{A}_{11}\bar{x}_1(t) + \bar{A}_{12}\bar{x}_2(t) + \bar{B}_1 u(t), \quad \bar{y}_1(t) = \bar{C}_1 \bar{x}_1(t)$$

另一部分是不能控的子系统，

$$\dot{\bar{x}}_2(t) = \bar{A}_{22}\bar{x}_2(t), \quad \bar{y}_2(t) = \bar{C}_2 \bar{x}_2(t)$$

且 $y(t) = \bar{y}_1(t) + \bar{y}_2(t)$。注意到

$$\det(sI - A) = \det(sI - \bar{A}) = \det \begin{bmatrix} sI - \bar{A}_{11} & -\bar{A}_{12} \\ 0 & sI - \bar{A}_{22} \end{bmatrix}$$

这表明 A 的特征值由两部分组成。一部分为 A_{11} 的特征值，输入 $u(t)$ 可以影响和改变这一部分的特征值，这些特征值通常被称为 A 的能控特征值。另一部分则是 A_{22} 的特征值，$u(t)$ 不能影响或改变这些特征值，它们通常被称为 A 的不能控特征值。

根据式 (3.45)，可以绘制出系统进行能控性结构分解后的框图，如图 3.6 所示。从图中可以看出，控制输入 $u(t)$ 不能直接或间接地影响不能控的状态。

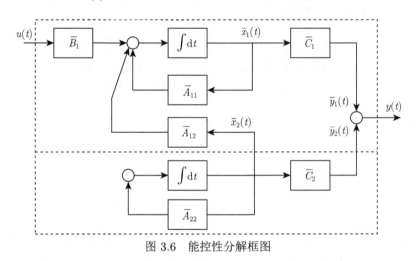

图 3.6 能控性分解框图

因为等价变换不改变系统的传递函数矩阵，系统方程 (3.45) 的传递函数矩阵和原系统方程 (3.43) 的相同。这种情况下，可知

$$\begin{aligned} G(s) &= \begin{bmatrix} \bar{C}_1 & \bar{C}_2 \end{bmatrix} \begin{bmatrix} sI - \bar{A}_{11} & -\bar{A}_{12} \\ 0 & sI - \bar{A}_{22} \end{bmatrix}^{-1} \begin{bmatrix} \bar{B}_1 \\ 0 \end{bmatrix} \\ &= \begin{bmatrix} \bar{C}_1 & \bar{C}_2 \end{bmatrix} \begin{bmatrix} (sI - \bar{A}_{11})^{-1} & (sI - \bar{A}_{11})^{-1}\bar{A}_{12}(sI - \bar{A}_{22})^{-1} \\ 0 & (sI - \bar{A}_{22})^{-1} \end{bmatrix} \begin{bmatrix} \bar{B}_1 \\ 0 \end{bmatrix} \\ &= \bar{C}_1(sI - \bar{A}_{11})^{-1}\bar{B}_1 \end{aligned}$$

因为变换矩阵 T_1 的选择是不唯一的，所以系统方程 (3.43) 存在不同的能控性结构分解。这些规范分解形式相同，但参数值不同。

3.4.3　线性定常系统的能观性结构分解

本节中，考虑系统方程 (3.43) 是不完全能观的。假设

$$\operatorname{rank} V = \operatorname{rank} \begin{bmatrix} C \\ CA \\ \vdots \\ CA^{n-1} \end{bmatrix} = k_2 < n$$

从 V 中任选 k_2 个线性无关的行并把它们记作 $h_1, \cdots, h_{k_2}$。此外，选取 $n - k_2$ 个行向量，并把它们记作 $h_{k_2+1}, \cdots, h_n$，使得 $h_1, \cdots, h_{k_2}, h_{k_2+1}, \cdots, h_n$ 是线性无关的。在此基础上，构造变换矩阵 T_2 为

$$T_2 = \begin{bmatrix} h_1 \\ \vdots \\ h_{k_2} \\ h_{k_2+1} \\ \vdots \\ h_n \end{bmatrix} \tag{3.47}$$

定理 3.23　对于不完全能观的线性定常系统方程 (3.43)，$\operatorname{rank} V = k_2 < n$，引入满足式 (3.47) 的等价变换矩阵 T_2，则 $\hat{x}(t) = T_2 x(t)$ 能够把系统方程 (3.43) 转化为

$$\begin{cases} \begin{bmatrix} \dot{\hat{x}}_1(t) \\ \dot{\hat{x}}_2(t) \end{bmatrix} = \begin{bmatrix} \hat{A}_{11} & 0 \\ \hat{A}_{21} & \hat{A}_{22} \end{bmatrix} \begin{bmatrix} \hat{x}_1(t) \\ \hat{x}_2(t) \end{bmatrix} + \begin{bmatrix} \hat{B}_1 \\ \hat{B}_2 \end{bmatrix} u(t) \\ y(t) = \begin{bmatrix} \hat{C}_1 & 0 \end{bmatrix} \begin{bmatrix} \hat{x}_1(t) \\ \hat{x}_2(t) \end{bmatrix} \end{cases} \tag{3.48}$$

其中，$\hat{x}_1(t) \in \mathbb{R}^{k_2}$ 是能观状态，$\hat{x}_2(t) \in \mathbb{R}^{n-k_2}$ 是不能观状态。

证明略。

根据定理 3.23，不完全能观的系统方程 (3.43) 可以分解为两部分，一部分是能观的子系统，

$$\dot{\hat{x}}_1(t) = \hat{A}_{11} \hat{x}_1(t) + \hat{B}_1 u(t), \quad \hat{y}_1(t) = \hat{C}_1 \hat{x}_1(t)$$

另一部分则是不能观的子系统

$$\dot{\hat{x}}_2(t) = \hat{A}_{21} \hat{x}_1(t) + \hat{A}_{22} \hat{x}_2(t) + \hat{B}_2 u(t), \quad \hat{y}_2(t) = 0$$

且 $y(t) = \hat{y}_1(t) + \hat{y}_2(t)$。注意到

$$\det(sI - A) = \det(sI - \hat{A}) = \det \begin{bmatrix} sI - \hat{A}_{11} & 0 \\ -\hat{A}_{11} & sI - \hat{A}_{22} \end{bmatrix}$$

$$= \det(sI - \hat{A}_{11}) \det(sI - \hat{A}_{22})$$

这表明 A 的特征值可以分为两部分。一部分是 $\hat{A}_{11}$ 的特征值，能够通过输出 $y(t)$ 反映，这些特征值一般被称作 A 的能观特征值。另一部分是 $\hat{A}_{22}$ 的特征值，$y(t)$ 不能反映这些特征值的信息，它们一般被称为 A 的不能观特征值。

根据式 (3.48)，可以绘制出系统经过能观性结构分解后的框图，如图 3.7 所示。从图中可以看出，不能观状态的信息不能从输出 $y(t)$ 中得到。

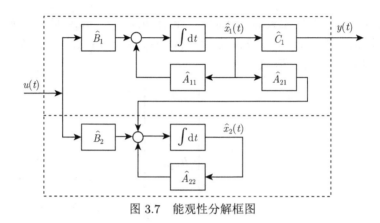

图 3.7 能观性分解框图

系统方程 (3.48) 的传递函数矩阵也是原系统方程 (3.43) 的传递函数矩阵，即

$$G(s) = \hat{C}_1(sI - \hat{A}_{11})^{-1}\hat{B}_1$$

系统方程 (3.43) 的能观性结构分解也不唯一。这些规范分解形式相同，但参数值不同。

3.4.4 线性定常系统的 Kalman 分解

本小节考虑系统方程 (3.43) 既不完全能控，也不完全能观的情况。这种情况下，下述结论成立，这一结论通常被称作规范分解定理或 Kalman 分解定理。

定理 3.24 对于不完全能控也不完全能观的线性定常系统方程 (3.43)，存在等价变换矩阵 T，使得 $\tilde{x}(t) = Tx(t)$ 能够把系统方程 (3.43) 转化为

$$\begin{cases} \begin{bmatrix} \dot{\tilde{x}}_1(t) \\ \dot{\tilde{x}}_2(t) \\ \dot{\tilde{x}}_3(t) \\ \dot{\tilde{x}}_4(t) \end{bmatrix} = \begin{bmatrix} \tilde{A}_{11} & 0 & \tilde{A}_{13} & 0 \\ \tilde{A}_{21} & \tilde{A}_{22} & \tilde{A}_{23} & \tilde{A}_{24} \\ 0 & 0 & \tilde{A}_{33} & 0 \\ 0 & 0 & \tilde{A}_{43} & \tilde{A}_{44} \end{bmatrix} \begin{bmatrix} \tilde{x}_1(t) \\ \tilde{x}_2(t) \\ \tilde{x}_3(t) \\ \tilde{x}_4(t) \end{bmatrix} + \begin{bmatrix} \tilde{B}_1 \\ \tilde{B}_2 \\ 0 \\ 0 \end{bmatrix} u(t) \\ y(t) = \begin{bmatrix} \tilde{C}_1 & 0 & \tilde{C}_3 & 0 \end{bmatrix} \begin{bmatrix} \tilde{x}_1(t) \\ \tilde{x}_2(t) \\ \tilde{x}_3(t) \\ \tilde{x}_4(t) \end{bmatrix} \end{cases} \quad (3.49)$$

其中，$\tilde{x}_1(t)$ 是能控且能观的状态，$\tilde{x}_2(t)$ 是能控但不能观的状态，$\tilde{x}_3(t)$ 是不能控但能观的状态，$\tilde{x}_4(t)$ 是不能控且不能观的状态。

证明 使用定理 3.22 可以把系统方程 (3.43) 分解为能控的部分和不能控的部分。在此基础上，应用定理 3.23，进一步把能控的子系统和不能控的子系统分别分解为能观的部分和不能观的部分。系统方程 (3.43) 可以最终被分解为方程 (3.49) 的形式。 □

系统结构规范分解方程 (3.49) 的框图如图 3.8 所示。需要指出的是这一分解也是不唯一的，但分解的结构形式是唯一的。

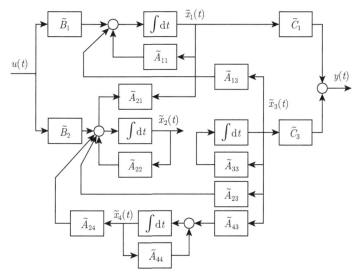

图 3.8 系统结构规范分解框图

另一方面，定理 3.24 中，首先进行了能控性结构分解。实际上，也可以先进行能观性结构分解，再对能观的子系统和不能观的子系统分别进行能控性结构分解。这种情况下，系统方程 (3.43) 也能够被分解为四部分。

子系统 $(\tilde{A}_{11}, \tilde{B}_1, \tilde{C}_1)$ 是能控且能观的，系统方程 (3.43) 的传递函数矩阵和 $(\tilde{A}_{11}, \tilde{B}_1, \tilde{C}_1)$ 的传递函数矩阵相同，即传递函数矩阵只能描述系统中能控且能观的子系统。所以，

$$G(s) = \tilde{C}_1(sI - \tilde{A}_{11})^{-1}\tilde{B}_1$$

例 3.17 考虑以下不完全能控也不完全能观的线性定常系统，

$$\dot{x}(t) = \begin{bmatrix} 0 & 0 & -1 \\ 1 & 0 & -3 \\ 0 & 1 & -3 \end{bmatrix} x(t) + \begin{bmatrix} 1 \\ 1 \\ 0 \end{bmatrix} u(t), \ y(t) = \begin{bmatrix} 0 & 1 & -2 \end{bmatrix} x(t)$$

对它进行结构分解。

首先，进行能控性结构分解。

可以计算出

$$\operatorname{rank} U = \operatorname{rank} \begin{bmatrix} b & Ab & A^2b \end{bmatrix} = \operatorname{rank} \begin{bmatrix} 1 & 0 & -1 \\ 1 & 1 & -3 \\ 0 & 1 & -2 \end{bmatrix} = 2 < n = 3$$

在能控性矩阵中选择线性无关的列 $q_1 = \begin{bmatrix} 1 & 1 & 0 \end{bmatrix}^{\mathrm{T}}$ 和 $q_2 = \begin{bmatrix} 0 & 1 & 1 \end{bmatrix}^{\mathrm{T}}$，同时选择能够保证 q_1，q_2 和 q_3 线性无关的列 $q_3 = \begin{bmatrix} 0 & 0 & 1 \end{bmatrix}^{\mathrm{T}}$。构造

$$T_1^{-1} = \begin{bmatrix} q_1 & q_2 & q_3 \end{bmatrix} = \begin{bmatrix} 1 & 0 & 0 \\ 1 & 1 & 0 \\ 0 & 1 & 1 \end{bmatrix}$$

可以进一步计算出

$$T_1 = \begin{bmatrix} 1 & 0 & 0 \\ -1 & 1 & 0 \\ 1 & -1 & 1 \end{bmatrix}$$

由此可知

$$\bar{A} = T_1 A T_1^{-1} = \begin{bmatrix} 0 & -1 & -1 \\ 1 & -2 & -2 \\ 0 & 0 & -1 \end{bmatrix}, \bar{B} = T_1 B = \begin{bmatrix} 1 \\ 0 \\ 0 \end{bmatrix}, \bar{C} = C T_1^{-1} = \begin{bmatrix} 1 & -1 & -2 \end{bmatrix}$$

接下来，进行能观性结构分解。

可以计算出

$$\operatorname{rank} V = \operatorname{rank} \begin{bmatrix} C \\ CA \\ CA^2 \end{bmatrix} = \operatorname{rank} \begin{bmatrix} 0 & 1 & -2 \\ 1 & -2 & 3 \\ -2 & 3 & -4 \end{bmatrix} = 2 < n = 3$$

在能观性矩阵中选择线性无关的行 $h_1 = \begin{bmatrix} 0 & 1 & -2 \end{bmatrix}$ 和 $h_2 = \begin{bmatrix} 1 & -2 & 3 \end{bmatrix}$，同时选择能够保证 h_1，h_2 和 h_3 线性无关的 $h_3 = \begin{bmatrix} 0 & 0 & 1 \end{bmatrix}$。构造

$$T_2 = \begin{bmatrix} h_1 \\ h_2 \\ h_3 \end{bmatrix} = \begin{bmatrix} 0 & 1 & -2 \\ 1 & -2 & 3 \\ 0 & 0 & 1 \end{bmatrix}$$

进一步可以计算出

$$T_2^{-1} = \begin{bmatrix} 2 & 1 & 1 \\ 1 & 0 & 2 \\ 0 & 0 & 1 \end{bmatrix}$$

能够得到

$$\hat{A} = T_2 A T_2^{-1} = \begin{bmatrix} 0 & 1 & 0 \\ -1 & -2 & 0 \\ 1 & 0 & -1 \end{bmatrix}, \quad \hat{B} = T_2 B = \begin{bmatrix} 1 \\ -1 \\ 0 \end{bmatrix}, \quad \hat{C} = C T_2^{-1} = \begin{bmatrix} 1 & 0 & 0 \end{bmatrix}$$

最后，进行结构规范分解。

能控性结构分解后，能控的子系统为

$$\dot{\bar{x}}(t) = \begin{bmatrix} 0 & -1 \\ 1 & -2 \end{bmatrix} \bar{x}_1(t) + \begin{bmatrix} -1 \\ -2 \end{bmatrix} \bar{x}_2(t) + \begin{bmatrix} 1 \\ 0 \end{bmatrix} u(t), \quad \bar{y}_1(t) = \begin{bmatrix} 1 & -1 \end{bmatrix} \bar{x}_1(t)$$

它的能观性矩阵是

$$V_1 = \begin{bmatrix} 1 & -1 \\ -1 & 1 \end{bmatrix}$$

可以看出 $\operatorname{rank} V_1 = 1 < 2$。

所以能控的子系统是不完全能观的。进行能观性结构分解，变换矩阵选择

$$T_{r1} = \begin{bmatrix} 1 & -1 \\ 0 & 1 \end{bmatrix}$$

另一方面，可以看出不能控的子系统是完全能观的。因此可以选择 $T_{r2} = 1$。T_{r1} 和 T_{r2} 构成了一个块对角矩阵，

$$T_r = \begin{bmatrix} T_{r1} & \\ & T_{r2} \end{bmatrix} = \begin{bmatrix} 1 & -1 & 0 \\ 0 & 1 & 0 \\ 0 & 0 & 1 \end{bmatrix}, \quad T_r^{-1} = \begin{bmatrix} 1 & 1 & 0 \\ 0 & 1 & 0 \\ 0 & 0 & 1 \end{bmatrix}$$

引入等价变换 $\tilde{x}(t) = T_r \bar{x}(t)$，能够得到

$$\tilde{A} = T_r \bar{A} T_r^{-1} = \begin{bmatrix} -1 & 0 & 1 \\ 1 & -1 & -2 \\ 0 & 0 & -1 \end{bmatrix}, \quad \tilde{B} = T_r \bar{B} = \begin{bmatrix} 1 \\ 0 \\ 0 \end{bmatrix}, \quad \tilde{C} = \bar{C} T_r^{-1} \begin{bmatrix} 1 & 0 & -2 \end{bmatrix}$$

可以看出，原系统最终被分解为能控且能观的部分、能控但不能观的部分和不能控但能观的部分。

3.4.5 极点、零点和特征值间的关系

在 1.2.4 小节中，基于 Smith-McMillan 形定义了传递函数矩阵 $G(s)$ 的极点和零点。下面，来介绍系统的极点和零点。

定义 3.12（Rosenbrock(罗森布罗克) 系统矩阵） 对于线性定常系统 (A,B,C,D)，矩阵

$$S(s) \triangleq \begin{bmatrix} sI-A & B \\ -C & D \end{bmatrix} \tag{3.50}$$

被称作系统的 Rosenbrock 系统矩阵。

考虑以下线性定常系统，

$$\dot{x}(t) = Ax(t) + Bu(t), \ y(t) = Cx(t) + Du(t) \tag{3.51}$$

其中，$x(t) \in \mathbb{R}^n$ 是状态向量，$u(t) \in \mathbb{R}^p$ 是输入向量，$y(t) \in \mathbb{R}^q$ 是输出向量，A，B，C 和 D 是具有相应维数的常值矩阵。结合式 (3.50) 和式 (3.51)，可以得到

$$S(s) \begin{bmatrix} -x(s) \\ u(s) \end{bmatrix} = \begin{bmatrix} 0 \\ y(s) \end{bmatrix}$$

其中，$x(s)$，$u(s)$ 和 $y(s)$ 分别表示 $x(t)$，$u(t)$ 和 $y(t)$ 的 Laplace 变换。可以看出系统矩阵 $S(s)$ 也可以用来描述线性定常系统。

定义 3.13（系统的极点） 对于线性定常系统 (A,B,C,D)，多项式

$$p_{\mathrm{s}}(s) = \det(sI-A)$$

被称作系统的极点多项式，且 $p_{\mathrm{s}}(s) = 0$ 的根被称作系统的极点。

可以看出系统 (A,B,C,D) 的极点与 A 的特征值完全相同。

定义 3.14（系统的零点） 对于线性定常系统 (A,B,C,D)，$\mathrm{rank}\,S(s) = r$，考虑以下形式的 $S(s)$ 的 r 阶非零子式：包含 $S(s)$ 的前 n 行和前 n 列，即包含 $sI-A$ 的所有行和列，然后添加 $[\,-C\ \ D\,]$ 中的 $r-n$ 行和 $[\,B^{\mathrm{T}}\ \ D^{\mathrm{T}}\,]^{\mathrm{T}}$ 中相应的 $r-n$ 列。这些子式的首一最大公因子被称作系统的零点多项式，记作 $z_{\mathrm{s}}(s)$。$z_{\mathrm{s}}(s) = 0$ 的根被称作系统的零点。

定义 3.15（系统的不变零点） 对于线性定常系统 (A,B,C,D)，$\mathrm{rank}\,S(s) = r$，$S(s)$ 的不变因子为 $s_1(s), \cdots, s_r(s)$。多项式

$$z_{\mathrm{s}}^{\mathrm{i}}(s) = \prod_{i=1}^{r} s_i(s)$$

被称作系统的不变零点多项式，且 $z_{\mathrm{s}}^{\mathrm{i}}(s) = 0$ 的根被称作系统的不变零点。

可以推导出 $S(s)$ 所有的最高阶非零子式的首一最大公因子等于 $z_{\mathrm{s}}^{\mathrm{i}}(s)$。一般地，

$$\{系统的不变零点\} \subset \{系统的零点\}$$

定义 3.16（系统的输入解耦零点） 对于线性定常系统 (A,B,C,D)，$[\,sI-A\ \ B\,]$ 的不变因子的乘积构成一个多项式，这一多项式的根被称作系统的输入解耦零点。

3.4 线性定常系统的结构分解

定义 3.17（系统的输出解耦零点） 对于线性定常系统 (A,B,C,D)，$[(sI-A)^{\mathrm{T}} - C^{\mathrm{T}}]^{\mathrm{T}}$ 的不变因子的乘积构成一个多项式，这一多项式的根被称作系统的输出解耦零点。

根据以上定义，不难推导出输入解耦零点和输出解耦零点都是 A 的特征值，也都是系统 (A,B,C,D) 的零点。同时，如果 λ 是一个输入解耦零点，那么 $\mathrm{rank}[\ \lambda I - A\quad B\ ] < n$。根据 PBH 秩判据，系统不完全能控。3.4.2 小节中指出，不完全能控的线性定常系统可以被分解为能控的部分和不能控的部分，实际上，系统的输入解耦零点与 A 的不能控的特征值完全相同。类似地，如果系统存在输出解耦零点，则系统是不完全能观的，且输出解耦零点与 A 的不能观的特征值完全相同。

3.4.4 节中指出，A 可能有既不能控也不能观的特征值。这些特征值反映在系统的零点上时，既是输入解耦零点，也是输出解耦零点，这类零点被称作系统的输入-输出解耦零点。

系统的零点确定后，移除输入解耦零点和输出解耦零点，剩余的零点就是传递函数矩阵 $G(s)$ 的零点。一般地，传递函数矩阵 $G(s)$ 的零点也被称为系统的传输零点。

根据上述讨论，可以得到以下结论，

$$\{\text{系统的零点}\} = \{G(s)\text{的零点}\} \cup \{\text{输入解耦零点}\} \cup \{\text{输出解耦零点}\}$$
$$- \{\text{输入输出解耦零点}\}$$

注意到系统的不变零点包含 $G(s)$ 的全部零点（传输零点），但不包含全部的解耦零点。当 $S(s)$ 是方阵且非奇异时，系统的零点与不变零点完全相同。进一步，当 (A,B,C) 完全能控能观时，系统的零点、不变零点和 $G(s)$ 的零点完全相同。

此外，A 的特征值构成的集合满足

$$\{A\text{的特征值（系统的极点）}\} = \{G(s)\text{的极点}\} \cup \{\text{输入解耦零点}\} \cup \{\text{输出解耦零点}\}$$
$$- \{\text{输入输出解耦零点}\}$$

3.4.4 节中指出，线性定常系统 (A,B,C) 可以被分解为能控且能观的部分，能控但不能观的部分，不能控但能观的部分和不能控也不能观的部分。(A,B,C) 的传递函数矩阵等于能控且能观子系统 $(\tilde{A}_{11}, \tilde{B}_1, \tilde{C}_1)$ 的传递函数矩阵。另一方面，1.2.4 小节中，基于 Smith-McMillan 形定义了传递函数矩阵 $G(s)$ 的极点和零点。实际上，对于线性定常系统 (A,B,C)，$G(s)$ 的极点就等于 $\tilde{A}_{11}$ 的特征值。

例 3.18 考虑 $A = \begin{bmatrix} 0 & -1 & 1 \\ 1 & -2 & 1 \\ 0 & 1 & -1 \end{bmatrix}$，$B = \begin{bmatrix} 1 & 0 \\ 1 & 1 \\ 1 & 2 \end{bmatrix}$ 且 $C = \begin{bmatrix} 0 & 1 & 0 \end{bmatrix}$，相应的传递函数矩阵为

$$G(s) = \begin{bmatrix} \dfrac{1}{s} & \dfrac{1}{s} \end{bmatrix}$$

$G(s)$ 的极点多项式为 $p(s) = s$，所以 $G(s)$ 只有一个极点 $s_1 = 0$。$\lambda_1 = 0$ 是 A 的唯一能控且能观的特征值。A 的另外两个特征值 $\lambda_2 = -1$，$\lambda_3 = -2$，不是完全能控且能观的，没有反映在 $G(s)$ 的极点中。

$G(s)$ 的零点多项式为 $z(s) = 1$，所以 $G(s)$ 没有零点。

考虑

$$S(s) = \begin{bmatrix} sI-A & B \\ -C & D \end{bmatrix} = \begin{bmatrix} s & 1 & -1 & 1 & 0 \\ -1 & s+2 & -1 & 1 & 1 \\ 0 & -1 & s+1 & 1 & 2 \\ 0 & -1 & 0 & 0 & 0 \end{bmatrix}$$

上述系统矩阵由两个包含 $sI-A$ 所有行和列的四阶子式，分别选择 $S(s)$ 的 1，2，3，4 列和 1，2，3，5 列，相应的子式分别为 $(s+1)(s+2)$ 和 $(s+1)(s+2)$。系统的零点多项式为 $z_s(s) = (s+1)(s+2)$，且系统的零点是 -1 和 -2。

为了确定系统的输入解耦零点，考虑 $[\,sI-A \;\; B\,]$ 的所有三阶非零子式，它们的最大公因子为 $s+2$，这表明这一系统有一个输入解耦零点 -2。类似地，考虑 $[\,(sI-A)^{\mathrm{T}} \;\; -C^{\mathrm{T}}\,]^{\mathrm{T}}$，$s+1$ 是其所有三阶非零子式的最大公因子，因此系统有一个输出解耦零点 -1。可以看出，这一系统没有输入-输出解耦零点。

也可以通过相关矩阵的 Smith 形和传递函数矩阵 $G(s)$ 的 Smith-McMillan 形来求解零点和极点。$S(s)$ 的 Smith 形为

$$\begin{bmatrix} 1 & 0 & 0 & 0 & 0 \\ 0 & 1 & 0 & 0 & 0 \\ 0 & 0 & 1 & 0 & 0 \\ 0 & 0 & 0 & s+2 & 0 \end{bmatrix}$$

$[\,sI-A \;\; B\,]$ 的 Smith 形为

$$\begin{bmatrix} 1 & 0 & 0 & 0 & 0 \\ 0 & 1 & 0 & 0 & 0 \\ 0 & 0 & s+2 & 0 & 0 \end{bmatrix}$$

$[\,(sI-A)^{\mathrm{T}} \;\; -C^{\mathrm{T}}\,]^{\mathrm{T}}$ 的 Smith 形为

$$\begin{bmatrix} 1 & 0 & 0 \\ 0 & 1 & 0 \\ 0 & 0 & s+1 \\ 0 & 0 & 0 \end{bmatrix}$$

$sI-A$ 的 Smith 形为

$$\begin{bmatrix} 1 & 0 & 0 \\ 0 & 1 & 0 \\ 0 & 0 & s(s+1)(s+2) \end{bmatrix}$$

同时，$G(s)$ 的 Smith-McMillan 形为

$$G_{\mathrm{M}}(s) = \begin{bmatrix} \dfrac{1}{s} & 0 \end{bmatrix}$$

基于此可以直接验证上述结论。注意到这种情况下，不变零点多项式为 $z_{\mathrm{s}}^{\mathrm{i}}(s) = s+2$，系统仅有一个不变零点为 -2。

第 4 章

线性定常系统的标准形和实现

前面章节中指出等价变换不会改变系统的结构和性质。为了更进一步揭示系统的特征和结构特性，更便捷地分析和设计控制系统，通常利用等价变换将系统转化为可以直接反映其能控性和能观性的结构形式。实现问题指的是把仅表征系统外部因果关系的传递函数矩阵转化为能揭示系统内部结构特性的状态空间描述。通过系统实现问题的研究，有助于深刻地揭示系统的结构和性质及其在不同描述下的标准形式。

4.1 线性定常系统的标准形

定理 3.10 和定理 3.20 指出，对线性定常系统进行等价变换后，系统的能控性和能观性保持不变。以此为依据，完全能控或能观的线性定常系统可以通过等价变换转化为相应的标准形。这些标准形在系统综合问题的研究中具有重要作用。

4.1.1 线性定常 SISO 系统的标准形

考虑以下线性定常 SISO 系统，

$$\dot{x}(t) = Ax(t) + bu(t), \quad y(t) = cx(t) \tag{4.1}$$

其中，$x(t) \in \mathbb{R}^n$ 是状态向量，$u(t) \in \mathbb{R}^1$ 是输入向量，$y(t) \in \mathbb{R}^1$ 是输出向量，A，b 和 c 是具有相应维数的常值矩阵。考虑

$$\alpha(s) = s^n + a_1 s^{n-1} + \cdots + a_{n-1} s + a_n \tag{4.2}$$

是式 (4.1) 的特征多项式。

首先，介绍系统方程 (4.1) 的能控标准形。

定理 4.1 如果线性定常 SISO 系统方程 (4.1) 是完全能控的，那么存在等价变换矩阵 T_c 使得 $\bar{x}(t) = T_c(t)$ 能够把系统方程 (4.1) 转化为能控标准形，

$$\dot{\bar{x}}(t) = \bar{A}_c \bar{x}(t) + \bar{b}_c u(t), \quad y(t) = \bar{c}_c \bar{x}(t) \tag{4.3}$$

其中，

4.1 线性定常系统的标准形

$$\bar{A}_c = \begin{bmatrix} 0 & 1 & 0 & \cdots & 0 \\ 0 & 0 & 1 & \cdots & 0 \\ \vdots & \vdots & \vdots & \ddots & \vdots \\ 0 & 0 & 0 & \cdots & 1 \\ -a_n & -a_{n-1} & -a_{n-2} & \cdots & -a_1 \end{bmatrix}, \bar{b}_c = \begin{bmatrix} 0 \\ 0 \\ \vdots \\ 0 \\ 1 \end{bmatrix}$$

$$\bar{c}_c = \begin{bmatrix} \beta_n & \beta_{n-1} & \beta_{n-2} & \cdots & \beta_1 \end{bmatrix}$$

$$= c \begin{bmatrix} b & Ab & \cdots & A^{n-1}b \end{bmatrix} \begin{bmatrix} a_{n-1} & a_{n-2} & \cdots & a_1 & 1 \\ a_{n-2} & \vdots & \ddots & 1 & \\ \vdots & a_1 & \ddots & & \\ a_1 & 1 & & & \\ 1 & & & & \end{bmatrix}$$

$a_1, \cdots, a_n$ 是式 (4.2) 中的系数。

证明 因为系统方程 (4.1) 完全能控，所以 $\operatorname{rank} U = \operatorname{rank} [\begin{array}{cccc} b & Ab & \cdots & A^{n-1}b \end{array}] = n$。构造

$$Q = \begin{bmatrix} q_1 & q_2 & \cdots & q_n \end{bmatrix}$$

$$= \begin{bmatrix} b & Ab & \cdots & A^{n-1}b \end{bmatrix} \begin{bmatrix} a_{n-1} & a_{n-2} & \cdots & a_1 & 1 \\ a_{n-2} & \vdots & \ddots & 1 & \\ \vdots & a_1 & \ddots & & \\ a_1 & 1 & & & \\ 1 & & & & \end{bmatrix} \quad (4.4)$$

显然，Q 是非奇异的。选择 Q^{-1} 作为变换矩阵 T_c，并引入等价变换 $\bar{x}(t) = T_c x(t)$，那么 $\bar{A}_c = T_c A T_c^{-1}$，即 $T_c^{-1} \bar{A}_c = A T_c^{-1}$，它可以被写作

$$\begin{bmatrix} q_1 & q_2 & \cdots & q_n \end{bmatrix} \bar{A}_c = A \begin{bmatrix} q_1 & q_2 & \cdots & q_n \end{bmatrix} \quad (4.5)$$

根据式 (4.4)，可以得到

$$\begin{cases} q_1 = a_{n-1} b + a_{n-2} Ab + \cdots + A^{n-1} b \\ q_2 = a_{n-2} b + a_{n-3} Ab + \cdots + A^{n-2} b \\ \vdots \\ q_{n-1} = a_1 b + Ab \\ q_n = b \end{cases}$$

这表明
$$Aq_{i+1} = q_i - a_{n-i}q_n, \quad i = 1, \cdots, n-1 \tag{4.6}$$

根据 Cayley-Hamilton 定理，$\alpha(A) = 0$。由此可以进一步得到
$$\begin{aligned} Aq_1 &= a_{n-1}Ab + a_{n-2}A^2b + \cdots + A^nb \\ &= (a_nI + a_{n-1}A + \cdots + A^n)b - a_nb = -a_nq_n \end{aligned} \tag{4.7}$$

结合式 (4.6) 和式 (4.7)，可得
$$A\begin{bmatrix} q_1 & q_2 & \cdots & q_n \end{bmatrix} = \begin{bmatrix} q_1 & q_2 & \cdots & q_n \end{bmatrix} \begin{bmatrix} 0 & 1 & 0 & \cdots & 0 \\ 0 & 0 & 1 & \cdots & 0 \\ \vdots & \vdots & \vdots & & \vdots \\ 0 & 0 & 0 & \cdots & 1 \\ -a_n & -a_{n-1} & a_{n-2} & \cdots & -a_1 \end{bmatrix}$$

与式 (4.5) 比较，可以得到 $\bar{A}_c$ 满足式 (4.3)。

与此同时，$\bar{b}_c = T_c b$，即 $T_c^{-1} \bar{b}_c = b$，它可以被写作
$$\begin{bmatrix} q_1 & q_2 & \cdots & q_n \end{bmatrix} \bar{b}_c = b$$

因为 $q_n = b$，可以推导出 $\bar{b}_c$ 满足式 (4.3)。

最后，$\bar{c}_c = cT_c^{-1}$。结合式 (4.4)，显然可得 $\bar{c}_c$ 也满足式 (4.3)。 □

为了求 SISO 系统方程 (4.1) 的能控标准形，需要首先确定变换矩阵 T_c。可以根据式 (4.4) 求取 T_c^{-1}，然后再求 T_c^{-1} 的逆以获得 T_c。实际上，也可以直接构造 T_c。考虑能控性矩阵的逆为

$$U^{-1} = \begin{bmatrix} b & Ab & \cdots & A^{n-1}b \end{bmatrix}^{-1} = \begin{bmatrix} p_1^{\mathrm{T}} \\ p_2^{\mathrm{T}} \\ \vdots \\ p_n^{\mathrm{T}} \end{bmatrix}$$

在此基础上，可以得到相应的变换矩阵为
$$T_c = \begin{bmatrix} p_n^{\mathrm{T}} \\ p_n^{\mathrm{T}} A \\ \vdots \\ p_n^{\mathrm{T}} A^{n-1} \end{bmatrix} \tag{4.8}$$

结合式 (4.4)，可以通过计算 $T_c T_c^{-1} = I$ 来验证式 (4.8)。

例 4.1 考虑以下线性定常系统,

$$\dot{x}(t) = \begin{bmatrix} 1 & 0 & 1 \\ 1 & 1 & 1 \\ 0 & 1 & 1 \end{bmatrix} x(t) + \begin{bmatrix} 1 \\ 0 \\ 1 \end{bmatrix} u(t), \ y(t) = \begin{bmatrix} 2 & 1 & 2 \end{bmatrix} x(t)$$

显然,能控性矩阵的秩为

$$\text{rank} \begin{bmatrix} b & Ab & A^2b \end{bmatrix} = \text{rank} \begin{bmatrix} 1 & 2 & 3 \\ 0 & 2 & 5 \\ 1 & 1 & 3 \end{bmatrix} = 3$$

这一系统是完全能控的,所以它可以被转化为能控标准形。这里,注意到

$$U^{-1} = \begin{bmatrix} b & Ab & A^2b \end{bmatrix}^{-1} = \begin{bmatrix} \frac{1}{5} & -\frac{3}{5} & \frac{4}{5} \\ 1 & 0 & -1 \\ -\frac{2}{5} & \frac{1}{5} & \frac{2}{5} \end{bmatrix}$$

使用 U^{-1} 的最后一行来构造变换矩阵 T_c,

$$T_c = \begin{bmatrix} p_3^{\mathrm{T}} \\ p_3^{\mathrm{T}} A \\ p_3^{\mathrm{T}} A^2 \end{bmatrix} = \begin{bmatrix} -\frac{2}{5} & \frac{1}{5} & \frac{2}{5} \\ -\frac{1}{5} & \frac{3}{5} & \frac{1}{5} \\ \frac{2}{5} & \frac{4}{5} & \frac{3}{5} \end{bmatrix}, \ T_c^{-1} = \begin{bmatrix} -1 & -1 & 1 \\ -1 & 2 & 0 \\ 2 & -2 & 1 \end{bmatrix}$$

进一步,可得

$$\bar{A}_c = T_c A T_c^{-1} = \begin{bmatrix} 0 & 1 & 0 \\ 0 & 0 & 1 \\ 1 & -2 & 3 \end{bmatrix}, \ \bar{b}_c = T_c b = \begin{bmatrix} 0 \\ 0 \\ 1 \end{bmatrix}, \ \bar{c}_c = c T_c^{-1} = \begin{bmatrix} 1 & -4 & 4 \end{bmatrix}$$

接下来,介绍系统方程 (4.1) 的能观标准形。

定理 4.2 如果线性定常 SISO 系统方程 (4.1) 是完全能观的,那么存在等价变换矩阵 T_o 使得 $\hat{x}(t) = T_o x(t)$ 能够把系统方程 (4.1) 转化为能观标准形,

$$\dot{\hat{x}}(t) = \hat{A}_o \hat{x}(t) + \hat{b}_o u(t), \ y(t) = \hat{c}_o \hat{x}(t) \tag{4.9}$$

其中,

$$\hat{A}_o = \begin{bmatrix} 0 & 0 & \cdots & 0 & -a_n \\ 1 & 0 & \cdots & 0 & -a_{n-1} \\ 0 & 1 & \cdots & 0 & -a_{n-2} \\ \vdots & \vdots & & \vdots & \vdots \\ 0 & 0 & \cdots & 1 & -a_1 \end{bmatrix}, \ \hat{b}_o = \begin{bmatrix} \beta_n \\ \beta_{n-1} \\ \beta_{n-2} \\ \vdots \\ \beta_1 \end{bmatrix}, \ \hat{c}_o = \begin{bmatrix} 0 & 0 & \cdots & 0 & 1 \end{bmatrix}$$

$a_1, \cdots, a_n$ 是式 (4.2) 中的系数，$\beta_1, \cdots, \beta_n$ 在定理 4.1中给出。

证明 这一定理可以基于定理 3.21 和定理 4.1 导出。因为 (A, b, c) 完全能观，所以它的对偶系统 $(A^{\mathrm{T}}, c^{\mathrm{T}}, b^{\mathrm{T}})$ 完全能控。这种情况下，系统方程 (4.1) 的能观标准形可以表示为

$$\hat{A}_{\mathrm{o}} = \bar{A}_{\mathrm{c}}^{\mathrm{T}}, \quad \hat{b}_{\mathrm{o}} = \bar{c}_{\mathrm{c}}^{\mathrm{T}}, \quad \hat{c}_{\mathrm{o}} = \bar{b}_{\mathrm{c}}^{\mathrm{T}}$$

这一形式满足式 (4.9)。 □

根据式 (4.4)，可知

$$T_{\mathrm{c}}^{-1} = \begin{bmatrix} c^{\mathrm{T}} & A^{\mathrm{T}}c^{\mathrm{T}} & \cdots & (A^{\mathrm{T}})^{n-1}c^{\mathrm{T}} \end{bmatrix} \begin{bmatrix} a_{n-1} & a_{n-2} & \cdots & a_1 & 1 \\ a_{n-2} & \vdots & \iddots & 1 & \\ \vdots & a_1 & \iddots & & \\ a_1 & 1 & & & \\ 1 & & & & \end{bmatrix}$$

因为

$$\hat{A}_{\mathrm{o}} = T_{\mathrm{o}} A T_{\mathrm{o}}^{-1} = \bar{A}_{\mathrm{c}}^{\mathrm{T}} = (T_{\mathrm{c}} A^{\mathrm{T}} T_{\mathrm{c}}^{-1})^{\mathrm{T}} = (T_{\mathrm{c}}^{-1})^{\mathrm{T}} A T_{\mathrm{c}}^{\mathrm{T}}$$

由此可以得到

$$T_{\mathrm{o}} = (T_{\mathrm{c}}^{-1})^{\mathrm{T}} = \begin{bmatrix} a_{n-1} & a_{n-2} & \cdots & a_1 & 1 \\ a_{n-2} & \vdots & \iddots & 1 & \\ \vdots & a_1 & \iddots & & \\ a_1 & 1 & & & \\ 1 & & & & \end{bmatrix} \begin{bmatrix} c \\ cA \\ \vdots \\ cA^{n-1} \end{bmatrix} \quad (4.10)$$

例 4.2 考虑例 4.1 中给出的系统。这一系统的特征多项式为

$$\alpha(s) = s^3 + a_1 s^2 + a_2 s + a_3 = s^3 - 3s^2 + 2s - 1$$

能观标准形的变换矩阵为

$$T_{\mathrm{o}} = \begin{bmatrix} a_2 & a_1 & 1 \\ a_1 & 1 & 0 \\ 1 & 0 & 0 \end{bmatrix} \begin{bmatrix} c \\ cA \\ cA^2 \end{bmatrix} = \begin{bmatrix} 2 & -3 & 1 \\ -3 & 1 & 0 \\ 1 & 0 & 0 \end{bmatrix} \begin{bmatrix} 2 & 1 & 2 \\ 3 & 3 & 5 \\ 6 & 8 & 11 \end{bmatrix} = \begin{bmatrix} 1 & 1 & 0 \\ -3 & 0 & -1 \\ 2 & 1 & 2 \end{bmatrix}$$

进一步可得

$$\hat{A}_{\mathrm{o}} = T_{\mathrm{o}} A T_{\mathrm{o}}^{-1} = \begin{bmatrix} 0 & 0 & 1 \\ 1 & 0 & -2 \\ 0 & 1 & 3 \end{bmatrix}, \quad \hat{b}_{\mathrm{o}} = T_{\mathrm{o}} b = \begin{bmatrix} 1 \\ -4 \\ 4 \end{bmatrix}^{\mathrm{T}}, \quad \hat{c}_{\mathrm{o}} = c T_{\mathrm{o}}^{-1} = \begin{bmatrix} 0 & 0 & 1 \end{bmatrix}$$

对于线性定常 SISO 系统，存在其他形式的标准形，这里不再详细介绍。

4.1.2 线性定常 MIMO 系统的标准形

考虑以下线性定常 MIMO 系统，

$$\dot{x}(t) = Ax(t) + Bu(t), \ y(t) = Cx(t) \tag{4.11}$$

其中，$x(t) \in \mathbb{R}^n$ 是状态向量，$u(t) \in \mathbb{R}^p$ 是输入向量，$y(t) \in \mathbb{R}^q$ 是输出向量，A，B 和 C 是具有相应维数的常值矩阵。

假设系统方程 (4.11) 完全能控且 $\text{rank}\, B = r \leqslant p$。能控性矩阵 U 是满秩的，U 的每一列可以写作

$$U = \begin{bmatrix} b_1 & \cdots & b_p & Ab_1 & \cdots & Ab_p & \cdots & A^{\mu-1}b_1 & \cdots & A^{\mu-1}b_p \end{bmatrix} \tag{4.12}$$

3.1.4 节中指出，可以从以上矩阵中从左至右依次选择 n 个线性无关的列向量。如果一个列向量不能表示为它左边的已选择的各列的线性组合，该列可以被选为 n 个线性无关的列向量中的一列。否则，不选取该列。因为 $\text{rank}\, B = r$，可以将选取的 n 个线性无关的列向量重新排列并构成系统方程 (4.11) 的等价变换矩阵，它满足

$$T_c^{-1} = \begin{bmatrix} b_1 & Ab_1 & \cdots & A^{\mu_1-1}b_1 & \cdots & b_r & Ab_r & \cdots & A^{\mu_r-1}b_r \end{bmatrix} \tag{4.13}$$

其中，$\{\mu_1, \mu_2, \cdots, \mu_r\}$ 是系统 (4.11) 的 Kronecker 不变量，且 $\mu_1 + \mu_2 + \cdots + \mu_r = n$。

后续内容中，系数矩阵中的 "$*$" 表示可能不为零的元素。

定理 4.3　如果线性定常系统 (4.11) 是完全能控的，那么存在满足式 (4.13) 的等价变换矩阵 T_c 使得 $\bar{x}(t) = T_c x(t)$ 能够把系统 (4.11) 转化为 Luenberger（龙伯格）第一能控标准形，

$$\dot{\bar{x}}(t) = \bar{A}_c \bar{x}(t) + \bar{B}_c u(t), \ y(t) = \bar{C}_c \bar{x}(t) \tag{4.14}$$

其中

$$\bar{A}_c = \begin{bmatrix} \bar{A}_{c11} & \cdots & \bar{A}_{c1r} \\ \vdots & & \vdots \\ \bar{A}_{cr1} & \cdots & \bar{A}_{crr} \end{bmatrix}, \ \bar{B}_c = \begin{bmatrix} \bar{B}_{c1} & * & \cdots & * \\ \vdots & \vdots & & \vdots \\ \bar{B}_{cr} & * & \cdots & * \end{bmatrix}$$

$\bar{C}_c = C T_c^{-1}$ 没有特殊形式，其中，$\bar{A}_{cii}$ 是 $\mu_i \times \mu_i$ 的矩阵，$\bar{A}_{cij}$ 是 $\mu_i \times \mu_j$ 的矩阵，且具有以下结构，

$$\bar{A}_{cii} = \begin{bmatrix} 0 & \cdots & 0 & * \\ 1 & \cdots & 0 & * \\ \vdots & & \vdots & \vdots \\ 0 & \cdots & 1 & * \end{bmatrix}, \ \bar{A}_{cij} = \begin{bmatrix} 0 & \cdots & 0 & * \\ 0 & \cdots & 0 & * \\ \vdots & & \vdots & \vdots \\ 0 & \cdots & 0 & * \end{bmatrix}$$

$i = 1, \cdots, r$，$j = 1, \cdots, r$，$i \neq j$，$\bar{B}_{ci}$ 是 $\mu_i \times r$ 的矩阵，且它第一行第 i 列的元为 1，其他元全为 0。

证明 可以验证

$$AT_c^{-1} = A \begin{bmatrix} b_1 & Ab_1 & \cdots & A^{\mu_1-1}b_1 & \cdots & b_r & Ab_r & \cdots & A^{\mu_r-1}b_r \end{bmatrix}$$

$$= \begin{bmatrix} Ab_1 & A^2b_1 & \cdots & A^{\mu_1}b_1 & \cdots & Ab_r & A^2b_r & \cdots & A^{\mu_r}b_r \end{bmatrix}$$

$$= \begin{bmatrix} b_1 & Ab_1 & \cdots & A^{\mu_1-1}b_1 & \cdots & b_r & Ab_r & \cdots & A^{\mu_r-1}b_r \end{bmatrix} \bar{A}_c$$

这表明，$\bar{A}_c = T_c A T_c^{-1}$。

与此同时，注意到 $\text{rank}\, B = r \leqslant p$。不失一般性地，假设 B 的前 r 列是线性独立的。这种情况下，可以得到

$$B = \begin{bmatrix} b_1 & \cdots & b_r & b_{r+1} & \cdots & b_p \end{bmatrix}$$

$$= \begin{bmatrix} b_1 & Ab_1 & \cdots & A^{\mu_1-1}b_1 & \cdots & b_r & Ab_r & \cdots & A^{\mu_r-1}b_r \end{bmatrix} \bar{B}_c$$

$$= T_c^{-1} \bar{B}_c$$

这表明，$\bar{B}_c = T_c B$。

$\bar{C}_c = C T_c^{-1}$ 没有特殊形式。 $\square$

更直观地，式 (4.14) 中的系数矩阵可以表示为

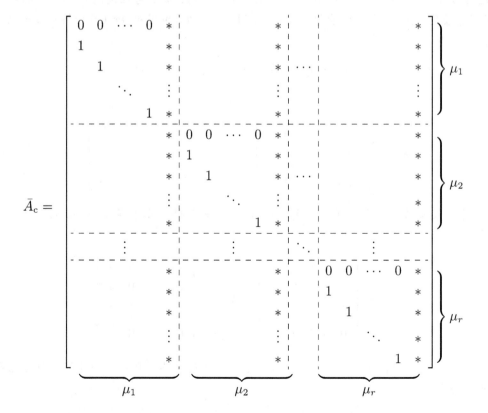

$$\bar{B}_{\mathrm{c}} = \begin{bmatrix} 1 & 0 & \cdots & 0 & * & \cdots & * \\ 0 & 0 & \cdots & 0 & * & \cdots & * \\ \vdots & \vdots & & \vdots & \vdots & & \vdots \\ 0 & 0 & \cdots & 0 & * & \cdots & * \\ \hline 0 & 1 & \cdots & 0 & * & \cdots & * \\ 0 & 0 & \cdots & 0 & * & \cdots & * \\ \vdots & \vdots & & \vdots & \vdots & & \vdots \\ 0 & 0 & \cdots & 0 & * & \cdots & * \\ \hline \vdots & \vdots & & \vdots & \vdots & & \vdots \\ \hline 0 & 0 & \cdots & 1 & * & \cdots & * \\ 0 & 0 & \cdots & 0 & * & \cdots & * \\ \vdots & \vdots & & \vdots & \vdots & & \vdots \\ 0 & 0 & \cdots & 0 & * & \cdots & * \end{bmatrix} \begin{matrix} \left.\vphantom{\begin{matrix}1\\0\\ \vdots \\0\end{matrix}}\right\}\mu_1 \\ \\ \left.\vphantom{\begin{matrix}0\\0\\ \vdots \\0\end{matrix}}\right\}\mu_2 \\ \\ \\ \left.\vphantom{\begin{matrix}0\\0\\ \vdots \\0\end{matrix}}\right\}\mu_r \end{matrix}$$

$$\underbrace{}_{r} \underbrace{}_{p-r}$$

假设系统方程 (4.11) 是完全能观的且 $\operatorname{rank} C = m \leqslant q$。结合 3.2.4 小节中的讨论，可以构造

$$T_{\mathrm{o}} = \begin{bmatrix} c_1 \\ c_1 A \\ \vdots \\ c_1 A^{\nu_1 - 1} \\ \vdots \\ c_m \\ c_m A \\ \vdots \\ c_m A^{\nu_m - 1} \end{bmatrix} \tag{4.15}$$

其中，$\{\nu_1, \nu_2, \cdots, \nu_m\}$ 是系统方程 (4.11) 的 Kronecker 不变量，且 $\nu_1 + \nu_2 + \cdots + \nu_m = n$。

定理 4.4 如果线性定常系统方程 (4.11) 是完全能观的，那么存在满足式 (4.15) 的等价变换矩阵 T_{o} 使得 $\hat{x}(t) = T_{\mathrm{o}} x(t)$ 能够把系统方程 (4.11) 转化为 Luenberger 第一能观标准形，

$$\dot{\hat{x}}(t) = \hat{A}_{\mathrm{o}} \hat{x}(t) + \hat{B}_{\mathrm{o}} u(t), \ y(t) = \hat{C}_{\mathrm{o}} \hat{x}(t) \tag{4.16}$$

其中，

$$\hat{A}_o = \begin{bmatrix} \hat{A}_{o11} & \cdots & \hat{A}_{o1m} \\ \vdots & & \vdots \\ \hat{A}_{om1} & \cdots & \hat{A}_{omm} \end{bmatrix}, \quad \hat{C}_o = \begin{bmatrix} \hat{C}_{o1} & \cdots & \hat{C}_{om} \\ * & \cdots & * \\ \vdots & & \vdots \\ * & \cdots & * \end{bmatrix}$$

$\hat{B}_o = T_o B$ 没有特殊形式,其中,$\hat{A}_{oii}$ 是 $\nu_i \times \nu_i$ 的矩阵,$\hat{A}_{oij}$ 是 $\nu_i \times \nu_j$ 的矩阵,且具有以下结构,

$$\hat{A}_{oii} = \begin{bmatrix} 0 & 1 & \cdots & 0 \\ \vdots & \vdots & & \vdots \\ 0 & 0 & \cdots & 1 \\ * & * & * & * \end{bmatrix}, \quad \hat{A}_{oij} = \begin{bmatrix} 0 & 0 & \cdots & 0 \\ \vdots & \vdots & & \vdots \\ 0 & 0 & \cdots & 0 \\ * & * & * & * \end{bmatrix}$$

$i = 1, \cdots, m, j = 1, \cdots, m, i \neq j$,$\hat{C}_{oi}$ 是 $m \times \nu_i$ 的矩阵,且它第 i 行第一列的元为 1,其他元全为 0。

证明略。

例 4.3 线性定常系统方程 (4.11) 的参数如下所示,

$$A = \begin{bmatrix} 0 & 0 & 0 & 1 \\ 1 & 0 & 0 & -2 \\ -22 & -11 & -4 & 0 \\ -23 & -6 & 0 & -6 \end{bmatrix}, \quad B = \begin{bmatrix} 0 & 0 \\ 0 & 0 \\ 0 & 1 \\ 1 & 3 \end{bmatrix}, \quad C = \begin{bmatrix} 0 & 0 & 0 & 1 \\ 0 & 0 & 1 & 0 \end{bmatrix}$$

(1) 求 Luenberger 第一能控标准形。

能控性矩阵为

$$U = \begin{bmatrix} 0 & 0 & 1 & 3 & -6 & -18 & 25 & 75 \\ 0 & 0 & -2 & -6 & 13 & 39 & -56 & -168 \\ 0 & 1 & 0 & -4 & 0 & 16 & -11 & -97 \\ 1 & 3 & -6 & -18 & 25 & 75 & -90 & -270 \end{bmatrix}$$
$$ b_1 \quad b_2 \quad Ab_1 \quad Ab_2 \quad A^2b_1 \quad A^2b_2 \quad A^3b_1 \quad A^3b_2$$

因为 $\text{rank}\, U = 4$,系统是完全能控的。构造

$$T_c^{-1} = \begin{bmatrix} b_1 & Ab_1 & A^2b_1 & b_2 \end{bmatrix} = \begin{bmatrix} 0 & 1 & -6 & 0 \\ 0 & -2 & 13 & 0 \\ 0 & 0 & 0 & 1 \\ 1 & -6 & 25 & 3 \end{bmatrix}$$

可以计算得到

4.1 线性定常系统的标准形

$$T_c = \begin{bmatrix} 28 & 11 & -3 & 1 \\ 13 & 6 & 0 & 0 \\ 2 & 1 & 0 & 0 \\ 0 & 0 & 1 & 0 \end{bmatrix}$$

进一步可得

$$\bar{A}_c = T_c A T_c^{-1} = \begin{bmatrix} 0 & 0 & 27 & 12 \\ 1 & 0 & -11 & 13 \\ 0 & 1 & -6 & 0 \\ 0 & 0 & -11 & -4 \end{bmatrix}, \quad \bar{B}_c = T_c B = \begin{bmatrix} 1 & 0 \\ 0 & 0 \\ 0 & 0 \\ 0 & 1 \end{bmatrix}$$

$$\bar{C}_c = C T_c^{-1} = \begin{bmatrix} 1 & -6 & 25 & 3 \\ 0 & 0 & 0 & 1 \end{bmatrix}$$

(2) 求 Luenberger 第一能观标准形。

能观性矩阵为

$$V = \begin{bmatrix} 0 & 0 & 0 & 1 \\ 0 & 0 & 1 & 0 \\ -23 & -6 & 0 & -6 \\ -22 & -11 & -4 & 0 \\ 132 & 36 & 0 & 25 \\ 77 & 44 & 16 & 0 \\ -539 & -150 & 0 & -90 \\ -308 & -176 & -64 & -11 \end{bmatrix} \begin{matrix} c_1 \\ c_2 \\ c_1 A \\ c_2 A \\ c_1 A^2 \\ c_2 A^2 \\ c_1 A^3 \\ c_2 A^3 \end{matrix}$$

因为 $\operatorname{rank} V = 4$,系统是完全能观的。构造

$$T_o = \begin{bmatrix} c_1 \\ c_1 A \\ c_2 \\ c_2 A \end{bmatrix} = \begin{bmatrix} 0 & 0 & 0 & 1 \\ -23 & -6 & 0 & -6 \\ 0 & 0 & 1 & 0 \\ -22 & -11 & -4 & 0 \end{bmatrix}$$

进一步可得

$$T_o^{-1} = \begin{bmatrix} -\dfrac{66}{121} & -\dfrac{11}{121} & \dfrac{24}{121} & \dfrac{6}{121} \\ \dfrac{132}{121} & \dfrac{22}{121} & -\dfrac{92}{121} & -\dfrac{23}{121} \\ 0 & 0 & 1 & 0 \\ 1 & 0 & 0 & 0 \end{bmatrix}.$$

所以,

$$\hat{A}_\text{o} = T_\text{o} A T_\text{o}^{-1} = \begin{bmatrix} 0 & 1 & 0 & 0 \\ -\dfrac{935}{121} & -\dfrac{660}{121} & -\dfrac{144}{121} & -\dfrac{36}{121} \\ 0 & 0 & 0 & 1 \\ 6 & 1 & -\dfrac{24}{11} & -\dfrac{550}{121} \end{bmatrix}, \quad \hat{B}_\text{o} = T_\text{o} B = \begin{bmatrix} 1 & 3 \\ -6 & -18 \\ 0 & 1 \\ 0 & -4 \end{bmatrix}$$

$$\hat{C}_\text{o} = C T_\text{o}^{-1} = \begin{bmatrix} 1 & 0 & 0 & 0 \\ 0 & 0 & 1 & 0 \end{bmatrix}$$

这里，只介绍了线性定常 MIMO 系统的 Luenberger 第一标准形。实际上，对于这类系统，还有 Luenberger 第二标准形，这里不再详细介绍。

另一方面，可以根据以下方法构造变换矩阵 T_wc。仍然从式 (4.12) 中的 U 内选择 n 个线性无关的列向量。从 b_1 开始，依次选择 $b_1, Ab_1, \cdots, A^{\bar{\mu}_1 - 1}b_1$，直到 $A^{\bar{\mu}_1} b_1$ 可以表示为 $b_1, Ab_1, \cdots, A^{\bar{\mu}_1 - 1}b_1$ 的线性组合为止。如果 $\bar{\mu}_1 = n$，即可结束选择。这种情况下，系统可以由 B 的第一列单独控制。如果 $\bar{\mu}_1 < n$，那么选择 $b_2, Ab_2, \cdots, A^{\bar{\mu}_2 - 1}b_2$，直到 $A^{\bar{\mu}_2}b_2$ 可以表示为所有已选择列的线性组合为止。如果 $\bar{\mu}_1 + \bar{\mu}_2 < n$，那么重复上述步骤，直到选择出全部的 n 个线性无关的列。应用选择的 n 个线性无关的列向量构造等价变换矩阵，它满足

$$T_\text{wc}^{-1} = \begin{bmatrix} b_1 & Ab_1 & \cdots & A^{\bar{\mu}_1 - 1}b_1 & \cdots & b_l & Ab_l & \cdots & A^{\bar{\mu}_l - 1}b_l \end{bmatrix} \tag{4.17}$$

其中，$\bar{\mu}_1 + \bar{\mu}_2 + \cdots + \bar{\mu}_l = n$。

定理 4.5 如果线性定常系统方程 (4.11) 是完全能控的，那么存在满足式 (4.17) 的等价变换矩阵 T_wc 使得 $\bar{x}(t) = T_\text{wc} \bar{x}(t)$ 能够把系统方程 (4.11) 转化为 Wonham（旺纳姆）第一能控标准形，

$$\dot{\bar{x}}(t) = \bar{A}_\text{wc} \bar{x}(t) + \bar{B}_\text{wc} u(t), \quad y(t) = \bar{C}_\text{wc} \bar{x}(t) \tag{4.18}$$

其中，

$$\bar{A}_\text{wc} = \begin{bmatrix} \bar{A}_{\text{wc}11} & \cdots & \bar{A}_{\text{wc}1l} \\ & \ddots & \vdots \\ & & \bar{A}_{\text{wc}ll} \end{bmatrix}, \quad \bar{B}_\text{wc} = \begin{bmatrix} \bar{B}_{\text{wc}1} & * & \cdots & * \\ \vdots & \vdots & & \vdots \\ \bar{B}_{\text{wc}l} & * & \cdots & * \end{bmatrix}$$

$\bar{C}_\text{wc} = C T_\text{wc}^{-1}$ 没有特殊形式，其中，$\bar{A}_{\text{wc}ii}$ 是 $\bar{\mu}_i \times \bar{\mu}_i$ 的矩阵，$\bar{A}_{\text{wc}ij}$ 是 $\bar{\mu}_i \times \bar{\mu}_j$ 的矩阵，且具有以下结构，

$$\bar{A}_{\text{wc}ii} = \begin{bmatrix} 0 & \cdots & 0 & * \\ 1 & \cdots & 0 & * \\ \vdots & & \vdots & \vdots \\ 0 & \cdots & 1 & * \end{bmatrix}, \quad \bar{A}_{\text{wc}ij} = \begin{bmatrix} 0 & \cdots & 0 & * \\ 0 & \cdots & 0 & * \\ \vdots & & \vdots & \vdots \\ 0 & \cdots & 0 & * \end{bmatrix}$$

$i = 1, \cdots, l$，$j = 1, \cdots, l$，$i \neq j$，$\bar{B}_{\text{wc}i}$ 是 $\mu_i \times l$ 的矩阵，且它第一行第 i 列的元为 1，其他元全为 0。

证明略。

更直观地，式 (4.18) 中的系数矩阵可以表示为

$$\bar{A}_{\mathrm{wc}} = \left[\begin{array}{ccccc|cccc|c|cccc}
0 & 0 & \cdots & 0 & * & & & & * & & & & & * \\
1 & & & & * & & & & * & & & & & * \\
 & 1 & & & * & & & & * & \cdots & & & & * \\
 & & \ddots & & \vdots & & & & \vdots & & & & & \vdots \\
 & & & 1 & * & & & & * & & & & & * \\
\hline
 & & & & & 0 & 0 & \cdots & 0 & * & & & & * \\
 & & & & & 1 & & & & * & & & & * \\
 & & & & & & 1 & & & * & \cdots & & & * \\
 & & & & & & & \ddots & & \vdots & & & & \vdots \\
 & & & & & & & & 1 & * & & & & * \\
\hline
 & & & & & & & & & & \ddots & & & \vdots \\
\hline
 & & & & & & & & & & & 0 & 0 & \cdots & 0 & * \\
 & & & & & & & & & & & 1 & & & * \\
 & & & & & & & & & & & & 1 & & * \\
 & & & & & & & & & & & & & \ddots & \vdots \\
 & & & & & & & & & & & & & 1 & *
\end{array}\right]\begin{array}{l}\left.\vphantom{\begin{array}{c}1\\1\\1\\1\\1\end{array}}\right\}\bar{\mu}_1\\ \\ \left.\vphantom{\begin{array}{c}1\\1\\1\\1\\1\end{array}}\right\}\bar{\mu}_2\\ \\ \\ \left.\vphantom{\begin{array}{c}1\\1\\1\\1\\1\end{array}}\right\}\bar{\mu}_l\end{array}$$

$$\underbrace{\qquad\qquad}_{\bar{\mu}_1}\underbrace{\qquad\qquad}_{\bar{\mu}_2}\underbrace{\qquad\qquad}_{\bar{\mu}_l}$$

$$\bar{B}_{\mathrm{wc}} = \left[\begin{array}{cccc|ccc}
1 & 0 & \cdots & 0 & * & \cdots & * \\
0 & 0 & \cdots & 0 & * & \cdots & * \\
\vdots & \vdots & & \vdots & \vdots & & \vdots \\
0 & 0 & \cdots & 0 & * & \cdots & * \\
\hline
0 & 1 & \cdots & 0 & * & \cdots & * \\
0 & 0 & \cdots & 0 & * & \cdots & * \\
\vdots & \vdots & & \vdots & \vdots & & \vdots \\
0 & 0 & \cdots & 0 & * & \cdots & * \\
\hline
\vdots & \vdots & & \vdots & \vdots & & \vdots \\
\hline
0 & 0 & \cdots & 1 & * & \cdots & * \\
0 & 0 & \cdots & 0 & * & \cdots & * \\
\vdots & \vdots & & \vdots & \vdots & & \vdots \\
0 & 0 & \cdots & 0 & * & \cdots & *
\end{array}\right]\begin{array}{l}\left.\vphantom{\begin{array}{c}1\\1\\1\\1\end{array}}\right\}\bar{\mu}_1\\ \\ \left.\vphantom{\begin{array}{c}1\\1\\1\\1\end{array}}\right\}\bar{\mu}_2\\ \\ \\ \left.\vphantom{\begin{array}{c}1\\1\\1\\1\end{array}}\right\}\bar{\mu}_l\end{array}.$$

$$\underbrace{\qquad\qquad}_{l}\underbrace{\qquad}_{p-l}$$

可以看出，在 Wonham 第一能控标准形中，状态系数矩阵 $\bar{A}_{\mathrm{wc}}$ 是上块三角矩阵。完

全能控的线性定常系统也有 Wonham 第二能控标准形，相应地，状态系数矩阵是下块三角矩阵。

仿照构造 T_{wc}^{-1} 类似的思路，可以构造

$$T_{\text{wo}} = \begin{bmatrix} c_1 \\ c_1 A \\ \vdots \\ c_1 A^{\bar{\nu}_1 - 1} \\ \vdots \\ c_k \\ c_k A \\ \vdots \\ c_k A^{\bar{\nu}_k - 1} \end{bmatrix} \tag{4.19}$$

其中，$\bar{\nu}_1 + \bar{\nu}_2 + \cdots + \bar{\nu}_k = n$。

定理 4.6 如果线性定常系统方程 (4.11) 是完全能观的，那么存在满足式 (4.19) 的等价变换矩阵 T_{wo} 使得 $\hat{x}(t) = T_{\text{wo}} x(t)$ 能够把系统方程 (4.11) 转化为 Wonham 第一能观标准形，

$$\dot{\hat{x}}(t) = \hat{A}_{\text{wo}} \hat{x}(t) + \hat{B}_{\text{wo}} u(t), \; y(t) = \hat{C}_{\text{wo}} \hat{x}(t) \tag{4.20}$$

其中，

$$\hat{A}_{\text{wo}} = \begin{bmatrix} \hat{A}_{\text{wo}11} & & \\ \vdots & \ddots & \\ \hat{A}_{\text{wo}k1} & \cdots & \hat{A}_{\text{wo}kk} \end{bmatrix}, \; \hat{C}_{\text{wo}} = \begin{bmatrix} \hat{C}_{\text{wo}1} & \cdots & \hat{C}_{\text{wo}k} \\ * & \cdots & * \\ \vdots & & \vdots \\ * & \cdots & * \end{bmatrix}$$

$\hat{B}_{\text{wo}} = T_{\text{wo}} B$ 没有特殊形式，其中，$\hat{A}_{\text{wo}ii}$ 是 $\bar{\nu}_i \times \bar{\nu}_i$ 的矩阵，$\hat{A}_{\text{wo}ij}$ 是 $\bar{\nu}_i \times \bar{\nu}_j$ 的矩阵，且具有以下结构，

$$\hat{A}_{\text{wo}ii} = \begin{bmatrix} 0 & 1 & \cdots & 0 \\ \vdots & \vdots & & \vdots \\ 0 & 0 & \cdots & 1 \\ * & * & * & * \end{bmatrix}, \; \hat{A}_{\text{wo}ij} = \begin{bmatrix} 0 & 0 & \cdots & 0 \\ \vdots & \vdots & & \vdots \\ 0 & 0 & \cdots & 0 \\ * & * & * & * \end{bmatrix}$$

$i = 1, \cdots, k$，$j = 1, \cdots, k$，$i \neq j$，$\hat{C}_{\text{wo}i}$ 是 $k \times \bar{\nu}_i$ 的矩阵，且它第 i 行第一列的元为 1，其他元全为 0。

证明略。

定理 4.6 可以根据对偶原理和定理 4.5 直接推得。类似地，完全能观的线性定常系统具有 Wonham 第二能观标准形，它是其对偶系统的 Wonham 第二能控标准形。

4.2 能控性和能观性的频域形式

第 3 章中，主要基于线性系统的状态空间描述来讨论它的能控性和能观性。事实上，传递函数矩阵 $G(s)$ 也能反映系统的这些性质。

1.3.3 节中指出，系统 (A,B,C,D) 的传递函数矩阵可以表示为

$$G(s) = \frac{C \operatorname{adj}(sI-A)B}{\alpha(s)} + D \tag{4.21}$$

其中，$\alpha(s) = \det(sI-A)$ 是 A 的特征多项式。

定理 4.7 如果 $C\operatorname{adj}(sI-A)B$ 和 $\alpha(s)$ 没有非常值的公因子，那么 (A,B,C) 是能控且能观的。

证明 使用反证法，假设 $C\operatorname{adj}(sI-A)B$ 和 $\alpha(s)$ 没有非常值的公因子，但是 (A,B,C) 不完全能控或者不完全能观。根据定理 3.24，必然存在维数为 $n_1 < n$ 的系统 (A_1, B_1, C_1)，$n = \dim A$，使得 (A_1, B_1, C_1) 和 (A,B,C) 具有相同的传递函数矩阵，即

$$\frac{C_1 \operatorname{adj}(sI-A_1)B_1}{\det(sI-A_1)} = \frac{C \operatorname{adj}(sI-A)B}{\alpha(s)}$$

这里，$\det(sI-A_1)$ 是关于 s 的 n_1 阶多项式。如果上式成立，那么 $C\operatorname{adj}(sI-A)B$ 和 $\alpha(s)$ 必然有非常值的公因子，这与假设矛盾。 □

定理 4.7 不是必要条件。接下来，给出一个必要条件。考虑 $d(s)$ 是 $\operatorname{adj}(sI-A)$ 的首一最大公因子，且满足 $\operatorname{adj}(sI-A) = d(s)P(s)$ 和 $d(s)\phi(s) = \alpha(s)$，其中，$\phi(s)$ 是 A 的最小多项式，$P(s)$ 是多项式矩阵。由此，式 (4.21) 可以表示为

$$G(s) = \frac{CP(s)B}{\phi(s)} + D$$

定理 4.8 如果 (A,B,C) 是能控且能观的，那么 $CP(s)B$ 和 $\phi(s)$ 没有非常值的公因子。

证明 使用反证法。假设 (A,B,C) 是能控且能观的，但是 $CP(s)B$ 和 $\phi(s)$ 具有非常值的公因子。这种情况下，存在 $s = s_0$ 使得 $CP(s_0)B = 0$，$\phi(s_0) = 0$。另一方面，由 $\phi(s)I = (sI-A)P(s)$，可以进一步得到

$$AP(s_0) = \begin{cases} s_0 P(s_0), & s_0 \neq 0 \\ 0, & s_0 = 0 \end{cases}$$

由此可得

$$CP(s_0)B = 0, \; CAP(s_0)B = s_0 CP(s_0)B = 0, \; \cdots, \; CA^{n-1}P(s_0)B = 0$$

这表明，

$$\begin{bmatrix} C \\ CA \\ \vdots \\ CA^{n-1} \end{bmatrix} P(s_0)B = 0$$

因为 (A,C) 是能观的，所以 $P(s_0)B = 0$。

另一方面，结合 1.3.3 节中的讨论，$\mathrm{adj}(s_0 I - A)$ 可以表示为 $\sum_{k=0}^{n-1} p_k(s_0) A^k$，由此可知，

$$d(s_0)P(s_0)B = \mathrm{adj}(s_0 I - A)B = \sum_{k=0}^{n-1} p_k(s_0) A^k B = 0$$

这表明，

$$\begin{bmatrix} B & AB & \cdots & A^{n-1}B \end{bmatrix} \begin{bmatrix} p_0(s_0) \\ p_1(s_0) \\ \vdots \\ p_{n-1}(s_0) \end{bmatrix} = 0$$

因为 $p_{n-1}(s_0) = 1$，所以 $\begin{bmatrix} B & AB & \cdots & A^{n-1}B \end{bmatrix}$ 必然是奇异的，也就是说，(A,B) 是不完全能控的，这与假设矛盾。 □

根据以上讨论，可以得到以下充分必要条件。

定理 4.9 当且仅当 $p(s) = \alpha(s)$ 时，其中，$p(s)$ 是 $G(s)$ 的极点多项式，$\alpha(s)$ 是 A 的特征多项式，(A,B,C) 是能控且能观的。

证明略。

4.3 线性定常系统的实现

这一节介绍如何根据线性定常系统的传递函数（矩阵）确定它的状态空间描述。

4.3.1 基本概念

定义 4.1（实现） 对于传递函数矩阵为 $G(s)$ 的线性定常系统，如果能够找到一个状态空间描述

$$\dot{x}(t) = Ax(t) + Bu(t), \ y(t) = Cx(t) + Du(t) \tag{4.22}$$

使得

$$G(s) = C(sI - A)^{-1}B + D \tag{4.23}$$

那么称状态空间描述式 (4.22) 是传递函数矩阵 $G(s)$ 的一个实现。

4.3 线性定常系统的实现

传递函数矩阵的实现具有以下性质。

(1) 传递函数矩阵实现的维数反映了系统的复杂程度，它取决于传递函数矩阵结构的复杂程度和获取实现的方法。

(2) 一个给定的传递函数矩阵 $G(s)$ 的实现是不唯一的。对于给定的 $G(s)$，不同的获取实现的方法可以得到具有不同维数的实现，或者具有相同维数但不同参数的实现。

(3) 在 $G(s)$ 的所有实现中，一定存在一个维数最低的实现，称之为最小实现。它是给定 $G(s)$ 的最简单的状态空间结构。在 4.4 节中，将详细介绍最小实现。

(4) 一般而言，给定 $G(s)$ 的各种实现之间没有代数等价关系，但它的最小实现之间存在代数等价关系。

(5) 实现的物理本质是对 $G(s)$ 内部状态结构的反映，最小实现则是对 $G(s)$ 内部状态结构的（运动规律意义上的）真实反映。

定理 4.10 当且仅当 $G(s)$ 是真有理分式矩阵时，$G(s)$ 可以由有限维动态方程 (4.22) 实现。

证明 首先，证明必要性。假设方程 (4.22) 是 $G(s)$ 的一个实现，根据 1.3.3 节中的讨论，它的传递函数矩阵满足式 (4.23)，是一个真有理分式矩阵。

后续介绍的构造传递函数矩阵实现的任何一种方法都可以用来说明这一定理的充分性，这里省略。 □

可以进一步得到当 $G(s)$ 是真有理分式矩阵时，它的实现具有 (A,B,C,D) 的形式，且

$$D = \lim_{s \to \infty} G(s) \tag{4.24}$$

当 $G(s)$ 是严格真有理分式矩阵时，它的实现则具有 (A,B,C) 的形式。

4.3.2 线性定常 SISO 系统的实现

考虑以下真有理传递函数，

$$\bar{g}(s) = \frac{\beta_1 s^{n-1} + \beta_2 s^{n-2} + \cdots + \beta_n}{s^n + a_1 s^{n-1} + \cdots + a_n} + d \triangleq g(s) + d \tag{4.25}$$

其中，a_i，β_i，$i = 1, \cdots, n$ 和 d 是实常数。因为 d 是实现中直接传递的部分，所以在研究实现问题时，只需要考虑严格真的部分 $g(s)$。

定理 4.11 考虑线性定常系统的传递函数 $g(s)$ 为式 (4.25)，它的能控形实现 (A,b,c) 满足

$$\begin{cases} A = \begin{bmatrix} 0 & 1 & 0 & \cdots & 0 \\ 0 & 0 & 1 & \cdots & 0 \\ \vdots & \vdots & \vdots & & \vdots \\ 0 & 0 & 0 & \cdots & 1 \\ -a_n & -a_{n-1} & -a_{n-2} & \cdots & -a_1 \end{bmatrix} \quad b = \begin{bmatrix} 0 \\ 0 \\ \vdots \\ 0 \\ 1 \end{bmatrix} \\ c = \begin{bmatrix} \beta_n & \beta_{n-1} & \beta_{n-2} & \cdots & \beta_1 \end{bmatrix} \end{cases} \tag{4.26}$$

证明　把式 (4.25) 中 $g(s)$ 的分子多项式和分母多项式分别记作 $\beta(s)$ 和 $\alpha(s)$。可以得到

$$y(s) = g(s)u(s) = \beta(s)\alpha^{-1}(s)u(s)$$

其中，$y(s) = \mathcal{L}[y(t)]$，$u(s) = \mathcal{L}[u(t)]$。

引入变换 $\bar{y}(s) = \alpha^{-1}(s)u(s)$，由此可得 $y(s) = \beta(s)\bar{y}(s)$。取 Laplace 反变换，可以得到

$$\frac{\mathrm{d}^n}{\mathrm{d}t^n}\bar{y}(t) + a_1 \frac{\mathrm{d}^{n-1}}{\mathrm{d}t^{n-1}}\bar{y}(t) + \cdots + a_{n-1}\frac{\mathrm{d}}{\mathrm{d}t}\bar{y}(t) + a_n\bar{y}(t) = u(t) \tag{4.27}$$

与

$$y(t) = \beta_1 \frac{\mathrm{d}^{n-1}}{\mathrm{d}t^{n-1}}\bar{y}(t) + \cdots + \beta_{n-1}\frac{\mathrm{d}}{\mathrm{d}t}\bar{y}(t) + \beta_n \bar{y}(t) \tag{4.28}$$

选择以下状态变量，

$$x_1(t) = \bar{y}(t),\ x_2(t) = \frac{\mathrm{d}}{\mathrm{d}t}\bar{y}(t),\ \cdots,\ x_n(t) = \frac{\mathrm{d}^{n-1}}{\mathrm{d}t^{n-1}}\bar{y}(t) \tag{4.29}$$

结合式 (4.27) 与式 (4.29)，可以得到

$$\begin{cases} \dot{x}_1(t) = x_2(t) \\ \vdots \\ \dot{x}_{n-1}(t) = x_n(t) \\ \dot{x}_n(t) = -a_1 x_n(t) - \cdots - a_{n-1}x_2(t) - a_n x_1(t) + u(t) \end{cases}$$

上式可以被写作 $\dot{x}(t) = Ax(t) + bu(t)$，其中 A 和 b 满足式 (4.26)。

结合式 (4.28) 与式 (4.29)，可以得到

$$y(t) = \beta_1 x_n(t) + \cdots + \beta_{n-1}x_2(t) + \beta_n x_1(t)$$

上式可以被写作 $y(t) = cx(t)$，其中 c 满足式 (4.26)。　□

可以看出定理 4.11 中的状态空间描述 (A, b, c) 是能控标准形。类似地，$g(s)$ 也存在能观形实现。

定理 4.12　考虑线性定常系统的传递函数 $g(s)$ 为式 (4.25)，它的能观形实现 (A, b, c) 满足

$$A = \begin{bmatrix} 0 & 0 & \cdots & 0 & -a_n \\ 1 & 0 & \cdots & 0 & -a_{n-1} \\ 0 & 1 & \cdots & 0 & -a_{n-2} \\ \vdots & \vdots & & \vdots & \vdots \\ 0 & 0 & \cdots & 1 & -a_1 \end{bmatrix},\ b = \begin{bmatrix} \beta_n \\ \beta_{n-1} \\ \beta_{n-2} \\ \vdots \\ \beta_1 \end{bmatrix},\ c = \begin{bmatrix} 0 & 0 & \cdots & 0 & 1 \end{bmatrix} \tag{4.30}$$

证明略。

例 4.4 给定传递函数

$$\bar{g}(s) = \frac{s^4 + 9s^3 + 30s^2 + 46s + 30}{s^4 + 9s^3 + 29s^2 + 39s + 18}$$

根据多项式除法，可以得到

$$\bar{g}(s) = g(s) + d = \frac{s^2 + 7s + 12}{s^4 + 9s^3 + 29s^2 + 39s + 18} + 1$$

$d = 1$ 是实现中直接传递的部分。根据定理 4.11，(A_1, b_1, c_1, d) 是 $\bar{g}(s)$ 的一个实现，其中，

$$A_1 = \begin{bmatrix} 0 & 1 & 0 & 0 \\ 0 & 0 & 1 & 0 \\ 0 & 0 & 0 & 1 \\ -18 & -39 & -29 & -9 \end{bmatrix}, \quad b_1 = \begin{bmatrix} 0 \\ 0 \\ 0 \\ 1 \end{bmatrix}, \quad c_1 = \begin{bmatrix} 12 & 7 & 1 & 0 \end{bmatrix}$$

根据定理 4.12，(A_2, b_2, c_2, d) 也是 $\bar{g}(s)$ 的一个实现，其中，

$$A_2 = \begin{bmatrix} 0 & 0 & 0 & -18 \\ 1 & 0 & 0 & -39 \\ 0 & 1 & 0 & -29 \\ 0 & 0 & 1 & -9 \end{bmatrix}, \quad b_2 = \begin{bmatrix} 12 \\ 7 \\ 1 \\ 0 \end{bmatrix}, \quad c_2 = \begin{bmatrix} 0 & 0 & 0 & 1 \end{bmatrix}$$

因为状态变量的选择具有不唯一性，所以实现也是不唯一的。需要说明的是，能控形实现一定是能控的，能观形实现一定是能观的。

4.3.3 线性定常 MIMO 系统的实现

考虑 $q \times p$ 的真有理传递函数矩阵 $\bar{G}(s) = G(s) + D$，其中 $D = \lim\limits_{s \to \infty} \bar{G}(s)$ 可以直接确定。下面，只需要考虑严格真的部分 $G(s)$ 的实现 (A, B, C)。$G(s)$ 可以表示为

$$G(s) = \begin{bmatrix} g_{11}(s) & \cdots & g_{1p}(s) \\ \vdots & & \vdots \\ g_{q1}(s) & \cdots & g_{qp}(s) \end{bmatrix} = \frac{1}{\alpha(s)} \begin{bmatrix} m_{11}(s) & \cdots & m_{1p}(s) \\ \vdots & & \vdots \\ m_{q1}(s) & \cdots & m_{qp}(s) \end{bmatrix}$$

$$= \frac{1}{\alpha(s)}(G_1 s^{r-1} + \cdots + G_{r-1} s + G_r) \tag{4.31}$$

其中，$\alpha(s) = s^r + a_1 s^{r-1} + \cdots + a_{r-1} s + a_r$ 是分母 $g_{ij}(s)$ 的首一最小公分母，$i = 1, \cdots, q$，$j = 1, \cdots, p$，$\alpha(s) g_{ij}(s) = m_{ij}(s)$，$G_k$ 是 $q \times p$ 的常值矩阵，$k = 1, \cdots, r$。

定理 4.13 考虑一个线性定常 MIMO 系统具有式 (4.31) 形式的严格真传递函数矩阵 $G(s)$，它的能控形实现 (A, B, C) 满足

$$\begin{cases} A = \begin{bmatrix} O_p & I_p & O_p & \cdots & O_p \\ O_p & O_p & I_p & \cdots & O_p \\ \vdots & \vdots & \vdots & & \vdots \\ O_p & O_p & O_p & \cdots & I_p \\ -a_r I_p & -a_{r-1} I_p & -a_{r-2} I_p & \cdots & -a_1 I_p \end{bmatrix}, \quad B = \begin{bmatrix} O_p \\ O_p \\ \vdots \\ O_p \\ I_p \end{bmatrix} \\ C = \begin{bmatrix} G_r & G_{r-1} & G_{r-2} & \cdots & G_1 \end{bmatrix} \end{cases} \quad (4.32)$$

其中，O_p 表示 $p \times p$ 的零矩阵，I_p 表示 $p \times p$ 的单位矩阵。

证明 考虑

$$(sI - A)^{-1} = \begin{bmatrix} & X_1 \\ & X_2 \\ * & \vdots \\ & X_r \end{bmatrix}$$

其中，"$*$" 表示无须求取的元素，X_k 是 $p \times p$ 的常值矩阵，$k = 1, \cdots, r$。在此基础上，可以得到

$$(sI - A) \begin{bmatrix} & X_1 \\ & X_2 \\ * & \vdots \\ & X_r \end{bmatrix} = \begin{bmatrix} & O_p \\ & O_p \\ I_{(r-1)p} & \vdots \\ & I_p \end{bmatrix}$$

把式 (4.32) 中的 A 代入到上述等式中，可以得到

$$\begin{cases} sX_1 = X_2 \\ \vdots \\ sX_{r-1} = X_r \\ sX_r = I_p - (a_r X_1 + a_{r-1} X_2 + \cdots + a_1 X_r) \end{cases} \quad (4.33)$$

这表明，$X_k = sX_{k-1}$，$k = 2, \cdots, r$。基于此，能够推导出

$$\begin{bmatrix} X_1 \\ X_2 \\ \vdots \\ X_r \end{bmatrix} = \frac{1}{\alpha(s)} \begin{bmatrix} I_p \\ sI_p \\ \vdots \\ s^{r-1} I_p \end{bmatrix}$$

4.3 线性定常系统的实现

由此可知，

$$C(sI-A)^{-1}B = \begin{bmatrix} G_r & G_{r-1} & G_{r-2} & \cdots & G_1 \end{bmatrix} \left[* \quad \frac{1}{\alpha(s)} \begin{bmatrix} I_p \\ sI_p \\ \vdots \\ s^{r-1}I_p \end{bmatrix} \right] \begin{bmatrix} O_p \\ O_p \\ \vdots \\ I_p \end{bmatrix}$$

$$= \frac{1}{\alpha(s)}(G_1 s^{r-1} + \cdots + G_{r-1} s + G_r) = G(s)$$

这表明满足式 (4.32) 的 (A, B, C) 是 $G(s)$ 的一个实现。 □

可以验证 $\operatorname{rank}[\begin{array}{cccc} B & AB & A^2B & \cdots & A^{r-1}B \end{array}] = rp$，所以定理 4.13 中给出的能控形实现一定是能控的，但它可能不完全能观。

类似地，可以给出式 (4.31) 中 $G(s)$ 的能观形实现，它一定是能观的但可能不完全能控。

定理 4.14 考虑一个线性定常 MIMO 系统具有式 (4.31) 形式的严格真传递函数矩阵 $G(s)$，它的能观形实现 (A, B, C) 满足

$$\begin{cases} A = \begin{bmatrix} O_q & O_q & \cdots & O_q & -a_r I_q \\ I_q & O_q & \cdots & O_q & -a_{r-1} I_q \\ O_q & I_q & \cdots & O_q & -a_{r-2} I_q \\ \vdots & \vdots & & \vdots & \vdots \\ O_q & O_q & \cdots & I_q & -a_1 I_q \end{bmatrix}, B = \begin{bmatrix} G_r \\ G_{r-1} \\ G_{r-2} \\ \vdots \\ G_1 \end{bmatrix} \\ C = \begin{bmatrix} O_q & O_q & \cdots & O_q & I_q \end{bmatrix} \end{cases} \quad (4.34)$$

其中，O_q 表示 $q \times q$ 的零矩阵，I_q 表示 $q \times q$ 的单位矩阵。

证明略。

例 4.5 给定以下严格真传递函数矩阵，

$$G(s) = \begin{bmatrix} \dfrac{2}{s+1} & \dfrac{1}{s+1} \\ \dfrac{1}{s+2} & \dfrac{1}{s+2} \end{bmatrix}$$

$G(s)$ 每个元的最小公分母为 $\alpha(s) = (s+1)(s+2) = s^2 + 3s + 2$。进一步，$G(s)$ 可以被表示为

$$G(s) = \frac{1}{s^2+3s+2} \begin{bmatrix} 2(s+2) & s+2 \\ s+1 & s+1 \end{bmatrix} = \frac{1}{s^2+3s+2} \left(\begin{bmatrix} 2 & 1 \\ 1 & 1 \end{bmatrix} s + \begin{bmatrix} 4 & 2 \\ 1 & 1 \end{bmatrix} \right)$$

根据定理 4.13，$G(s)$ 的能控形实现为

$$A_1 = \begin{bmatrix} O_2 & I_2 \\ -a_2 I_2 & -a_1 I_2 \end{bmatrix} = \begin{bmatrix} 0 & 0 & 1 & 0 \\ 0 & 0 & 0 & 1 \\ -2 & 0 & -3 & 0 \\ 0 & -2 & 0 & -3 \end{bmatrix}, \quad B_1 = \begin{bmatrix} O_2 \\ I_2 \end{bmatrix} = \begin{bmatrix} 0 & 0 \\ 0 & 0 \\ 1 & 0 \\ 0 & 1 \end{bmatrix}$$

$$C_1 = \begin{bmatrix} G_2 & G_1 \end{bmatrix} = \begin{bmatrix} 4 & 2 & 2 & 1 \\ 1 & 1 & 1 & 1 \end{bmatrix}$$

根据定理 4.14，$G(s)$ 的能观形实现为

$$A_2 = \begin{bmatrix} O_2 & -a_2 I_2 \\ I_2 & -a_1 I_2 \end{bmatrix} = \begin{bmatrix} 0 & 0 & -2 & 0 \\ 0 & 0 & 0 & -2 \\ 1 & 0 & -3 & 0 \\ 0 & 1 & 0 & -3 \end{bmatrix}, \quad B_2 = \begin{bmatrix} G_2 \\ G_1 \end{bmatrix} = \begin{bmatrix} 4 & 2 \\ 1 & 1 \\ 2 & 1 \\ 1 & 1 \end{bmatrix}$$

$$C_2 = \begin{bmatrix} O_2 & I_2 \end{bmatrix} = \begin{bmatrix} 0 & 0 & 1 & 0 \\ 0 & 0 & 0 & 1 \end{bmatrix}$$

4.4 线性定常系统的最小实现

本节将详细介绍最小实现的性质及获取最小实现的方法。

4.4.1 基本概念

4.3 节中指出，同一个传递函数矩阵可以有不同维数的实现。这些实现里，一定存在一类维数最小的实现。

由于真有理分式矩阵 $\bar{G}(s) = G(s) + D$，且 $D = \lim_{s \to \infty} G(s)$ 可以直接得到，接下来，只讨论严格真有理分式矩阵 $G(s)$ 的最小实现。

定义 4.2（最小实现） 传递函数矩阵 $G(s)$ 的所有实现中，具有最小维数的实现被称为最小实现或不可约实现。

定理 4.15 当且仅当 (A,B) 能控且 (A,C) 能观时，(A,B,C) 是严格真传递函数矩阵 $G(s)$ 的最小实现。

证明 首先，证明必要性。

使用反证法。假设 (A,B,C) 是 $G(s)$ 的最小实现，但 (A,B) 不完全能控或 (A,C) 不完全能观。这种情况下，(A,B,C) 进行结构分解后存在既能控也能观的部分 (A_1,B_1,C_1)，(A,B,C) 和 (A_1,B_1,C_1) 具有相同的传递函数矩阵 $G(s)$，但是 $\dim A > \dim A_1$。这表明 (A,B,C) 不是 $G(s)$ 的最小实现，这与假设矛盾。因此，最小实现 (A,B,C) 既能控也能观。

接下来，证明充分性。

使用反证法，假设 (A,B,C) 不是 $G(s)$ 的最小实现，但是它完全能控且能观。这种情况下，一定存在最小实现 $(\bar{A},\bar{B},\bar{C})$，使得 $n=\dim A>\dim \bar{A}=\bar{n}$。

因为 (A,B,C) 和 $(\bar{A},\bar{B},\bar{C})$ 具有相同的传递函数矩阵，所以在零初始条件下，任意输入 $u(t)$ 会激励出相同的输出，也就是说，

$$\int_0^t Ce^{A(t-\tau)}Bu(\tau)d\tau = \int_0^t \bar{C}e^{\bar{A}(t-\tau)}\bar{B}u(\tau)d\tau$$

上述等式中，$u(\tau)$ 和 t 具有任意性，由此可得对于任意的 t 和 τ，

$$Ce^{A(t-\tau)}Bu(\tau)d\tau = \bar{C}e^{\bar{A}(t-\tau)}\bar{B}u(\tau)$$

对上式两端关于 t 取 k 阶导，可以得到

$$CA^k e^{A(t-\tau)}B = \bar{C}\bar{A}^k e^{\bar{A}(t-\tau)}\bar{B}, \ k=0,1,2,\cdots$$

令 $t-\tau=0$，可以进一步得到

$$CA^k B = \bar{C}\bar{A}^k \bar{B}, \ k=0,1,2,\cdots \tag{4.35}$$

上式可以被表示为

$$VU = \begin{bmatrix} C \\ CA \\ \vdots \\ CA^{n-1} \end{bmatrix} \begin{bmatrix} B & AB & \cdots & A^{n-1}B \end{bmatrix}$$

$$= \begin{bmatrix} \bar{C} \\ \bar{C}\bar{A} \\ \vdots \\ \bar{C}\bar{A}^{n-1} \end{bmatrix} \begin{bmatrix} \bar{B} & \bar{A}\bar{B} & \cdots & \bar{A}^{n-1}\bar{B} \end{bmatrix} = \bar{V}\bar{U} \tag{4.36}$$

因为 (A,B,C) 完全能控且能观，所以 $\text{rank}\,V=n, \text{rank}\,U=n$，且 $\text{rank}\,VU=n$。根据式 (4.36) 可知，$n=\text{rank}\,VU=\text{rank}\,\bar{V}\bar{U} \leqslant \min\{\text{rank}\,\bar{V},\text{rank}\,\bar{U}\}$。另一方面，$\text{rank}\,\bar{V} \leqslant \bar{n}$ 且 $\text{rank}\,\bar{U} \leqslant \bar{n}$，所以 $n \leqslant \bar{n}$，这与假设 $n>\bar{n}$ 矛盾。因此，(A,B,C) 是 $G(s)$ 的最小实现。 □

结合定理 4.9 和定理 4.15，可以得到以下结论。

推论 4.1 如果 (A,B,C) 是 $G(s)$ 的最小实现，那么 $\dim A$ 等于 $G(s)$ 的极点多项式 $p(s)$ 的阶次。

$G(s)$ 的最小实现是不唯一的。$G(s)$ 的任意两个最小实现具有以下关系。

定理 4.16 如果 (A,B,C) 和 $(\bar{A},\bar{B},\bar{C})$ 是传递函数矩阵 $G(s)$ 的两个最小实现，那么它们一定是代数等价的，即一定存在非奇异的常值矩阵 T 使得 $\bar{A}=TAT^{-1}$，$\bar{B}=TB$ 且 $\bar{C}=CT^{-1}$，同时，$T=(\bar{V}^{\mathrm{T}}\bar{V})^{-1}\bar{V}^{\mathrm{T}}V$，$T^{-1}=U\bar{U}^{\mathrm{T}}(\bar{U}\bar{U}^{\mathrm{T}})^{-1}$，其中 U 和 V 分别是 (A,B,C) 的能控性矩阵和能观性矩阵，$\bar{U}$ 和 $\bar{V}$ 分别是 $(\bar{A},\bar{B},\bar{C})$ 的能控性矩阵和能观性矩阵。

证明 (A, B, C) 和 $(\bar{A}, \bar{B}, \bar{C})$ 是最小实现。根据定理 4.15，它们是完全能控且能观的。所以，$\operatorname{rank} U = \operatorname{rank} V = \operatorname{rank} \bar{U} = \operatorname{rank} \bar{V} = n$。进一步，根据定理 4.15 的证明，可以得到

$$VU = \bar{V}\bar{U} \tag{4.37}$$

这表明

$$\bar{U} = (\bar{V}^{\mathrm{T}}\bar{V})^{-1}\bar{V}^{\mathrm{T}}\bar{V}\bar{U} = (\bar{V}^{\mathrm{T}}\bar{V})^{-1}\bar{V}^{\mathrm{T}}VU = TU$$

且

$$\bar{V} = \bar{V}\bar{U}\bar{U}^{\mathrm{T}}(\bar{U}\bar{U}^{\mathrm{T}})^{-1} = VU\bar{U}^{\mathrm{T}}(\bar{U}\bar{U}^{\mathrm{T}})^{-1} = V\bar{T}$$

结合式 (4.37)，可以验证

$$T\bar{T} = (\bar{V}^{\mathrm{T}}\bar{V})^{-1}\bar{V}^{\mathrm{T}}VU\bar{U}^{\mathrm{T}}(\bar{U}\bar{U}^{\mathrm{T}})^{-1} = (\bar{V}^{\mathrm{T}}\bar{V})^{-1}\bar{V}^{\mathrm{T}}\bar{V}\bar{U}\bar{U}^{\mathrm{T}}(\bar{U}\bar{U}^{\mathrm{T}})^{-1} = I$$

也就是说，$T^{-1} = \bar{T} = U\bar{U}^{\mathrm{T}}(\bar{U}\bar{U}^{\mathrm{T}})^{-1}$。

进一步，$\bar{U} = TU$ 和 $\bar{V} = VT^{-1}$ 可以分别表示为

$$\begin{bmatrix} \bar{B} & \bar{A}\bar{B} & \cdots & \bar{A}^{n-1}\bar{B} \end{bmatrix} = \begin{bmatrix} TB & TAB & \cdots & TA^{n-1}B \end{bmatrix}$$

和

$$\begin{bmatrix} \bar{C} \\ \bar{C}\bar{A} \\ \vdots \\ \bar{C}^{n-1}\bar{A} \end{bmatrix} = \begin{bmatrix} CT^{-1} \\ CAT^{-1} \\ \vdots \\ CA^{n-1}T^{-1} \end{bmatrix}$$

这表明 $\bar{B} = TB$ 且 $\bar{C} = CT^{-1}$。

另一方面，当 (A, B, C) 和 $(\bar{A}, \bar{B}, \bar{C})$ 都是最小实现时，式 (4.35) 成立。基于此，可得

$$VAU = \begin{bmatrix} C \\ CA \\ \vdots \\ CA^{n-1} \end{bmatrix} A \begin{bmatrix} B & AB & \cdots & A^{n-1}B \end{bmatrix}$$

$$= \begin{bmatrix} \bar{C} \\ \bar{C}\bar{A} \\ \vdots \\ \bar{C}\bar{A}^{n-1} \end{bmatrix} \bar{A} \begin{bmatrix} \bar{B} & \bar{A}\bar{B} & \cdots & \bar{A}^{n-1}\bar{B} \end{bmatrix} = \bar{V}\bar{A}\bar{U}$$

在 $VAU = \bar{V}\bar{A}\bar{U}$ 两端分别左乘 $\bar{V}^{\mathrm{T}}$ 右乘 $\bar{U}^{\mathrm{T}}$，可以得到

$$\bar{V}^{\mathrm{T}}VAU\bar{U}^{\mathrm{T}} = \bar{V}^{\mathrm{T}}\bar{V}\bar{A}\bar{U}\bar{U}^{\mathrm{T}}$$

这表明 $\bar{A} = (\bar{V}^{\mathrm{T}}\bar{V})^{-1}\bar{V}^{\mathrm{T}}VAU\bar{U}^{\mathrm{T}}(\bar{U}\bar{U}^{\mathrm{T}})^{-1} = TAT^{-1}$。 □

4.4 线性定常系统的最小实现

事实上，$\bar{V}^+ = (\bar{V}^{\mathrm{T}}\bar{V})^{-1}\bar{V}^{\mathrm{T}}$ 和 $\bar{U}^+ = \bar{U}^{\mathrm{T}}(\bar{U}\bar{U}^{\mathrm{T}})^{-1}$ 分别是 $\bar{V}$ 和 $\bar{U}$ 的 Moore-Penrose（穆尔-彭罗斯）逆。基于此，$T = \bar{V}^+ V$，$T^{-1} = U\bar{U}^+$。

$q \times p$ 的真有理传递函数矩阵 $\bar{G}(s)$ 可以表示为 Laurent（洛朗）级数展开，

$$\bar{G}(s) = \sum_{i=0}^{\infty} h_i s^{-i} = h_0 + h_1 s^{-1} + h_2 s^{-2} + h_3 s^{-3} + \cdots \tag{4.38}$$

其中，$h_i \in \mathbb{R}^{q \times p}$ 被称作 Markov（马尔可夫）参数，$i = 0, 1, 2, \cdots$，可以根据下式确定，

$$h_0 = \lim_{s \to \infty} \bar{G}(s)$$
$$h_1 = \lim_{s \to \infty} s(\bar{G}(s) - h_0)$$
$$h_2 = \lim_{s \to \infty} s^2(\bar{G}(s) - h_0 - h_1 s^{-1})$$
$$\vdots$$

以此类推。

定义 4.3（Hankel(汉克尔) 矩阵） 阶次为 (i_q, j_p) 的 Hankel 矩阵 $H(i, j)$ 与 Markov 参数序列相关，它被定义为

$$H(i, j) \triangleq \begin{bmatrix} h_1 & h_2 & \cdots & h_j \\ h_2 & h_3 & \cdots & h_{j+1} \\ \vdots & \vdots & & \vdots \\ h_i & h_{i+1} & \cdots & h_{i+j+1} \end{bmatrix} \tag{4.39}$$

定理 4.17 假设 (A, B, C) 是具有式 (4.38) 形式的 $G(s)$ 的最小实现，那么这一最小实现的维数 n 为

$$\dim A = \operatorname{rank} H(n, n) \tag{4.40}$$

证明 考虑 (A, B, C) 是给定的 $G(s)$ 的最小实现。将

$$(sI - A)^{-1} = \sum_{k=0}^{\infty} \frac{A^k}{s^{k+1}}$$

代入到 $G(s) = C(sI - A)^{-1}B$ 中，可以得到

$$G(s) = \sum_{k=0}^{\infty} CA^k B s^{-(k+1)} = \sum_{i=1}^{\infty} h_i s^{-i}$$

这表明

$$h_i = CA^{i-1}B, \; i = 1, 2, \cdots$$

在此基础上，$H(n,n)$ 可以表示为

$$H(n,n) = \begin{bmatrix} C \\ CA \\ \vdots \\ CA^{n-1} \end{bmatrix} \begin{bmatrix} B & AB & \cdots & A^{n-1}B \end{bmatrix} = VU$$

因为 (A,B,C) 是 $G(s)$ 的最小实现，所以 (A,B,C) 能控且能观的。由此可知 $\operatorname{rank} V = \operatorname{rank} U = n$。进一步可得

$$\operatorname{rank} V + \operatorname{rank} U - n \leqslant \operatorname{rank} H(n,n) \leqslant \min\{\operatorname{rank} V, \operatorname{rank} U\}$$

这表明 $\operatorname{rank} H(n,n) = n$。

根据 Cayley-Hamilton 定理，可以得到

$$\operatorname{rank} H(n+k, n+k) = \operatorname{rank} H(n,n) = n, \ k = 0, 1, 2, \cdots \qquad \square$$

事实上，随着 Hankel 矩阵的维数增加，一旦 Hankel 矩阵的秩不再改变，那么随着它的维数继续增加，它的秩不会再改变。式 (4.17) 可以被修改为

$$\dim A = \operatorname{rank} H(r,r) \tag{4.41}$$

其中，r 是 $G(s)$ 所有元的最小公分母的阶次。这里不详细讨论这一结果的证明。一般情况下，式 (4.41) 可以更有效地确定最小实现的维数。

4.4.2 线性定常 SISO 系统的最小实现

在 4.3.2 小节中，介绍了两种线性定常 SISO 系统的实现的方法。在此基础上，本小节讨论求取线性定常 SISO 系统的最小实现的方法。

如果给定的传递函数 $\bar{g}(s)$ 是真的但不是严格真的，那么 $\bar{g}(s)$ 可以表示为 $g(s) + d$，d 可以直接利用多项式除法计算得到。因此，仅需要讨论严格真的部分 $g(s)$ 的最小实现。

对于给定的 $g(s)$，如果它的分子多项式和分母多项式是互质的，那么定理 4.11 和定理 4.12 给出的实现就是 $g(s)$ 的最小实现。这是由于它们的特征多项式都等于 $g(s)$ 的极点多项式，根据推论 4.1 可知它们都是 $g(s)$ 的最小实现。

接下来讨论 $g(s)$ 的分子多项式和分母多项式存在非常值公因子的情况。此时，为了获得 $g(s)$ 的最小实现，可以先根据定理 4.11 得到能控形实现，然后根据定理 3.23 对这一实现进行能观性分解，得到的能观的子系统即为 $g(s)$ 的一个最小实现，这是由于这一子系统是完全能控且能观的。类似地，也可以先根据定理 4.12 得到能观形实现，然后利用定理 3.22 进行能控性分解以获得最小实现。

例 4.6 考虑以下传递函数，

$$\bar{g}(s) = g(s) + d = \frac{s^2 + 7s + 12}{s^4 + 9s^3 + 29s^2 + 39s + 18} + 1$$

4.4 线性定常系统的最小实现

例 4.4 给出了它的能控形实现，

$$A_1 = \begin{bmatrix} 0 & 1 & 0 & 0 \\ 0 & 0 & 1 & 0 \\ 0 & 0 & 0 & 1 \\ -18 & -39 & -29 & -9 \end{bmatrix}, \quad b_1 = \begin{bmatrix} 0 \\ 0 \\ 0 \\ 1 \end{bmatrix}, \quad c_1 = \begin{bmatrix} 12 & 7 & 1 & 0 \end{bmatrix}, \quad d = 1$$

(A_1, c_1) 是不完全能观的，这是由于

$$\operatorname{rank} V_1 = \operatorname{rank} \begin{bmatrix} 12 & 7 & 1 & 0 \\ 0 & 12 & 7 & 1 \\ -18 & -39 & -17 & -2 \\ 36 & 60 & 19 & 1 \end{bmatrix} = 3$$

从 V_1 中选取 3 个线性无关的行，然后另选一个行向量，构成一个等价变换矩阵，

$$T_1 = \begin{bmatrix} 12 & 7 & 1 & 0 \\ 0 & 12 & 7 & 1 \\ -18 & -39 & -17 & -2 \\ 0 & 1 & 0 & 0 \end{bmatrix}, \quad T_1^{-1} = \begin{bmatrix} \dfrac{1}{6} & \dfrac{1}{9} & \dfrac{1}{18} & -\dfrac{1}{3} \\ 0 & 0 & 0 & 1 \\ -1 & -\dfrac{4}{3} & -\dfrac{2}{3} & -3 \\ 7 & \dfrac{31}{3} & \dfrac{14}{3} & 9 \end{bmatrix}$$

进一步，可以计算出

$$\hat{A}_1 = T_1 A_1 T_1^{-1} = \begin{bmatrix} 0 & 1 & 0 & 0 \\ 0 & 0 & 1 & 0 \\ -6 & -11 & -6 & 0 \\ -1 & -\dfrac{4}{3} & -\dfrac{2}{3} & -3 \end{bmatrix}, \quad \hat{b}_1 = T_1 b_1 = \begin{bmatrix} 0 \\ 1 \\ -2 \\ 0 \end{bmatrix}$$

$$\hat{c}_1 = c_1 T_1^{-1} = \begin{bmatrix} 1 & 0 & 0 & 0 \end{bmatrix}$$

所以，$\bar{g}(s)$ 的最小实现为

$$(A_{m1}, b_{m1}, C_{m1}, d) = \left(\begin{bmatrix} 0 & 1 & 0 \\ 0 & 0 & 1 \\ -6 & -11 & -6 \end{bmatrix}, \begin{bmatrix} 0 \\ 1 \\ -2 \end{bmatrix}, \begin{bmatrix} 1 & 0 & 0 \end{bmatrix}, 1 \right)$$

另一种方法是使用能观形实现

$$A_2 = \begin{bmatrix} 0 & 0 & 0 & -18 \\ 1 & 0 & 0 & -39 \\ 0 & 1 & 0 & -29 \\ 0 & 0 & 1 & -9 \end{bmatrix}, \quad b_2 = \begin{bmatrix} 12 \\ 7 \\ 1 \\ 0 \end{bmatrix}, \quad c_2 = \begin{bmatrix} 0 & 0 & 0 & 1 \end{bmatrix}$$

(A_2, b_2) 不完全能控，这是由于

$$\operatorname{rank} U_2 = \operatorname{rank} \begin{bmatrix} 12 & 0 & -18 & 36 \\ 7 & 12 & -39 & 60 \\ 1 & 7 & -17 & 19 \\ 0 & 1 & -2 & 1 \end{bmatrix} = 3$$

在 U_2 中选择 3 个线性无关的列，然后另选一个列向量，构成一个等价变换矩阵，

$$T_2^{-1} = \begin{bmatrix} 12 & 0 & -18 & 0 \\ 7 & 12 & -39 & 1 \\ 1 & 7 & -17 & 0 \\ 0 & 1 & -2 & 0 \end{bmatrix}, \quad T_2 = \begin{bmatrix} \frac{1}{6} & 0 & -1 & 7 \\ \frac{1}{9} & 0 & -\frac{4}{3} & \frac{31}{3} \\ \frac{1}{18} & 0 & -\frac{2}{3} & \frac{14}{3} \\ -\frac{1}{3} & 1 & -3 & 9 \end{bmatrix}$$

可以计算得到

$$\bar{A}_2 = T_2 A_2 T_2^{-1} = \begin{bmatrix} 0 & 0 & -6 & -1 \\ 1 & 0 & -11 & -\frac{4}{3} \\ 0 & 1 & -6 & -\frac{2}{3} \\ 0 & 0 & 0 & -3 \end{bmatrix}, \quad \bar{b}_2 = T_2 b_2 = \begin{bmatrix} 1 \\ 0 \\ 0 \\ 0 \end{bmatrix}$$

$$\bar{c}_2 = c_2 T_2^{-1} = \begin{bmatrix} 0 & 1 & -2 & 0 \end{bmatrix}$$

$\bar{g}(s)$ 的最小实现为

$$(A_{\mathrm{m}2}, b_{\mathrm{m}2}, c_{\mathrm{m}2}, d) = \left(\begin{bmatrix} 0 & 0 & -6 \\ 1 & 0 & -11 \\ 0 & 1 & -6 \end{bmatrix}, \begin{bmatrix} 1 \\ 0 \\ 0 \end{bmatrix}, \begin{bmatrix} 0 & 1 & -2 \end{bmatrix}, 1 \right)$$

事实上，对于线性定常 SISO 系统 (A, b, c)，它完全能控且能观的充要条件是它的传递函数

$$g(s) = \frac{c \operatorname{adj}(sI - A) b}{\det(sI - A)}$$

没有零极点对消。

注意到

$$g(s) = \frac{s^2 + 7s + 12}{s^4 + 9s^3 + 29s^2 + 39s + 18} = \frac{s + 4}{s^3 + 6s^2 + 11s + 6} = \tilde{g}(s)$$

在 $\tilde{g}(s)$ 中，不存在零极点对消，所以 $\tilde{g}(s)$ 的能控标准形

$$(A_c, b_c, c_c, d) = \left(\begin{bmatrix} 0 & 1 & 0 \\ 0 & 0 & 1 \\ -6 & -11 & -6 \end{bmatrix}, \begin{bmatrix} 0 \\ 0 \\ 1 \end{bmatrix}, \begin{bmatrix} 4 & 1 & 0 \end{bmatrix}, 1 \right)$$

和 $\tilde{g}(s)$ 的能观标准形

$$(A_o, b_o, c_o, d) = \left(\begin{bmatrix} 0 & 0 & -6 \\ 1 & 0 & -11 \\ 0 & 1 & -6 \end{bmatrix}, \begin{bmatrix} 4 \\ 1 \\ 0 \end{bmatrix}, \begin{bmatrix} 0 & 0 & 1 \end{bmatrix}, 1 \right)$$

都是 $\tilde{g}(s)$ 的最小实现。

4.4.3 线性定常 MIMO 系统的最小实现

考虑 $q \times p$ 的真有理传递函数矩阵 $\bar{G}(s) = G(s) + D$，其中，$D = \lim\limits_{s \to \infty} \bar{G}(s)$ 可以直接得到。根据定理 4.13，可以得到它的满足式 (4.32) 的能控形实现 (A, B, C)。如果 (A, C) 是能观的，那么 (A, B, C) 是 $G(s)$ 的最小实现，因为它完全能控且能观。否则，可以根据定理 3.23 对它进行能观性结构分解，得到的能观的子系统 $(\hat{A}_{11}, \hat{B}_1, \hat{C}_1)$ 就是一个最小实现。

类似地，可以根据定理 4.14 得到能观形实现，然后根据定理 3.22 进行能控性分解，以得到最小实现 $(\bar{A}_{11}, \bar{B}_1, \bar{C}_1)$。

例 4.7 对于以下传递函数矩阵，

$$G(s) = \begin{bmatrix} \dfrac{2}{s+1} & \dfrac{1}{s+1} \\ \dfrac{1}{s+2} & \dfrac{1}{s+2} \end{bmatrix}$$

例 4.5 中给出了它的能控形实现，

$$A_1 = \begin{bmatrix} 0 & 0 & 1 & 0 \\ 0 & 0 & 0 & 1 \\ -2 & 0 & -3 & 0 \\ 0 & -2 & 0 & -3 \end{bmatrix}, B_1 = \begin{bmatrix} 0 & 0 \\ 0 & 0 \\ 1 & 0 \\ 0 & 1 \end{bmatrix}, C_1 = \begin{bmatrix} 4 & 2 & 2 & 1 \\ 1 & 1 & 1 & 1 \end{bmatrix}$$

因为 $\operatorname{rank} V_1 = 2$，$(A_1, C_1)$ 是不完全能观的。注意到 $\operatorname{rank} C_1 = 2$，应用 C_1 的行来构造以下变换矩阵，

$$T_1 = \begin{bmatrix} 4 & 2 & 2 & 1 \\ 1 & 1 & 1 & 1 \\ 1 & 0 & 0 & 0 \\ 0 & 1 & 0 & 0 \end{bmatrix}, T_1^{-1} = \begin{bmatrix} 0 & 0 & 1 & 0 \\ 0 & 0 & 0 & 1 \\ 1 & -1 & -3 & -1 \\ -1 & 2 & 2 & 0 \end{bmatrix}$$

进一步可以得到

$$\hat{A}_1 = T_1 A_1 T_1^{-1} = \begin{bmatrix} -1 & 0 & 0 & 0 \\ 0 & -2 & 0 & 0 \\ 1 & -1 & -3 & -1 \\ -1 & 2 & 2 & 0 \end{bmatrix}, \quad \hat{B}_1 = T_1 B_1 = \begin{bmatrix} 2 & 1 \\ 1 & 1 \\ 0 & 0 \\ 0 & 0 \end{bmatrix}$$

$$\hat{C}_1 = C_1 T_1^{-1} = \begin{bmatrix} 1 & 0 & 0 & 0 \\ 0 & 1 & 0 & 0 \end{bmatrix}$$

所以，$G(s)$ 的最小实现为

$$(A_{m1}, B_{m1}, C_{m1}) = \left(\begin{bmatrix} -1 & 0 \\ 0 & -2 \end{bmatrix}, \begin{bmatrix} 2 & 1 \\ 1 & 1 \end{bmatrix}, \begin{bmatrix} 1 & 0 \\ 0 & 1 \end{bmatrix} \right)$$

另一种方法是使用 $G(s)$ 的能观形实现，

$$A_2 = \begin{bmatrix} 0 & 0 & -2 & 0 \\ 0 & 0 & 0 & -2 \\ 1 & 0 & -3 & 0 \\ 0 & 1 & 0 & -3 \end{bmatrix}, \quad B_2 = \begin{bmatrix} 4 & 2 \\ 1 & 1 \\ 2 & 1 \\ 1 & 1 \end{bmatrix}, \quad C_2 = \begin{bmatrix} 0 & 0 & 1 & 0 \\ 0 & 0 & 0 & 1 \end{bmatrix}$$

因为 $\operatorname{rank} U_2 = 2$，$(A_2, B_2)$ 是不完全能控的。注意到 $\operatorname{rank} B_2 = 2$。应用 B_2 的列来构造以下变换矩阵。

$$T_2^{-1} = \begin{bmatrix} 4 & 2 & 0 & 0 \\ 1 & 1 & 0 & 0 \\ 2 & 1 & 1 & 0 \\ 1 & 1 & 0 & 1 \end{bmatrix}, \quad T_2 = \begin{bmatrix} \frac{1}{2} & -1 & 0 & 0 \\ -\frac{1}{2} & 2 & 0 & 0 \\ -\frac{1}{2} & 0 & 1 & 0 \\ 0 & -1 & 0 & 1 \end{bmatrix}$$

进一步，可以得到

$$\bar{A}_2 = T_2 A_2 T_2^{-1} = \begin{bmatrix} 0 & 1 & -1 & 2 \\ -2 & -3 & 1 & -4 \\ 0 & 0 & -2 & 0 \\ 0 & 0 & 0 & -1 \end{bmatrix}, \quad \bar{B}_2 = T_2 B_2 = \begin{bmatrix} 1 & 0 \\ 0 & 1 \\ 0 & 0 \\ 0 & 0 \end{bmatrix}$$

$$\bar{C}_2 = C_2 T_2^{-1} = \begin{bmatrix} 2 & 1 & 1 & 0 \\ 1 & 1 & 0 & 1 \end{bmatrix}$$

所以，$G(s)$ 的最小实现为

$$(A_{m2}, B_{m2}, C_{m2}) = \left(\begin{bmatrix} 0 & 1 \\ -2 & -3 \end{bmatrix}, \begin{bmatrix} 1 & 0 \\ 0 & 1 \end{bmatrix}, \begin{bmatrix} 2 & 1 \\ 1 & 1 \end{bmatrix} \right)$$

4.4 线性定常系统的最小实现

接下来，介绍求取最小实现的 Hankel 矩阵法。

4.4.1 小节中指出，一个 $q\times p$ 的严格真有理分式传递函数矩阵 $G(s)$ 可以表示为 Laurent 级数展开式 (4.38)。在此基础上，定义了 Hankel 矩阵式 (4.39)。考虑 Hankel 矩阵的秩 $\text{rank}\, H(n,n)$ 为 n，构造以下矩阵：

(1) $F\in\mathbb{R}^{n\times n}$，它是由 $H(n,n)$ 中前 n 个线性无关的行和前 n 个线性无关的列构成的非奇异矩阵；

(2) $F^*\in\mathbb{R}^{n\times n}$，它是由 $H(n,n)$ 中和 F 相同的列，但低 q 行的行构成的矩阵；

(3) $F_1\in\mathbb{R}^{q\times n}$，它是由 $H(n,n)$ 中的前 q 行以及和 F 相同的列构成的矩阵；

(4) $F_2\in\mathbb{R}^{n\times p}$，它是由 $H(n,n)$ 中前 n 个线性无关的行以及前 p 列构成的矩阵。

进一步，考虑 $A=F^*F^{-1}$，$B=F_2$ 且 $C=F_1F^{-1}$，可以验证这种方法得到的 (A,B,C) 完全能控且能观，即 (A,B,C) 是 $G(s)$ 的一个最小实现。

例 4.8 考虑以下严格真有理分式传递函数矩阵

$$G(s)=\begin{bmatrix} \dfrac{1}{s(s+1)} & \dfrac{1}{s} \\ \dfrac{1}{s} & \dfrac{s-1}{s(s+1)} \end{bmatrix}$$

它可以表示为 Laurent 级数展开，其中

$$h_1=\begin{bmatrix} 0 & 1 \\ 1 & 1 \end{bmatrix},\ h_2=\begin{bmatrix} 1 & 0 \\ 0 & -2 \end{bmatrix},\ h_3=-h_2,\ h_4=h_2$$

构造 Hankel 矩阵，

$$H(3,2)=\begin{bmatrix} 0 & 1 & 1 & 0 \\ 1 & 1 & 0 & -2 \\ 1 & 0 & -1 & 0 \\ 0 & -2 & 0 & 2 \\ -1 & 0 & 1 & 0 \\ 0 & 2 & 0 & -2 \end{bmatrix}$$

可以验证 Hankel 矩阵的秩为 4。在此基础上，构造

$$F=\begin{bmatrix} 0 & 1 & 1 & 0 \\ 1 & 1 & 0 & -2 \\ 1 & 0 & -1 & 0 \\ 0 & -2 & 0 & 2 \end{bmatrix},\ F^*=\begin{bmatrix} 1 & 0 & -1 & 0 \\ 0 & -2 & 0 & 2 \\ -1 & 0 & 1 & 0 \\ 0 & 2 & 0 & -2 \end{bmatrix}$$

$$F_1=\begin{bmatrix} 0 & 1 & 1 & 0 \\ 1 & 1 & 0 & -2 \end{bmatrix},\ F_2=\begin{bmatrix} 0 & 1 \\ 1 & 1 \\ 1 & 0 \\ 0 & -2 \end{bmatrix}$$

进一步，可以得到

$$A = F^*F^{-1} = \begin{bmatrix} 1 & 0 & -1 & 0 \\ 0 & -2 & 0 & 2 \\ -1 & 0 & 1 & 0 \\ 0 & 2 & 0 & -2 \end{bmatrix} \cdot \frac{1}{2} \begin{bmatrix} 1 & 1 & 1 & 1 \\ 1 & -1 & 1 & -1 \\ 1 & 1 & -1 & 1 \\ 1 & -1 & 1 & 0 \end{bmatrix} = \begin{bmatrix} 0 & 0 & 1 & 0 \\ 0 & 0 & 0 & 1 \\ 0 & 0 & -1 & 0 \\ 0 & 0 & 0 & -1 \end{bmatrix}$$

$$B = F_2 = \begin{bmatrix} 0 & 1 \\ 1 & 1 \\ 1 & 0 \\ 0 & -2 \end{bmatrix}, \quad C = F_1 F^{-1} = \begin{bmatrix} 1 & 0 & 0 & 0 \\ 0 & 1 & 0 & 0 \end{bmatrix}$$

可以验证，(A, B, C) 完全能控且能观，且 $C(sI - A)^{-1}B = G(s)$。所以，(A, B, C) 是 $G(s)$ 的最小实现。

第 5 章

线性系统的稳定性

稳定性是系统另一种重要的结构性质,它和系统的能控性与能观性一样,是系统的一种定性性质。对系统稳定性的分析与综合是系统控制理论的重要组成部分。一个实际的系统必须是稳定的,只有稳定的系统才能在实际工程中应用。尽管稳定还不是系统达到成功工程应用的全部,但是它是系统实际应用的前提和先决条件。本章首先讨论 Lyapunov(李雅普诺夫) 意义下稳定性的基本概念。在此基础上,进一步讨论系统的内部稳定性、外部稳定性,以及各类稳定性之间的关系和判据。

5.1 Lyapunov 稳定理论

5.1.1 基本概念

在研究系统的稳定性时,通常考虑没有外部输入的系统,这类系统通常被称作自治系统。本节考虑由以下一阶常微分方程描述的自治系统,

$$\dot{x}(t) = f(x,t), \ x(t_0) = x_0, \ t \geqslant t_0 \tag{5.1}$$

其中,$x(t) \in \mathbb{R}^n$ 是状态向量,$f(x,t)$ 是 n 维函数向量。

定义 5.1(平衡点) 如果 $x_e \in \mathbb{R}^n$ 对于 $\forall t \geqslant t_0$,都使得

$$f(x_e,t) = 0$$

成立,那么称之为式 (5.1) 的平衡点。

大多数情况下,$x_e = 0$,即状态空间的原点是系统的一个平衡点。除此之外,系统也可能有一些非零的平衡点。如果系统的平衡点在状态空间中呈现为彼此分离的孤立点,则称其为孤立平衡点。对于孤立平衡点,总是可以通过移动坐标系而将其转换为状态空间的原点,所以在后续讨论中,假设平衡点 x_e 就是原点。

假设系统方程 (5.1) 满足唯一解条件,那么由初始条件 x_0 引起的运动 $x(t)$ 可以表示为 $x(t) = x(t;x_0,t_0)$, $t \geqslant t_0$,一般称之为系统的受扰运动。事实上,受扰运动等价于系统状态的零输入响应。

系统运动的稳定性就是研究其平衡点的稳定性,或者说,偏离平衡点的受扰运动能否仅仅依靠系统内部的结构特性而返回到平衡点,或者限制在平衡点的一个有限邻域内。

定义 5.2（稳定） 如果对于任意 $\varepsilon > 0$，都对应地存在一个实数 $\delta(t_0, \varepsilon) > 0$，使得由满足

$$\|x_0 - x_e\| \leqslant \delta(t_0, \varepsilon) \tag{5.2}$$

的任意初始状态出发的受扰运动都满足

$$\|x(t; x_0, t_0) - x_e\| \leqslant \varepsilon, \ \forall t \geqslant t_0 \tag{5.3}$$

那么称平衡点 x_e 是稳定的。

接下来解释这一定义的几何意义。对于任意的正实数 ε，在状态空间内以平衡点 x_e 为球心，构造半径为 ε 的超球体，相应的球域记为 $S(\varepsilon)$。若存在一个同时依赖于 ε 和初始时刻 t_0 的正实数 $\delta(t_0, \varepsilon)$，那么可以仍以 x_e 为球心，以 $\delta(t_0, \varepsilon)$ 为半径构造另一个超球体，相应的球域记为 $S(\delta)$。如果从球域 $S(\delta)$ 内任意一点出发的运动，它的运动轨迹 $x(t; x_0, t_0)$ 对于所有的 $t \geqslant t_0$ 始终都不脱离球域 $S(\varepsilon)$，那么，平衡点 x_e 被称为是（Lyapunov 意义下）稳定的。例如，考虑 $x(t) \in \mathbb{R}^2$ 且 $x_e = 0$，上述几何意义如图 5.1 所示。

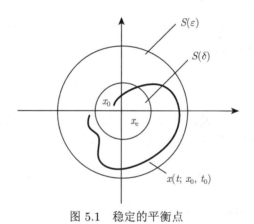

图 5.1 稳定的平衡点

定义 5.3（一致稳定） 如果对于任意 $\varepsilon > 0$，都对应地存在一个实数 $\delta(\varepsilon) > 0$，使得由满足

$$\|x_0 - x_e\| \leqslant \delta(\varepsilon) \tag{5.4}$$

的任意初始状态出发的受扰运动都满足方程 (5.3)，那么称平衡点 x_e 是一致稳定的。

可以看出，在定义 5.2 中，如果 δ 的选取和 t_0 无关，即得到了一致稳定的定义。对于定常系统，其稳定和一致稳定等价；但是对于时变系统，则不存在这样的等价关系。

定义 5.4（渐近稳定） 如果平衡点 x_e 是渐近稳定的，
(1) 平衡点是稳定的，
(2) 对于 $\delta(t_0, \varepsilon) > 0$ 和 $\varepsilon > 0$，存在 $T(\varepsilon, \delta, t_0) > 0$，使得任意满足方程 (5.2) 的初始状态 x_0 引起的受扰运动满足

$$\|x(t; x_0, t_0) - x_e\| \leqslant \varepsilon, \ \forall t \geqslant t_0 + T(\varepsilon, \delta, t_0) \tag{5.5}$$

那么称平衡点 x_e 是渐近稳定的。

特别地，如果 δ 和 T 的选取都和初始时刻 t_0 无关，那么称平衡点 x_e 是一致渐近稳定的。

考虑 $x(t)\in\mathbb{R}^2$，渐近稳定的几何意义如图 5.2 所示。图 5.2(a) 反映了运动的有界性，图 5.2(b) 反映了运动随时间变化过程的渐近性。显然，随着 $\varepsilon\to 0$，$T\to\infty$，因此，当位于原点的平衡点 x_e 渐近稳定时，对于任意满足方程 (5.2) 的初始状态引起的运动，总是满足

$$\lim_{t\to\infty} x(t;x_0,t_0)=0 \tag{5.6}$$

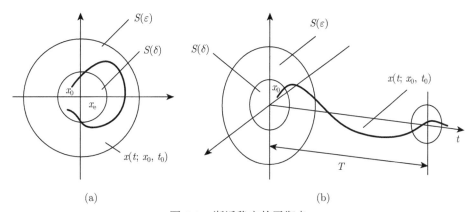

图 5.2 渐近稳定的平衡点

当 $t_0\geqslant 0$，随着 $\varepsilon\to 0$，使得 $x(t;x_0,t_0)\to 0$ 的所有状态 $x_0\in\mathbb{R}^n$ 构成的集合被称为平衡点 $x_e=0$ 的吸引域。

定义 5.5（指数稳定） 如果对于 $\delta(\varepsilon)>0$ 和 $\varepsilon>0$，存在 $\nu>0$，使得任意满足方程 (5.4) 的初始状态 x_0 引起的受扰运动满足

$$\|x(t;x_0,t_0)-x_e\|\leqslant \varepsilon\mathrm{e}^{-\nu(t-t_0)},\ \forall t\geqslant t_0 \tag{5.7}$$

那么称平衡点 x_e 是指数稳定的。

定义 5.5 给出了"渐近"的数量概念。一个平衡点如果是指数稳定的，那么它必然是渐近稳定的。

上述概念都仅反映了平衡点的局部稳定性质。接下来，考虑平衡点的全局稳定性。

定义 5.6（全局渐近稳定） 如果对于任意有限值的非零初始状态 x_0，相应的受扰运动 $x(t;x_0,t_0)$ 是有界的，且方程 (5.6) 成立，那么称平衡点 $x_e=0$ 是全局渐近稳定的。

如果平衡点 $x_e=0$ 是全局渐近稳定的，也称之为是大范围渐近稳定的。如果一个平衡点 x_e 是渐近稳定的，但不是全局渐近稳定的，则称之为是局部渐近稳定的。当平衡点 $x_e=0$ 全局渐近稳定时，相应的吸引域为 $\mathbb{R}^2$。这种情况下，$x_e=0$ 是系统方程 (5.1) 唯一的平衡点。也就是说，系统全局渐近稳定的必要前提是除了位于原点的平衡点外，系统不存在其他任何孤立平衡点。如果一个线性系统是渐近稳定的，那么它一定是全局渐近稳定的。一般来说，在工程应用中，通常期望系统具有全局渐近稳定的特性。

定义 5.7（全局指数稳定） 如果对于任意 $\delta > 0$，总是存在 $k(\delta) > 0$ 和 $\nu > 0$，使得任意 $x_0 < \delta$ 引起的受扰运动总是满足

$$\|x(t; x_0, t_0) - x_e\| \leqslant k(\delta) \|x_0\| e^{-\nu(t-t_0)}, \ \forall t \geqslant t_0 \tag{5.8}$$

那么称平衡点 $x_e = 0$ 是全局指数稳定的。

定义 5.8（不稳定） 无论取多大的有界实数 $\varepsilon > 0$，都不能找到相应的实数 $\delta(t_0, \varepsilon) > 0$，使得由满足方程 (5.2) 初始状态引起的受扰运动满足方程 (5.3)，那么称平衡点 x_e 是不稳定的。

对于 $x(t) \in \mathbb{R}^2$，图 5.3 给出了不稳定的物理意义。显然，如果平衡点 x_e 不稳定，那么无论 $S(\varepsilon)$ 多大，$S(\delta)$ 多小，总是存在非零状态 $x_0 \in S(\delta)$，使得由其引起的受扰运动轨迹越出球域 $S(\varepsilon)$。

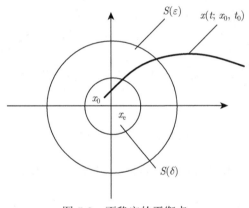

图 5.3 不稳定的平衡点

例 5.1 考虑一个系统 $\dot{x}_1(t) = -x_2(t)$，$\dot{x}_2(t) = -x_1(t)$。初始状态分别为 $x_1(t_0)$ 和 $x_2(t_0)$。可以看出原点是这一系统唯一的平衡点，进一步可以求解得到

$$\begin{bmatrix} x_1(t) \\ x_2(t) \end{bmatrix} = \begin{bmatrix} \cos t & -\sin t \\ \sin t & \cos t \end{bmatrix} \begin{bmatrix} x_1(t_0) \\ x_2(t_0) \end{bmatrix}$$

在此基础上，可以进一步推得

$$\begin{aligned} \|x(t)\| &= \sqrt{x_1^2(t) + x_2^2(t)} \\ &= \sqrt{(\cos t x_1(t_0) - \sin t x_2(t_0))^2 + (\sin t x_1(t_0) + \cos t x_2(t_0))^2} \\ &= \sqrt{x_1^2(t_0) + x_2^2(t_0)} = \|x(t_0)\| \end{aligned}$$

显然，对于任意 $\varepsilon > 0$，总是存在 $\delta(\varepsilon) = \varepsilon$ 使得 $\|x(t_0)\| \leqslant \delta(\varepsilon)$ 和 $\|x(t)\| \leqslant \varepsilon$ 成立，$\forall t \geqslant t_0$。所以，平衡点 $x_e = 0$ 是一致稳定的。

另一方面，因为 $\lim\limits_{t\to\infty}\|x(t)\| = \sqrt{x_1^2(t_0)+x_2^2(t_0)} \neq 0$，所以平衡点 $x_e = 0$ 不是渐近稳定的。

5.1.2 Lyapunov 第二法的主要定理

1892 年，Lyapunov 发表了著名的论文《运动稳定性的一般问题》，提出并建立了稳定性理论研究体系和基础，对稳定性理论的研究具有里程碑式的意义。一百多年来，Lyapunov 提出的理论和方法已经渗透到数学、力学、控制与系统理论等众多领域中，形成了从理论到应用的丰富体系。

Lyapunov 研究稳定性的方法主要包括以他名字命名的第一法和第二法。第一法，也被称作间接法，主要通过求系统的微分方程的解来分析系统的稳定性。第二法，也被称作直接法，直接从系统的方程出发，构造一个类似于能量函数的 Lyapunov 函数，并分析这一函数及其一阶导数的正负来获取系统稳定性的相关信息。这一方法概念直观，物理意义清晰，适用性很广。该方法在 1960 年前后被引入系统和控制理论的研究中后，得到了迅速发展和广泛应用，在理论和应用方面都具有重要意义。

考虑以下连续非线性时变系统

$$\dot{x}(t) = f(x,t),\; t \geqslant t_0 \tag{5.9}$$

其中，对于任意 t，$f(0,t) = 0$，即状态空间的原点是这一系统的平衡点。

对于系统方程 (5.9)，首先其给出全局一致渐近稳定判据，这一结论一般也被称为 Lyapunov 主稳定性定理。

定理 5.1 对于系统方程 (5.9)，如果存在一个标量函数 $V(x,t)$，使得

(1) $V(x,t)$，$\dfrac{\partial}{\partial x}V(x,t)$ 和 $\dfrac{\partial}{\partial t}V(x,t)$，对于 x 和 t 是连续的，且 $V(0,t) = 0$；

(2) $V(x,t)$ 正定且有界，即存在两个连续非减的标量函数 $\alpha(\|x\|)$ 与 $\beta(\|x\|)$，满足 $\alpha(0) = 0$ 与 $\beta(0) = 0$，使得对于 $\forall t \geqslant t_0$ 和 $x \neq 0$，

$$0 < \alpha(\|x\|) \leqslant V(x,t) \leqslant \beta(\|x\|) \tag{5.10}$$

(3) $\dfrac{\partial}{\partial t}V(x,t)$ 负定且有界，即存在一个连续非减的标量函数 $\gamma(\|x\|)$ 满足 $\gamma(0) = 0$，使得对于 $\forall t \geqslant t_0$，$x \neq 0$，

$$\frac{\partial}{\partial t}V(x,t) \leqslant -\gamma(\|x\|) < 0 \tag{5.11}$$

(4) 当 $\|x\| \to \infty$，$\alpha(\|x\|) \to \infty$，即 $V(x,t) \to \infty$，

那么，$x_e = 0$ 是全局渐近稳定的。

证明 分三步进行证明。第一步，证明平衡点 $x_e = 0$ 是一致稳定的；第二步，证明这种稳定是渐近的；第三步，证明这种稳定是全局的。

首先证明位于原点的平衡点 $x_e = 0$ 是一致稳定的。定理 5.1 中条件 (1) 和 (2) 的几何意义如图 5.4 所示。

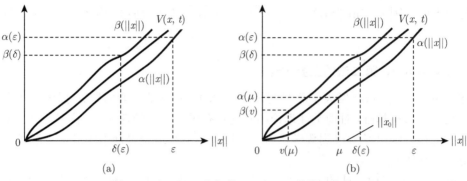

图 5.4　定理 5.1 中条件 (1) 和 (2) 的物理意义

因为 $\beta(\|x\|)$ 是一个连续非减的函数，且 $\beta(0)=0$，所以对于任意实数 $\varepsilon>0$，总是存在相应的实数 $\delta(\varepsilon)>0$ 使得 $\beta(\delta)\leqslant\alpha(\varepsilon)$。

由于 $\dfrac{\partial}{\partial t}V(x,t)$ 是负定的，所以

$$V\left(x(t;x_0,t_0),t\right)-V(x_0,t_0)=\int_{t_0}^{t}\frac{\partial}{\partial\tau}V\left(x(\tau;x_0,t_0),\tau\right)\mathrm{d}\tau\leqslant 0,\ \forall t\geqslant t_0$$

在此基础上，对于任意 t_0 和满足 $\|x_0\|\leqslant\delta(\varepsilon)$ 的 x_0，以及任意 $t\geqslant t_0$，均有

$$\alpha(\varepsilon)\geqslant\beta(\delta)\geqslant V(x_0,t_0)\geqslant V(x(t;x_0,t_0),t)\geqslant\alpha(\|x(t;x_0,t_0)\|)$$

因为 $\alpha(\|x\|)$ 是一个连续非减的函数，且 $\alpha(0)=0$，对于任意 $t\geqslant t_0$ 和满足 $\|x_0\|\leqslant\delta(\varepsilon)$ 的 x_0，可以得到

$$\|x(t;x_0,t_0)\|\leqslant\varepsilon,\ \forall t\geqslant t_0$$

这表明，对于任意实数 $\varepsilon>0$，总是存在实数 $\delta(\varepsilon)>0$ 使得任意从满足 $\|x_0\|\leqslant\delta(\varepsilon)$ 的初始状态 x_0 出发的受扰运动都满足上述不等式。这里，$\delta(\varepsilon)$ 和初始时刻 t_0 的选取无关。根据定义 5.3，位于原点的平衡点 $x_e=0$ 是一致稳定的。

接下来，证明 $x_e=0$ 是一致渐近稳定的。

给定一个实数 $\mu>0$，对于 $\delta(\varepsilon)>0$，找出一个相应的实数 $T(\mu,\delta)>0$。假设初始时刻 t_0 为任意，同时 x_0 是满足 $\|x_0\|\leqslant\delta(\varepsilon)$ 的任意初始状态。不失一般性地，假设 $\mu\leqslant\|x_0\|\leqslant\delta(\varepsilon)$。因为 $V(x,t)$ 是有界的，如图 5.4 所示，对于给定的 $\mu>0$，能够找到相应的实数 $0<\nu(\mu)<\mu$ 使得 $\beta(\nu)\leqslant\alpha(\mu)$。同时，$\gamma(\|x\|)$ 是连续非减的。考虑 $\rho(\mu,\delta)$ 是 $\gamma(\|x\|)$ 在区间 $0\leqslant\|x\|\leqslant\varepsilon$ 上的极小值，进一步选取

$$T(\mu,\delta)=\frac{\beta(\delta)}{\rho(\mu,\delta)}$$

这种情况下，对于任意的 $\mu>0$，总是能够找到对应的与初始时刻 t_0 无关的 $T(\mu,\delta)>0$。进一步，证明存在满足 $t_0\leqslant t_1\leqslant t_2$ 的时刻 t_1，其中 $t_2=t_0+T(\mu,\delta)$，能够使得

$\|x(t_1; x_0, t_0)\| = \nu(\mu)$。这里使用反证法。假设对于 $\forall t \in [t_0, t_2]$，$\|x(t; x_0, t_0)\| > \nu(\mu)$。根据条件 (3)，可以推导出

$$0 < \alpha(\nu) \leqslant \alpha(\|x(t_2; x_0, t_0)\|) \leqslant V(x(t_2; x_0, t_0), t_2)$$
$$\leqslant V(x_0, t_0) - (t_2 - t_0)\rho(\mu, \delta) \leqslant \beta(\delta) - T(\mu, \delta)\rho(\mu, \delta) = 0$$

这一结果存在矛盾。所以，在 $[t_0, t_2]$ 的区间内，总是存在 t_1 使得 $x(t_1; x_0, t_0) = \nu(\mu)$。对于 $\forall t \geqslant t_1$，可以得到

$$\alpha(\|x(t; x_0, t_0)\|) \leqslant V(x(t; x_0, t_0), t) \leqslant V(x(t_1; x_0, t_0), t_1) \leqslant \beta(\nu) \leqslant \alpha(\mu)$$

这表明对于 $\forall t \geqslant t_1, \|x(t; x_0, t_0)\| \leqslant \mu$ 总是成立。因为 $t_0 + T(\mu, \delta) \geqslant t_1$，所以 $\|x(t; x_0, t_0)\| \leqslant \mu$ 对于 $\forall t \geqslant t_0 + T(\mu, \delta)$ 也成立。同时，随着 $\mu \to 0$，$T \to \infty$。所以，$x_e = 0$ 是一致渐近稳定的。

最后，证明 $x_e = 0$ 是全局一致渐近稳定的，即证明由任意非零初始状态 $x_0 \in \mathbb{R}^n$ 引起的受扰运动都是一致有界的。

注意到如果 $\|x\| \to \infty$，那么 $\alpha(\|x\|) \to \infty$。由此可知，对于任意大的有界的实数 $\delta > 0$，总是存在有界实数 $\varepsilon(\delta) > 0$ 使得 $\beta(\delta) < \alpha(\varepsilon)$。所以，对于状态空间内任意的非零状态 x_0 和 $t \geqslant t_0$，总是有

$$\alpha(\varepsilon) > \beta(\delta) \geqslant V(x_0, t_0) \geqslant V(x(t; x_0, t_0), t) \geqslant \alpha(\|x(t; x_0, t_0)\|)$$

因为 $\alpha(\|x\|)$ 是连续非减的，进一步可得 $\|x(t; x_0, t_0)\| < \varepsilon(\delta)$ 对于 $\forall t \geqslant t_0$ 和 $\forall x_0 \in \mathbb{R}^n$ 都成立，其中，$\varepsilon(\delta)$ 与初始时刻 t_0 的选取无关。这表明 $x(t; x_0, t_0)$ 对于任意 $x_0 \in \mathbb{R}^n$ 是一致有界的。 □

需要强调的是定理 5.1 仅给出了判断系统方程 (5.9) 是全局一致渐近稳定的充分条件。对于给定的系统，如果不能找到合适的标量函数 $V(x, t)$ 使得定理 5.1 中的条件都成立，那么对判断系统的稳定性不能提供任何信息。

如果定理 5.1 中的条件 (4) 不满足，且条件 (1)~ 条件 (3) 仅在 $x \in \Omega$ 内满足，其中 Ω 是原点的一个邻域，那么可以得到局部一致渐近稳定性判据。

如果所讨论的问题限于定常系统，即

$$\dot{x}(t) = f(x), \ t \geqslant 0 \tag{5.12}$$

其中，对于任意 $t \geqslant 0$，$f(0) = 0$，那么定理 5.1 可以表述为以下形式。

定理 5.2 对于定常系统方程 (5.12)，如果存在一个标量函数 $V(x)$，使得

(1) $V(x)$ 及其一阶导数连续；

(2) $V(x) \geqslant 0$ 且 $V(x) = 0$ 仅当 $x = 0$ 时成立；

(3) $\dfrac{\mathrm{d}}{\mathrm{d}t} V(x) < 0$ 对于 $\forall x \neq 0$ 都成立；

(4) 当 $\|x\| \to \infty, V(x) \to \infty$，

那么，$x_e = 0$ 是全局一致渐近稳定的。

例 5.2 考虑以下连续定常系统，

$$\begin{cases} \dot{x}_1 = x_2 - x_1(x_1^2 + x_2^2) \\ \dot{x}_2 = -x_1 - x_2(x_1^2 + x_2^2) \end{cases}$$

显然，$x_1 = x_2 = 0$ 是该系统唯一的平衡点。考虑 $V(x) = x_1^2 + x_2^2$，那么

$$\begin{aligned} \frac{\mathrm{d}}{\mathrm{d}t}V(x) &= \frac{\partial V(x)}{\partial x_1}\frac{\mathrm{d}x_1}{\mathrm{d}t} + \frac{\partial V(x)}{\partial x_2}\frac{\mathrm{d}x_2}{\mathrm{d}t} \\ &= \begin{bmatrix} \dfrac{\partial V(x)}{\partial x_1} & \dfrac{\partial V(x)}{\partial x_2} \end{bmatrix} \begin{bmatrix} \dot{x}_1 \\ \dot{x}_2 \end{bmatrix} \\ &= \begin{bmatrix} 2x_1 & 2x_2 \end{bmatrix} \begin{bmatrix} x_2 - x_1(x_1^2 + x_2^2) \\ -x_1 - x_2(x_1^2 + x_2^2) \end{bmatrix} \\ &= -2(x_1^2 + x_2^2)^2 \end{aligned}$$

从 $V(x)$ 和 $\dfrac{\mathrm{d}}{\mathrm{d}t}V(x)$ 的表达式可以看出，选取的标量函数满足定理 5.2 中的所有条件。所以 $x_e = 0$ 是全局一致渐近稳定的。

定理 5.3 对于定常系统方程 (5.12)，如果存在一个标量函数 $V(x)$，使得

(1) $V(x)$ 及其一阶导数连续；

(2) $V(x) \geqslant 0$ 且 $V(x) = 0$ 仅当 $x = 0$ 时成立；

(3) $\dfrac{\mathrm{d}}{\mathrm{d}t}V(x) \leqslant 0$；

(4) 对于任意非零的 $x_0 \in \mathbb{R}^n$，$\dfrac{\mathrm{d}}{\mathrm{d}t}V(x(t; x_0, 0)) \neq 0$；

(5) 当 $\|x\| \to \infty$，$V(x) \to \infty$，

那么，$x_e = 0$ 是全局一致渐近稳定的。

例 5.3 考虑以下连续定常系统，

$$\begin{cases} \dot{x}_1 = x_2 \\ \dot{x}_2 = -x_1 - (1 + x_2)^2 x_2 \end{cases}$$

显然，$x_1 = x_2 = 0$ 是该系统唯一的平衡点。考虑 $V(x) = x_1^2 + x_2^2$，那么

$$\begin{aligned} \frac{\mathrm{d}}{\mathrm{d}t}V(x) &= \begin{bmatrix} \dfrac{\partial V(x)}{\partial x_1} & \dfrac{\partial V(x)}{\partial x_2} \end{bmatrix} \begin{bmatrix} \dot{x}_1 \\ \dot{x}_2 \end{bmatrix} \\ &= \begin{bmatrix} 2x_1 & 2x_2 \end{bmatrix} \begin{bmatrix} x_2 \\ -x_1 - (1+x_2)^2 x_2 \end{bmatrix} \\ &= -2x_2^2(1 + x_2)^2 \end{aligned}$$

从 $V(x)$ 和 $\dfrac{\mathrm{d}}{\mathrm{d}t}V(x)$ 的表达式可以看出，定理 5.3 中的条件 (1)，(2)，(3) 和 (5) 都满足，下边需要验证 (4) 是否满足。

除了以下两种情况：(a) x_1 为任意且 $x_2 = 0$，(b) x_1 为任意且 $x_2 = -1$，$\dfrac{\mathrm{d}}{\mathrm{d}t}V(x) < 0$ 均成立。这两种情况下，$\dfrac{\mathrm{d}}{\mathrm{d}t}V(x) = 0$。

对于情况 (a)，$x_2 = 0$，所以

$$\begin{cases} \dot{x}_1 = x_2 = 0 \\ 0 = \dot{x}_2 = -x_1 - (1 + x_2)^2 x_2 = -x_1 \end{cases}$$

这表明，除了 $x_1 = x_2 = 0$，$\bar{x}(t; x_0, 0) = [\ x_1(t)\ \ 0\ ]^{\mathrm{T}}$ 不是受扰运动的解。

对于情况 (b)，$x_2(t) = -1$，所以

$$\begin{cases} \dot{x}_1 = x_2 = -1 \\ 0 = \dot{x}_2 = -x_1 - (1 + x_2)^2 x_2 = -x_1 \end{cases}$$

显然这是一个矛盾的结果。这表明 $\tilde{x}(t; x_0, 0) = [\ x_1(t)\ \ -1\ ]^{\mathrm{T}}$ 也不是受扰运动的解。因此，定理 5.3 中的条件 (4) 也满足。由此可知，位于原点的平衡点是全局一致渐近稳定的。

如果很难判断系统的一致渐近稳定性，应当转而判断系统的一致稳定性。

定理 5.4 对于系统方程 (5.9)，如果存在一个连续的标量函数 $V(x, t)$ 满足 $V(0, t) = 0$，且对于 x 和 t 具有连续一阶偏导，同时存在一个围绕原点的吸引域 Ω，使得对于任意 $x \in \Omega$ 和 $t \geqslant t_0$，

(1) $V(x, t)$ 正定且有界；

(2) $\dfrac{\partial}{\partial t}V(x, t)$ 半负定且有界，

那么，$x_{\mathrm{e}} = 0$ 在 Ω 域内一致稳定。

定理 5.5 对于系统方程 (5.12)，如果存在一个连续的标量函数 $V(x)$ 满足 $V(0) = 0$，且对于 x 存在连续的一阶导数，同时存在一个围绕原点的吸引域 Ω，使得对于任意 $x \in \Omega$ 和 $t \geqslant 0$，

(1) $V(x, t)$ 正定且有界；

(2) $\dfrac{\partial}{\partial t}V(x, t)$ 半负定，

那么，$x_{\mathrm{e}} = 0$ 在 Ω 域内一致稳定。

最后，给出不稳定判据。

定理 5.6 对于系统方程 (5.9)，如果存在一个连续的标量函数 $V(x, t)$ 满足 $V(0, t) = 0$，且对于 x 和 t 具有连续一阶偏导，同时存在一个围绕原点的吸引域 Ω，使得对于任意 $x \in \Omega$ 和 $t \geqslant t_0$，

(1) $V(x, t)$ 正定且有界；

(2) $\frac{\partial}{\partial t}V(x,t)$ 也正定且有界,

那么, $x_e = 0$ 是不稳定的。

定理 5.7 对于系统方程 (5.12),如果存在一个连续的标量函数 $V(x)$ 满足 $V(0) = 0$,且对于 x 存在连续的一阶导数,同时存在一个围绕原点的吸引域 Ω,使得对于任意 $x \in \Omega$ 和 $t \geqslant 0$,

(1) $V(x,t)$ 正定;

(2) $\frac{\partial}{\partial t}V(x,t)$ 也正定,

那么, $x_e = 0$ 是不稳定的。

5.2 线性系统的内部稳定性

这一节中主要讨论线性系统的稳定性判据。

5.2.1 线性时变系统的内部稳定性

考虑以下线性定常系统,

$$\dot{x}(t) = A(t)x(t), \ x(0) = x_0, \ t \geqslant t_0 \tag{5.13}$$

其中, $x(t) \in \mathbb{R}^n$ 是状态变量, $A(t)$ 是时变矩阵。假设方程 (5.13) 具有唯一解。$x_e = 0$ 是系统方程 (5.13) 的平衡点。

定理 5.8 对于线性时变系统方程 (5.13),当且仅当

$$\|\Phi(t,t_0)\| \leqslant k(t_0) < \infty$$

平衡点 $x_e = 0$ 是稳定的,其中, $\Phi(t,t_0)$ 是方程 (5.13) 的状态转移矩阵, $k(t_0)$ 是一个常值,可能与 t_0 的选取相关。

证明 首先,证明充分性。

注意到

$$\|x(t;x_0,t_0)\| = \|\Phi(t,t_0)x_0\| \leqslant \|\Phi(t,t_0)\|\|x_0\|$$

对于任意实数 $\varepsilon > 0$,取 $\delta(t_0,\varepsilon) = \varepsilon/k(t_0)$。如果 $\|x_0\| < \delta(t_0,\varepsilon)$,可以得到

$$\|x(t;x_0,t_0)\| \leqslant \delta(t_0,\varepsilon)k(t_0) = \varepsilon, \ \forall t \geqslant t_0$$

这表明 $x_e = 0$ 是稳定的。

接下来,证明必要性。

假设 $x_e = 0$ 是稳定的,但是 $\|\Phi(t,t_0)\|$ 是无界的。这种情况下,对于 $t \geqslant t_0$, $\|\Phi(t,t_0)\|$ 的至少一个元是无界的。不失一般性地,假设对于 $t \geqslant t_0$, $\Phi_{ij}(t,t_0)$ 满足

$$|\Phi_{ij}(t,t_0)| > k > 0$$

其中，$\Phi_{ij}(t,t_0)$ 是 $\Phi(t,t_0)$ 的第 i 行第 j 列的元，$i \in \{1,\cdots,n\}$，$j \in \{1,\cdots,n\}$。考虑 $x_0 = [\ 0\ \cdots\ 0\ \delta\ 0\ \cdots\ 0\ ]^{\mathrm{T}}$，其中，$x_0$ 的第 j 个元是 δ。可以进一步推导出

$$x(t;x_0,t_0) = \Phi_j(t,t_0)\delta$$

其中，$\Phi_j(t,t_0)$ 是 $\Phi(t,t_0)$ 的第 j 列。注意到 $x_\mathrm{e}=0$，选择 $k=\varepsilon/\delta$，可以进一步得到

$$\|x(t;x_0,t_0) - x_\mathrm{e}\| = \|\Phi_j(t,t_0)\delta\| > k\delta = \varepsilon,\ \forall t \geqslant t_0$$

这表明，无论 δ 取多小，总是有 $\|x(t;x_0,t_0) - x_\mathrm{e}\| > \varepsilon$，这与平衡点是稳定的相矛盾。 □

定理 5.9 对于线性时变系统方程 (5.13)，当且仅当

$$\|\Phi(t,t_0)\| \leqslant k < \infty$$

平衡点 $x_\mathrm{e}=0$ 是一致稳定的，其中，$\Phi(t,t_0)$ 是方程 (5.13) 的状态转移矩阵，k 是一个常值，与 t_0 的选取无关。

证明过程与定理 5.8 的证明类似。

定理 5.10 对于线性时变系统方程 (5.13)，以下描述是等价的：
(1) 平衡点 $x_\mathrm{e}=0$ 是渐近稳定的；
(2) 平衡点 $x_\mathrm{e}=0$ 是全局渐近稳定的；
(3) $\lim\limits_{t\to\infty} \|\Phi(t,t_0)\| = 0$，其中，$\Phi(t,t_0)$ 是方程 (5.13) 的状态转移矩阵。

证明 假设描述 (1) 成立，那么存在一个常值 $\eta(t_0)>0$，使得当 $\|x_0\| \leqslant \eta(t_0)$ 时，$\lim\limits_{t\to\infty} x(t;x_0,t_0) = 0$ 成立。事实上，对于任意的 $x_0 \neq 0$，根据线性系统的齐次性，可以得到

$$\lim_{t\to\infty} x(t;x_0,t_0) = \lim_{t\to\infty} x\left(t; \frac{\eta(t_0)}{\|x_0\|}x_0,t_0\right) \bigg/ \left(\frac{\eta(t_0)}{\|x_0\|}\right) = 0$$

这表明描述 (2) 成立。

接下来，假设描述 (2) 成立。根据定义 5.6，可以得到

$$\lim_{t\to\infty} x(t;x_0,t_0) = \lim_{t\to\infty} \Phi(t,t_0)x_0 = 0$$

对于任意 $x_0 \neq 0$ 均成立。所以，$\lim\limits_{t\to\infty} \Phi(t,t_0) = 0$，进一步可知描述 (3) 成立。

最后，假设描述 (3) 成立，那么 $\|\Phi(t,t_0)\|$ 对于任意 $t \geqslant t_0$ 都是有界的。根据定理 5.8，$x_\mathrm{e}=0$ 是稳定的。进一步，对于 $t_0 \geqslant 0$，如果存在常数 $\eta(t_0)>0$ 使得 $\|x_0\| \leqslant \eta(t_0)$，那么 $\|x(t;x_0,t_0)\| \leqslant \|x_0\|\|\Phi(t,t_0)\| \leqslant \eta(t_0)\|\Phi(t,t_0)\|$。这种情况下，$\lim\limits_{t\to\infty}\|x(t;x_0,t_0)\| = 0$。由此可知，描述 (1) 成立。 □

定理 5.11 对于线性时变系统方程 (5.13)，以下描述是等价的：
(1) 平衡点 $x_\mathrm{e}=0$ 是一致渐近稳定的；
(2) 平衡点 $x_\mathrm{e}=0$ 是指数稳定的；
(3) 对于任意 t_0 和 $t \geqslant t_0$，存在 $\mu>0$ 和 $\nu>0$，使得

$$\|\Phi(t,t_0)\| \leqslant \mu \mathrm{e}^{-\nu(t-t_0)} \tag{5.14}$$

成立，其中，$\Phi(t,t_0)$ 是方程 (5.13) 的状态转移矩阵。

证明 假设描述 (1) 成立。这种情况下，对于任意给定的 $\varepsilon>0$，存在 $\delta>0$ 和 $T>0$，使得对于任意满足 $\|x_0\|\leqslant\delta$ 的初始状态 x_0，由其引起的受扰运动在 $t\geqslant t_0+T$ 时总是满足 $\|x(t;x_0,t_0)\|\leqslant\varepsilon$。不失一般性地，假设当 $t=t_0+T$ 时，$\|x_0\|=\delta$ 且 $\varepsilon=\delta/2$，那么可以得到

$$\|x(t;x_0,t_0)\|=\|\Phi(t,t_0)x_0\|=\|\Phi(t_0+T,t_0)x_0\|=\|\Phi(t_0+T,t_0)\|\delta\leqslant\frac{\delta}{2}$$

对任意 t_0 均成立，这表明

$$\|\Phi(t_0+T,t_0)\|\leqslant\frac{1}{2}$$

因为 $x_e=0$ 是一致稳定的，根据定理 5.9，存在常数 k 使得对于 $t\geqslant t_0$，$\|\Phi(t,t_0)\|\leqslant k$。进一步，对于 $\forall t\in[t_0+T,t_0+2T]$，可以得到

$$\|\Phi(t,t_0)\|=\|\Phi(t,t_0+T)\Phi(t_0+T,t_0)\|\leqslant\|\Phi(t,t_0+T)\|\|\Phi(t_0+T,t_0)\|\leqslant\frac{k}{2}$$

在此基础上，对于 $\forall t\in[t_0+2T,t_0+3T]$，可以推得

$$\|\Phi(t,t_0)\|=\|\Phi(t,t_0+2T)\Phi(t_0+2T,t_0+T)\Phi(t_0+T,t_0)\|\leqslant\frac{k}{2^2}$$

如此规律，可知对于 $\forall t\in[t_0+nT,t_0+(n+1)T]$，有

$$\|\Phi(t,t_0)\|\leqslant\frac{k}{2^n}$$

构造函数 $\mu e^{-\nu(t-t_0)}$，使得对于 $t=t_0+nT$，$n=1,2,\cdots$，$\mu e^{-\nu(t-t_0)}=\dfrac{k}{2^{n-1}}$，由此可以得到 $\mu(e^{-\nu T})^n=2k\left(\dfrac{1}{2}\right)^n$。考虑 $\nu=\dfrac{1}{T}\ln 2$ 且 $\mu=2k$，可知式 (5.14) 成立，这表明描述 (3) 成立。

进一步，可以得到

$$\|x(t;x_0,t_0)\|\leqslant\|\Phi(t,t_0)\|\|x_0\|\leqslant\mu\delta e^{-\nu(t-t_0)}$$

这表明 $x_e=0$ 是指数稳定的，即描述 (2) 成立。

对于线性系统，描述 (2) 成立时描述 (1) 总是成立，这里不再详细证明。

最后，考虑描述 (3) 成立。式 (5.14) 表明 $\|\Phi(t,t_0)\|\leqslant\mu$。根据定理 5.9，$x_e=0$ 是一致稳定的。同时，可知

$$\|x(t;x_0,t_0)\|\leqslant\|\Phi(t,t_0)\|\|x_0\|\leqslant\mu e^{-\nu(t-t_0)}\|x_0\|$$

对于任意实数 $\varepsilon>0$ 和 $\delta>0$，总是存在 $T=-\dfrac{1}{\nu}\ln\dfrac{\varepsilon}{\mu\delta}$，它与 t_0 的选取无关。对于 $\|x_0\|\leqslant\delta$，$t\geqslant t_0+T$，有

$$\|x(t;x_0,t_0)\|\leqslant\mu e^{-\nu T}\|x_0\|\leqslant\varepsilon$$

这表明 $x_e=0$ 是一致渐近稳定的。也就是说如果描述 (3) 成立，那么描述 (1) 成立。 □

5.2 线性系统的内部稳定性

例 5.4 考虑以下系统，

$$\dot{x}(t) = -\mathrm{e}^{2t}x(t), \ x(t_0) = x_0$$

它的解为 $x(t;x_0,t_0) = \Phi(t,t_0)x_0$，其中，

$$\Phi(t,t_0) = \mathrm{e}^{\frac{1}{2}(\mathrm{e}^{2t_0} - \mathrm{e}^{2t})}$$

因为 $\lim_{t\to\infty} \Phi(t,t_0) = 0$，所以平衡点 $x_\mathrm{e} = 0$ 是（全局）渐近稳定的。

因为 $\mathrm{e}^{2t} > 2t$，可以得到

$$\|x(t;x_0,t_0)\| = \left|x_0 \mathrm{e}^{\frac{1}{2}\mathrm{e}^{2t_0}} \mathrm{e}^{-\frac{1}{2}\mathrm{e}^{2t}}\right| \leqslant |x_0| \mathrm{e}^{\frac{1}{2}\mathrm{e}^{2t_0}} \mathrm{e}^{-t}, \ t \geqslant t_0 \geqslant 0$$

所以平衡点 $x_\mathrm{e} = 0$ 是全局一致渐近稳定的，也是全局指数稳定的。

接下来，根据 Lyapunov 第二法给出线性时变系统的稳定性判据。

考虑 $Q(t)$ 是一个分段连续矩阵函数，$t \in [t_0, \infty)$，如果它对称、正定，且满足

$$0 < c_1 I \leqslant Q(t) \leqslant c_2 I$$

其中，$c_2 > c_1 > 0$ 是两个正实数，那么 $Q(t)$ 被称作是一致有界且一致正定的。

为了导出稳定性判据，首先介绍以下引理。

引理 5.1 假设线性时变系统方程 (5.13) 的平衡点 $x_\mathrm{e} = 0$ 是一致渐近稳定的，那么对于任意一致有界且一致正定的矩阵函数 $Q(t)$，积分

$$0 \leqslant P(t) = \int_t^\infty \Phi^\mathrm{T}(\tau,t) Q(\tau) \Phi(\tau,t) \mathrm{d}\tau \tag{5.15}$$

对于任意 t 是收敛的，且它是

$$\dot{P}(t) + A^\mathrm{T}(t)P(t) + P(t)A(t) + Q(t) = 0, \ \forall t \geqslant t_0 \tag{5.16}$$

的唯一解，其中，$\Phi(\tau,t)$ 是方程 (5.13) 的状态转移矩阵。

证明 因为平衡点是一致渐近稳定的，根据定理 5.11，存在 $\mu > 0$ 和 $\nu > 0$，使得对于 $\forall \tau \geqslant t$，有

$$\|\Phi(\tau,t)\| \leqslant \mu \mathrm{e}^{-\nu(\tau-t)}$$

结合式 (5.15)，可以得到

$$P(t) \leqslant \int_t^\infty c_2 \mu^2 \mathrm{e}^{-2\nu(\tau-t)} I \mathrm{d}\tau = \frac{c_2 \mu^2}{2\nu} I$$

很容易验证式 (5.15) 是式 (5.16) 的解。接下来，证明这一解是唯一的。假设 $P_1(t)$ 和 $P_2(t)$ 都是式 (5.16) 的解，结合式 (5.15) 和式 (5.16)，可以得到

$$P_2(t) = \int_t^\infty \Phi^\mathrm{T}(\tau,t) Q(\tau) \Phi(\tau,t) \mathrm{d}\tau$$

$$= -\int_t^\infty \Phi^{\mathrm{T}}(\tau,t)\left(\dot{P}_1(t) + A^{\mathrm{T}}(t)P_1(t) + P_1(t)A(t)\right)\Phi(\tau,t)\mathrm{d}\tau$$

$$= -\int_t^\infty \frac{\mathrm{d}}{\mathrm{d}\tau}\left(\Phi^{\mathrm{T}}(\tau,t)P_1(t)\Phi(\tau,t)\right)\mathrm{d}\tau$$

$$= -\left.\Phi^{\mathrm{T}}(\tau,t)P_1(t)\Phi(\tau,t)\right|_t^\infty = P_1(t)$$

由此可知式 (5.16) 具有唯一解。□

定理 5.12 对于线性时变系统方程 (5.13),当且仅当对于任意一致有界且一致正定的 $Q(t)$,矩阵微分方程 (5.16) 都存在唯一的一致有界且一致正定的解 $P(t)$,平衡点 $x_e = 0$ 是一致渐近稳定的。

证明 定理的必要性可以根据引理 5.1 直接证得。接下来,证明充分性。构造 $V(x,t) = x^{\mathrm{T}}P(t)x$。可以验证 $V(x,t)$ 正定且有界。同时

$$\frac{\partial}{\partial t}V(x,t) = x^{\mathrm{T}}\left(\dot{P}(t) + A^{\mathrm{T}}(t)P(t) + P(t)A(t)\right)x = -x^{\mathrm{T}}Q(t)x$$

这表明 $\frac{\partial}{\partial t}V(x,t)$ 负定且有界。由此可以进一步证得 $x_e = 0$ 是一致渐近稳定的。□

5.2.2 线性定常系统的内部稳定性

考虑以下线性定常系统,

$$\dot{x}(t) = Ax(t),\ x(0) = x_0,\ t \geqslant 0 \tag{5.17}$$

其中,$x(t) \in \mathbb{R}^n$ 是状态向量,A 是常值矩阵。$x_e = 0$ 是系统方程 (5.17) 的平衡点。

定理 5.13 对于线性定常系统方程 (5.17),当且仅当 A 的所有特征值具有非正实部,且每个实部为零的特征值是 A 的最小多项式的单根,或者说每个实部为零的特征值对应的 Jordan 块是一阶的,那么 $x_e = 0$ 是稳定的,事实上是一致稳定的。

证明 首先证明 $\|\mathrm{e}^{At}\| \leqslant k < \infty$ 是保证 $x_e = 0$ 稳定的充要条件。
可知

$$x(t;x_0,0) = \mathrm{e}^{At}x_0,\ \forall t \geqslant 0$$

另一方面,平衡点 $x_e = 0$ 满足

$$x_e = \mathrm{e}^{At}x_e,\ \forall t \geqslant 0$$

进一步,可得

$$x(t;x_0,0) - x_e = \mathrm{e}^{At}(x_0 - x_e), \forall t \geqslant 0$$

对于任意给定的 $\varepsilon > 0$,选择 $\delta(\varepsilon) = \varepsilon/k$,那么对于任意满足

$$\|x_0 - x_e\| \leqslant \delta(\varepsilon)$$

5.2 线性系统的内部稳定性

的初始状态，从它出发的受扰运动满足

$$\|x(t;x_0,0) - x_{\mathrm{e}}\| \leqslant \|\mathrm{e}^{At}\|\|x_0 - x_{\mathrm{e}}\| \leqslant \frac{\varepsilon}{k}\|\mathrm{e}^{At}\|$$

如果 $\|\mathrm{e}^{At}\| \leqslant k < \infty$，那么有

$$\|x(t;x_0,0) - x_{\mathrm{e}}\| \leqslant \varepsilon$$

这表明 $x_{\mathrm{e}} = 0$ 是稳定的。

接下来，假设 $x_{\mathrm{e}} = 0$ 稳定，但是 $\|\mathrm{e}^{At}\|$ 无界。这种情况下，e^{At} 至少有一个元是无界的。不失一般性地，假设 $\varphi_{ij}(t)$ 在 $t \geqslant 0$ 时满足

$$|\varphi_{ij}(t)| > k > 0$$

其中，$\varphi_{ij}(t)$ 表示 e^{At} 第 i 行第 j 列的元，$i \in \{1,\cdots,n\}$，$j \in \{1,\cdots,n\}$。选取 $x_0 = [\ 0\ \cdots\ 0\ \delta\ 0\ \cdots\ 0\ ]^{\mathrm{T}}$，其中，$x_0$ 的第 j 个元是 δ。可以得到

$$x(t;x_0,0) = \varphi_j(t)\delta$$

其中，$\varphi_j(t)$ 是 e^{At} 的第 j 列。注意到 $x_{\mathrm{e}} = 0$，选择 $k = \varepsilon/\delta$，可以得到

$$\|x(t;x_0,0) - x_{\mathrm{e}}\| = \|\varphi_j(t)\delta\| > k\delta = \varepsilon,\ \forall t \geqslant 0$$

这表明，无论 δ 取多小，总是有 $\|x(t;x_0,0) - x_{\mathrm{e}}\| > \varepsilon$。这与 $x_{\mathrm{e}} = 0$ 稳定的假设矛盾。因此，当且仅当 $\|\mathrm{e}^{At}\| \leqslant k < \infty$ 时，$x_{\mathrm{e}} = 0$ 是稳定的。

接下来，讨论 $\|\mathrm{e}^{At}\|$ 的有界性和 A 的特征值之间的关系。考虑 P 是一个非奇异矩阵，能够使 $J = PAP^{-1}$ 化为 Jordan 标准形。此时，有 $\mathrm{e}^{Jt} = P\mathrm{e}^{At}P^{-1}$。这表明，

$$\|\mathrm{e}^{Jt}\| \leqslant \|P\|\|\mathrm{e}^{At}\|\|P^{-1}\|,\ \|\mathrm{e}^{At}\| \leqslant \|P^{-1}\|\|\mathrm{e}^{Jt}\|\|P\|$$

因此可知 $\|\mathrm{e}^{At}\|$ 的有界性与 $\|\mathrm{e}^{Jt}\|$ 的有界性等价。

2.3.1 小节中指出，$\|\mathrm{e}^{Jt}\|$ 的元具有 $ct^m\mathrm{e}^{(\alpha+j\beta)t}$ 的形式，$0 \leqslant m \leqslant p-1$，其中 c 是一个常值，$\lambda = \alpha + j\beta$ 是 A 的特征值，p 是 λ 对应的 Jordan 块的阶次。由此可知，当且仅当 $\alpha < 0$ 或 $\alpha = 0$ 且 $p = 1$ 时，e^{Jt} 是有界的。当且仅当 $\|\mathrm{e}^{Jt}\|$ 有界，即当且仅当 $\|\mathrm{e}^{At}\|$ 有界，e^{Jt} 是有界的。因此，当且仅当 A 的所有特征值具有非正实部，且每个实部为零的特征值是 A 的最小多项式的单根，$\|\mathrm{e}^{At}\|$ 是有界的。 □

注意到定理 5.13 是判断线性定常系统稳定性的充要条件。在此基础上，可以总结出当且仅当 A 至少有一个特征值具有正实部，或至少有一个实部为零的特征值是 A 的最小多项式的重根，那么 $x_{\mathrm{e}} = 0$ 是不稳定的。

定理 5.14 对于线性定常系统方程 (5.17)，当且仅当 A 的所有特征值都具有负实部时，$x_{\mathrm{e}} = 0$ 是渐近稳定的。

证明 根据定理 5.13 的证明,可知
$$\lim_{t\to\infty} x(t;x_0,0) = \lim_{t\to\infty} e^{At} = 0$$
成立,当且仅当 $\lim_{t\to\infty} e^{Jt} = 0$,它等价于 $\lim_{t\to\infty} ct^m e^{(\alpha+j\beta)t} = 0$。当且仅当 $\alpha < 0$,即 A 的所有特征值都具有负实部时,上式成立。□

接下来,基于 Lyapunov 第二法,给出线性定常系统的稳定性判据。

定理 5.15 对于线性定常系统方程 (5.17),当且仅当对于任意对称且正定的矩阵 Q,总是存在对称且正定的矩阵 P 满足
$$A^{\mathrm{T}}P + PA = -Q \tag{5.18}$$
且这一方程的解 P 是唯一的,那么 $x_e = 0$ 是渐近稳定的。

式 (5.18) 被称作 Lyapunov 矩阵方程。

证明 首先,证明充分性。

选取 $V(x) = x^{\mathrm{T}}Px$。因为 $P > 0$,所以 $V(x) = x^{\mathrm{T}}Px \geqslant 0$,且 $V(x) = 0$ 仅当 $x = 0$ 时成立。同时可得
$$\dot{V}(x) = \dot{x}^{\mathrm{T}}Px + x^{\mathrm{T}}P\dot{x} = x^{\mathrm{T}}A^{\mathrm{T}}Px + x^{\mathrm{T}}PAx$$
$$= x^{\mathrm{T}}\left(A^{\mathrm{T}}P + PA\right)x = -x^{\mathrm{T}}Qx < 0$$

所以,平衡点 $x_e = 0$ 是渐近稳定的。

接下来,证明必要性。

考虑矩阵方程
$$\dot{X}(t) = A^{\mathrm{T}}X(t) + X(t)A, \ X(0) = Q \tag{5.19}$$

可以得到矩阵方程的解为
$$X(t) = e^{A^{\mathrm{T}}t}Qe^{At} \tag{5.20}$$

对式 (5.19) 进行积分,可以得到
$$X(\infty) - X(0) = A^{\mathrm{T}}\left(\int_0^\infty X(t)\mathrm{d}t\right) + \left(\int_0^\infty X(t)\mathrm{d}t\right)A$$

因为 $x_e = 0$ 是渐近稳定的,所以 $\lim_{t\to\infty} e^{At} = 0$。结合式 (5.20),可以得到 $X(\infty) = 0$。在此基础上,有
$$A^{\mathrm{T}}\left(\int_0^\infty e^{A^{\mathrm{T}}t}Qe^{At}\mathrm{d}t\right) + \left(\int_0^\infty e^{A^{\mathrm{T}}t}Qe^{At}\mathrm{d}t\right)A = -Q$$

这表明
$$P = \int_0^\infty e^{A^{\mathrm{T}}t}Qe^{At}\mathrm{d}t$$

是方程 (5.18) 的解。因为方程 (5.19) 的解 $X(t)$ 是唯一的，且 $X(\infty) = 0$，所以 $P = \int_0^\infty X(t)\mathrm{d}t$ 是唯一的。

进一步可知，$P = P^\mathrm{T}$，$x^\mathrm{T}Px = \int_0^\infty (\mathrm{e}^{At}x)^\mathrm{T}Q(\mathrm{e}^{At}x)\mathrm{d}t \geqslant 0$，且仅当 $x=0$ 时等号成立。所以，P 对称且正定。$\square$

例 5.5 考虑线性定常系统，

$$\dot{x} = Ax = \begin{bmatrix} a_{11} & a_{12} \\ a_{21} & a_{22} \end{bmatrix} x$$

确定这一系统渐近稳定时其相应参数应满足的条件。

对于方程 (5.18)，考虑 $Q = I$，记

$$P = \begin{bmatrix} p_{11} & p_{12} \\ p_{21} & p_{22} \end{bmatrix}$$

那么，可以得到

$$\begin{bmatrix} 2a_{11} & 2a_{21} & 0 \\ a_{12} & a_{11}+a_{22} & a_{21} \\ 0 & 2a_{12} & 2a_{22} \end{bmatrix} \begin{bmatrix} p_{11} \\ p_{12} \\ p_{22} \end{bmatrix} \triangleq A_1 \begin{bmatrix} p_{11} \\ p_{12} \\ p_{22} \end{bmatrix} = \begin{bmatrix} -1 \\ 0 \\ -1 \end{bmatrix}$$

上述线性方程组中，系数矩阵 A_1 的行列式为

$$\det A_1 = 4(a_{11}+a_{22})(a_{11}a_{22} - a_{12}a_{21})$$

如果 $\det A_1 \neq 0$，那么上述方程组会有以下形式的唯一解，

$$P = \frac{-2}{\det A_1} \begin{bmatrix} \det A + a_{21}^2 + a_{22}^2 & -(a_{12}a_{22}+a_{21}a_{11}) \\ -(a_{12}a_{22}+a_{21}a_{11}) & \det A + a_{21}^2 + a_{22}^2 \end{bmatrix}$$

因为 P 是正定的，所以有

$$p_{11} = \frac{\det A + a_{21}^2 + a_{22}^2}{-2(a_{11}+a_{22})(a_{11}a_{22}-a_{12}a_{21})} > 0, \quad \det P = \frac{(a_{11}+a_{22})^2 + (a_{12}-a_{21})^2}{4(a_{11}+a_{22})^2 \det A} > 0$$

为了保证系统是渐近稳定的，系统的参数应该满足

$$\det A = a_{11}a_{22} - a_{12}a_{21} > 0, \quad a_{11}+a_{22} < 0$$

为了简化计算过程，在使用方程 (5.18) 时，Q 通常被选为 I。

推论 5.1 对于任意对称且正定的矩阵 Q，当且仅当 A 的所有特征值都具有负实部时，矩阵方程 (5.19) 具有唯一解对称且正定的解 P。

这一推论可以根据定理 5.14 和定理 5.15 导出，证明略。

推论 5.2 对于任意对称且正定的矩阵 Q 和任意实数 $\sigma \geqslant 0$，当且仅当

$$\mathrm{Re}\lambda_i < -\sigma, \ i = 1, \cdots, n$$

矩阵方程

$$2\sigma P + A^{\mathrm{T}} P + PA = -Q \tag{5.21}$$

具有唯一解对称且正定的解 P，其中 λ_i 是 A 的特征值。

证明 式 (5.21) 可以被写作

$$(A + \sigma I)^{\mathrm{T}} P + P(A + \sigma I) = -Q$$

根据推论 5.1，式 (5.21) 具有唯一正定且对称解的充要条件是 $A + \sigma I$ 的所有特征值都具有负实部。可以验证，当且仅当 $\lambda_i + \sigma$ 是 $A + \sigma I$ 的特征值时，λ_i 是 $A + \sigma I$ 的特征值。$\mathrm{Re}(\lambda_i + \sigma) < 0$ 等价于 $\mathrm{Re}\lambda_i < -\sigma$。 $\square$

5.3 线性系统的外部稳定性

前两节中，主要讨论了平衡点的稳定性，或者说，系统自由运动的稳定性。一般地，平衡点的稳定性与外部输入无关，也被称作内部稳定性。本节则主要考虑外部稳定性，外部稳定性要求每个有界的输入引起的输出都应该是有界的，即有界输入有界输出（bounded-input, bounded-output，BIBO）稳定。在此基础上，进一步研究在有界输入下状态是否是有界的，即有界输入有界状态（bounded-input, bounded-state，BIBS）稳定。

5.3.1 线性系统的 BIBO 稳定性

考虑一个因果的、在 t_0 时刻松弛的线性系统，它的输入 $u(t) \in \mathbb{R}^p$ 和输出 $y(t) \in \mathbb{R}^q$ 之间满足

$$y(t) = \int_{t_0}^{t} G(t, \tau) u(\tau) \mathrm{d}\tau, \ t \geqslant t_0 \tag{5.22}$$

其中，$G(t, \tau)$ 是系统的脉冲响应矩阵。

如果系统是定常的，可以考虑 $t_0 = 0$，相应的脉冲响应矩阵可以被记作 $G(t - \tau)$。此时输入 $u(t)$ 和输出 $y(t)$ 之间满足

$$y(t) = \int_{0}^{t} G(t - \tau) u(\tau) \mathrm{d}\tau, \ t \geqslant 0 \tag{5.23}$$

定义 5.9（BIBO 稳定） 对于在 t_0 时刻松弛的线性系统，在任意有界输入 $u(t)$ 作用下，或者说，在任意满足

$$\|u(t)\| \leqslant \bar{k} < \infty, \ \forall t \geqslant t_0$$

的输入 $u(t)$ 作用下，相应的输出 $y(t)$ 满足

$$\|y(t)\| \leqslant c(t_0, u) < \infty, \ \forall t \geqslant t_0$$

其中，$\bar{k}$ 和 $c(t_0, u)$ 是两个常值，那么系统被称为是有界输入有界输出（BIBO）稳定的。进一步，如果 $c(t_0, u)$ 与 t_0 的选取无关，此时它可以被记作 $c(u)$，那么系统被称为是一致 BIBO 稳定的。

需要注意的是，定义 5.9 中限制了系统具有零初始条件，因为仅在这种情况下，系统的输入-输出描述才有意义。

首先，给出线性时变系统方程 (5.22) 的稳定性判据。

定理 5.16 当且仅当存在一个常值 k 使得

$$\int_{t_0}^{t} |g_{ij}(t, \tau)| \mathrm{d}\tau \leqslant k < \infty, \ i = 1, 2, \cdots, q, \ j = 1, 2, \cdots, p \tag{5.24}$$

线性时变系统方程 (5.22) 是 BIBO 稳定的，其中 $g_{ij}(t, \tau)$ 是 $G(t, \tau)$ 第 i 行第 j 列的元。

证明 首先，证明充分性。

同 $y_i(t)$ 表示第 i 个输出，$u_j(t)$ 表示第 j 个输入。这里，考虑 $|u_j(t)| \leqslant k_j < \infty$，那么

$$|y_i(t)| = \left| \int_{t_0}^{t} \sum_{j=1}^{p} g_{ij}(t, \tau) u_j(\tau) \mathrm{d}\tau \right| \leqslant \sum_{j=1}^{p} \int_{t_0}^{t} |g_{ij}(t, \tau)| |u_j(\tau)| \mathrm{d}\tau$$

$$\leqslant \sum_{j=1}^{p} k_j \int_{t_0}^{t} |g_{ij}(t, \tau)| \mathrm{d}\tau \leqslant p k_j k$$

这表明系统的输出是有界的。所以，系统是 BIBO 稳定的。

接下来，证明必要性。

使用反证法，假设式 (5.22) 是 BIBO 稳定的，但存在 $g_{ij}(t, \tau)$ 和 t_1 使得 $\int_{0}^{t_1} |g_{ij}(t, \tau)| \mathrm{d}\tau$ 是无界的。

选择有界输入 $u_j(\tau) = \mathrm{sign}[g_{ij}(t, \tau)]$，其中，$\mathrm{sign}[g_{ij}(t, \tau)]$ 是 $g_{ij}(t, \tau)$ 的符号，可以得到

$$y_i(t_1) = \int_{t_0}^{t_1} \sum_{j=1}^{p} g_{ij}(t, \tau) u_j(\tau) \mathrm{d}\tau = \sum_{j=1}^{p} \int_{t_0}^{t_1} |g_{ij}(t, \tau)| \mathrm{d}\tau$$

这一输出是无界的，这和系统 BIBO 稳定的假设矛盾。 □

接下来，考虑线性定常系统。

定理 5.17 当且仅当存在一个常值 k 使得

$$\int_{0}^{\infty} |g_{ij}(t - \tau)| \mathrm{d}\tau \leqslant k < \infty, \ i = 1, 2, \cdots, q, \ j = 1, 2, \cdots, p \tag{5.25}$$

线性定常系统方程 (5.23) 是 BIBO 稳定的，其中，$g_{ij}(t - \tau)$ 是 $G(t - \tau)$ 第 i 行第 j 列的元。

证明与定理 5.16 的证明类似，这里不再给出。

定理 5.18 当且仅当线性定常系统方程 (5.23) 的传递函数矩阵 $G(s)$ 是真有理分式矩阵，且 $G(s)$ 所有元的每个极点都具有负实部时，这一系统是 BIBO 稳定的。

证明 考虑 $g_{ij}(s)$ 是 $G(s)$ 第 i 行第 j 列的元。当 $g_{ij}(s)$ 是真有理分式时，它可以展开为有限项之和的形式，且每一项可以表示为 $\beta_l/(s-\lambda_l)^{\alpha_l}$，其中，$\lambda_l$ 是 $g_{ij}(s)$ 的极点，α_l 和 β_l 是常值。因为 λ_l 具有负实部，所以 $g_{ij}(s)$ 的 Laplace 反变换是 $t^{\alpha_l-1}e^{\lambda_l t}$ 与 $\delta(t)$ 的和的形式。因此，式 (5.25) 成立，根据定理 5.17，可以证明这一结论。 □

5.3.2 线性系统的 BIBS 稳定性

考虑以下线性时变系统，

$$\dot{x}(t) = A(t)x(t) + B(t)u(t), \ y(t) = C(t)x(t), \ x(t_0) = x_0, \ t \in [t_0, t_f] \tag{5.26}$$

其中，$x(t) \in \mathbb{R}^n$ 是状态向量，$u(t) \in \mathbb{R}^p$ 是输入向量，$y(t) \in \mathbb{R}^q$ 是输出向量，$A(t)$，$B(t)$ 和 $C(t)$ 是时变矩阵。式 (5.26) 省略了系统中直接传递的部分，因为它对系统的稳定性没有影响。

接下来，讨论有界的输入是否能保证状态有界。

定义 5.10（BIBS 稳定） 对于线性系统方程 (5.26)，在任意有界输入 $u(t)$ 和任意非零初始状态 x_0 的作用下，如果相应的状态响应 $x(t)$ 满足

$$\|x(t)\| \leqslant c(t_0, x_0, u) < \infty \tag{5.27}$$

其中，$c(t_0, x_0, u)$ 是一个常值，那么系统被称为是有界输入有界状态（BIBS）稳定的。进一步，如果 $c(t_0, x_0, u)$ 与 t_0 的选取无关，此时它可以被记作 $c(x_0, u)$，那么系统被称为是一致 BIBS 稳定的。

定理 5.19 当且仅当存在两个常值 $k_1(t_0)$ 和 $k_2(t_0)$ 使得

$$\|\Phi(t, t_0)\| \leqslant k_1(t_0) \tag{5.28}$$

和

$$\int_{t_0}^{t} \|\Phi(t, \tau)B(\tau)\| \mathrm{d}\tau \leqslant k_2(t_0) \tag{5.29}$$

那么线性时变系统方程 (5.26) 是 BIBS 稳定的，其中，$\Phi(t, t_0)$ 是系统方程 (5.26) 的状态转移矩阵。

证明 首先，证明充分性。

假设 $\|u(t)\| \leqslant \bar{k} < \infty$，可以得到

$$\|x(\tau)\| = \left\| \Phi(t, t_0)x_0 + \int_{t_0}^{t} \Phi(t, \tau)B(\tau)u(\tau)\mathrm{d}\tau \right\|$$

$$\leqslant \|\Phi(t, \tau_0)\| \|x_0\| + \int_{t_0}^{t} \|\Phi(t, \tau)B(\tau)\| \|u(\tau)\| \mathrm{d}\tau$$

$$\leqslant k_1(t_0)\|x_0\| + k_2(t_0)\bar{k}$$

这表明式 (5.27) 成立，所以系统方程 (5.26) 是 BIBS 稳定的。

接下来，证明必要性。

使用反证法。假设系统方程 (5.26) 是 BIBS 稳定的，但是存在初始状态 x_0 或有界输入 $u(t)$ 使得 $\|\Phi(t,t_0)\|$ 或 $\int_{t_0}^{t}\|\Phi(t,\tau)B(\tau)\|\mathrm{d}\tau$ 是无界的。此时，$\|x(t)\|$ 一定是无界的，这与假设矛盾。 □

事实上，定理 5.19 中，如果 $k_1(t_0)$ 和 $k_2(t_0)$ 都与 t_0 的选取无关，那么，式 (5.28) 和式 (5.29) 是系统方程 (5.26) 一致 BIBS 稳定的充要条件。定理 5.19 对线性定常系统也成立。特别地，对于线性定常系统 (A,B,C)，下述结论成立。

定理 5.20 当且仅当 $(sI-A)^{-1}B$ 所有元的每个极点都具有负实部时，线性定常系统 (A,B,C) 是一致 BIBS 稳定的。

证明略。

5.4 外部稳定和内部稳定的关系

如果 (A,B,C) 是 BIBS 稳定的，那么显然，它也是 BIBO 稳定的，但是反之不成立。在 3.4.3 节中指出，如果 (A,B,C) 不完全能观，那么它可以被分解为一个能观的子系统和一个不能观的子系统。事实上，输出 $y(t)$ 不能反映不能观子系统的状态信息。由此，可以得到以下结论。

定理 5.21 如果线性定常系统 (A,B,C) 是完全能观的，那么它的 BIBS 稳定性等价于它的 BIBO 稳定性。

接下来，讨论线性定常系统 (A,B,C) 的外部稳定性与内部稳定性之间的关系。

定理 5.22 如果线性定常系统 (A,B,C) 是渐近稳定的，那么它是 BIBO 稳定的，也是 BIBS 稳定的。

证明 系统 (A,B,C) 的脉冲响应矩阵为

$$G(t-\tau) = C\mathrm{e}^{A(t-\tau)}B$$

当系统是渐近稳定的，5.2.1 节中指出，

$$\lim_{t\to\infty}\mathrm{e}^{At} = 0$$

在此基础上，不难推导出式 (5.24) 成立。所以，系统是 BIBO 稳定的。

另一方面，对于线性定常系统，它的状态转移矩阵是 e^{At}。因此，不难验证定理 5.19 成立，所以系统也是 BIBS 稳定的。 □

值得一提的是，线性定常系统的 BIBS 稳定性不能保证它的渐近稳定性。3.4.2 小节中指出，如果一个线性系统不完全能控，它可以被分解为一个能控的子系统和一个不能控的子系统。输入 $u(t)$ 仅能影响能控部分的特征值，不能影响不能控部分的特征值。基于此，可以得到以下结论。

定理 5.23 如果线性定常系统 (A, B, C) 是完全能控的，那么它的 BIBS 稳定性等价于它的渐近稳定性。

显然，线性定常系统的 BIBO 稳定性也不能保证渐近稳定性。3.4.4 小节中指出，一个线性定常系统可以被分解为能控且能观的部分，能控但不能观的部分，不能控但能观的部分，不能控也不能观的部分。系统的输入-输出特性仅能反映能控且能观的子系统的性质。所以，一个线性定常系统是 BIBO 稳定的，只能说明它能控且能观的子系统是渐近稳定的。由此，可以得到以下结论。

定理 5.24 如果线性定常系统 (A, B, C) 是完全能控且完全能观的，那么它的 BIBO 稳定性等价于它的渐近稳定性。

线性定常系统的外部稳定性与内部稳定性的关系如图 5.5 所示。

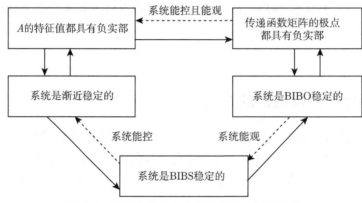

图 5.5　外部稳定性与内部稳定性之间的关系

例 5.6 考虑以下系统，

$$\dot{x}(t) = \begin{bmatrix} 1 & 0 \\ 1 & -1 \end{bmatrix} x(t) + \begin{bmatrix} 0 \\ 1 \end{bmatrix} u(t), \quad y(t) = \begin{bmatrix} 1 & 1 \end{bmatrix} x(t)$$

状态系数矩阵的特征值为 1 和 -1，系统不是渐近稳定的。

同时，可以计算出

$$(sI - A)^{-1}b = \begin{bmatrix} s-1 & 0 \\ -1 & s+1 \end{bmatrix}^{-1} \begin{bmatrix} 0 \\ 1 \end{bmatrix} = \begin{bmatrix} 0 \\ \dfrac{1}{s+1} \end{bmatrix}$$

$(sI - A)^{-1}b$ 的每个元的极点为 -1，所以系统是 BIBS 稳定的。

进一步，可以得到

$$c(sI - A)^{-1}b = \begin{bmatrix} 1 & 1 \end{bmatrix} \begin{bmatrix} 0 \\ \dfrac{1}{s+1} \end{bmatrix} = \dfrac{1}{s+1}$$

它的极点为 -1，所以系统是 BIBO 稳定的。

5.4 外部稳定和内部稳定的关系

可以进一步计算得到

$$U = \begin{bmatrix} b & Ab \end{bmatrix} = \begin{bmatrix} 0 & 0 \\ 1 & -1 \end{bmatrix}, V = \begin{bmatrix} c \\ cA \end{bmatrix} = \begin{bmatrix} 1 & 1 \\ 2 & -1 \end{bmatrix}$$

因为 $\operatorname{rank} U = 1$，$\operatorname{rank} V = 2$。这一系统完全能观但不完全能控。可以看出这一系统的外部稳定性与内部稳定性满足图 5.5 中的关系。

第 6 章

线性系统的时域综合与反馈控制

系统的分析与综合是研究线性系统的两个重要任务,两者密切相关。分析的目的是研究系统的定性行为(如能控性、能观性、稳定性等)和定量变化规律(如动态响应)。综合的目的则是设计闭环系统,使之具有期望的性能。本章重点讨论线性定常系统的综合问题。

6.1 状态反馈控制

反馈在实际系统中是非常常见的。动态过程的模型本身以及与环境交互过程中都具有不确定性,反馈在其自动控制过程中是必不可少的。经典控制理论中,常选用系统的输出作为反馈信号来镇定系统或改善系统的性能。当使用状态空间模型来描述系统时,状态变量包含了系统的所有信息,状态响应可以揭示系统的所有运动行为,而输出信号只能反映系统能够直接被观测到的部分运动特性。因此,选择状态变量作为反馈信号可以更有效地提高系统的性能。本节将详细讨论状态反馈。

6.1.1 状态反馈系统的能控性和能观性

考虑线性定常系统

$$\dot{x}(t) = Ax(t) + Bu(t), \quad y(t) = Cx(t) + Du(t) \tag{6.1}$$

其中,$x(t) \in \mathbb{R}^n$ 是状态向量,$u(t) \in \mathbb{R}^p$ 是输入向量,$y(t) \in \mathbb{R}^q$ 是输出向量,A,B,C 和 D 是具有相应维数的实常数矩阵。

引入状态反馈控制律

$$u(t) = Kx(t) + v(t) \tag{6.2}$$

其中,$v(t) \in \mathbb{R}^p$ 是参考输入向量,K 是常数矩阵,通常被称作反馈增益。把式 (6.2) 代入式 (6.1) 中,可以得到相应的闭环系统,

$$\dot{x}(t) = (A + BK)x(t) + Bv(t), \quad y(t) = (C + DK)x(t) + Dv(t) \tag{6.3}$$

它的框图如图 6.1 所示。

6.1 状态反馈控制

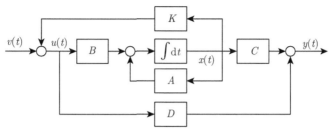

图 6.1 状态反馈闭环系统的框图

定理 6.1 对于线性定常系统方程 (6.1)，引入状态反馈控制律 (6.2) 不会改变它的能控性，但是可能改变它的能观性。

证明 系统方程 (6.1) 的能控性矩阵为 $U = [\ B\ \ AB\ \ \cdots\ \ A^{n-1}B\ ]$，相应的闭环系统方程 (6.3) 的能控性矩阵为 $U_k = [\ B\ \ (A+BK)B\ \ \cdots\ \ (A+BK)^{n-1}B\ ]$。这里只需要证明 $\operatorname{rank} U = \operatorname{rank} U_k$，即可说明两者的能控性相同。

记 B 的第 i 列为 b_i，b_1 的第 j 个元素为 b_{j1}，K 第 i 行第 j 列的元素为 k_{ij}，$i = 1, \cdots, p$，$j = 1, \cdots, n$。由此，$(A+BK)B$ 的第一列可以表示为

$$(A+BK)b_1 = Ab_1 + \begin{bmatrix} b_1 & b_2 & \cdots & b_p \end{bmatrix} \begin{bmatrix} k_{11} & k_{12} & \cdots & k_{1n} \\ k_{21} & k_{22} & \cdots & k_{2n} \\ \vdots & \vdots & & \vdots \\ k_{p1} & k_{p2} & \cdots & k_{pn} \end{bmatrix} \begin{bmatrix} b_{11} \\ b_{21} \\ \vdots \\ b_{n1} \end{bmatrix}$$

$$= Ab_1 + \begin{bmatrix} b_1 & b_2 & \cdots & b_p \end{bmatrix} \begin{bmatrix} c_1 \\ c_2 \\ \vdots \\ c_p \end{bmatrix} = Ab_1 + c_1 b_1 + c_2 b_2 + \cdots + c_p b_p$$

其中，$c_l = \sum_{j=1}^{n} k_{lj} b_{j1}$，$l = 1, \cdots, p$。这表明 $(A+BK)b_1$ 是 $Ab_1, b_1, \cdots, b_p$ 的线性组合。类似地，可以推出 $(A+BK)B$ 的第 i 列 $(A+BK)b_i$ 能够表示为 $Ab_i, b_1, \cdots, b_p$ 的线性组合。所以，$(A+BK)B$ 的每一列都可以表示为 $[\ B\ \ AB\ ]$ 的列的线性组合。

类似地，可以推导出 $(A+BK)^m B$ 的每一列都可以表示为 $[\ B\ \ AB\ \ \cdots\ \ A^m B\ ]$ 的列的线性组合，$m = 1, \cdots, n-1$。也就是说，U_k 的每一列都可以表示为 U 的列的线性组合，因此，

$$\operatorname{rank} U \geqslant \operatorname{rank} U_k$$

另一方面，注意到

$$\dot{x}(t) = Ax(t) + Bu(t) = [(A+BK) - BK]x(t) + Bu(t)$$

即原系统可以被看作系统方程 (6.3) 中引入状态反馈后得到的闭环系统，且反馈增益为 $-K$。此时，可以得到

$$\operatorname{rank} U \leqslant \operatorname{rank} U_{\mathrm{k}}$$

所以，$\operatorname{rank} U = \operatorname{rank} U_{\mathrm{k}}$，即引入状态反馈后不会改变原系统的能控性。 □

引入状态反馈后系统的能观性可能发生变化，下面通过一个例子来说明这一结论。

例 6.1 考虑如下动态系统，

$$\dot{x}(t) = \begin{bmatrix} 3 & 4 \\ 4 & 6 \end{bmatrix} x(t) + \begin{bmatrix} 0 \\ 1 \end{bmatrix} u(t),\ y(t) = \begin{bmatrix} 3 & 4 \end{bmatrix} x(t)$$

可以验证这一系统完全能控能观，引入状态反馈

$$u(t) = \begin{bmatrix} -4 & -6 \end{bmatrix} x(t) + v(t)$$

相应的闭环系统为

$$\dot{x}(t) = \begin{bmatrix} 3 & 4 \\ 0 & 0 \end{bmatrix} x(t) + \begin{bmatrix} 0 \\ 1 \end{bmatrix} v(t),\ y(t) = \begin{bmatrix} 3 & 4 \end{bmatrix} x(t)$$

显然，这一系统仍然能控，但是不能观。

值得注意的是在引入状态反馈后，系统的能观性也可能不发生改变。同时，能控性不变意味着如果原系统 (A, B) 完全能控，那么相应的闭环系统 $(A + BK, B)$ 也完全能控。如果原系统 (A, B) 不完全能控，则相应的闭环系统 $(A + BK, B)$ 也不完全能控。

6.1.2 SISO 系统的状态反馈特征值配置

上一小节中指出，在原系统中引入状态反馈，可以把系统的状态系数矩阵改变为 $A + BK$。这一小节讨论如何设计反馈增益 K 能够把闭环系统的特征值配置在期望的位置。特征值配置问题可以表述为给定一个线性定常系统 (A, B)，设计合适的 K 把 $A + BK$ 的 n 个特征值配置为任意的实数或共轭复数。这一问题也常被称作极点配置问题。

在这一小节，首先讨论 SISO 系统的特征值配置问题。考虑线性定常 SISO 系统

$$\dot{x}(t) = Ax(t) + bu(t),\ y(t) = cx(t) + du(t) \tag{6.4}$$

其中 $x(t) \in \mathbb{R}^n$ 是状态向量，$u(t) \in \mathbb{R}^1$ 是输入量，$y(t) \in \mathbb{R}^1$ 是输出量，A, b, c 和 d 是具有相应维数的实常数矩阵。

引入状态反馈控制律 (6.2)，相应的闭环系统是

$$\dot{x}(t) = (A + bK)x(t) + bv(t),\ y(t) = (c + dK)x(t) + dv(t) \tag{6.5}$$

定理 6.2 对于线性定常 SISO 系统方程 (6.4)，当且仅当它完全能控时，它的所有特征值可以通过状态反馈控制律 (6.2) 任意配置。

6.1 状态反馈控制

证明

首先，证明必要性。

使用反证法，假设系统方程 (6.4) 的所有极点可以任意配置，但是 (A,b) 不完全能控，对其进行结构分解，可以得到

$$\bar{A} = TAT^{-1} = \begin{bmatrix} \bar{A}_{11} & \bar{A}_{12} \\ \bar{A}_{21} & \bar{A}_{22} \end{bmatrix}, \quad \bar{b} = Tb = \begin{bmatrix} \bar{b}_1 \\ 0 \end{bmatrix}$$

其中 T 是非奇异矩阵。

引入状态反馈控制律 (6.2) 且反馈增益为 $K = \begin{bmatrix} K_1 & K_2 \end{bmatrix}$，考虑 $\bar{K} = KT^{-1} = \begin{bmatrix} \bar{K}_1 & \bar{K}_2 \end{bmatrix}$，可以推导出

$$\begin{aligned}\det(sI - A - bK) &= \det(sI - \bar{A} - \bar{b}\bar{K}) \\ &= \det \begin{bmatrix} sI - \bar{A}_{11} - \bar{b}_1\bar{K}_1 & -\bar{A}_{12} - \bar{b}_1\bar{K}_2 \\ 0 & sI - \bar{A}_{22} \end{bmatrix} \\ &= \det(sI - \bar{A}_{11} - \bar{b}_1\bar{K}_1)\det(sI - \bar{A}_{22})\end{aligned}$$

这表明不能控部分的特征值不能被改变，与假设矛盾，必要性得证。

接下来，证明充分性。

由于 (A,b) 完全能控，那么必然存在一个变换矩阵 T 可以把 (A,b) 转化为能控标准形 $(\bar{A}, \bar{b})$，

$$\bar{A} = TAT^{-1} = \begin{bmatrix} 0 & 1 & 0 & \cdots & 0 \\ 0 & 0 & 1 & \cdots & 0 \\ \vdots & \vdots & \vdots & & \vdots \\ 0 & 0 & 0 & \cdots & 1 \\ -a_n & -a_{n-1} & -a_{n-2} & \cdots & -a_1 \end{bmatrix}, \quad \bar{b} = Tb = \begin{bmatrix} 0 \\ 0 \\ \vdots \\ 0 \\ 1 \end{bmatrix}$$

对 $(\bar{A}, \bar{b})$ 引入状态反馈，选择状态反馈矩阵为

$$\bar{K} = KT^{-1} = \begin{bmatrix} a_n - \bar{a}_n & a_{n-1} - \bar{a}_{n-1} & \cdots & a_1 - \bar{a}_1 \end{bmatrix}$$

其中，$\bar{a}_1, \cdots, \bar{a}_n$ 是可以任意选取的系数。这种情况下，相应的闭环系统为

$$\bar{A} + \bar{b}\bar{K} = \begin{bmatrix} 0 & 1 & 0 & \cdots & 0 \\ 0 & 0 & 1 & \cdots & 0 \\ \vdots & \vdots & \vdots & & \vdots \\ 0 & 0 & 0 & \cdots & 1 \\ -\bar{a}_n & -\bar{a}_{n-1} & -\bar{a}_{n-2} & \cdots & -\bar{a}_1 \end{bmatrix}, \quad \bar{b} = \begin{bmatrix} 0 \\ \vdots \\ 0 \\ 1 \end{bmatrix}$$

上述闭环系统的特征方程为

$$\alpha_c(s) = s^n + \bar{a}_1 s^{n-1} + \cdots + \bar{a}_n = 0$$

由于系数 $\bar{a}_1, \cdots, \bar{a}_n$ 可以任意选取，所以闭环系统的特征值可以任意配置。 □
下边给出 SISO 系统的特征值配置算法。

算法 6.1 考虑完全能控的线性定常 SISO 系统方程 (6.4)，引入状态反馈控制律 (6.2)，$A + bK$ 的特征值可以被配置在 $\lambda_1^*, \cdots, \lambda_n^*$。状态反馈增益 K 可以根据下列步骤计算。

(1) 计算 A 的特征多项式，

$$\alpha(s) = \det(sI - A) = s^n + a_1 s^{n-1} + \cdots + a_n$$

(2) 根据期望的特征值计算期望的特征多项式，

$$\alpha_c(s) = (s - \lambda_1^*)(s - \lambda_2^*) \cdots (s - \lambda_n^*) = s^n + \bar{a}_1 s^{n-1} + \cdots + \bar{a}_n$$

(3) 构造变换矩阵 T，

$$T = \left\{ \begin{bmatrix} b & Ab & \cdots & A^{n-1}b \end{bmatrix} \begin{bmatrix} a_{n-1} & a_{n-2} & \cdots & a_1 & 1 \\ a_{n-2} & \cdots & a_1 & 1 & \\ \vdots & \ddots & \ddots & & \\ a_1 & 1 & & & \\ 1 & & & & \end{bmatrix} \right\}^{-1}$$

(4) 计算系统的原状态反馈增益 K，

$$K = \bar{K}T = \begin{bmatrix} a_n - \bar{a}_n & a_{n-1} - \bar{a}_{n-1} & \cdots & a_1 - \bar{a}_1 \end{bmatrix} T$$

基于定理 6.2 的证明可以归纳出这一算法。接下来，给出另一种 SISO 系统的极点配置算法。

算法 6.2 问题与算法 6.1 相同。状态反馈增益 $K = \begin{bmatrix} k_1 & k_2 & \cdots & k_n \end{bmatrix}$ 可以通过下述步骤得到。

(1) 计算特征多项式，

$$\alpha_k(s) = \det(sI - (A + bK)) = s^n + \tilde{a}_1 s^{n-1} + \cdots + \tilde{a}_n$$

其中 $\tilde{a}_1, \cdots, \tilde{a}_n$ 是和 $k_1, \cdots, k_n$ 相关的系数。

(2) 根据期望的特征值计算期望的特征多项式，

$$\alpha_c(s) = (s - \lambda_1^*)(s - \lambda_2^*) \cdots (s - \lambda_n^*) = s^n + \bar{a}_1 s^{n-1} + \cdots + \bar{a}_n$$

(3) 令 $\tilde{a}_i = \bar{a}_i$, $i = 1, \cdots, n$，可以得到 n 个方程。
(4) 联立求解上述方程，可以解得 $K = \begin{bmatrix} k_1 & k_2 & \cdots & k_n \end{bmatrix}$。

6.1 状态反馈控制

例 6.2 考虑如下 SISO 系统,

$$\dot{x}(t) = \begin{bmatrix} 1 & 0 & -1 \\ 1 & 2 & 1 \\ 2 & 2 & 3 \end{bmatrix} x(t) + \begin{bmatrix} 1 \\ 0 \\ 1 \end{bmatrix} u(t), \ y(t) = \begin{bmatrix} 1 & 0 & 0 \end{bmatrix} x(t)$$

设计状态反馈使得闭环系统的特征值为 -1,$-1\pm 2\mathrm{j}$。

可以验证这一系统完全能控。

首先,使用算法 6.1 来设计控制律。

计算 A 的特征多项式,

$$\det(sI - A) = s^3 - 6s^2 + 11s - 6$$

根据给定的期望特征值,计算闭环系统的期望特征多项式,

$$\alpha_c(s) = (s+1)(s+1+\mathrm{j}2)(s+1-\mathrm{j}2) = s^3 + 3s^2 + 7s + 5$$

构造变换矩阵 T,

$$T = \left\{ \begin{bmatrix} 1 & 0 & -5 \\ 0 & 2 & 9 \\ 1 & 5 & 19 \end{bmatrix} \begin{bmatrix} 11 & -6 & 1 \\ -6 & 1 & 0 \\ 1 & 0 & 0 \end{bmatrix} \right\}^{-1} = \begin{bmatrix} 6 & -6 & 1 \\ -3 & 2 & 0 \\ 0 & -1 & 1 \end{bmatrix}^{-1} = \begin{bmatrix} -\frac{2}{3} & -\frac{5}{3} & \frac{2}{3} \\ -1 & -2 & 1 \\ -1 & -2 & 2 \end{bmatrix}$$

计算系统的原状态反馈增益 K,

$$K = \bar{K}T = \begin{bmatrix} -6-5 & 11-7 & -6-3 \end{bmatrix} \begin{bmatrix} -\frac{2}{3} & -\frac{5}{3} & \frac{2}{3} \\ -1 & -2 & 1 \\ -1 & -2 & 2 \end{bmatrix} = \begin{bmatrix} \frac{37}{3} & \frac{85}{3} & -\frac{64}{3} \end{bmatrix}$$

接下来,应用算法 6.2 来设计控制律。

考虑 $K = \begin{bmatrix} k_1 & k_2 & k_3 \end{bmatrix}$,可以得到

$$\det(sI - (A+bK)) = \det\left(\begin{bmatrix} s & & \\ & s & \\ & & s \end{bmatrix} - \begin{bmatrix} 1 & 0 & -1 \\ 1 & 2 & 1 \\ 2 & 2 & 3 \end{bmatrix} - \begin{bmatrix} k_1 & k_2 & k_3 \\ 0 & 0 & 0 \\ k_1 & k_2 & k_3 \end{bmatrix} \right)$$

$$= \det\left(\begin{bmatrix} s-1-k_1 & -k_2 & 1-k_3 \\ -1 & s-2 & -1 \\ -k_1-2 & -k_2-2 & s-k_3-3 \end{bmatrix} \right)$$

$$= s^3 - (k_1+k_3+6)s^2 + (11+6k_1-2k_2+k_3)s - 6 - 6k_1 + 3k_2$$

由此，可以得到下述方程，

$$\begin{cases} -(k_1+k_3+6)=3 \\ 11+6k_1-2k_2+k_3=7 \\ -6-6k_1+3k_2=5 \end{cases}$$

通过求解上述联立方程，可以得到 K，和使用算法 6.1 计算得到的结果相同。

6.1.3　MIMO 系统的状态反馈特征值配置

这一小节讨论 MIMO 系统的特征值配置问题。

考虑线性定常 MIMO 系统方程 (6.1)，首先介绍下述引理。

1.3.2 小节中指出，当且仅当矩阵 $A\in\mathbb{R}^{n\times n}$ 的特征多项式等于它的最小多项式，那么它被称作是循环的。

引理 6.1　当且仅当 A 的 Jordan 标准形中，每个特征值只对应一个 Jordan 块，那么 A 是循环的。

证明　1.3.4 小节中指出，A 的最小多项式可以表示为

$$\phi(s)=\prod_{i=1}^{l}(s-\lambda_i)^{\sigma_{i\max}}$$

其中，$\sigma_{i\max}=\max_{1\leqslant k\leqslant\alpha_i}\sigma_{ik}$，$\sigma_{ik}$ 表示与 λ_i 相关的第 k 个 Jordan 块的阶次，α_i 表示 λ_i 的几何重数，$i=1,2,\cdots,l$。

另一方面，A 的特征多项式可以表示为

$$\alpha(s)=\prod_{i=1}^{l}(s-\lambda_i)^{\sigma_i}$$

其中，σ_i 表示 λ_i 的代数重数。

注意到 $\sigma_i=\sum_{k=1}^{\alpha_i}\sigma_{ik}$。因此，当且仅当 $\alpha_i=1$ 时，$\phi(s)=\alpha(s)$，这表明每个特征值仅存在一个 Jordan 块。　□

根据上述讨论，引理 6.1 表明当且仅当 A 的每个特征值的几何重数都是 1 时，A 是循环的。特别地，如果 A 的 n 个特征值都互异，A 是循环的。

引理 6.2　如果 A 是循环的，那么存在 $b\in\mathbb{R}^n$ 使得

$$\mathrm{rank}\begin{bmatrix} b & Ab & \cdots & A^{n-1}b \end{bmatrix}=n \tag{6.6}$$

即 (A,b) 是完全能控的。

证明略。

基于引理 6.2 也可以定义矩阵的循环性，即如果存在 $b\in\mathbb{R}^n$ 使得式 (6.6) 成立，那么 $A\in\mathbb{R}^{n\times n}$ 被称作是循环的。

6.1 状态反馈控制

引理 6.3 对于线性定常系统方程 (6.1)，如果 A 是循环的，(A,B) 完全能控，那么对于几乎所有的 $l \in \mathbb{R}^p$，(A,b) 完全能控，其中 $b = Bl$。

下面给出一个例子来说明这一引理。

例 6.3 考虑状态方程的系数矩阵为

$$A = \begin{bmatrix} 2 & 1 & 0 & 0 & 0 \\ 0 & 2 & 1 & 0 & 0 \\ 0 & 0 & 2 & 0 & 0 \\ 0 & 0 & 0 & 3 & 1 \\ 0 & 0 & 0 & 0 & 3 \end{bmatrix}, \quad B = \begin{bmatrix} 0 & 1 \\ 0 & 0 \\ 1 & 1 \\ 1 & 2 \\ 1 & 0 \end{bmatrix}$$

可以看出 (A,B) 完全能控，A 是循环的。对于任意 $l = [\begin{array}{cc} l_1 & l_2 \end{array}]^{\mathrm{T}}$，

$$Bl = \begin{bmatrix} l_2 \\ 0 \\ l_1 + l_2 \\ l_1 + 2l_2 \\ l_1 \end{bmatrix}$$

根据 Jordan 标准形判据，除 $l_1 + l_2 = 0$ 或 $l_1 = 0$ 的情形外，(A, Bl) 是完全能控的。

引理 6.4 对于线性定常 MIMO 系统，$\operatorname{rank} B = p$，如果系统完全能控，那么对于任意 b_i，总是存在实常数矩阵 K_1，使得 $(A + BK_1, b_i)$ 完全能控，其中，b_i 指 B 的第 i 列，$i = 1, \cdots, p$。

证明 由于 (A,B) 是能控的，所以可以从能控性矩阵中选择出 n 个线性无关的列向量。首先，选择 $b_1, Ab_1, \cdots, A^{\mu_1 - 1}b_1$，直到 $A^{\mu_1}b_1$ 可以表示为 $b_1, Ab_1, \cdots, A^{\mu_1 - 1}b_1$ 的线性组合。接下来，继续选择 $b_2, Ab_2, \cdots, A^{\mu_2 - 1}b_2$，直到 $A^{\mu_2}b_2$ 可以表示为已选择的列向量的线性组合。重复这一过程直到选择出 n 个线性无关的列向量为止。其中，$\mu_i \geqslant 0$，$i = 1, 2, \cdots, p$。使用选择的 n 个线性无关的列向量，可以构造如下矩阵，

$$W = \begin{bmatrix} b_1 & Ab_1 & \cdots & A^{\mu_1-1}b_1 & \cdots & b_p & Ab_p & \cdots & A^{\mu_p-1}b_p \end{bmatrix} \tag{6.7}$$

它满足 $\operatorname{rank} W = n$。

构造 $p \times n$ 的矩阵 S，

$$S = [\underbrace{0 \quad \cdots \quad 0}_{\mu_1 列} \quad e_2 \quad 0 \quad \cdots \quad 0 \quad \cdots \quad \underset{\sum\limits_{i=1}^{p-1}\mu_i 列}{e_p} \quad 0 \quad \cdots \quad 0]$$

其中 e_i 表示 $p \times p$ 的单位矩阵的第 i 列。S 的第 μ_1 列是 e_2，第 $\mu_1 + \mu_2$ 列是 e_3，以此类推，第 $\sum\limits_{i=1}^{p-1} \mu_i$ 列是 e_p，其他列都是零向量。

选择 $K_1 = SW^{-1}$，可以进一步推导出

$$(A + BK_1)b_1 = Ab_1 + BK_1b_1 = Ab_1$$
$$(A + BK_1)^2 b_1 = (A + BK_1)Ab_1 = A^2 b_1$$
$$\vdots$$
$$(A + BK_1)^{\mu_1 - 1} b_1 = A^{\mu_1 - 1} b_1$$
$$(A + BK_1)^{\mu_1} b_1 = (A + BK_1)(A + BK_1)^{\mu_1 - 1} b_1 = A^{\mu_1} b_1 + Be_2 = A^{\mu_1} b_1 + b_2$$

由于 $A^{\mu_1} b_1$ 是 $b_1, Ab_1, \cdots, A^{\mu_1 - 1} b_1$ 的线性组合，所以它可以被表示为

$$A^{\mu_1} b_1 = \alpha_0 b_1 + \alpha_1 Ab_1 + \cdots + \alpha_{\mu_1 - 1} A^{\mu_1 - 1} b_1$$

其中 $\alpha_0, \alpha_1, \cdots, \alpha_{\mu_1 - 1}$ 是常数系数。进一步，可以得到

$$(A + BK_1)^{\mu_1 + 1} b_1 = (A + BK_1)(A^{\mu_1} b_1 + b_2)$$
$$= Ab_2 + A^{\mu_1 + 1} b_1 + BK_1 A^{\mu_1} b_1 + BK_1 b_2$$
$$= Ab_2 + A^{\mu_1 + 1} b_1 + BK_1 (\alpha_0 b_1 + \alpha_1 Ab_1 + \cdots + \alpha_{\mu_1 - 1} A^{\mu_1 - 1} b_1)$$
$$= Ab_2 + A^{\mu_1 + 1} b_1 + \alpha_{\mu_1 - 1} b_2$$

重复这一过程，可以得到

$$U_k = \begin{bmatrix} b_1 & (A + BK_1)b_1 & \cdots & (A + BK_1)^{n-1} b_1 \end{bmatrix}$$
$$= W \begin{bmatrix} I_{\mu_1 \times \mu_1} & * & \cdots & * \\ & I_{\mu_2 \times \mu_2} & \ddots & \vdots \\ & & \ddots & * \\ & & & I_{\mu_p \times \mu_p} \end{bmatrix}$$

其中，系数矩阵中的"$*$"代表可能不为零的元。

可以看出，$\operatorname{rank} U_k = n$，所以 $(A + BK_1, b_1)$ 是完全能控的。采用类似的方法，可以构造 K_{1i} 使得 $(A + BK_{1i}, b_i)$ 是能控的，$i = 2, \cdots, p$。 □

引理 6.4 指出，当 (A, B) 完全能控但 A 不是循环的时，总是可以找到一个实常数矩阵 K_1 使得 $A + BK_1$ 是循环的。实际上，如果 (A, B) 完全能控但 A 不是循环的，那么几乎对于所有的 $K_1 \in \mathbb{R}^{p \times n}$，$A + BK_1$ 是循环的。

定理 6.3 对于线性定常 MIMO 系统方程 (6.1)，当且仅当它完全能控时，它的所有特征值可以通过状态反馈控制律 (6.2) 任意配置。

6.1 状态反馈控制

证明 必要性的证明与定理 6.2 的证明相同。接下来仅证明充分性。根据引理 6.4，可以引入一个反馈增益为 K_1 的状态反馈控制律使得 $(A+BK_1, b_1)$ 能控。在此基础上，可以引入另一个反馈增益为 K_2 的状态反馈控制律，

$$K_2 = \begin{bmatrix} k_2 \\ 0 \\ \vdots \\ 0 \end{bmatrix} \tag{6.8}$$

其中 k_2 是 K_2 的第一行。

由于 $(A+BK_1, b_1)$ 完全能控，根据定理 6.2 的充分性证明，可以选择合适的行向量 k_2 使得 $A+BK_1+b_1k_2$ 的特征值任意配置。根据叠加原理，选择反馈增益为 $K = K_1 + K_2$ 的状态反馈控制律，即可实现系统方程 (6.1) 所有特征值的任意配置。 □

上述控制框图如图 6.2 所示。根据上述结论，可以归纳出以下算法，来实现系统方程 (6.1) 所有特征值的任意配置。

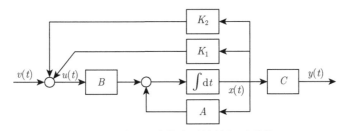

图 6.2 引入两次状态反馈的闭环系统

算法 6.3 对于完全能控的 MIMO 系统方程 (6.1)，使用状态反馈控制律 (6.2)，$A+BK$ 的特征值可以被配置到 $\lambda_1^*, \cdots, \lambda_n^*$，状态反馈增益 K 可以根据下述步骤确定。

(1) 构造 $K_1 = SW^{-1}$ 使得 $(A+BK_1, b_1)$ 能控，其中 S 和 W 可以根据引理 6.4 选取，b_1 是 B 的第一个列向量。

(2) 对于能控对 $(A+BK_1, b_1)$，应用算法 6.1 或算法 6.2 确定 k_2，使得 $A+BK_1+b_1k_2$ 的特征值被配置到 $\lambda_1^*, \cdots, \lambda_n^*$。

(3) 构造满足式 (6.8) 的 K_2。

(4) 计算反馈增益 $K = K_1 + K_2$。

如果 B 中存在一个列向量 b，或者 B 的一个线性组合（这里把这一线性组合也记作 b），使得 (A, b) 能控，那么算法 6.3 中的第 (1) 步可以省略，事实上，根据引理 6.3，这样的列向量 b 往往是存在的。

例 6.4 考虑下述完全能控的系统

$$\dot{x}(t) = \begin{bmatrix} 1 & 1 & 0 & 0 \\ 0 & 2 & 0 & 0 \\ 1 & 0 & 0 & 0 \\ 0 & 1 & 0 & 0 \end{bmatrix} x(t) + \begin{bmatrix} 1 & 2 \\ 1 & 0 \\ 0 & 0 \\ 0 & 0 \end{bmatrix} u(t)$$

设计状态反馈增益矩阵 K 使得闭环系统的所有特征值为 $-1, -2, -3$ 和 -4。构造

$$W = \begin{bmatrix} b_1 & Ab_1 & b_2 & Ab_2 \end{bmatrix} = \begin{bmatrix} 1 & 2 & 2 & 2 \\ 1 & 2 & 0 & 0 \\ 0 & 1 & 0 & 2 \\ 0 & 1 & 0 & 0 \end{bmatrix}, \quad S = \begin{bmatrix} 0 & 0 & 0 & 0 \\ 0 & 1 & 0 & 0 \end{bmatrix}$$

由此,可以计算出

$$K_1 = SW^{-1} = \begin{bmatrix} 0 & 0 & 0 & 0 \\ 0 & 0 & 0 & 1 \end{bmatrix}$$

$A + BK_1$ 的特征多项式为

$$\det(sI - A - BK_1) = s^4 - 3s^3 + 2s^2$$

期望的极点多项式为

$$\alpha_c(s) = s^4 + 10s^3 + 35s^2 + 50s + 24$$

对于能控对 $(A + BK_1, b_1)$,构造变换矩阵 T,

$$T = \left\{ \begin{bmatrix} 1 & 2 & 6 & 14 \\ 1 & 2 & 4 & 8 \\ 0 & 1 & 2 & 6 \\ 0 & 1 & 2 & 4 \end{bmatrix} \begin{bmatrix} 0 & 2 & -3 & 1 \\ 2 & -3 & 1 & \\ -3 & 1 & & \\ 1 & & & \end{bmatrix} \right\}^{-1} = \begin{bmatrix} 0 & 0 & 0.5 & -0.5 \\ 0.5 & -0.5 & 0 & 0 \\ 0.5 & -0.5 & 0 & 1 \\ 0.5 & -0.5 & 0 & 1 \end{bmatrix}$$

进一步,可以计算出

$$k_2 = \begin{bmatrix} 0-24 & 0-50 & 2-35 & -3-10 \end{bmatrix} \begin{bmatrix} 0 & 0 & 0.5 & -0.5 \\ 0.5 & -0.5 & 0 & 0 \\ 0.5 & -0.5 & 0 & 1 \\ 0.5 & -0.5 & 0 & 1 \end{bmatrix}$$

$$= \begin{bmatrix} -48 & -35 & -12 & -34 \end{bmatrix}$$

最后,可以计算得到

$$K = K_1 + K_2 = \begin{bmatrix} -48 & -35 & -12 & -34 \\ 0 & 0 & 0 & 1 \end{bmatrix}$$

基于线性定常 MIMO 系统的能控标准形,可以推导出一些特征值配置算法。这里,仅给出一种基于 Luenberger 第二能控标准形的特征值配置算法。

6.1 状态反馈控制

算法 6.4 问题与算法 6.3中的相同。状态反馈增益矩阵 K 可以根据下列步骤确定。
(1) 将原系统转化为 Luenberger 第二能控标准形，变换矩阵为 T。
(2) 计算期望的特征多项式 $\alpha_c(s)$。
(3) 设计 $\bar{K}$ 使得 $\bar{A}+\bar{B}\bar{K}$ 的特征多项式为 $\alpha_c(s)$。
(4) 计算 $K=\bar{K}T$。

例 6.5 考虑一个系统可以被转化为下述 Luenberger 第二能控标准形，

$$\dot{\bar{x}}(t)=\left[\begin{array}{ccc:ccc} 0 & 1 & 0 & 0 & 0 & 0 \\ 0 & 0 & 1 & 0 & 0 & 0 \\ 3 & 2 & 0 & 4 & 7 & 9 \\ \hdashline 0 & 0 & 0 & 0 & 1 & 0 \\ 0 & 0 & 0 & 0 & 0 & 1 \\ 2 & 3 & 4 & -5 & -3 & -4 \end{array}\right]\bar{x}(t)+\left[\begin{array}{cc} 0 & 0 \\ 0 & 0 \\ 1 & 1 \\ \hdashline 0 & 0 \\ 0 & 0 \\ 0 & 1 \end{array}\right]u(t).$$

期望的特征值为 -1，-2，-3，-4，-5 和 -6，可以计算出

$$\alpha_{c1}(s)=(s+1)(s+2)(s+3)=s^3+6s^2+11s+6$$
$$\alpha_{c2}(s)=(s+4)(s+5)(s+6)=s^3+15s^2+74s+120$$

或者

$$\alpha_c(s)=\alpha_{c1}(s)\alpha_{c2}(s)=s^6+21s^5+175s^4+735s^3+1624s^2+1764s+720$$

选择状态反馈增益矩阵 $\bar{K}$，

$$\bar{K}=\left[\begin{array}{cccccc} k_{11} & k_{12} & k_{13} & k_{14} & k_{15} & k_{16} \\ k_{21} & k_{22} & k_{23} & k_{24} & k_{25} & k_{26} \end{array}\right]$$

那么，

$$\bar{A}+\bar{B}\bar{K}=\left[\begin{array}{ccc:ccc} 0 & 1 & 0 & 0 & 0 & 0 \\ 0 & 0 & 1 & 0 & 0 & 0 \\ 3+k_{11}+k_{21} & 2+k_{12}+k_{22} & k_{13}+k_{23} & 4+k_{14}+k_{24} & 7+k_{15}+k_{25} & 9+k_{16}+k_{26} \\ \hdashline 0 & 0 & 0 & 0 & 1 & 0 \\ 0 & 0 & 0 & 0 & 0 & 1 \\ 2+k_{21} & 3+k_{22} & 4+k_{23} & -5+k_{24} & -3+k_{25} & -4+k_{26} \end{array}\right]$$

同时，根据 $\bar{A}+\bar{B}\bar{K}$ 的期望的特征多项式，可以得到

$$\bar{A}+\bar{B}\bar{K}=\left[\begin{array}{ccc:ccc} 0 & 1 & 0 & 0 & 0 & 0 \\ 0 & 0 & 1 & 0 & 0 & 0 \\ -6 & -11 & -6 & 0 & 0 & 0 \\ \hdashline 0 & 0 & 0 & 0 & 1 & 0 \\ 0 & 0 & 0 & 0 & 0 & 1 \\ 0 & 0 & 0 & -120 & -74 & -15 \end{array}\right]$$

或者

$$\bar{A}+\bar{B}\bar{K}=\left[\begin{array}{cccccc} 0 & 1 & 0 & 0 & 0 & 0 \\ 0 & 0 & 1 & 0 & 0 & 0 \\ 0 & 0 & 0 & 1 & 0 & 0 \\ 0 & 0 & 0 & 0 & 1 & 0 \\ 0 & 0 & 0 & 0 & 0 & 1 \\ -720 & -1764 & -1264 & -735 & -175 & -21 \end{array}\right]$$

进一步，可以计算出

$$\bar{K}=\left[\begin{array}{cccccc} -7 & -10 & -2 & 111 & 64 & 2 \\ -2 & -3 & -4 & -115 & -71 & -11 \end{array}\right]$$

或者

$$\bar{K}=\left[\begin{array}{cccccc} 719 & 1765 & 1268 & 727 & 165 & 8 \\ -722 & -1767 & -1268 & -730 & -172 & -17 \end{array}\right]$$

最后，应用 $K=\bar{K}T$ 可以计算出原系统的反馈增益矩阵。值得注意的是，这里设计出的状态反馈增益矩阵是不唯一的。

如果所有期望的特征值和原系统的特征值都不相同，可以使用下一算法解决特征值配置问题。

算法 6.5 问题与算法 6.3 中的相同，并额外要求期望的特征值与原系统的特征值都不相同。状态反馈增益矩阵 K 可以根据下列步骤确定。

(1) 选择 $n\times n$ 的常数矩阵 F，它的特征值与期望的特征值相同。

(2) 选择 $p\times n$ 的常数矩阵 $\bar{K}$ 使得 $(F,\bar{K})$ 完全能观。

(3) 解矩阵方程 $AT-TF=-B\bar{K}$。

(4) 判断 T 是否非奇异。如果 T 是非奇异的，继续下一步，否则，返回第 (1) 步或第 (2) 步重新选择 F 或 $\bar{K}$。

(5) 计算 $K=\bar{K}T^{-1}$。

6.1 状态反馈控制

形如 $XT+TY=Z$ 的矩阵方程被称为 Sylvester 方程,其中,$X \in \mathbb{R}^{n \times n}$, $Y \in \mathbb{R}^{m \times m}$, $Z \in \mathbb{R}^{n \times m}$ 是已知的,$T \in \mathbb{R}^{n \times m}$ 是这一方程的解。对于任意的 Z,当且仅当 X 和 $-Y$ 没有公共的特征值时,矩阵方程的解 T 是唯一的。正因如此,要求 A 的所有特征值都和 F 的特征值不同,以确保 $AT - TF = -B\bar{K}$ 的解 T 是唯一的。

在第 (2) 步中,要求 $(F, \bar{K})$ 完全能观,是因为 (A, B) 完全能控且 $(F, \bar{K})$ 完全能观是确保 T 非奇异的必要条件。如果 (A, B) 是 SISO 系统,这一条件也是充分的。所以,这一选择有助于得到非奇异的 T。

根据 $AT - TF = -B\bar{K}$ 和 $K = \bar{K}T^{-1}$,可知 $TFT^{-1} = A + B\bar{K}T^{-1} = A + BK$。所以 $A + BK$ 具有和 F 相同的特征值。

如果期望的 n 个特征值互异,可以推导出另一个特征值配置算法。

假设式 (6.2) 中的反馈增益矩阵 K 可以表示为两个向量的外积,即 $K = lm$,其中,$l \in \mathbb{R}^p$,$m^\mathrm{T} \in \mathbb{R}^n$。由此,相应的闭环系统为

$$\dot{x}(t) = (A + BK)x(t) + Bv(t) = (A + Blm)x(t) + Bv(t) = (A + bm)x(t) + Bv(t)$$

其中,$b = Bl$ 是 B 的列向量的线性组合。如果存在向量 l,使得 (A, b) 完全能控,那么 MIMO 系统的特征值配置问题可以转化为 SISO 系统的特征值配置问题。

对于上述闭环系统,可以得到

$$\begin{aligned} \det(sI - A - bm) &= \det\left((sI - A)(I - (sI - A)^{-1}bm)\right) \\ &= \det(sI - A)\det\left(I - (sI - A)^{-1}bm\right) \\ &= \det(sI - A)\left(1 - m(sI - A)^{-1}b\right) \\ &= \det(sI - A) - m\,\mathrm{adj}(sI - A)b \end{aligned}$$

记 $\alpha_c(s) = \det(sI - A - bm)$,$\alpha(s) = \det(sI - A)$,$R(s) = \mathrm{adj}(sI - A)$,可以得到

$$\alpha_c(s) = \alpha(s) - mR(s)b$$

把期望的特征值 λ_1^*,$\cdots$,λ_n^* 代入到上述方程中,可以得到

$$m \begin{bmatrix} R(\lambda_1^*)b & \cdots & R(\lambda_n^*)b \end{bmatrix} = \begin{bmatrix} \alpha(\lambda_1^*) & \cdots & \alpha(\lambda_n^*) \end{bmatrix} \tag{6.9}$$

为了保证 $\begin{bmatrix} R(\lambda_1^*)b & \cdots & R(\lambda_n^*)b \end{bmatrix}$ 是非奇异的,λ_1^*,$\cdots$,λ_n^* 应该是互异的。

算法 6.6 问题与算法 6.3 中的相同,并额外要求期望的特征值互异。状态反馈增益矩阵 $K = lm$ 可以根据下列步骤确定,其中,$l \in \mathbb{R}^p$,$m^\mathrm{T} \in \mathbb{R}^n$。

(1) 选择任意的 l 使得 (A, b) 完全能控,其中 $b = Bl$。
(2) 根据算法 1.1 确定 $\alpha(s) = \det(sI - A)$ 和 $R(s) = \mathrm{adj}(sI - A)$。
(3) 根据期望的特征值,构造方程 (6.9),进一步可以计算得到

$$m = \begin{bmatrix} \alpha(\lambda_1^*) & \cdots & \alpha(\lambda_n^*) \end{bmatrix} \begin{bmatrix} R(\lambda_1^*)b & \cdots & R(\lambda_n^*)b \end{bmatrix}^{-1}$$

(4) 计算 $K = lm$。

例 6.6 某飞行控制系统的纵向姿态运动方程为

$$\dot{x}(t) = Ax(t) + Bu(t) = \begin{bmatrix} -0.605 & 0.023 & -5.816 \\ 9.29 & -0.343 & -33.6 \\ 0 & 0 & -20 \end{bmatrix} x(t) + \begin{bmatrix} 0 & 0 \\ 1 & 0 \\ 0 & 1 \end{bmatrix} u(t)$$

设计状态反馈增益矩阵 K 使得闭环系统的特征值为 -5，-10 和 -1。

选择 $l = [33.6 \quad 20]^T$，则可以计算出 $b = Bl = [0 \quad 33.6 \quad 20]^T$，进一步，可以验证 (A, b) 是能控的。

根据算法 1.1，可以得到

$$R_2 = I, \ a_1 = -\operatorname{tr}(R_2 A) = 20.94$$

$$R_1 = AR_2 + a_1 I = \begin{bmatrix} 20.343 & 0.023 & -5.816 \\ 9.29 & 20.605 & -33.6 \\ 0 & 0 & 0.948 \end{bmatrix}, \ a_2 = -\frac{1}{2}\operatorname{tr}(R_1 A) = 18.954$$

$$R_0 = AR_1 + a_2 I = \begin{bmatrix} 6.86 & 0.46 & -2.768 \\ 185.8 & 12.1 & -74.359 \\ 0 & 0 & -0.006 \end{bmatrix}, \ a_3 = -\frac{1}{3}\operatorname{tr}(R_0 A) = -0.122$$

由此可得 $\alpha(s) = s^3 + a_1 s^2 + a_2 s + a_3$，$R(s) = R_2 s^2 + R_1 s + R_0$。进一步，可知

$$m = \begin{bmatrix} \alpha(\lambda_1^*) & \alpha(\lambda_2^*) & \alpha(\lambda_3^*) \end{bmatrix} \begin{bmatrix} R(\lambda_1^*)b & R(\lambda_2^*)b & R(\lambda_3^*)b \end{bmatrix}^{-1}$$

$$= \begin{bmatrix} 303.808 & 905.138 & 1053.868 \end{bmatrix} \begin{bmatrix} 537.831 & 1115.566 & 1693.301 \\ -342.26 & 2076.1 & 6174.46 \\ 405.08 & 1810.28 & 4215.48 \end{bmatrix}^{-1}$$

$$= \begin{bmatrix} 2.080 & 0.6175 & -1.49 \end{bmatrix}$$

最后，计算

$$K = lm = \begin{bmatrix} 33.6 \\ 20 \end{bmatrix} \begin{bmatrix} 2.080 & 0.618 & -1.49 \end{bmatrix} = \begin{bmatrix} 69.89 & 20.75 & -50.06 \\ 41.6 & 12.35 & -29.8 \end{bmatrix}$$

6.1.4 状态反馈镇定

这一小节关注的问题是确定状态反馈控制律 (6.2)，使得在 $v = 0$ 时相应的闭环系统是渐近稳定的。这一问题通常被称作镇定问题。

考虑线性定常系统方程 (6.1)，存在变换矩阵 T 能够把系统分解为

$$\begin{bmatrix} \dot{x}_{\mathrm{c}}(t) \\ \dot{x}_{\mathrm{nc}}(t) \end{bmatrix} = \begin{bmatrix} A_{11} & A_{12} \\ 0 & A_{22} \end{bmatrix} \begin{bmatrix} x_{\mathrm{c}}(t) \\ x_{\mathrm{nc}}(t) \end{bmatrix} + \begin{bmatrix} B_1 \\ 0 \end{bmatrix} u(t) \quad (6.10)$$

6.1 状态反馈控制

其中 (A_{11}, B_1) 是完全能控的。系统 (6.10) 的特征值由 A_{11} 和 A_{22} 的特征值组成，其中 A_{11} 的特征值能控，A_{22} 的特征值不能控。

选择

$$u(t) = \begin{bmatrix} K_c & K_{nc} \end{bmatrix} \begin{bmatrix} x_c(t) \\ x_{nc}(t) \end{bmatrix} \quad (6.11)$$

把式 (6.11) 代入式 (6.10) 中，可以推导出相应的闭环系统为

$$\begin{bmatrix} \dot{x}_c(t) \\ \dot{x}_{nc}(t) \end{bmatrix} = \begin{bmatrix} A_{11} + B_1 K_c & A_{12} + B_1 K_{nc} \\ 0 & A_{22} \end{bmatrix} \begin{bmatrix} x_c(t) \\ x_{nc}(t) \end{bmatrix} \quad (6.12)$$

闭环系统方程 (6.12) 的特征值为 $A_{11} + B_1 K_c$ 和 A_{22} 的特征值，上一小节中指出，$A_{11} + B_1 K_c$ 的特征值可以任意配置。由此，可以得到下述结论。

定理 6.4 线性定常系统可以被状态反馈控制律 $u(t) = Kx(t)$ 镇定的充要条件是系统不能控的特征值都具有负实部。

证明略。

显然，如果方程 (6.1) 完全能控，它可以被状态反馈控制律镇定。如果系统不完全能控，可以通过下述算法解决镇定问题。

算法 6.7 对于不完全能控的系统方程 (6.1)，使用状态反馈控制律 (6.2) 镇定该系统。状态反馈增益矩阵可以根据下列步骤确定。

(1) 进行能控性分解，变换矩阵是 T，系统转化为式 (6.10) 的形式。A_{22} 的特征值应具有负实部，否则，系统方程 (6.1) 不能被镇定。

(2) 把 A_{11} 的特征值通过反馈增益为 K_c 的状态反馈控制律配置在期望的位置，期望的特征值可以任意配置，只要它们都具有负实部。

(3) 计算 $K = \begin{bmatrix} K_c & K_{nc} \end{bmatrix} T$，其中具有合适维数的 K_{nc} 可以任意选择。

例 6.7 考虑下述不完全能控的系统

$$\dot{x}(t) = \begin{bmatrix} 1 & 0 & -1 \\ 0 & -2 & 0 \\ -1 & 0 & 2 \end{bmatrix} x(t) = \begin{bmatrix} 0 \\ 0 \\ 1 \end{bmatrix} u(t)$$

利用变换矩阵

$$T = \begin{bmatrix} 0 & 0 & 1 \\ 1 & 0 & 0 \\ 0 & 1 & 0 \end{bmatrix}$$

系统可以被分解分

$$\dot{\bar{x}}(t) = \begin{bmatrix} 2 & -1 & 0 \\ -1 & 1 & 0 \\ 0 & 0 & -2 \end{bmatrix} \bar{x}(t) + \begin{bmatrix} 1 \\ 0 \\ 0 \end{bmatrix} u(t)$$

显然，这一系统可以被镇定。

选择 $\bar{K} = \begin{bmatrix} -9 & 17 & 0 \end{bmatrix}$，闭环系统为

$$\dot{\bar{x}}(t) = \begin{bmatrix} -7 & 16 & 0 \\ -1 & 1 & 0 \\ 0 & 0 & -2 \end{bmatrix} \bar{x}(t)$$

它是渐近稳定的。

反馈增益是

$$K = \bar{K}T = \begin{bmatrix} 17 & 0 & -9 \end{bmatrix}$$

定理 5.15 也可以用来解决镇定问题。只需要求解矩阵方程

$$(A + BK)^{\mathrm{T}} P + P(A + BK) = -Q \tag{6.13}$$

来获得 K，这里 P 和 Q 是对称正定矩阵，为了计算方便，Q 和 P 越简单越好。例如，可以选择 $P = Q = I$。

例 6.8 考虑下述完全能控的系统，

$$\dot{x}(t) = \begin{bmatrix} 0 & 1 \\ 2 & 2 \end{bmatrix} x(t) + \begin{bmatrix} 1 & 1 \\ -2 & 1 \end{bmatrix} u(t)$$

选取 $u(t) = Kx(t)$，其中

$$K = \begin{bmatrix} k_{11} & k_{12} \\ k_{21} & k_{22} \end{bmatrix}$$

可以计算得到

$$A + BK = \begin{bmatrix} k_{11} + k_{21} & 1 + k_{12} + k_{22} \\ 2 - 2k_{11} + k_{21} & 2 - 2k_{12} + k_{22} \end{bmatrix}$$

选择 $P = Q = I$，可以进一步得到

$$\begin{bmatrix} k_{11} + k_{21} & 2 - 2k_{11} + k_{21} \\ 1 + k_{12} + k_{22} & 2 - 2k_{12} + k_{22} \end{bmatrix} + \begin{bmatrix} k_{11} + k_{21} & 1 + k_{12} + k_{22} \\ 2 - 2k_{11} + k_{21} & 2 - 2k_{12} + k_{22} \end{bmatrix} = -I$$

由此可以得到下述方程，

$$\begin{cases} 2k_{11} + 2k_{21} = -1 \\ -2k_{11} + k_{12} + k_{21} + k_{22} = -3 \\ -4k_{12} + 2k_{22} = -5 \end{cases}$$

选择 $k_{11} = 1$，可以解得 $k_{12} = 1$，$k_{21} = -1.5$ 和 $k_{22} = -0.5$。可以验证 $A + BK$ 的特征值具有负实部，即闭环系统是渐近稳定的。

6.1 状态反馈控制

此外,对于线性定常系统方程 (6.1),在最优调节问题中经常使用下述二次型指标,

$$J = \int_0^\infty \left(x^{\mathrm{T}}(t)Qx(t) + u^{\mathrm{T}}(t)Ru(t) \right) \mathrm{d}t \tag{6.14}$$

其中,Q 和 R 是两个对称正定矩阵。这一指标与误差和输入信号所消耗的能量相关。设计要求是确定状态反馈控制律 $u(t) = Kx(t)$ 使得性能指标 J 达到最小。

应用状态反馈,式 (6.14) 可以转化为

$$J = \int_0^\infty x^{\mathrm{T}}(t)(Q + K^{\mathrm{T}}RK)x(t)\mathrm{d}t$$

令

$$x^{\mathrm{T}}(t)(Q + K^{\mathrm{T}}RK)x(t) = -\frac{\mathrm{d}}{\mathrm{d}t}x^{\mathrm{T}}(t)Px(t)$$

其中,P 是对称正定矩阵。上述等式可以被写作

$$(A + BK)^{\mathrm{T}}P + P(A + BK) = -(Q + K^{\mathrm{T}}RK) \tag{6.15}$$

同时可以得到

$$J = \int_0^\infty -\frac{\mathrm{d}}{\mathrm{d}t}x^{\mathrm{T}}(t)Px(t)\mathrm{d}t = x^{\mathrm{T}}(0)Px(0) - x^{\mathrm{T}}(\infty)Px(\infty)$$

由于 $A + BK$ 被设计为渐近稳定的,所以 $x(\infty) = 0$,基于此,二次型指标 J 可以被进一步表示为

$$J = x^{\mathrm{T}}(0)Px(0) \tag{6.16}$$

这种情况下,求解式 (6.15),可以得到 P 的元为 K 的元的函数。在此基础上,把 P 代入式 (6.16) 中,通过使 J 极小来确定 K。

例 6.9 考虑如下的状态方程和性能指标,

$$\dot{x}(t) = \begin{bmatrix} 0 & 1 \\ 0 & 0 \end{bmatrix} x(t) + \begin{bmatrix} 0 \\ 1 \end{bmatrix} u(t)$$

$$J = \int_0^\infty (x^{\mathrm{T}}(t)x(t) + u^{\mathrm{T}}(t)u(t))\mathrm{d}t$$

在状态方程中引入 $u(t) = \begin{bmatrix} k_1 & k_2 \end{bmatrix} x(t)$,可以得到

$$\dot{x}(t) = \begin{bmatrix} 0 & 1 \\ k_1 & k_2 \end{bmatrix} x(t)$$

根据式 (6.15) 和性能指标 J，可知 $Q=I$, $R=I$。此时，式 (6.15) 可以被写作

$$\begin{bmatrix} 0 & k_1 \\ 1 & k_2 \end{bmatrix} P + P \begin{bmatrix} 0 & 1 \\ k_1 & k_2 \end{bmatrix} = -\begin{bmatrix} 1+k_1^2 & k_1 k_2 \\ k_1 k_2 & 1+k_2^2 \end{bmatrix}$$

这表明矩阵 P 是关于 k_1 和 k_2 的函数。

结合式 (6.16)，性能指标 J 是关于 k_1 和 k_2 的函数。使 $\dfrac{\partial J}{\partial k_1}=0$ 和 $\dfrac{\partial J}{\partial k_2}=0$ 可以使 J 对于任意的 $x(0)$ 为极小，可以解得 $k_1=-1$, $k_2=-\sqrt{3}$。所以，

$$P = \begin{bmatrix} \sqrt{3} & 1 \\ 1 & \sqrt{3} \end{bmatrix}$$

且

$$J = x^{\mathrm{T}}(0) \begin{bmatrix} \sqrt{3} & 1 \\ 1 & \sqrt{3} \end{bmatrix} x(0)$$

6.2 输出反馈控制

考虑状态反馈时，总是假设状态值都可以通过合适的传感器测量得到。实际上，对于一些系统，测量其所有状态值是不可能或不实际的。与状态信息相比，输出值更容易被测量。这一节来讨论输出反馈。

6.2.1 输出反馈系统的能控性和能观性

考虑线性定常系统

$$\dot{x}(t) = Ax(t) + Bu(t), \quad y(t) = Cx(t) \tag{6.17}$$

其中 $x(t) \in \mathbb{R}^n$ 是状态向量，$u(t) \in \mathbb{R}^p$ 是输入向量，$y(t) \in \mathbb{R}^q$ 是输出向量，A, B 和 C 是具有相应维数的实常数矩阵。

引入输出反馈控制律

$$u(t) = Ky(t) + v(t) \tag{6.18}$$

其中，$v(t) \in \mathbb{R}^p$ 是参考输入向量，K 是常数矩阵。一般地，式 (6.18) 被称作静态输出反馈控制律。

把式 (6.18) 代入式 (6.17) 中，相应的闭环系统是

$$\dot{x}(t) = (A+BKC)x(t) + Bv(t), \quad y(t) = Cx(t) \tag{6.19}$$

结构图如图 6.3 所示。

6.2 输出反馈控制

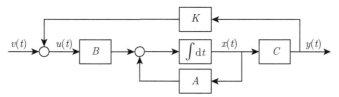

图 6.3 静态输出反馈闭环系统的结构图

定理 6.5 对于线性定常系统方程 (6.17), 引入输出反馈控制律 (6.18) 不会改变它的能控性和能观性。

证明 输出反馈控制律 (6.18) 可以被看作状态反馈增益是 KC 的状态反馈控制律。根据定理 6.1, 系统的能控性不会改变。

接下来, 证明系统的能观性也不会发生改变。

系统方程 (6.17) 和 (6.19) 的能观性矩阵分别是 $V = [\ C^\mathrm{T}\ \ A^\mathrm{T}C^\mathrm{T}\ \ \cdots\ \ (A^{n-1})^\mathrm{T}C^\mathrm{T}\ ]^\mathrm{T}$ 和 $V_\mathrm{ok} = [\ C^\mathrm{T}\ \ (A+BKC)^\mathrm{T}C^\mathrm{T}\ \ \cdots\ \ ((A+BKC)^{n-1})^\mathrm{T}C^\mathrm{T}\ ]^\mathrm{T}$, 只需证明 $\mathrm{rank}\,V = \mathrm{rank}\,V_\mathrm{ok}$ 即可。

与定理 6.1 的证明过程类似, 可以推导出 $C(A+BKC)^m$ 的每一行都可以表示为 $[\ C^\mathrm{T}\ \ A^\mathrm{T}C^\mathrm{T}\ \ \cdots\ \ (A^m)^\mathrm{T}C^\mathrm{T}\ ]^\mathrm{T}$ 的行的线性组合, $m = 1, \cdots, n-1$。所以, V_ok 的每一行都可以表示为 V 的行的线性组合, 这表明 $\mathrm{rank}\,V \geqslant \mathrm{rank}\,V_\mathrm{ok}$。

另一方面, 注意到

$$\dot{x}(t) = Ax(t) + Bu(t) = [(A+BKC) - BKC]x(t) + Bu(t)$$

即原系统可以被看作是系统方程 (6.17) 中应用了反馈增益为 $-K$ 的输出反馈控制律。由此可以推出 $\mathrm{rank}\,V \leqslant \mathrm{rank}\,V_\mathrm{ok}$。

所以, $\mathrm{rank}\,V = \mathrm{rank}\,V_\mathrm{ok}$, 输出反馈不会改变系统的能观性。□

6.2.2 输出反馈特征值配置

6.1.3 小节中详细讨论了矩阵的循环性, 这里, 再引入两个引理。

引理 6.5 对于线性定常系统方程 (6.17), 如果 A 是循环的, (A, C) 完全能观, 那么几乎所有的 $m^\mathrm{T} \in \mathbb{R}^q$ 都可以使得 (A, c) 完全能观, 其中 $c = mC$。

引理 6.6 对于线性定常系统方程 (6.17), 如果 A 不是循环的, (A, B, C) 完全能控能观, 那么存在常值矩阵 K 使得 $(A+BKC, B, C)$ 完全能控能观, 且 $A+BKC$ 是循环的。

上述两个引理的证明省略。

定理 6.6 对于线性定常系统方程 (6.17), $\mathrm{rank}\,B = p$, $\mathrm{rank}\,C = q$, (A, B, C) 完全能控能观, 那么总是存在常值反馈矩阵 K 使得 $A+BKC$ 的 $\max(p, q)$ 个特征值可以任意接近期望的位置。

证明 不失一般性地, 假设 A 是循环的。如果 A 不是循环的, 由引理 6.6 可知, 总是能够找到一个常值矩阵 K 使得 $A+BKC$ 循环。

考虑 $K = lm$, 其中, $l \in \mathbb{R}^p$, $m \in \mathbb{R}^q$, 代入式 (6.19) 中, 可以得到

$$\dot{x}(t) = (A+BlmC)x(t) + Bv(t)$$

注意到

$$\begin{aligned}\det(sI-A-BlmC) &= \det\left((sI-A)(I-(sI-A)^{-1}BlmC)\right)\\ &= \det(sI-A)\det\left(I-(sI-A)^{-1}BlmC\right)\\ &= \det(sI-A)(1-mC(sI-A)^{-1}Bl)\\ &= \det(sI-A)-mC\operatorname{adj}(sI-A)Bl\end{aligned}$$

记 $\alpha_c(s)=\det(sI-A-BlmC)$, $\alpha(s)=\det(sI-A)$, $R(s)=\operatorname{adj}(sI-A)$, 可以得到

$$\alpha_c(s)=\alpha(s)-mCR(s)Bl$$

把 q 个期望的特征值 $\lambda_1^*,\cdots,\lambda_q^*$ 代入上述等式中, 可以得到

$$m\begin{bmatrix} CR(\lambda_1^*)Bl & \cdots & CR(\lambda_q^*)Bl \end{bmatrix} = \begin{bmatrix} \alpha(\lambda_1^*) & \cdots & \alpha(\lambda_q^*) \end{bmatrix} \tag{6.20}$$

如果 $H \triangleq \begin{bmatrix} CR(\lambda_1^*)Bl & \cdots & CR(\lambda_q^*)Bl \end{bmatrix}$ 是非奇异的, 那么对于任意预先选取的向量 l, l 可以确保 (A,Bl) 是能控的, 总是存在

$$m = \begin{bmatrix} \alpha(\lambda_1^*) & \cdots & \alpha(\lambda_q^*) \end{bmatrix} H^{-1} \tag{6.21}$$

且相应的 $K=lm$ 可以把 q 个特征值准确配置到期望的位置。

如果 H 是奇异的, 可以对期望的特征值引入小扰动 $\Delta\lambda_i^*$, 并用 $\lambda_i^*+\Delta\lambda_i^*$ 替换 λ_i^*, $i\in\{1,\cdots,q\}$, 以使得 H 非奇异。由于 $\operatorname{rank}C=q$, 这一方法总是有效的。进一步, 选择满足式 (6.21) 的 m, $K=lm$ 可以把 q 个特征值近似配置到期望的位置。

考虑系统 (A,B,C) 的对偶系统 $(A^{\mathrm{T}},C^{\mathrm{T}},B^{\mathrm{T}})$, 采用类似的方法, 可以找到一个输出反馈矩阵 K 使得 A^{T} 的 $p=\operatorname{rank}B^{\mathrm{T}}$ 个特征值精确或近似地配置到期望的位置。

综上所示, 利用输出反馈控制律 (6.18), 可以把 $\max\{p,q\}$ 个特征值精确或近似地配置到期望的位置。 □

根据定理 6.6, 参考算法 6.6, 可以归纳出利用输出反馈控制律 (6.18) 实现的特征值配置算法, 这里不详细展开, 仅通过一个例子来说明。

例 6.10 考虑一个线性定常系统方程 (6.17), 其系数矩阵为

$$A=\begin{bmatrix} 0 & 1 & 0 & 0 \\ 0 & 0 & 1 & 0 \\ 0 & 0 & 0 & 1 \\ -4 & 12 & -13 & 6 \end{bmatrix},\ B=\begin{bmatrix} 0 & 0 \\ 0 & 1 \\ 0 & 0 \\ 1 & 0 \end{bmatrix},\ C=\begin{bmatrix} 1 & 0 & 0 & 0 \\ 0 & 0 & 1 & 0 \\ 0 & 1 & 0 & 1 \end{bmatrix}$$

尝试把闭环系统的特征值配置在 -1, -2, -3 和 -4。

可以验证 A 是循环的且 (A,B,C) 是能控能观的。

6.2 输出反馈控制

根据算法 1.1，可以得到

$$R_3 = I,\ a_1 = -\operatorname{tr}(R_3 A) = -6$$

$$R_2 = AR_3 + a_1 I = \begin{bmatrix} -6 & 1 & 0 & 0 \\ 0 & -6 & 1 & 0 \\ 0 & 0 & -6 & 1 \\ -4 & 12 & -13 & 0 \end{bmatrix},\ a_2 = -\frac{1}{2}\operatorname{tr}(R_2 A) = 13$$

$$R_1 = AR_2 + a_2 I = \begin{bmatrix} 13 & -6 & 1 & 0 \\ 0 & 13 & -6 & 1 \\ -4 & 12 & 0 & 0 \\ 0 & -4 & 12 & 0 \end{bmatrix},\ a_3 = -\frac{1}{3}\operatorname{tr}(R_1 A) = -12$$

$$R_0 = AR_1 + a_3 I = \begin{bmatrix} -12 & 13 & -6 & 1 \\ -4 & 0 & 0 & 0 \\ 0 & -4 & 0 & 0 \\ 0 & 0 & -4 & 0 \end{bmatrix},\ a_4 = -\frac{1}{4}\operatorname{tr}(R_0 A) = 4$$

由此可以进一步得到 $\alpha(s) = s^4 - 6s^3 + 13s^2 - 12s + 4$，$R(s) = R_3 s^3 + R_2 s^2 + R_1 s + R_0$。选择 $l = [\ 1\ \ 0\ ]^{\mathrm{T}}$，可以验证 $(A, b = Bl)$ 是能控的，同时，

$$CR(s)b = \begin{bmatrix} 1 & 0 & 0 & 0 \\ 0 & 0 & 1 & 0 \\ 0 & 1 & 0 & 1 \end{bmatrix}(R_3 s^3 + R_2 s^2 + R_1 s + R_0)\begin{bmatrix} 0 \\ 0 \\ 0 \\ 1 \end{bmatrix}$$

$$= \begin{bmatrix} 0 \\ 0 \\ 1 \end{bmatrix} s^3 + \begin{bmatrix} 0 \\ 1 \\ 0 \end{bmatrix} s^2 + \begin{bmatrix} 0 \\ 0 \\ 1 \end{bmatrix} s + \begin{bmatrix} 1 \\ 0 \\ 0 \end{bmatrix}$$

因为 $\operatorname{rank} C = 3$，选择 $\lambda_1^* = -1$，$\lambda_2^* = -2$，$\lambda_3^* = -3$。可以计算得到

$$m = \begin{bmatrix} \alpha(\lambda_1^*) & \alpha(\lambda_2^*) & \alpha(\lambda_3^*) \end{bmatrix}\begin{bmatrix} CR(\lambda_1^*)b & CR(\lambda_2^*)b & CR(\lambda_3^*)b \end{bmatrix}^{-1}$$

$$= \begin{bmatrix} 36 & 144 & 400 \end{bmatrix}\begin{bmatrix} 1 & 1 & 1 \\ 1 & 4 & 9 \\ -2 & -10 & -30 \end{bmatrix}^{-1} = \begin{bmatrix} 7.6 & 5.6 & -11.4 \end{bmatrix}$$

进一步，可得

$$K = lm = \begin{bmatrix} 1 \\ 0 \end{bmatrix}\begin{bmatrix} 7.6 & 5.6 & -11.4 \end{bmatrix} = \begin{bmatrix} 7.6 & 5.6 & -11.4 \\ 0 & 0 & 0 \end{bmatrix}$$

且
$$\alpha_c(s) = \det(sI - A - BKC) = s^4 + 5.4s^3 + 7.4s^2 - 0.6s - 3.6$$

可以看出仅有三个特征值可以配置到期望的位置，另一个特征值是 0.6，它移动到了 s 右半平面。

另一方面，可知

$$A^{\rm T} = \begin{bmatrix} 0 & 0 & 0 & -4 \\ 1 & 0 & 0 & 12 \\ 0 & 1 & 0 & -13 \\ 0 & 0 & 1 & 6 \end{bmatrix}, \ C^{\rm T} = \begin{bmatrix} 1 & 0 & 0 \\ 0 & 0 & 1 \\ 0 & 1 & 0 \\ 0 & 0 & 1 \end{bmatrix}, \ B^{\rm T} = \begin{bmatrix} 0 & 0 & 0 & 1 \\ 0 & 1 & 0 & 0 \end{bmatrix}$$

采用类似的方法，可以计算得到

$$R_3^{\rm T} = I, \ a_1 = -\operatorname{tr}(R_3^{\rm T} A^{\rm T}) = -6$$

$$R_2^{\rm T} = A^{\rm T} R_3^{\rm T} + a_1 I = \begin{bmatrix} -6 & 0 & 0 & -4 \\ 1 & -6 & 0 & 12 \\ 0 & 1 & -6 & -13 \\ 0 & 0 & 1 & 0 \end{bmatrix}, \ a_2 = -\frac{1}{2}\operatorname{tr}(R_2^{\rm T} A^{\rm T}) = 13$$

$$R_1^{\rm T} = A^{\rm T} R_2^{\rm T} + a_2 I = \begin{bmatrix} 13 & 0 & -4 & 0 \\ -6 & 13 & 12 & -4 \\ 1 & -6 & 0 & 12 \\ 0 & 1 & 0 & 0 \end{bmatrix}, \ a_3 = -\frac{1}{3}\operatorname{tr}(R_1^{\rm T} A^{\rm T}) = -12$$

$$R_0^{\rm T} = A^{\rm T} R_1^{\rm T} + a_3 I = \begin{bmatrix} -12 & -4 & 0 & 0 \\ 13 & 0 & -4 & 0 \\ -6 & 0 & 0 & -4 \\ 1 & 0 & 0 & 0 \end{bmatrix}, \ a_4 = -\frac{1}{4}\operatorname{tr}(R_0^{\rm T} A^{\rm T}) = 4$$

由此可得 $\alpha(s) = s^4 - 6s^3 + 13s^2 - 12s + 4$，$R^{\rm T}(s) = R_3^{\rm T} s^3 + R_2^{\rm T} s^2 + R_1^{\rm T} s + R_0^{\rm T}$。

选择 $l = [\ 1\ \ 0\ \ 0\ ]^{\rm T}$，可以验证 $(A^{\rm T}, C^{\rm T} l)$ 完全能控，且

$$B^{\rm T} R^{\rm T}(s) C^{\rm T} l = \begin{bmatrix} 0 & 0 & 0 & 1 \\ 0 & 1 & 0 & 0 \end{bmatrix} [R_3^{\rm T} s^3 + R_2^{\rm T} s^2 + R_1^{\rm T} s + R_0^{\rm T}] \begin{bmatrix} 1 & 0 & 0 \\ 0 & 0 & 1 \\ 0 & 1 & 0 \\ 0 & 0 & 1 \end{bmatrix} \begin{bmatrix} 1 \\ 0 \\ 0 \end{bmatrix}$$

$$= \begin{bmatrix} 0 \\ 1 \end{bmatrix} s^2 + \begin{bmatrix} 0 \\ -6 \end{bmatrix} s^1 + \begin{bmatrix} 1 \\ 13 \end{bmatrix}$$

6.2 输出反馈控制

由于 $\operatorname{rank} B^{\mathrm{T}} = 2$，选择 $\lambda_1^* = -1$，$\lambda_2^* = -2$，可以计算得到

$$m = \begin{bmatrix} 36 & 144 \end{bmatrix} \begin{bmatrix} 1 & 1 \\ 20 & 29 \end{bmatrix}^{-1} = \begin{bmatrix} -204 & 12 \end{bmatrix}$$

且

$$\alpha_{\mathrm{c}}(s) = \det(sI - A - BKC) = s^4 - 6s^3 + s^2 + 60s + 52$$

这种情况下仅有两个特征值可以配置到期望的位置，剩余两个特征值为 $43.5 \pm \mathrm{j}\sqrt{11.5}$，位于 s 右半平面。

定理 6.6 表明，$\max\{p,q\}$ 个特征值可以近似配置到期望的位置，但是剩余的 $n - \max\{p,q\}$ 个特征值的去向并未被指明。

定理6.7 对于线性定常系统方程 (6.17)，$\operatorname{rank} B = p$，$\operatorname{rank} C = q$，$(A,B,C)$ 完全能控能观，那么对于几乎所有的 (A,B,C)，总是存在矩阵 K 使得 $A+BKC$ 的 $\min\{p+q-1,n\}$ 个特征值可以任意接近期望的位置。如果 $p+q-1 \geqslant n$，那么对于几乎所有的 (A,B,C)，$A+BKC$ 的所有特征值都可以任意接近期望的位置。

下述算法可以说明这一定理，这里不详细讨论它的证明。

算法 6.8 对于完全能控能观的系统方程 (6.17)，$\operatorname{rank} B = p$，$\operatorname{rank} C = q$，使用输出反馈控制律 (6.18)，$A+BK$ 的 $\min\{p+q-1,n\}$ 个特征值可以被配置到接近期望特征值的位置。输出反馈增益矩阵 K 可以根据下列步骤确定。

(1) 考虑 $K_1 = l_1 m_1$，其中，$l_1 = \begin{bmatrix} l_{11} & \cdots & l_{1p} \end{bmatrix}^{\mathrm{T}}$，$m_1 = \begin{bmatrix} m_{11} & \cdots & m_{1q} \end{bmatrix}$。记 $W_1(s) = C\operatorname{adj}(sI - A)B$，可以导出 $\alpha_{\mathrm{c}1}(s) = \alpha(s) - m_1 W_1(s) l_1$，其中 $\alpha_{\mathrm{c}1}(s) = \det(sI - A - BK_1 C)$，$\alpha(s) = \det(sI - A)$。把 $p-1$ 个特征值配置到 $\lambda_1^*, \cdots, \lambda_{p-1}^*$ 处，可以得到

$$\begin{cases} \alpha_{\mathrm{c}1}(\lambda_1^*) = \alpha(\lambda_1^*) - m_1 W_1(\lambda_1^*) l_1 = 0 \\ \vdots \\ \alpha_{\mathrm{c}1}(\lambda_{p-1}^*) = \alpha(\lambda_{p-1}^*) - m_1 W_1(\lambda_{p-1}^*) l_1 = 0 \end{cases}$$

选定 m_1 和 l_{11}，然后求解上述联立方程以得到 $l_{12}, \cdots, l_{1p}$。进一步，解得 K_1。

(2) 保持第一步中配置好的 $p-1$ 个特征值不变，然后把剩余的 q 个特征值配置到 λ_p^*，$\cdots, \lambda_{p+q-1}^*$。经过第 (1) 步后，系统变为 $(A+BK_1 C, B, C)$。考虑 $K_2 = l_2 m_2$，其中，$l_2 = \begin{bmatrix} l_{21} & \cdots & l_{2p} \end{bmatrix}^{\mathrm{T}}$，$m_2 = \begin{bmatrix} m_{21} & \cdots & m_{2q} \end{bmatrix}$。记 $W_2(s) = C\operatorname{adj}(sI - A - BK_1 C)B$。可以导出 $\alpha_{\mathrm{c}2}(s) = \alpha_{\mathrm{c}1}(s) - m_2 W_2(s) l_2$，其中 $\alpha_{\mathrm{c}2}(s) = \det(sI - A - BK_1 C - BK_2 C)$。

为了保证 $\lambda_1^*, \cdots, \lambda_{p-1}^*$ 不变，不论 m_2 为何值，总是需要保证

$$\begin{cases} m_2 W_2(\lambda_1^*) l_2 = 0 \\ \vdots \\ m_2 W_2(\lambda_{p-1}^*) l_2 = 0 \end{cases}$$

成立，所以
$$\begin{cases} W_2(\lambda_1^*)l_2 = 0 \\ \vdots \\ W_2(\lambda_{p-1}^*)l_2 = 0 \end{cases}$$

选择 l_2 使上述联立方程成立，进一步，为了把 q 个特征值配置到 $\lambda_p^*, \cdots, \lambda_{p+q-1}^*$，应有

$$\begin{cases} \alpha_{c1}(\lambda_p^*) - m_2 W_2(\lambda_p^*)l_2 = 0 \\ \vdots \\ \alpha_{c1}(\lambda_{p+q-1}^*) - m_2 W_2(\lambda_{p+q-1}^*)l_2 = 0 \end{cases}$$

求解上述联立方程以获得 m_2。然后，计算 K_2。

(3) 计算 $K = K_1 + K_2$。

算法 6.8 中假设了 $p + q - 1 \leqslant n$。如果 $p + q - 1 > n$，第 (2) 步只需要配置剩余的 $n - p + 1$ 个特征值。

例 6.11 考虑一个线性定常系统方程 (6.17)，其系数矩阵为

$$A = \begin{bmatrix} 0 & -1 & 1 & 0 \\ 0 & 1 & 1 & 0 \\ 0 & 0 & 2 & 0 \\ 0 & 0 & 0 & -3 \end{bmatrix}, B = \begin{bmatrix} 1 & 0 & 0 \\ 0 & 1 & 0 \\ 0 & 0 & 1 \\ 1 & 0 & 0 \end{bmatrix}, C = \begin{bmatrix} 0 & 1 & 0 & 0 \\ 1 & 0 & 0 & 1 \end{bmatrix}$$

尝试把闭环系统的特征值配置在 -1，-2，-3 和 -4。

可以验证 A 是循环的且 (A, B, C) 是能控能观的。因为 $\operatorname{rank} B + \operatorname{rank} C - 1 = 4$，所有特征值都可以通过输出反馈配置。

根据算法 1.1，可以得到 $\alpha(s) = s^4 - 7s^2 + 6s$，且

$$W_1(s) = C \operatorname{adj}(sI - A)B = \begin{bmatrix} 0 & s(s-2)(s+3) & s(s+3) \\ (s-1)(s-2)(2s+3) & -(s-2)(s+3) & (s-2)(s+3) \end{bmatrix}$$

设计输出反馈增益矩阵 K_1 把 $p - 1 = 2$ 个特征值配置到 -1 和 -3。选择 $m_1 = \begin{bmatrix} -9 & 1 \end{bmatrix}$，$l_1 = \begin{bmatrix} 0 & l_{12} & l_{13} \end{bmatrix}^{\mathrm{T}}$，求解

$$\begin{cases} \alpha(-1) - m_1 W_1(-1) l_1 = 0 \\ \alpha(-3) - m_1 W_1(-3) l_1 = 0 \end{cases}$$

$l_1 = \begin{bmatrix} 0 & 0 & 1 \end{bmatrix}^{\mathrm{T}}$ 是上述联立方程的一个解，所以

$$K_1 = l_1 m_1 = \begin{bmatrix} 0 \\ 0 \\ -1 \end{bmatrix} \begin{bmatrix} -9 & 1 \end{bmatrix} = \begin{bmatrix} 0 & 0 \\ 0 & 0 \\ 9 & -1 \end{bmatrix}$$

6.2 输出反馈控制

接下来，可以计算出 $\alpha_{c1}(s) = s^4 - 15s^2 - 20s - 6$，且

$$W_2(s) = C\operatorname{adj}(sI - A - BK_1C)$$
$$= \begin{bmatrix} -(2s+3) & (s+3)(s-1)^2 & s(s+3) \\ (2s+3)(s^2-3s-7) & -(s+3)(s-11) & (s+3)(s-2) \end{bmatrix}$$

为了保持特征值 -1 和 -3 不变，需满足

$$\begin{cases} W_2(-1)l_2 = [\ -1\ \ 8\ \ -2\ ]l_2 = 0 \\ W_2(-3)l_2 = [\ 3\ \ 0\ \ 0\ ]l_2 = 0 \end{cases}$$

$l_2 = [\ 0\ \ 1\ \ 4\ ]^\mathrm{T}$ 是上述联立方程的一个解。

为了把剩余的两个极点配置到 -2 和 -4，需满足

$$\begin{cases} \alpha_{c1}(-2) - m_2 W_2(-2)l_2 = -10 - m_2 \begin{bmatrix} 1 & 9 & -2 \\ -3 & 13 & -4 \end{bmatrix} \begin{bmatrix} 0 \\ 1 \\ 4 \end{bmatrix} = 0 \\ \\ \alpha_{c1}(-4) - m_2 W_2(-4)l_2 = 90 - m_2 \begin{bmatrix} 5 & -25 & 4 \\ -105 & -15 & 6 \end{bmatrix} \begin{bmatrix} 0 \\ 1 \\ 4 \end{bmatrix} = 0 \end{cases}$$

$m_2 = [\ -10\ \ 0\ ]$ 是上述方程的一个解，所以

$$K_2 = l_2 m_2 = \begin{bmatrix} 0 \\ 1 \\ 4 \end{bmatrix} \begin{bmatrix} -10 & 0 \end{bmatrix} = \begin{bmatrix} 0 & 0 \\ -10 & 0 \\ -40 & 0 \end{bmatrix}$$

最后，可以计算出

$$K = K_1 + K_2 = \begin{bmatrix} 0 & 0 \\ -10 & 0 \\ -31 & -1 \end{bmatrix}$$

并且

$$A + B(K_1 + K_2)C = \begin{bmatrix} 0 & -1 & 1 & 0 \\ 0 & -9 & 1 & 0 \\ -1 & -31 & 2 & -1 \\ 0 & 0 & 0 & -3 \end{bmatrix}$$

它的特征值为 -1，-2，-3 和 -4。

6.2.3 动态输出反馈补偿器

这一小节来讨论线性定常系统的动态输出反馈控制律。

考虑线性定常系统方程 (6.17)，引入动态输出补偿器，

$$\dot{z}(t) = A_1 z(t) + B_1 y(t), \ w(t) = C_1 z(t) + D_1 y(t) \tag{6.22}$$

其中，$z(t) \in \mathbb{R}^l$，$w(t) \in \mathbb{R}^p$，A_1，B_1，C_1 和 D_1 是具有相应维数的常值矩阵。

考虑

$$u(t) = v(t) - w(t) = v(t) - C_1 z(t) - D_1 y(t) \tag{6.23}$$

其中 $v(t) \in \mathbb{R}^p$。相应的闭环系统为

$$\begin{bmatrix} \dot{x}(t) \\ \dot{z}(t) \end{bmatrix} = \begin{bmatrix} A - BD_1C & -BC_1 \\ B_1 C & A_1 \end{bmatrix} \begin{bmatrix} x(t) \\ z(t) \end{bmatrix} + \begin{bmatrix} B \\ 0 \end{bmatrix} v(t), \ y(t) = Cx(t) \tag{6.24}$$

它的结构图如图 6.4 所示。

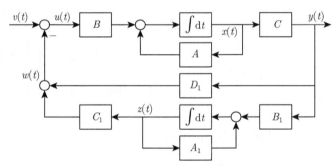

图 6.4 动态输出反馈闭环系统的结构图

记

$$x_c(t) = \begin{bmatrix} x(t) \\ z(t) \end{bmatrix}, \ v_c(t) = \begin{bmatrix} -v(t) \\ 0 \end{bmatrix}, \ y_c(t) = \begin{bmatrix} y(t) \\ z(t) \end{bmatrix} = \begin{bmatrix} C & 0 \\ 0 & I \end{bmatrix} x_c(t)$$

系统方程 (6.24) 可以写作

$$\dot{x}_c(t) = (A_e + B_e K_e C_e) x_c(t) + B_e v_c(t), \ y_c(t) = C_e x_c(t) \tag{6.25}$$

其中，

$$A_e = \begin{bmatrix} A & 0 \\ 0 & 0 \end{bmatrix}, \ B_e = \begin{bmatrix} -B & 0 \\ 0 & I \end{bmatrix}, \ C_e = \begin{bmatrix} C & 0 \\ 0 & I \end{bmatrix}, \ K_e = \begin{bmatrix} D_1 & C_1 \\ B_1 & A_1 \end{bmatrix}$$

这种情况下，闭环系统方程 (6.25) 可以被看作是一个等效的开环系统 (A_e, B_e, C_e) 应用了等效的静态输出反馈控制律而构成的系统，且反馈增益为 K_e。结构图如图 6.5 所示。

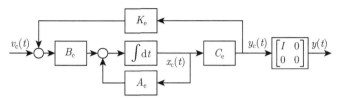

图 6.5 具有动态输出反馈补偿器的等效闭环系统

定理 6.8 当且仅当系统方程 (6.17) 完全能控能观时，闭环系统方程 (6.25) 完全能控能观。

证明略。

定理 6.9 如果系统 (A_e, B_e, C_e) 是完全能控能观的，其中 $\operatorname{rank} B_e = p+l$，$\operatorname{rank} C_e = q+l$，那么对于几乎所有的 (A_e, B_e, C_e)，存在常值矩阵 K_e 使得 $A_e + B_e K_e C_e$ 的 $\min\{p+q+2l-1, n+l\}$ 个特征值可以任意接近期望的位置。如果 $p+q+l-1 \geqslant n$，那么对于几乎所有的 (A_e, B_e, C_e)，$A_e + B_e K_e C_e$ 的所有特征值可以任意接近期望的位置。

这一定理可以拓展定理 6.9 得到，证明略。根据定理 6.9，可以得到下述结论。

推论 6.1 如果系统方程 (6.17) 完全能控能观，那么对于几乎所有的 (A, B, C)，存在动态输出反馈补偿器方程 (6.22) 使得闭环系统方程 (6.24) 的特征值可以任意接近期望的位置。

以上讨论表明，对于任意完全能控能观的系统，如果需要设计动态补偿器来构成输出反馈闭环系统，那么总可以把动态补偿器的参数转化为等效的静态输出反馈增益矩阵来设计，且动态补偿器的维数 l 可以利用下式设计，

$$l \geqslant n+1-p-q$$

如果 $l = 0$，动态输出反馈就退化为静态输出反馈。动态补偿器相较于单纯的静态输出反馈可以更有效地改善系统的性能。例如，考虑完全能控能观的系统 (A, B, C)，$\dim A = 5$，$\operatorname{rank} B = 2$，$\operatorname{rank} C = 2$，并且 A 的所有特征值都具有正实部。因为 $p+q-1 = 3 < n = 5$，不能通过静态输出反馈来镇定系统。但是当应用动态输出补偿器时，只需要 $\dim A_1 = l = n+1-p-q = 2$，选择合适的动态补偿器 (A_1, B_1, C_1, D_1) 就能够镇定系统。

例 6.12 考虑线性定常系统方程 (6.17)，其中，

$$A = \begin{bmatrix} -2 & 1 \\ 0 & -1 \end{bmatrix}, B = \begin{bmatrix} 0 \\ 1 \end{bmatrix}, C = \begin{bmatrix} 1 & 0 \end{bmatrix}$$

尝试设计动态输出反馈补偿器使得闭环系统的所有特征值都为 -3。

可以验证 (A, B, C) 完全能控能观。需要设计维数为 $l = n+1-p-q = 1$ 的动态输出反馈补偿器。

动态输出反馈补偿器的形式为

$$\dot{z}(t) = a_1 z(t) + b_1 y(t), \ w(t) = c_1 z(t) + d_1 y(t)$$

等效的开环系统为

$$A_e = \begin{bmatrix} -2 & 1 & 0 \\ 0 & -1 & 0 \\ 0 & 0 & 0 \end{bmatrix}, B_e = \begin{bmatrix} 0 & 0 \\ -1 & 0 \\ 0 & 1 \end{bmatrix}, C_e = \begin{bmatrix} 1 & 0 & 0 \\ 0 & 0 & 1 \end{bmatrix}$$

等效的静态输出反馈增益矩阵为

$$K_e = \begin{bmatrix} d_1 & c_1 \\ b_1 & a_1 \end{bmatrix}$$

相应的闭环系统的系数矩阵为

$$A_e + B_e K_e C_e = \begin{bmatrix} -2 & 1 & 0 \\ 0 & -1 & 0 \\ 0 & 0 & 0 \end{bmatrix} + \begin{bmatrix} 0 & 0 \\ -1 & 0 \\ 0 & 1 \end{bmatrix} \begin{bmatrix} d_1 & c_1 \\ b_1 & a_1 \end{bmatrix} \begin{bmatrix} 1 & 0 & 0 \\ 0 & 0 & 1 \end{bmatrix}$$

$$= \begin{bmatrix} -2 & 1 & 0 \\ -d_1 & -1 & -c_1 \\ b_1 & 0 & a_1 \end{bmatrix}$$

相应的特征多项式为

$$\alpha_k(s) = \det(sI - A_e - B_e K_e C_e)$$
$$= s^3 + (3 - a_1)s^2 + (2 - 3a_1 + d_1)s - d_1 a_1 + b_1 c_1 - 2a_1$$

期望的特征多项式为

$$\alpha_c(s) = (s+3)^3 = s^3 + 9s^2 + 27s + 27$$

进一步，可以计算出

$$K_e = \begin{bmatrix} d_1 & c_1 \\ b_1 & a_1 \end{bmatrix} = \begin{bmatrix} 7 & \dfrac{27}{b_1} \\ b_1 & -6 \end{bmatrix}$$

K_e 中存在一个自由参数，这表明 K_e 是不唯一的，也就是说，动态补偿器是不唯一的。动态补偿器的传递函数为

$$\frac{w(s)}{y(s)} = \frac{7s + 15}{s + 6}$$

它不含有自由参数。

需要说明的是，动态输出反馈补偿器自身应该是稳定的。也就是说，需要保证增广系统方程 (6.24) 的所有特征根都具有负实部。

6.3 解耦控制

一般情况下，对于 MIMO 系统，每个输入会同时激励出多个输出，每个输出也会同时受到多个输入的影响。这一现象被称作耦合，它非常不利于系统的设计。如果系统输入的数量与输出的数量相同，可以尝试设计控制律使得每个输入仅控制一个输出，每个输出也仅受到一个输入的影响，这就是解耦控制问题。

6.3.1 基本概念

考虑线性定常系统

$$\dot{x}(t) = Ax(t) + Bu(t), \ y(t) = Cx(t) \tag{6.26}$$

其中，$x(t) \in \mathbb{R}^n$ 是状态向量，$u(t) \in \mathbb{R}^p$ 是输入向量，$y(t) \in \mathbb{R}^p$ 是输出向量，A，B 和 C 是具有相应维数的实常值矩阵。考虑在零初始条件下，系统的输入和输出关系可以用传递函数矩阵描述，

$$y(s) = G(s)u(s) = C(sI-A)^{-1}Bu(s) = \begin{bmatrix} g_{11}(s) & \cdots & g_{1p}(s) \\ \vdots & & \vdots \\ g_{p1}(s) & \cdots & g_{pp}(s) \end{bmatrix} u(s)$$

系统的第 i 个输出 $y_i(s)$，$i = 1, \cdots, p$，可以表示为

$$y_i(s) = \sum_{j=1}^{p} g_{ij}(s) u_j(s)$$

其中，$u_j(s)$ 表示系统的第 j 个输入，$j = 1, \cdots, p$。可以看出如果 $G(s)$ 每行有超过一个的非零元时，每个输出都受到多个输入的影响。

定义 6.1（解耦系统） 如果系统的传递函数矩阵是对角非奇异矩阵，则称该系统是解耦的。当系统是解耦的，第 i 个输入仅控制第 i 个输出，第 i 个输出也仅受到第 i 个输入的影响，$i = 1, \cdots, p$。

最常使用的解耦控制律为

$$u(t) = Kx(t) + Hv(t) \tag{6.27}$$

其中，$v(t) \in \mathbb{R}^p$，K 是常值矩阵，H 是非奇异常值矩阵，它可以被看作是状态反馈和输入变换的结合。

把式 (6.27) 代入式 (6.26) 中，可以得到相应的闭环系统为

$$\dot{x}(t) = (A + BK)x(t) + BHv(t), \ y(t) = Cx(t) \tag{6.28}$$

它的传递函数矩阵是

$$G_{\mathrm{KH}}(s) = C(sI - A - BK)^{-1}BH$$

系统方程 (6.28) 的结构图如图 6.6 所示。由于解耦后的系统每个变量的控制可以单独进行，所以解耦控制很大程度上简化了控制过程。这类解耦控制也被称作 Morgan 受限解耦问题。

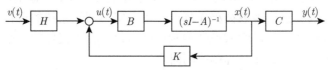

图 6.6 状态反馈解耦的结构图

接下来，首先讨论开环传递函数矩阵 $G(s)$ 和闭环传递函数矩阵 $G_{KH}(s)$ 之间的关系。

引理 6.7 对于线性定常系统方程 (6.26)，引入控制律 (6.27)，它的开环传递函数矩阵 $G(s)$ 和闭环传递函数矩阵 $G_{KH}(s)$ 满足

$$G_{KH}(s) = G(s)[I + K(sI - A - BK)^{-1}B]H = G(s)[I - K(sI - A)^{-1}B]^{-1}H \quad (6.29)$$

证明 可以验证

$$\begin{aligned}
G_{KH}(s) &= C(sI - A - BK)^{-1}BH \\
&= C(sI - A)^{-1}[(sI - A - BK) + BK](sI - A - BK)^{-1}BH \\
&= C(sI - A)^{-1}[B + BK(sI - A - BK)^{-1}B]H \\
&= C(sI - A)^{-1}B[I + K(sI - A - BK)^{-1}B]H \\
&= G(s)[I + K(sI - A - BK)^{-1}B]H
\end{aligned}$$

同时，也可以验证

$$\begin{aligned}
&[I + K(sI - A - BK)^{-1}B][I - K(sI - A)^{-1}B] \\
&= I + K(sI - A - BK)^{-1}B - K(sI - A)^{-1}B - K(sI - A - BK)^{-1}BK(sI - A)^{-1}B \\
&= I + K(sI - A - BK)^{-1}B - K(sI - A - BK)^{-1}(sI - A - BK)(sI - A)^{-1}B \\
&\quad - K(sI - A - BK)^{-1}BK(sI - A)^{-1}B \\
&= I + K(sI - A - BK)^{-1}B - K(sI - A - BK)^{-1}(sI - A)(sI - A)^{-1}B \\
&\quad + K(sI - A - BK)^{-1}BK(sI - A)^{-1}B - K(sI - A - BK)^{-1}BK(sI - A)^{-1}B \\
&= I
\end{aligned}$$

所以，可以得到

$$I + K(sI - A - BK)^{-1}B = [I - K(sI - A)^{-1}B]^{-1}$$

这表明式 (6.29) 成立。 □

6.3 解耦控制

接下来，引入传递函数矩阵的两个特征量。

定义 6.2（传递函数矩阵的特征量） 把 $g_{ij}(s)$ 的分子多项式和分母多项式的阶次分别记作 m_{ij} 和 n_{ij}，同时记 $\sigma_{ij}=n_{ij}-m_{ij}$，$i=1,\cdots,p$，$j=1,\cdots,p$。定义传递函数矩阵 $G(s)$ 的两个特征量，

$$d_i \triangleq \min\{\sigma_{i1},\sigma_{i2},\cdots,\sigma_{ip}\}-1$$

$$E_i \triangleq \lim_{s\to\infty} s^{d_i+1}G_i(s)$$

其中，$G_i(s)$ 是 $G(s)$ 的第 i 行。

例 6.13 对于给定的传递函数矩阵

$$G(s)=\begin{bmatrix} \dfrac{s+2}{s^2+2s+1} & \dfrac{1}{s^2+s+2} \\ \dfrac{1}{s^2+2s+1} & \dfrac{3}{s^2+s+4} \end{bmatrix}$$

可以计算出 $d_1=\min\{1,2\}-1=0$，$E_1=\lim_{s\to\infty}s^{0+1}G_1(s)=\begin{bmatrix} 1 & 0 \end{bmatrix}$，$d_2=\min\{2,2\}-1=1$，$E_2=\lim_{s\to\infty}s^{1+1}G_2(s)=\begin{bmatrix} 1 & 3 \end{bmatrix}$。

1.3.3小节中指出，传递函数矩阵 $G(s)$ 可以表示为

$$G(s)=\frac{C\,\mathrm{adj}(sI-A)B}{\det(sI-A)}=\frac{1}{\alpha(s)}C(R_{n-1}s^{n-1}+R_{n-2}s^{n-2}+\cdots+R_1 s+R_0)B$$

其中，$\alpha(s)=s^n+a_1 s^{n-1}+\cdots+a_{n-1}s+a_n$ 是 A 的特征多项式，$R_0,\cdots,R_{n-1}$ 是常值矩阵，可以根据算法 1.1 计算。

在此基础上，可以得到

$$G_i(s)=\frac{c_i\,\mathrm{adj}(sI-A)B}{\det(sI-A)}=\frac{1}{\alpha(s)}c_i(R_{n-1}s^{n-1}+R_{n-2}s^{n-2}+\cdots+R_1 s+R_0)B$$

其中，c_i 是 C 的第 i 行，$i=1,\cdots,p$。

根据定义 6.2，$G_i(s)$ 每个元的分母与分子的最小阶次差为 d_i+1。所以，$s^{n-1},\cdots,s^{n-d_i}$ 的系数为零，即

$$\begin{cases} c_i R_{n-1}B = c_i B = 0 \\ c_i R_{n-2}B = c_i(AR_{n-1}+a_1 I)B = c_i AB = 0 \\ \vdots \\ c_i R_{n-d_i}B = c_i A^{d_i-1}B = 0 \\ E_i = c_i R_{n-d_i-1}B = c_i A^{d_i}B \neq 0 \end{cases} \quad (6.30)$$

把闭环系统传递函数矩阵 $G_{\mathrm{KH}}(s)$ 的两个特征量记为 $\bar{d}_i$ 和 $\bar{E}_i$，类似地，可以得到

$$\begin{cases} c_i BH = 0 \\ c_i(A+BK)BH = 0 \\ \vdots \\ c_i(A+BK)^{\bar{d}_i-1}BH = 0 \\ \bar{E}_i = c_i(A+BK)^{\bar{d}_i}BH \neq 0 \end{cases} \tag{6.31}$$

引理 6.8 对于引入了控制律 (6.27) 的线性定常系统方程 (6.26)，开环传递函数矩阵 $G(s)$ 和闭环传递函数矩阵 $G_{\mathrm{KH}}(s)$ 的特征量满足

$$\bar{d}_i = d_i,\ \bar{E}_i = E_i H \tag{6.32}$$

证明 首先证明

$$c_i(A+BK)^k = c_i A^k,\ k = 1,2,\cdots,d_i \tag{6.33}$$

这里使用数学归纳法来证明式 (6.33)。

当 $k=1$ 时，根据式 (6.30) 可知 $c_i B = 0$，所以 $c_i(A+BK) = c_i A$。

当 $0 < L \leqslant d_i - 1$ 时，假设 $c_i(A+BK)^L = c_i A^L$。根据式 (6.30) 可知 $c_i A^L B = 0$，进一步，可以得到

$$c_i(A+BK)^{L+1} = c_i A^L(A+BK) = c_i A^{L+1} + c_i A^L BK = c_i A^{L+1}$$

所以，式 (6.33) 成立。

结合式 (6.30)，式 (6.31) 和式 (6.33)，可以得到

$$c_i(A+BK)^k BH = c_i A^k BH = 0,\ k=1,2,\cdots,d_i - 1$$

同时，当 $k = d_i$ 时，可以得到

$$c_i(A+BK)^{d_i} BH = c_i A^{d_i} BH \neq 0$$

由此可以证明式 (6.32) 成立。 □

6.3.2 状态反馈解耦

定理 6.10 线性定常系统方程 (6.26) 可以通过控制律 (6.27) 实现解耦，当且仅当

$$E \triangleq \begin{bmatrix} E_1 \\ \vdots \\ E_p \end{bmatrix} \tag{6.34}$$

是非奇异的，其中，E_i 在定义 6.2 中给出，$i = 1,\cdots,p$。

6.3 解耦控制

证明 首先，证明必要性。

如果 $G(s)$ 可以被解耦，那么 $G_{\text{KH}}(s)$ 是非奇异的对角阵，因此

$$\bar{E} \triangleq \begin{bmatrix} \bar{E}_1 \\ \vdots \\ \bar{E}_p \end{bmatrix}$$

一定是对角阵。同时，$\bar{E}_i \neq 0, i = 1, \cdots, p$。所以 $\bar{E}$ 是非奇异的。另一方面，由于 $\bar{E}_i = E_i H$，且 H 是非奇异的，所以 E 是非奇异的。

接下来，证明充分性。

根据式 (6.30)，$G_i(s)$ 可以被表示为

$$G_i(s) = c_i(sI - A)^{-1}B = c_i \left(\sum_{k=0}^{\infty} \frac{A^k}{s^{k+1}} \right) B = \sum_{k=d_i}^{\infty} \frac{c_i A^k B}{s^{k+1}}$$

$$= \frac{1}{s^{d_i+1}} \left(c_i A^{d_i} B + c_i A^{d_i+1} \left(\frac{I}{s} + \frac{A}{s^2} + \frac{A^2}{s^3} + \cdots \right) B \right)$$

$$= \frac{1}{s^{d_i+1}}(E_i + F_i(sI - A)^{-1}B)$$

其中，$F_i \triangleq c_i A^{d_i+1}, \ i = 1, \cdots, p$。

进一步，$G(s)$ 可以被表示为

$$G(s) = \begin{bmatrix} \frac{1}{s^{d_1+1}} & & & \\ & \frac{1}{s^{d_2+1}} & & \\ & & \ddots & \\ & & & \frac{1}{s^{d_p+1}} \end{bmatrix} \begin{bmatrix} E_1 + F_1(sI - A)^{-1}B \\ E_2 + F_2(sI - A)^{-1}B \\ \vdots \\ E_p + F_p(sI - A)^{-1}B \end{bmatrix}$$

$$= \begin{bmatrix} \frac{1}{s^{d_1+1}} & & & \\ & \frac{1}{s^{d_2+1}} & & \\ & & \ddots & \\ & & & \frac{1}{s^{d_p+1}} \end{bmatrix} \begin{bmatrix} E + F(sI - A)^{-1}B \end{bmatrix}$$

其中，

$$F \triangleq \begin{bmatrix} F_1 \\ \vdots \\ F_p \end{bmatrix}$$

选择

$$u(t) = Kx(t) + Hv(t) = -E^{-1}Fx(t) + E^{-1}v(t) \tag{6.35}$$

根据引理 6.7，可以得到

$$\begin{aligned}G_{\mathrm{KH}}(s) &= G(s)[I - K(sI-A)^{-1}B]^{-1}H \\ &= G(s)[I + E^{-1}F(sI-A)^{-1}B]^{-1}E^{-1} \\ &= G(s)[E + F(sI-A)^{-1}B]^{-1}\end{aligned}$$

由此可以获得

$$G_{\mathrm{KH}}(s) = \begin{bmatrix} \dfrac{1}{s^{d_1+1}} & & & \\ & \dfrac{1}{s^{d_2+1}} & & \\ & & \ddots & \\ & & & \dfrac{1}{s^{d_p+1}} \end{bmatrix} \tag{6.36}$$

这表明控制律 (6.35) 能够解耦 $G(s)$。 □

可以看出应用 (6.35)，$G(s)$ 可以被解耦为具有式 (6.36) 形式的 $G_{\mathrm{KH}}(s)$。这里，$G_{\mathrm{KH}}(s)$ 对角线上的元都是积分项，所以这种方法也被称作积分解耦。

例 6.14 考虑线性定常系统方程 (6.26)，其中，

$$A = \begin{bmatrix} 0 & 0 & 0 \\ 0 & 0 & 1 \\ -1 & -2 & -3 \end{bmatrix}, B = \begin{bmatrix} 1 & 0 \\ 0 & 0 \\ 0 & 1 \end{bmatrix}, C = \begin{bmatrix} 1 & 1 & 0 \\ 0 & 0 & 1 \end{bmatrix}$$

设计解耦控制律 (6.27) 以解耦这一系统。

传递函数矩阵为

$$G(s) = C(sI-A)^{-1}B = \begin{bmatrix} \dfrac{s^2+3s+1}{s(s+1)(s+2)} & \dfrac{1}{(s+1)(s+2)} \\ \dfrac{-1}{(s+1)(s+2)} & \dfrac{s}{(s+1)(s+2)} \end{bmatrix}$$

相应的特征量是 $d_1 = 0$，$E_1 = \lim\limits_{s\to\infty} sG_1(s) = \begin{bmatrix} 1 & 0 \end{bmatrix}$，$d_2 = 0$，$E_2 = \lim\limits_{s\to\infty} sG_2(s) = \begin{bmatrix} 0 & 1 \end{bmatrix}$。$E = \begin{bmatrix} 1 & 0 \\ 0 & 1 \end{bmatrix}$ 是非奇异的，所以这一系统可以被解耦。

计算

$$F = \begin{bmatrix} c_1 A^{d_1+1} \\ c_2 A^{d_2+1} \end{bmatrix} = \begin{bmatrix} 0 & 0 & 1 \\ -1 & -2 & -3 \end{bmatrix}$$

6.3 解耦控制

进一步可得

$$u(t) = Kx(t) + Hv(t) = -E^{-1}Fx(t) + E^{-1}v(t) = \begin{bmatrix} 0 & 0 & 1 \\ 1 & 2 & 3 \end{bmatrix} x(t) + \begin{bmatrix} 1 & 0 \\ 0 & 1 \end{bmatrix} v(t)$$

解耦系统的系数矩阵为

$$\bar{A} = A + BK = \begin{bmatrix} 0 & 0 & -1 \\ 0 & 0 & 1 \\ 0 & 0 & 0 \end{bmatrix}, \quad \bar{B} = BH = \begin{bmatrix} 1 & 0 \\ 0 & 0 \\ 0 & 1 \end{bmatrix}, \quad \bar{C} = C = \begin{bmatrix} 1 & 1 & 0 \\ 0 & 0 & 1 \end{bmatrix}$$

相应的传递函数矩阵为

$$G_{\mathrm{KH}}(s) = \begin{bmatrix} \dfrac{1}{s} & 0 \\ 0 & \dfrac{1}{s} \end{bmatrix}$$

应用积分解耦后，系统传递函数矩阵的对角元为至少一阶的积分项，这不能被应用到实际系统中。所以，需要进行极点配置来改变极点位置。

定理 6.11 对于线性定常系统方程 (6.26)，如果式 (6.34) 中定义的 E 是非奇异的，控制律 (6.27) 可以使得

$$G_{\mathrm{KH}}(s) = \begin{bmatrix} \dfrac{1}{s^{d_1+1} + a_{11}s^{d_1} + \cdots + a_{1d_1+1}} & & \\ & \ddots & \\ & & \dfrac{1}{s^{d_p+1} + a_{p1}s^{d_p} + \cdots + a_{pd_p+1}} \end{bmatrix}$$

其中，$a_{i1}, \cdots, a_{id_i+1}$ 是 $G_{\mathrm{KH}}(s)$ 第 i 个对角元期望的极点多项式系数，$i = 1, \cdots, p$，$H = E^{-1}$，$K = -E^{-1}\tilde{F}$，

$$\tilde{F} \triangleq \begin{bmatrix} \tilde{F}_1 \\ \vdots \\ \tilde{F}_p \end{bmatrix}$$

且 $\tilde{F}_i \triangleq c_i A^{d_i+1} + a_{i1} c_i A^{d_i} + a_{i2} c_i A^{d_i-1} + \cdots + a_{id_i+1} c_i$。

证明 假设 $G_{\mathrm{KH}}(s)$ 第 i 个对角元期望的极点多项式是

$$\alpha_{ci}(s) = s^{d_i+1} + a_{i1}s^{d_i} + a_{i2}s^{d_i-1} + \cdots + a_{id_i}s + a_{id_i+1}$$

用 $G_i(s)$ 左乘 $\alpha_{ci}(s)$，$G_i(s)$ 是 $G(s)$ 的第 i 行，可以得到

$$\alpha_{ci}(s) G_i(s) = \alpha_{ci}(s) c_i (sI - A)^{-1} B = \alpha_{ci}(s) \sum_{k=0}^{\infty} \dfrac{c_i A^k B}{s^{k+1}}$$

$$= c_i A^{d_i} B + (c_i A^{d_i+1} + a_{i1} c_i A^{d_i} + \cdots + a_{id_i+1} c_i) \left(\frac{I}{s} + \frac{A}{s^2} + \frac{A^2}{s^3} + \cdots \right) B$$

$$= c_i A^{d_i} B + (c_i A^{d_i+1} + a_{i1} c_i A^{d_i} + \cdots + a_{id_i+1} c_i)(sI - A)^{-1} B$$

$$= E_i + \tilde{F}_i (sI - A)^{-1} B$$

由此可得

$$G_i(s) = \frac{1}{\alpha_{ci}(s)} [E_i + \tilde{F}_i (sI - A)^{-1} B]$$

这表明

$$G(s) = \begin{bmatrix} \frac{1}{\alpha_{c1}(s)} & & & \\ & \frac{1}{\alpha_{c2}(s)} & & \\ & & \ddots & \\ & & & \frac{1}{\alpha_{cp}(s)} \end{bmatrix} [E + \tilde{F}(sI - A)^{-1} B]$$

选择控制律

$$u(t) = Kx(t) + Hv(t) = -E^{-1}\tilde{F}x(t) + E^{-1}v(t) \tag{6.37}$$

根据引理 6.7，可以推导得到

$$G_{\text{KH}}(s) = G(s)[I - K(sI - A)^{-1}B]^{-1} H = G(s)[I + E^{-1}\tilde{F}(sI - A)^{-1}B]^{-1} E^{-1}$$

$$= G(s)[E + \tilde{F}(sI - A)^{-1}B]^{-1} = \begin{bmatrix} \frac{1}{\alpha_{c1}(s)} & & & \\ & \frac{1}{\alpha_{c2}(s)} & & \\ & & \ddots & \\ & & & \frac{1}{\alpha_{cp}(s)} \end{bmatrix}$$

这表明 $G_{\text{KH}}(s)$ 的极点可以通过式 (6.37) 任意配置。□

例 6.15 考虑线性定常系统方程 (6.26)，其中，

$$A = \begin{bmatrix} 0 & 0 & 0 \\ 0 & 0 & 1 \\ -1 & -2 & -3 \end{bmatrix}, \quad B = \begin{bmatrix} 0 & 0 \\ 1 & 0 \\ 0 & 1 \end{bmatrix}, \quad C = \begin{bmatrix} 1 & 1 & 0 \\ 0 & 0 & 1 \end{bmatrix}$$

设计控制律 (6.27) 以解耦这一系统并把所有极点配置在 s 左半平面。

传递函数矩阵为

$$G(s) = C(sI - A)^{-1}B = \begin{bmatrix} \dfrac{s+3}{(s+1)(s+2)} & \dfrac{1}{(s+1)(s+2)} \\ \dfrac{-2}{(s+1)(s+2)} & \dfrac{s}{(s+1)(s+2)} \end{bmatrix}$$

这一传递函数矩阵的特征量为 $d_1 = 0$,$E_1 = \lim\limits_{s \to \infty} sG_1(s) = \begin{bmatrix} 1 & 0 \end{bmatrix}$,$d_2 = 0$,$E_2 = \lim\limits_{s \to \infty} sG_2(s) = \begin{bmatrix} 0 & 1 \end{bmatrix}$。$E = \begin{bmatrix} 1 & 0 \\ 0 & 1 \end{bmatrix}$ 是非奇异的,所以这一系统可以被解耦。

选择 $\alpha_{c1}(s) = s+4$,$\alpha_{c2}(s) = s+5$。计算

$$\tilde{F}_1 = c_1 A + a_{11} c_1 = \begin{bmatrix} 1 & 1 & 0 \end{bmatrix} \begin{bmatrix} 0 & 0 & 0 \\ 0 & 0 & 1 \\ -1 & -2 & -3 \end{bmatrix} + 4 \begin{bmatrix} 1 & 1 & 0 \end{bmatrix} = \begin{bmatrix} 4 & 4 & 1 \end{bmatrix}$$

$$\tilde{F}_2 = c_2 A + a_{21} c_2 = \begin{bmatrix} 0 & 0 & 1 \end{bmatrix} \begin{bmatrix} 0 & 0 & 0 \\ 0 & 0 & 1 \\ -1 & -2 & -3 \end{bmatrix} + 5 \begin{bmatrix} 0 & 0 & 1 \end{bmatrix} = \begin{bmatrix} -1 & -2 & 2 \end{bmatrix}$$

所以,

$$\tilde{F} = \begin{bmatrix} \tilde{F}_1 \\ \tilde{F}_2 \end{bmatrix} = \begin{bmatrix} 4 & 4 & 1 \\ -1 & -2 & 2 \end{bmatrix}$$

进一步可以得到

$$\begin{aligned} u(t) &= Kx(t) + Hv(t) = -E^{-1}\tilde{F}x(t) + E^{-1}v(t) \\ &= \begin{bmatrix} -4 & -4 & -1 \\ 1 & 2 & -2 \end{bmatrix} x(t) + \begin{bmatrix} 1 & 0 \\ 0 & 1 \end{bmatrix} v(t) \end{aligned}$$

解耦系统的系数矩阵为

$$\bar{A} = A + BK = \begin{bmatrix} 0 & 0 & 0 \\ -4 & -4 & 0 \\ 0 & 0 & -5 \end{bmatrix},\ \bar{B} = BH = \begin{bmatrix} 0 & 0 \\ 1 & 0 \\ 0 & 1 \end{bmatrix},\ \bar{C} = C = \begin{bmatrix} 1 & 1 & 0 \\ 0 & 0 & 1 \end{bmatrix}$$

相应的传递函数矩阵为

$$G_{KH}(s) = \begin{bmatrix} \dfrac{1}{s+4} & 0 \\ 0 & \dfrac{1}{s+5} \end{bmatrix}$$

上述例子表明解耦和极点配置后，一些零点和极点可能对消，系统从能观的变为不能观的。另一方面，状态反馈不改变系统的能控性，本例中系统不完全能控的性质并未改变。但是，不完全能控的系统仍然能够解耦并配置极点，但不能配置系统的全部极点，或任意配置系统的极点。

6.3.3 稳态解耦

考虑系统方程 (6.26)，如果存在控制律 (6.27) 使得闭环系统方程 (6.28) 满足下述性质，
(1) 闭环系统方程 (6.28) 是渐近稳定的；
(2) $G_{\mathrm{KH}}(s) = C(sI - A - BK)^{-1}BH$ 在 $s \to 0$ 时是对角的。

那么这一方法被称作稳态解耦，相应地，前一小节介绍的解耦方法被称作动态解耦。应当指出，当考虑稳态解耦时，$G_{\mathrm{KH}}(s)$ 除了在 $s \to 0$ 以外，一般都是非对角矩阵。同时，稳态解耦只适用于参考输入是阶跃输入的情况，即对于 $t \geqslant 0$，

$$v(t) = \begin{bmatrix} b_1 & \cdots & b_p \end{bmatrix}^{\mathrm{T}}$$

其中，$b_1, \cdots, b_p$ 是常值。

这种情况下，可以得到

$$\lim_{t \to \infty} y(t) = \lim_{s \to 0} sG_{\mathrm{KH}}(s) \begin{bmatrix} b_1 \\ \vdots \\ b_p \end{bmatrix} \frac{1}{s} = \begin{bmatrix} g_{11}(0) & & \\ & \ddots & \\ & & g_{pp}(0) \end{bmatrix} \begin{bmatrix} b_1 \\ \vdots \\ b_p \end{bmatrix}$$

$$= \begin{bmatrix} g_{11}(0)b_1 \\ \vdots \\ g_{pp}(0)b_p \end{bmatrix}$$

这表明

$$\lim_{t \to \infty} y_i(t) = g_{ii}(0)b_i \tag{6.38}$$

其中，$y_i(t)$ 是第 i 个输出，$g_{ii}(0)$ 是 $G_{\mathrm{KH}}(0)$ 的第 i 个对角元。

式 (6.38) 表明在阶跃输入作用下，稳态解耦可以使得稳态下每个输出仅受到相应的一个输入控制。但是在动态过程中，输入和输出之间的耦合关系不能被消除。

定理 6.12 线性定常系统方程 (6.26) 可以通过控制律 (6.27) 实现稳态解耦，当且仅当系统可以被状态反馈控制律镇定，并且

$$\mathrm{rank} \begin{bmatrix} A & B \\ C & 0 \end{bmatrix} = n + p \tag{6.39}$$

证明 首先证明充分性。

如果反馈增益为 K 的状态反馈控制律可以镇定系统，那么 $A + BK$ 的所有特征值都具有负实部，所以 $(A + BK)^{-1}$ 存在。

6.3 解耦控制

由于式 (6.39) 成立，可以得到

$$\begin{bmatrix} A+BK & B \\ C & 0 \end{bmatrix} = \begin{bmatrix} A & B \\ C & 0 \end{bmatrix} \begin{bmatrix} I & 0 \\ K & I \end{bmatrix} \tag{6.40}$$

是非奇异的。在此基础上，可以推导出

$$\det \begin{bmatrix} A+BK & B \\ C & 0 \end{bmatrix} = \det(A+BK)\det(-C(A+BK)^{-1}B) \neq 0 \tag{6.41}$$

这表明 $C(A+BK)^{-1}B$ 也是非奇异的。

选择

$$H = (-C(A+BK)^{-1}B)^{-1}M \tag{6.42}$$

其中，M 是期望设计的对角非奇异矩阵。由此，可以进一步得到

$$G_{\text{KH}}(0) = C(sI - A - BK)^{-1}B(-C(A+BK)^{-1}B)^{-1}M\big|_{s=0} = M$$

这表明系统可以被稳态解耦。

接下来，证明必要性。

如果一个系统需要被稳态解耦，那么 $G_{\text{KH}}(s)$ 应该能够进入稳态，并且 $G_{\text{KH}}(0) = -C(A+BK)^{-1}BH$ 应为对角的非奇异矩阵。前一条件表明系统可以被状态反馈控制律稳定。后一条件表明 $C(A+BK)^{-1}B$ 和 $A+BK$ 都是非奇异的。根据式 (6.40) 和式 (6.41) 可知，式 (6.39) 成立。

定理 6.12 的证明给出了实现稳态解耦的方法。利用控制律 (6.27)，可以任意选择 K，只要确保闭环系统渐近稳定即可，H 可以根据式 (6.42) 选取。 □

例 6.16 考虑线性定常系统方程 (6.26)，其中

$$A = \begin{bmatrix} 0 & 0 & 1 \\ 1 & 0 & 0 \\ 1 & 1 & 1 \end{bmatrix}, B = \begin{bmatrix} 1 & 1 \\ -1 & 1 \\ 0 & -1 \end{bmatrix}, C = \begin{bmatrix} 0 & 1 & 1 \\ -1 & 1 & 0 \end{bmatrix}$$

传递函数矩阵为

$$G(s) = \begin{bmatrix} \dfrac{-s^2+2s+1}{s^3-s^2-s-1} & \dfrac{2s-2}{s^3-s^2-s-1} \\ \dfrac{-2s^2+3s-1}{s^3-s^2-s-1} & \dfrac{-2}{s^3-s^2-s-1} \end{bmatrix}$$

可以计算得到

$$E = \begin{bmatrix} -1 & 0 \\ -2 & 0 \end{bmatrix}$$

是奇异的。所以，系统不能被动态解耦。

但是，系统完全能控并且满足式 (6.39)，它可以被稳态解耦。可以选择 K 使得 $A+BK$ 的特征值都具有负实部，在此基础上进一步根据式 (6.42) 选择 H，以实现稳态解耦。

6.4 状态观测器

考虑状态反馈时，总是假设系统的状态可以通过合适的传感器测量得到。一般情况下，获得系统的全部状态信息可能很困难。实际上，一些状态可能完全不能被测量。所以，需要根据可以获得的测量值，如输出量和输入量，来估计系统的状态值。

6.4.1 全维状态观测器

考虑线性定常系统

$$\dot{x}(t) = Ax(t) + Bu(t), \ y(t) = Cx(t), \ x(0) = x_0 \tag{6.43}$$

其中，$x(t) \in \mathbb{R}^n$ 是状态向量，$u(t) \in \mathbb{R}^p$ 是输入向量，$y(t) \in \mathbb{R}^p$ 是输出向量，A，B 和 C 是具有相应维数的实常值矩阵。

全维观测器是以 $u(t)$ 和 $y(t)$ 作为输入构造的 n 维模型系统，它的输出 $\hat{x}(t) \in \mathbb{R}^n$ 满足

$$\lim_{t \to \infty} \hat{x}(t) = \lim_{t \to \infty} x(t) \tag{6.44}$$

接下来，介绍两种方法来为系统方程 (6.43) 构造全维观测器。

$x(t)$ 的全维观测器可以按照以下形式构造，

$$\dot{\hat{x}}(t) = A\hat{x}(t) + Bu(t) + G\left(y(t) - \hat{y}(t)\right), \ \hat{x}(0) = \hat{x}_0 \tag{6.45}$$

其中，$\hat{y}(t) \triangleq C\hat{x}(t)$，$G$ 是一个常值矩阵。

观测器 (6.45) 中包含了反馈修正项 $G(y(t) - \hat{y}(t))$，如果移除这一部分，观测器的结构和原系统 (6.43) 相同。此时很难保证观测器的初始条件 $\hat{x}_0$ 与原系统的初始条件 x_0 相同，所以式 (6.44) 很难满足。特别地，如果原系统不稳定，那么 $\hat{x}_0$ 和 x_0 之间任意微小的差异都会导致 $\hat{x}(t) - x(t)$ 发散。

注意到式 (6.45) 可以被写作

$$\dot{\hat{x}}(t) = (A - GC)\hat{x}(t) + Bu(t) + Gy(t)$$

它的框图如图 6.7 所示。定义观测误差 $e(t) \triangleq x(t) - \hat{x}(t)$，结合式 (6.43) 与式 (6.45)，可以得到

$$\dot{e}(t) = (A - GC)e(t), \ e(0) = e_0 = x_0 - \hat{x}_0 \tag{6.46}$$

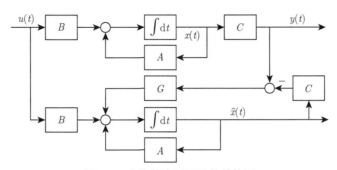

图 6.7 全维状态观测器的结构图

如果 $A - GC$ 的特征值全部具有负实部，那么随着 $t \to \infty$，$e(t) \to 0$，且与初始条件 e_0 无关，即式 (6.44) 成立。这一渐近状态观测器被称作 Luenberger 观测器。可以看出，如果能够找到 G 使得误差系统方程 (6.46) 渐近稳定，那么 Luenberger 观测器 (6.45) 存在。显然，如果原系统方程 (6.43) 完全能观，误差系统方程 (6.46) 可以被镇定。如果 (A, C) 不完全能观，则存在变换矩阵 T，可以把系统分解为

$$\begin{bmatrix} \dot{x}_o(t) \\ \dot{x}_{no}(t) \end{bmatrix} = \begin{bmatrix} A_{11} & 0 \\ A_{21} & A_{22} \end{bmatrix} \begin{bmatrix} x_o(t) \\ x_{no}(t) \end{bmatrix} + \begin{bmatrix} B_1 \\ B_2 \end{bmatrix} u(t), \ y(t) = [C_1 \ 0] \begin{bmatrix} x_o(t) \\ x_{no}(t) \end{bmatrix} \quad (6.47)$$

其中，(A_{11}, C_1) 完全能观，系统方程 (6.47) 的特征值由 A_{11} 和 A_{22} 的特征值组成。如果 A_{22} 的所有特征值都具有负实部，那么 (A, C) 被称为是可检测的。

定理 6.13 对于线性定常系统方程 (6.43)，当且仅当 (A, C) 可检测时，系统存在全维观测器 (6.45)。

证明 对于系统方程 (6.47)，应用 Luenberger 观测器 (6.45)。记 $G = [\ G_1^T \ \ G_2^T \ ]^T$，相应的误差系统可以表示为

$$\dot{e}(t) = \begin{bmatrix} \dot{e}_1(t) \\ \dot{e}_2(t) \end{bmatrix} = \begin{bmatrix} A_{11} - G_1 C_1 & 0 \\ A_{21} - G_2 C_1 & A_{22} \end{bmatrix} \begin{bmatrix} e_1(t) \\ e_2(t) \end{bmatrix}$$

因为 (A_{11}, C_1) 完全能观，所以可以选择 G_1 使得 $A_{11} - G_1 C_1$ 的所有特征值都具有负实部。同时，由于 (A, C) 可检测，A_{22} 的所有特征值都具有负实部。所以，$\lim_{t \to \infty} e(t) = 0$ 成立，即相应的 Luenberger 观测器存在。

必要性的证明略。 □

定理 6.14 对于线性定常系统方程 (6.43)，当且仅当 (A, C) 完全能观时，全维滤波器 (6.45) 的特征值可以被任意配置。

证明略。

算法 6.9 对于可检测的线性定常系统方程 (6.43)，Luenberger 观测器 (6.45) 可以根据下述步骤来设计。

(1) 根据对偶系统 (A^T, C^T, B^T)，来计算 $A^T - C^T G^T$。
(2) 选择反馈增益 G^T 把 $A^T - C^T G^T$ 的特征值配置到期望的位置。
(3) 写出 Luenberger 观测器 (6.45) 的方程。

接下来，介绍为系统方程 (6.43) 设计全维观测器的另一种方法。这一方法要求系统方程 (6.43) 是完全能控且能观的。全维观测器具有下述形式，

$$\dot{z}(t) = Fz(t) + Gy(t) + Hu(t), \ z(0) = z_0, \ \hat{x}(t) = T^{-1}z(t) \qquad (6.48)$$

其中，$z(t) \in \mathbb{R}^n$，F，G，H 和 T 是具有相应维数的实常值矩阵，且 T 是非奇异的。

定理 6.15 对于完全能控且能观的系统方程 (6.43)，当且仅当 F 的所有特征值都具有负实部，$TA - FT = GC$，且 $H = TB$ 时，式 (6.48) 是系统方程 (6.43) 的全维滤波器。

证明 首先，证明充分性。

观测误差仍定义为 $e(t) \triangleq x(t) - \hat{x}(t)$，结合式 (6.43) 与式 (6.48)，可以得到

$$\begin{aligned}\dot{e}(t) &= \dot{x}(t) - \dot{\hat{x}}(t) = \dot{x}(t) - T^{-1}\dot{z}(t) \\ &= Ax(t) + Bu(t) - T^{-1}(Fz(t) + Hu(t) + Gy(t)) \\ &= Ax(t) + Bu(t) - T^{-1}FT\hat{x}(t) - T^{-1}Hu(t) - T^{-1}GCx(t) \\ &= T^{-1}FTe(t) + T^{-1}(TA - FT - GC)x(t) + T^{-1}(TB - H)u(t) \end{aligned} \qquad (6.49)$$

如果 F 的所有特征值都具有负实部，$TA - FT = GC$，且 $H = TB$，那么可以推导 $\lim_{t \to \infty} e(t) = 0$，即式 (6.44) 成立。所以，式 (6.48) 是式 (6.43) 的全维观测器。

接下来，证明必要性。

利用反证法，假设 $\lim_{t \to \infty} e(t) = 0$，但是定理中的三个条件不同时满足。这种情况下，根据式 (6.49) 可知，$t \to \infty$ 时不能保证 $e(t) \to 0$，这与假设矛盾。 $\square$

算法 6.10 对于完全能控且能观的线性定常系统方程 (6.43)，全维观测器 (6.48) 可以根据下述步骤来设计。

(1) 选择矩阵 F，使得它的特征值都具有负实部，并且和 A 的特征值都不相同。

(2) 选择矩阵 G，使得 (F, G) 完全能控。

(3) 求解矩阵方程 $TA - FT = GC$ 来获得它的唯一解 T。

(4) 如果 T 是非奇异的，计算 $H = TB$，进一步可以得到全维观测器 (6.48)。否则，重新选择 F 或 G 并重复上述步骤。

例 6.17 考虑以下线性定常系统，

$$\dot{x}(t) = \begin{bmatrix} -1 & 2 \\ 0 & 1 \end{bmatrix} x(t) + \begin{bmatrix} 0 \\ 1 \end{bmatrix} u(t), \ y(t) = \begin{bmatrix} 1 & 0 \end{bmatrix} x(t)$$

为这一系统设计全维滤波器。

选择

$$F = \begin{bmatrix} -2 & 1 \\ 0 & -3 \end{bmatrix}$$

可以看出 F 的两个特征值都具有负实部，并且和 A 的特征值都不相同。

选择 $G = [\ 0\ \ 1\ ]^T$ 它可以保证 (F, G) 是完全能控的。

求解矩阵方程 $TA - FT = GC$,可以得到

$$T = \begin{bmatrix} -\dfrac{1}{12} & -\dfrac{5}{12} \\ -\dfrac{1}{4} & -\dfrac{1}{4} \end{bmatrix}$$

它是非奇异的,所以,

$$H = TB = \begin{bmatrix} -\dfrac{5}{12} \\ -\dfrac{1}{4} \end{bmatrix}$$

进一步可以写出全维观测器的方程,

$$\dot{z}(t) = \begin{bmatrix} -2 & 1 \\ 0 & -3 \end{bmatrix} z(t) + \begin{bmatrix} -\dfrac{5}{12} \\ -\dfrac{1}{4} \end{bmatrix} u(t) + \begin{bmatrix} 0 \\ 1 \end{bmatrix} y(t)$$

$$\hat{x}(t) = T^{-1} z(t) = \begin{bmatrix} 3 & -5 \\ -3 & 1 \end{bmatrix} z(t)$$

6.4.2 降维状态观测器

因为系统的输出 $y(t)$ 包含了系统的部分状态信息,所以可以考虑为系统设计一个低阶的状态观测器,即降维状态观测器。

对于线性定常系统方程 (6.43),$\operatorname{rank} C = q$,降维状态观测器的最小维数是 $n - q$。

考虑 $C = [\ C_1\ \ C_2\ ]$,其中 $C_1 \in \mathbb{R}^{q \times q}$,$C_2 \in \mathbb{R}^{q \times (n-q)}$,且 $\operatorname{rank} C_1 = q$。如果 $\operatorname{rank} C_1 \neq q$,总是能够找到列变换使得 $\operatorname{rank} C_1 = q$。

选择变换矩阵

$$T = \begin{bmatrix} C_1 & C_2 \\ 0 & I \end{bmatrix}$$

并对系统方程 (6.43) 进行等价变换,$\bar{x}(t) = Tx(t)$,可以得到

$$\begin{bmatrix} \dot{\bar{x}}_1(t) \\ \dot{\bar{x}}_2(t) \end{bmatrix} = \begin{bmatrix} A_{11} & A_{12} \\ A_{21} & A_{22} \end{bmatrix} \begin{bmatrix} \bar{x}_1(t) \\ \bar{x}_2(t) \end{bmatrix} + \begin{bmatrix} B_1 \\ B_2 \end{bmatrix} u(t),\ y(t) = \begin{bmatrix} I & 0 \end{bmatrix} \begin{bmatrix} \dot{\bar{x}}_1(t) \\ \dot{\bar{x}}_2(t) \end{bmatrix} \quad (6.50)$$

能够看出 $\bar{x}_1(t) \in \mathbb{R}^q$ 可以通过 $y(t)$ 直接获得,所以,只需要设计状态观测器来估计 $\bar{x}_2(t) \in \mathbb{R}^{n-q}$。

根据式 (6.50),可以得到

$$\dot{\bar{x}}_2(t) = A_{22} \bar{x}_2(t) + (A_{21} y(t) + B_2 u(t)),\ \dot{y}(t) - A_{11} y(t) - B_1 u(t) = A_{12} \bar{x}_2(t) \quad (6.51)$$

定义 $\bar{u}(t) \triangleq A_{21}y(t) + B_2u(t)$, $w(t) \triangleq \dot{y}(t) - A_{11}y(t) - B_1u(t)$，并把它们代入式 (6.51) 中，可以得到

$$\dot{\bar{x}}_2(t) = A_{22}\bar{x}_2(t) + \bar{u}(t), \; w(t) = A_{12}\bar{x}_2(t) \tag{6.52}$$

引理 6.9 当且仅当 (A, C) 完全能观时，(A_{22}, A_{12}) 完全能观。

证明 因为 (A, C) 完全能观，所以系统 (6.50) 完全能观，系统 (6.50) 的能观性矩阵满足

$$\operatorname{rank} V = \operatorname{rank} \begin{bmatrix} I & 0 \\ A_{11} & A_{12} \\ A_{11}^2 + A_{12}A_{21} & A_{11}A_{12} + A_{12}A_{22} \\ \vdots & \vdots \end{bmatrix}$$

$$= \operatorname{rank} \begin{bmatrix} I & 0 \\ 0 & A_{12} \\ 0 & A_{11}A_{12} + A_{12}A_{22} \\ \vdots & \vdots \end{bmatrix} = \operatorname{rank} \begin{bmatrix} I & 0 \\ 0 & A_{12} \\ 0 & A_{12}A_{22} \\ \vdots & \vdots \end{bmatrix} = n$$

这表明

$$\operatorname{rank} \begin{bmatrix} A_{12} \\ A_{12}A_{22} \\ \vdots \\ A_{12}A_{22}^{n-q-1} \end{bmatrix} = n - q$$

所以，(A_{22}, A_{12}) 完全能观。

必要性的证明略。 □

如果原系统完全能观，那么系统方程 (6.52) 也是完全能观的。接下来，可以设计 Luenberger 观测器来估计 $\bar{x}_2(t)$，

$$\dot{\hat{\bar{x}}}_2(t) = (A_{22} - G_2A_{12})\hat{\bar{x}}_2(t) + G_2w(t) + \bar{u}(t) \tag{6.53}$$

其中 G_2 是待设计的常值矩阵，通过选取 G_2，$A_{22} - G_2A_{12}$ 的特征值可以任意配置。

把 $\bar{u}(t)$ 和 $w(t)$ 代入式 (6.53) 中，可以得到

$$\dot{\hat{\bar{x}}}_2(t) = (A_{22} - G_2A_{12})\hat{\bar{x}}_2(t) + G_2\left(\dot{y}(t) - A_{11}y(t) - B_1u(t)\right) + A_{21}y(t) + B_2u(t) \tag{6.54}$$

上式包含 $y(t)$ 的微分，这对于扰动抑制来说是不利的。

引入变换

$$z(t) \triangleq \hat{\bar{x}}_2(t) - G_2y(t)$$

6.4 状态观测器

并代入式 (6.54) 中，可以得到

$$\dot{z}(t) = (A_{22} - G_2 A_{12})z(t) + \bar{G}_2 y(t) + (B_2 - G_2 B_1)u(t) \tag{6.55}$$

其中，$\bar{G}_2 = (A_{22} - G_2 A_{12})G_2 + (A_{21} - G_2 A_{11})$。

进一步，状态 $\bar{x}_2(t)$ 的估计可以表示为

$$\hat{\bar{x}}_2(t) = z(t) + G_2 y(t) \tag{6.56}$$

把 $\bar{x}_1(t)$ 的观测值记作 $\hat{\bar{x}}_1(t)$，可知 $\hat{\bar{x}}_1(t) = y(t)$，进一步，把 $x(t)$ 的观测值记作 $\hat{x}(t)$，可以得到

$$\begin{aligned}
\hat{x}(t) &= T^{-1} \begin{bmatrix} \hat{\bar{x}}_1(t) \\ \hat{\bar{x}}_2(t) \end{bmatrix} = \begin{bmatrix} C_1^{-1} & -C_1^{-1} C_2 \\ 0 & I_q \end{bmatrix} \begin{bmatrix} y(t) \\ z(t) + G_2 y(t) \end{bmatrix} \\
&= \begin{bmatrix} -C_1^{-1} C_2 \\ I \end{bmatrix} z(t) + \begin{bmatrix} C_1^{-1}(I - C_2 G_2) \\ G_2 \end{bmatrix} y(t)
\end{aligned} \tag{6.57}$$

定理 6.16 对于完全能观的线性定常系统 (6.43)，$\operatorname{rank} C = q$，可以为系统 (6.43) 构造 $n-q$ 维的状态观测器 (6.55) 和 (6.57)，且观测器的所有特征值可以任意配置。

证明略。

根据以上讨论，可以总结出降维状态观测器的设计方法。

算法 6.11 对于完全能观的线性定常系统 (6.43)，$\operatorname{rank} C = q$，$n-q$ 维的状态观测器 (6.55) 和 (6.57) 可以根据下述步骤来设计。

(1) 变换 $C = [\, C_1 \ \ C_2 \,]$，选择 $T = \begin{bmatrix} C_1 & C_2 \\ 0 & I \end{bmatrix}$，把原系统 (6.43) 通过 $\bar{x}(t) = Tx(t)$ 转换为式 (6.50)。

(2) 选择 $G_2 \in \mathbb{R}^{(n-q) \times q}$ 把 $A_{22} - G_2 A_{12}$ 的特征值配置到期望的位置。

(3) 计算式 (6.55) 和式 (6.57) 中的参数，进一步可以得到降维状态观测器。

例 6.18 考虑以下线性定常系统，

$$\dot{x}(t) = \begin{bmatrix} -1 & 0 & 0 \\ 0 & 1 & 1 \\ 0 & 0 & 1 \end{bmatrix} x(t) + \begin{bmatrix} 1 & 0 \\ 0 & 1 \\ 0 & 1 \end{bmatrix} u(t), \ y(t) = \begin{bmatrix} 1 & 0 & 0 \\ 0 & 1 & 1 \end{bmatrix} x(t)$$

可以验证系统完全能观且 $\operatorname{rank} C = 2$。接下来为这一系统设计一个 $n - q = 1$ 的降维状态观测器。

选择

$$T = \begin{bmatrix} C_1 & C_2 \\ 0 & I \end{bmatrix} = \begin{bmatrix} 1 & 0 & 0 \\ 0 & 1 & 1 \\ 0 & 0 & 1 \end{bmatrix}, \ T^{-1} = \begin{bmatrix} 1 & 0 & 0 \\ 0 & 1 & -1 \\ 0 & 0 & 1 \end{bmatrix}$$

原系统转化为

$$\begin{bmatrix} \dot{\bar{x}}_1(t) \\ \dot{\bar{x}}_2(t) \end{bmatrix} = \begin{bmatrix} -1 & 0 & 0 \\ 0 & 1 & 1 \\ 0 & 0 & 1 \end{bmatrix} \begin{bmatrix} \bar{x}_1(t) \\ \bar{x}_2(t) \end{bmatrix} + \begin{bmatrix} 1 & 0 \\ 0 & 2 \\ 0 & 1 \end{bmatrix} u(t)$$

$$y(t) = \begin{bmatrix} y_1(t) \\ y_2(t) \end{bmatrix} = \begin{bmatrix} 1 & 0 & 0 \\ 0 & 1 & 0 \end{bmatrix} \begin{bmatrix} \bar{x}_1(t) \\ \bar{x}_2(t) \end{bmatrix}$$

选择 $G_2 = \begin{bmatrix} g_1 & g_2 \end{bmatrix}$，一维状态观测器可以写作

$$\dot{z}(t) = (1 - g_2)z(t) + \begin{bmatrix} -g_1 & 1 - 2g_2 \end{bmatrix} u(t) + \begin{bmatrix} 2g_1 - g_1 g_2 & -g_2^2 \end{bmatrix} y(t)$$

且重构的状态为

$$\hat{x}(t) = \begin{bmatrix} 0 \\ -1 \\ 1 \end{bmatrix} z(t) + \begin{bmatrix} 1 & 0 \\ -g_1 & 1 - g_2 \\ g_1 & g_2 \end{bmatrix} y(t)$$

这里，选择 $G_2 = \begin{bmatrix} 0 & 5 \end{bmatrix}$ 可以把两个特征值都配置到 -4。由此，可以进一步得到相应的降维状态观测器是

$$\dot{z}(t) = -4z(t) + \begin{bmatrix} 0 & -9 \end{bmatrix} u(t) + \begin{bmatrix} 0 & -25 \end{bmatrix} y(t)$$

$$\hat{x}(t) = \begin{bmatrix} 0 \\ -1 \\ 1 \end{bmatrix} z(t) + \begin{bmatrix} 1 & 0 \\ 0 & -4 \\ 0 & 5 \end{bmatrix} y(t)$$

接下来，介绍为系统 (6.43) 设计降维观测器的另一种方法。这一方法要求系统 (6.43) 完全能控能观，且 $\text{rank}\, C = q$。该方法可以设计一个具有下述形式的 $n - q$ 维的降维观测器。

构造一个线性定常系统，

$$\dot{z}(t) = Fz(t) + Gy(t) + Hu(t), \quad z(0) = z_0 \tag{6.58}$$

其中 $z(t) \in \mathbb{R}^{n-q}$，$F$，$G$ 和 H 是具有相应维数的实常值矩阵。

定理 6.17 对于完全能控能观的线性定常系统 (6.43)，$\text{rank}\, C = q$，式 (6.58) 是系统 (6.43) 的 $n - q$ 维的降维观测器，当且仅当存在行满秩矩阵 $T \in \mathbb{R}^{(n-q) \times n}$ 使得

$$P \triangleq \begin{bmatrix} C \\ T \end{bmatrix}$$

是非奇异的，F 的所有特征值都具有负实部，$TA - FT = GC$，且 $H = TB$。同时，估计的状态 $\hat{x}(t)$ 为

$$\hat{x}(t) = P^{-1} \begin{bmatrix} y(t) \\ z(t) \end{bmatrix} = \begin{bmatrix} C \\ T \end{bmatrix}^{-1} \begin{bmatrix} y(t) \\ z(t) \end{bmatrix} = \begin{bmatrix} Q_1 & Q_2 \end{bmatrix} \begin{bmatrix} y(t) \\ z(t) \end{bmatrix} \tag{6.59}$$

其中，$Q = \begin{bmatrix} Q_1 & Q_2 \end{bmatrix} \triangleq P^{-1}$。

仿照定理 6.15 的证明，可以推导出这一定理。这里省略证明过程。

根据定理 6.17，可以总结出下述算法。

算法 6.12 对于完全能控且能观的线性定常系统 (6.43)，$\text{rank}\,C = q$，$n - q$ 维的状态观测器 (6.55) 和 (6.57) 可以根据下述步骤来设计。

(1) 选择矩阵 $F \in \mathbb{R}^{(n-q) \times (n-q)}$ 使得它的所有特征值都具有负实部，并且和 A 的特征值都不相同。

(2) 选择矩阵 $G \in \mathbb{R}^{(n-q) \times q}$ 使得 (F, G) 完全能控。

(3) 求解 $TA - FT = GC$ 以获得它的唯一解 T。

(4) 构造矩阵 P 并确定 P 是否是非奇异的。如果 P 非奇异，那么计算 $H = TB$，进一步可以得到降维观测器 (6.58) 和 (6.59)。否则，重新选择 F 或 G。

应当指出由于降维状态观测器的维数更低，所以它的结构更简单，也更加可靠。这使得降维状态观测器更实用。与全维状态观测器不同，降维状态观测器把系统的输出通过一个常值矩阵 Q_1 直接反映到 $\hat{x}(t)$ 中，而全维观测器则需通过积分器反映到 $\hat{x}(t)$ 中。所以，如果 $y(t)$ 中存在扰动或噪声，选取降维观测器是不利的。如何取舍两者的优缺点，应根据具体情况来确定。

6.4.3 函数状态观测器

一些应用中，只需关注状态的线性组合。构造状态观测器的主要目的是为了实现状态反馈，可以直接对状态函数 Kx 进行重构，这有助于进一步降低观测器的维数。一般地，这类观测器被称为函数状态观测器。

考虑线性定常系统 (6.43)，假设它是完全能控能观的。构造一个完全能观的系统，

$$\dot{z}(t) = Fz(t) + Gy(t) + Hu(t),\ w(t) = Ez(t) + My(t),\ z(0) = z_0 \qquad (6.60)$$

其中，$z(t) \in \mathbb{R}^r$ 是这一系统的状态向量，$w(t) \in \mathbb{R}^p$ 是这一系统的输出向量，F，G，H，E 和 M 是具有相应维数的常值矩阵。这里的目的是在式 (6.60) 中选取合适的矩阵参数，使得

$$\lim_{t \to \infty} w(t) = \lim_{t \to \infty} Kx(t) \qquad (6.61)$$

其中，K 是反馈增益。这种情况下，式 (6.61) 是系统 (6.43) 的一个函数状态观测器。

定理 6.18 对于完全能控且能观的线性定常系统 (6.43)，$\text{rank}\,C = q$，当且仅当 F 的所有特征值都具有负实部，并存在一个满秩矩阵 $T \in \mathbb{R}^{r \times n}$ 使得 $TA - FT = GC$，$H = TB$，且 $K = ET + MC$，那么式 (6.60) 是系统 (6.43) 的函数状态观测器。

证明 首先，证明充分性。

可知

$$\dot{z}(t) - T\dot{x}(t) = (Fz(t) + Gy(t) + Hu(t)) - T(Ax(t) + Bu(t))$$
$$= (GC - TA)x(t) + (H - TB)u(t) + Fz(t)$$

$$= Fz(t) - FTx(t) = F(z(t) - Tx(t))$$

这表明对于任意的 $u(t)$，x_0 和 z_0，总是存在

$$z(t) - Tx(t) = \mathrm{e}^{Ft}(z_0 - Tx_0)$$

由于 F 的所有特征值都具有负实部，所以 $t \to \infty$ 时，$\mathrm{e}^{Ft} \to 0$。进一步，可得

$$\lim_{t\to\infty} z(t) = \lim_{t\to\infty} Tx(t)$$

在此基础上，可以进一步推导出

$$\lim_{t\to\infty} w(t) = \lim_{t\to\infty} (Ez(t) + My(t)) = \lim_{t\to\infty} (ET + MC)x(t) = \lim_{t\to\infty} Kx(t)$$

这表明式 (6.60) 是系统 (6.43) 的函数观测器。

接下来，证明必要性。

假设对于任意 $u(t)$，x_0 和 z_0，都有式 (6.61) 成立，在此基础上，可以得到

$$\lim_{t\to\infty} (Kx(t) - w(t)) = 0$$

$$\lim_{t\to\infty} (K\dot{x}(t) - \dot{w}(t)) = 0$$

$$\vdots$$

$$\lim_{t\to\infty} \left(Kx^{(r-1)}(t) - w^{(r-1)}(t)\right) = 0$$

结合式 (6.60)，可以得到

$$Kx(t) - w(t) = (K - MC)x(t) - Ez(t) \triangleq M_0 x(t) - Ez(t)$$

$$K\dot{x}(t) - \dot{w}(t) = [(K - MC)A - EGC]x(t) - EFz(t) + [(K - MC)B - EH]u(t)$$

$$\triangleq M_1 x(t) - EFz(t) + N_1 u(t)$$

$$K\ddot{x}(t) - \ddot{w}(t) \triangleq M_2 x(t) - EF^2 z(t) + N_1 \dot{u}(t) + N_2 u(t)$$

$$\vdots$$

$$Kx^{(r-1)}(t) - w^{(r-1)}(t) \triangleq M_{r-1} x(t) - EF^{r-1} z(t) + \sum_{j=1}^{r-1} N_j u^{(r-1-j)}(t)$$

进一步，可知

$$\lim_{t\to\infty} (M_0 x(t) - Ez(t)) = 0$$

$$\lim_{t\to\infty} (M_1 x(t) - EFz(t) + N_1 u(t)) = 0$$

6.4 状态观测器

$$\lim_{t\to\infty} \left(M_2 x(t) - EF^2 z(t) + N_1 \dot{u}(t) + N_2 u(t)\right) = 0$$

$$\vdots$$

$$\lim_{t\to\infty} \left(M_{r-1} x(t) - EF^{r-1} z(t) + \sum_{j=1}^{r-1} N_j u^{(r-1-j)}(t)\right) = 0$$

由于控制输入的任意性,可以选取 $u(t) = \dot{u}(t) = \cdots = u^{(r-1)}(t) = 0$。这种情况下,可以得到

$$\lim_{t\to\infty} \left(M_0 x(t) - E z(t)\right) = 0$$

$$\lim_{t\to\infty} \left(M_1 x(t) - EF z(t)\right) = 0$$

$$\lim_{t\to\infty} \left(M_2 x(t) - EF^2 z(t)\right) = 0$$

$$\vdots$$

$$\lim_{t\to\infty} \left(M_{r-1} x(t) - EF^{r-1} z(t)\right) = 0$$

定义

$$R \triangleq \begin{bmatrix} M_0 \\ M_1 \\ \vdots \\ M_{r-1} \end{bmatrix}, \quad Q \triangleq \begin{bmatrix} E \\ EF \\ \vdots \\ EF^{r-1} \end{bmatrix}$$

由此可以得到

$$\lim_{t\to\infty} \left(R x(t) - Q z(t)\right) = 0$$

因为 (F, E) 完全能观且 Q 是列满秩的,所以它的广义逆 Q^+ 存在,令 $T = Q^+ R$,可以得到

$$\lim_{t\to\infty} \left(Q^+ R x(t) - z(t)\right) = \lim_{t\to\infty} \left(T x(t) - z(t)\right) = 0$$

为了保证上式成立,根据充分性的推导过程,需要满足 $TA - FT = GC$, $H = TB$ 且 F 的所有特征值具有负实部。进一步,对于任意的 $u(t)$, x_0 和 z_0,可以得到

$$\lim_{t\to\infty} \left(K x(t) - w(t)\right) = \lim_{t\to\infty} \left[K x(t) - (E z(t) + M y(t))\right] = \lim_{t\to\infty} \left[K - (ET + MC)\right] x(t)$$

$$= \left[K - (ET + MC)\right] \lim_{t\to\infty} x(t)$$

为了保证式 (6.61),需要满足 $K - (ET + MC) = 0$。 $\square$

如何确定函数状态观测器的最小维数是一个非常复杂的问题。如果 $K \in \mathbb{R}^{1 \times n}$,那么观测器的维数 $r = \nu - 1$,其中,ν 是 (A, C) 的能观性指数。一般地,当 q 的值相对较大时,$\nu - 1$ 会比 $n - q$ 更小,此时函数状态观测器的维数会比降维状态观测器的维数更低。

6.4.4 带状态观测器的动态系统

状态观测器的存在为在工程中实现状态反馈带来了便利。然而，利用状态观测器进行反馈和直接利用实际系统的状态进行反馈不同。因此，有必要研究在具有状态观测器的动态系统中引入状态反馈后的影响。

考虑完全能控且能观的线性定常系统方程 (6.43)。不失一般性地，考虑在这一系统中引入降维观测器 (6.58) 和 (6.59)，在此基础上，进一步引入状态反馈控制律

$$u(t) = K\hat{x}(t) + v(t) \tag{6.62}$$

从式 (6.43)，式 (6.58)，式 (6.59) 和式 (6.62) 可以推导出下述增广系统，

$$\begin{cases} \begin{bmatrix} \dot{x}(t) \\ \dot{z}(t) \end{bmatrix} = \begin{bmatrix} A + BKQ_1C & BKQ_2 \\ GC + HKQ_1C & F + HKQ_2 \end{bmatrix} \begin{bmatrix} x(t) \\ z(t) \end{bmatrix} + \begin{bmatrix} B \\ H \end{bmatrix} v(t) \\ y(t) = \begin{bmatrix} C & 0 \end{bmatrix} \begin{bmatrix} x(t) \\ z(t) \end{bmatrix} \end{cases} \tag{6.63}$$

系统结构如图 6.8 所示。

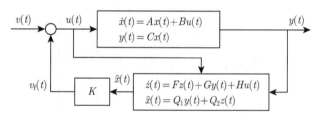

图 6.8 具有状态观测器的状态反馈系统

引入变换矩阵

$$\bar{T} = \begin{bmatrix} I & 0 \\ -T & I \end{bmatrix}$$

可以计算出

$$\bar{T}^{-1} = \begin{bmatrix} I & 0 \\ T & I \end{bmatrix}$$

其中，T 在式 (6.59) 中给出。

进行等价变换

$$\begin{bmatrix} \bar{x}(t) \\ \bar{z}(t) \end{bmatrix} = \bar{T} \begin{bmatrix} x(t) \\ z(t) \end{bmatrix}$$

系统 (6.63) 变换为

$$\begin{bmatrix} \dot{\bar{x}}(t) \\ \dot{\bar{z}}(t) \end{bmatrix} = \begin{bmatrix} A+BK & BKQ_2 \\ 0 & F \end{bmatrix} \begin{bmatrix} \bar{x}(t) \\ \bar{z}(t) \end{bmatrix} + \begin{bmatrix} B \\ 0 \end{bmatrix} v(t), \ y(t) = [C \ \ 0] \begin{bmatrix} \bar{x}(t) \\ \bar{z}(t) \end{bmatrix} \quad (6.64)$$

可以看出，闭环系统 (6.64) 的特征多项式是 $A+BK$ 和 F 的特征多项式的乘积。所以，引入状态观测器并不影响通过状态反馈设计的 $A+BK$ 的特征值。同时，状态反馈对观测器的特征值也没有影响。因此，控制律 (6.62) 的设计与观测器 (6.58) 和 (6.59) 的设计可以独立进行。这一性质通常被称为分离特性。当引入本节介绍的其他观测器时，这一结论仍然成立。

观测器的引入不会改变系统的闭环传递函数矩阵。实际上，直接应用状态反馈 $u(t) = Kx(t) + v(t)$ 时，闭环系统的传递函数矩阵是

$$G_\text{K}(s) = C(sI - A - BK)^{-1}B$$

等价变换不会改变系统的传递函数矩阵。所以，系统 (6.63) 的传递函数矩阵

$$\hat{G}_\text{K}(s) = [C \ \ 0] \begin{bmatrix} sI - A - BK & -BKQ_2 \\ 0 & sI - F \end{bmatrix}^{-1} \begin{bmatrix} B \\ 0 \end{bmatrix} = C(sI - A - BK)^{-1}B = G_\text{K}(s)$$

一般地，对具有观测器的系统进行状态反馈，它的鲁棒性比直接应用状态反馈的差。实际应用中，通常会将观测器特征值的负实部设计为闭环系统 $A+BK$ 特征值负实部的 $2 \sim 3$ 倍。但需要注意的是，如果观测器特征值的负实部过大，将会起到微分作用。这将会带来高频干扰，是不希望的。

例 6.19 考虑以下线性定常系统，

$$\dot{x}(t) = \begin{bmatrix} 0 & 1 & 0 & 0 \\ 0 & 0 & -2 & 0 \\ 0 & 0 & -3 & 1 \\ 0 & 0 & 4 & 0 \end{bmatrix} x(t) + \begin{bmatrix} 0 & 0 \\ 0 & 1 \\ 0 & 1 \\ -1 & 0 \end{bmatrix} u(t), \ y(t) = \begin{bmatrix} 1 & 0 & 0 & 0 \\ 0 & 1 & 0 & 0 \end{bmatrix} x(t)$$

(1) 设计 $K\hat{x}(t) + v(t)$ 使得 $\lambda_1^* = -1$, $\lambda_{2,3}^* = -1 \pm \text{j}$, $\lambda_4^* = -2$。

期望的特征多项式为

$$\alpha_\text{c}(s) = (s+1)(s+1-j)(s+1+j)(s+2) = s^4 + 5s^3 + 10s^2 + 10s + 4$$

假设状态反馈增益为

$$K = \begin{bmatrix} k_1 & k_2 & k_3 & k_4 \\ k_5 & k_6 & k_7 & k_8 \end{bmatrix}$$

由此可以得到

$$A + BK = \begin{bmatrix} 0 & 1 & 0 & 0 \\ k_5 & k_6 & k_7 - 2 & k_8 \\ k_5 & k_6 & k_7 - 3 & k_8 + 1 \\ -k_1 & -k_2 & -k_3 + 4 & -k_4 \end{bmatrix} = \begin{bmatrix} 0 & 1 & 0 & 0 \\ 0 & 0 & 1 & 0 \\ 0 & 0 & 0 & 1 \\ -4 & -10 & -10 & -5 \end{bmatrix}$$

进一步可以计算出

$$K = \begin{bmatrix} 4 & 10 & 14 & 5 \\ 0 & 0 & 3 & 0 \end{bmatrix}$$

(2) 设计降维观测器。

注意到系统具有式 (6.50) 的形式，且 $\text{rank}\,C = 2$，这种情况下，

$$A_{11} = \begin{bmatrix} 0 & 1 \\ 0 & 0 \end{bmatrix}, \; A_{12} = \begin{bmatrix} 0 & 0 \\ -2 & 0 \end{bmatrix}, \; A_{21} = \begin{bmatrix} 0 & 0 \\ 0 & 0 \end{bmatrix}, \; A_{22} = \begin{bmatrix} -3 & 1 \\ 4 & 0 \end{bmatrix}$$

$$B_1 = \begin{bmatrix} 0 & 0 \\ 0 & 1 \end{bmatrix}, \; B_2 = \begin{bmatrix} 0 & 1 \\ -1 & 0 \end{bmatrix}, \; C_1 = \begin{bmatrix} 1 & 0 \\ 0 & 1 \end{bmatrix}, \; C_2 = \begin{bmatrix} 0 & 0 \\ 0 & 0 \end{bmatrix}$$

假设观测器增益为

$$G_2 = \begin{bmatrix} g_1 & g_2 \\ g_3 & g_4 \end{bmatrix}$$

进一步可得

$$A_{22} - G_2 A_{12} = \begin{bmatrix} -3 + 2g_2 & 1 \\ 4 + 2g_4 & 0 \end{bmatrix}$$

把观测器的特征值配置在 -5 和 -6，可以选择

$$G_2 = \begin{bmatrix} 0 & -4 \\ 0 & -17 \end{bmatrix}$$

计算得到

$$A_{22} - G_2 A_{12} = \begin{bmatrix} -11 & 1 \\ -30 & 0 \end{bmatrix}$$

进一步可得

$$(A_{22} - G_2 A_{12})G_2 + (A_{21} - G_2 A_{11}) = (A_{22} - G_2 A_{12})G_2 = \begin{bmatrix} 0 & 27 \\ 0 & 120 \end{bmatrix}$$

且

$$B_2 - G_2 B_1 = \begin{bmatrix} 0 & 1 \\ -1 & 0 \end{bmatrix} - \begin{bmatrix} 0 & -4 \\ 0 & -17 \end{bmatrix} \begin{bmatrix} 0 & 0 \\ 0 & 1 \end{bmatrix} = \begin{bmatrix} 0 & 5 \\ -1 & 17 \end{bmatrix}$$

6.4 状态观测器

所以，期望的降维观测器为

$$\dot{z}(t) = \begin{bmatrix} -11 & 1 \\ -30 & 0 \end{bmatrix} z(t) + \begin{bmatrix} 0 & 27 \\ 0 & 120 \end{bmatrix} y(t) + \begin{bmatrix} 0 & 5 \\ -1 & 17 \end{bmatrix} u(t)$$

$$\hat{x}(t) = \begin{bmatrix} y(t) \\ z(t) + G_2 y(t) \end{bmatrix} = \begin{bmatrix} I & 0 \\ G_2 & I \end{bmatrix} \begin{bmatrix} y(t) \\ z(t) \end{bmatrix} = \begin{bmatrix} 1 & 0 & 0 & 0 \\ 0 & 1 & 0 & 0 \\ 0 & -4 & 1 & 0 \\ 0 & -17 & 0 & 1 \end{bmatrix} \begin{bmatrix} y(t) \\ z(t) \end{bmatrix}$$

由 $\hat{x}(t)$ 构成的状态反馈律为

$$u(t) = \begin{bmatrix} 4 & 10 & 14 & 5 \\ 0 & 0 & 3 & 0 \end{bmatrix} \hat{x}(t) + v(t)$$

事实上，带观测器的状态反馈系统，就输入与输出的关系而言，可以看作是同时带有串联补偿和并联补偿的输出反馈系统。考虑如图 6.8 所示的具有降维观测器的状态反馈系统，把观测器看作是一个以 $y(t)$ 和 $u(t)$ 作为输入，以 $v_f(t)$ 作为输出的线性定常系统，那么这一系统可以表示为图 6.9 所示的形式，其中 $G_1(s)$ 表示从 $u(t)$ 到 $v_f(t)$ 的传递函数，$G_2(s)$ 表示从 $y(t)$ 到 $v_f(t)$ 的传递函数。根据 $v_f(t) = K\hat{x}(t)$，可知

$$G_1(s) = KQ_2(sI - F)^{-1}H$$

且

$$G_2(s) = KQ_1 + KQ_2(sI - F)^{-1}G$$

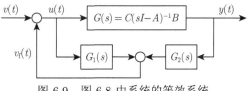

图 6.9 图 6.8 中系统的等效系统

简化图 6.9 中的结构图，可以得到如图 6.10 所示的等效系统，这一等效系统可以被进一步简化为如图 6.11 所示的系统。它是一个具有串联补偿 $G_b(s)$ 和并联补偿 $G_f(s)$ 的输出反馈系统，其中，

$$G_f(s) = G_2(s) = KQ_1 + KQ_2(sI - F)^{-1}G$$

且

$$G_b(s) = (I - G_1(s))^{-1}$$

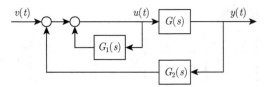

图 6.10　图 6.9 中系统的等效系统

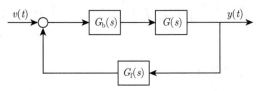

图 6.11　图 6.10 中系统的等效系统

可以进一步验证上述带观测器的闭环系统，其传递函数矩阵不变, 即

$$\hat{G}_{\text{K}}(s) = C(sI - A - BK)^{-1}B$$

参 考 文 献
References

陈复扬, 姜斌. 2022. 自动控制原理 [M]. 北京: 高等教育出版社.

段广仁. 2016. 线性系统理论 [M]. 3 版. 北京: 科学出版社.

胡寿松. 2007. 自动控制原理 [M]. 5 版. 北京: 科学出版社.

姜长生, 吴庆宪, 江驹, 等. 2008. 线性系统理论与设计（中英文版）[M]. 北京: 科学出版社.

郑大钟. 2002. 线性系统理论 [M]. 2 版. 北京: 清华大学出版社.

Antsaklis P J, Michel A N. 1997.Linear Systems[M]. New York: McGraw-Hill Book Company.

Bavafa-Toosi Y. 2017. Introduction to Linear Control Systems[M]. London: Academic Press.

Blyth T S, Robertson E F. 2013. Further Linear Algebra[M]. London: Springer Science & Business Media.

Callier F M, Desoer C A. 1991. Linear System Theory[M]. New York: Springer Science & Business Media.

Chen C T. 2019. 线性系统理论与设计 [M]. 4 版. 高飞, 王俊, 孙进平, 译. 北京: 北京航空航天大学出版社.

Dorf R C, Bishop R H. 2013. Modern Control Systems[M]. New York: Pearson.

Hespanha J P. 2018. Linear Systems Theory[M]. New Jersey: Princeton University Press.

Khalil H K. 2014. Nonlinear Control[M]. New York: Pearson.

Zadeh L A, Desoer C A. 1963. Linear System Theory: The State Space Approach[M]. New York: McGraw-Hill Book Company.

Zhang F Z. 2011. Matrix Theory: Basic Results and Techniques[M]. New York: Springer.

Chapter 1

Mathematical Description of Linear Systems

Based on physical or chemical principles of system kinetic processes, mathematical equations can be obtained to describe dynamical and static actions of systems. The mathematical equations are called mathematical description or mathematical model of systems. However, the mathematical model of a real system is usually nonlinear. For convenience, it can be regarded as a linear system near some working states. In other words, nonlinear mathematical model of a system can be linearized at a working state. In many cases, this approximation is accurate enough for resolving real project problems. The linearized mathematical model for a system is called a linear model. A class of systems, which can be described by linear model, is called linear systems. Linear system theory does not study real physical or chemical systems; it only studies the structural properties of a system and system analysis and design techniques based on the linear mathematical model of a system.

In general, system descriptions include two classes: the real variable description and the complex variable description. The most commonly used real variable description is the state space description, which takes time t as real variable and directly studies system actions changing with time. It is called the time domain method. The most commonly used complex variable method is the transfer function matrix description. It takes s as variable, which is associated with the frequency. It investigates system properties by studying system action changing with the input frequency, and is called the frequency domain method.

1.1 Basic Concepts

A system is called a continuous-time system if it applies continuous-time signals as input while generates continuous-time signals as output. This text only considers the continuous-time systems. A system is called a *single-variable system* or a *single-input single-output* (SISO) *system* if it only has one input terminal and one output terminal. The study object in the classical control theory is such a system. A system with more than one input terminal and/or more than one output terminal is called a *multivariable system*. Specifically, a system with two or more input terminals and output terminals can be called a *multi-input multi-output* (MIMO) *system*.

Consider a system with p inputs and q outputs, shown in Fig.1.1, where $u =$

$[\ u_1\ \ u_2\ \ \cdots\ \ u_p\ ]^\mathrm{T} \in \mathbb{R}^{p\times 1}$ denotes the input vector, and $y = [\ y_1\ \ y_2\ \ \cdots\ \ y_q\ ]^\mathrm{T} \in \mathbb{R}^{q\times 1}$ denotes the output.

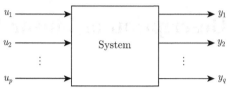

Fig.1.1　A MIMO system

1.1.1　Causality

A system is called a *memoryless system* if its output $y(t_0)$ depends only on the input applied at time t_0. It is independent of the input applied before or after t_0. A circuit network that consists of only resistors is an example of memoryless system. Most systems, however, are not memoryless ones. In other words, the output at time t_0 depends not only on the input applied at time t_0, but also on the inputs before t_0 and/or after t_0. That is to say, the current output of a system may depend on past, current, and future inputs. Such a system is called a *memory system*.

A system is called a *causal system* if its current output only depends on its past and current inputs but not on the future input. The current output of a noncausal system also depends on the future input. That is, a noncausal system is capable to predict what will be applied in the future. In fact, no physical system has such capability. In a word, for any real physical process, the result cannot take place before the cause. Therefore, the causality is required for a real physical system. This text only studies causal systems.

Generally, the input from $-\infty$ to the current time t has an influence on the current output $y(t)$ of a casual system. The current output of a casual system is affected by the past input. Tracking $u(t)$ from $t = -\infty$ is very inconvenient. The concept of the state can deal with this problem. The *state* $x(t_0)$ of a system at t_0 is the information at t_0 that, together with the input $u(t)$ for $t \geqslant t_0$, determines uniquely the output $y(t)$ for all $t \geqslant t_0$. In this way, it is not necessary to know the input $u(t)$ before t_0 in determining the output $y(t)$ after t_0 when knowing the state $x(t_0)$. For instance, consider a circuit network shown in Fig.1.2, if the voltages $x_1(t_0)$ and $x_2(t_0)$ across the two capacitors and the current $x_3(t_0)$ passing through the inductor is known, then for any input applied at and after t_0, the output for $t \geqslant t_0$ can be determined uniquely. Therefore, the state of the network at t_0 is $x(t_0) = [\ x_1(t_0)\ \ x_2(t_0)\ \ x_3(t_0)\ ]^\mathrm{T} \in \mathbb{R}^3$. The entry of x is called the *state variable*. In general, the initial state can be simply considered as a set of initial conditions.

Utilizing the state at t_0, the input and output of a system can be described as

$$\left. \begin{array}{r} x(t_0) \\ u(t), t \geqslant t_0 \end{array} \right\} \to y(t),\ t \geqslant t_0 \qquad (1.1)$$

which indicates that the output is partly excited by the initial state $x(t_0)$ and partly by the input at and after t_0. In this way, it is no more need to know the input applied

1.1 Basic Concepts

before t_0 all the way back to $-\infty$, which is more convenient. In general, (1.1) is called a state-input-output pair.

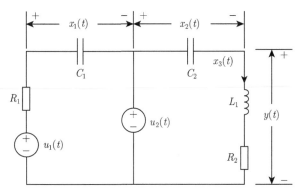

Fig.1.2 A circuit network with 3 state variables

1.1.2 Linearity

A system is called a *linear system* if for every t_0 and any two state-input-output pairs

$$\left.\begin{array}{r} x_1(t_0) \\ u_1(t), t \geqslant t_0 \end{array}\right\} \to y_1(t),\ t \geqslant t_0$$

and

$$\left.\begin{array}{r} x_2(t_0) \\ u_2(t), t \geqslant t_0 \end{array}\right\} \to y_2(t),\ t \geqslant t_0$$

it holds that

$$\left.\begin{array}{r} x_1(t_0) + x_2(t_0) \\ u_1(t) + u_2(t), t \geqslant t_0 \end{array}\right\} \to y_1(t) + y_2(t),\ t \geqslant t_0$$

and

$$\left.\begin{array}{r} \alpha x_1(t_0) \\ \alpha u_1(t), t \geqslant t_0 \end{array}\right\} \to \alpha y_1(t),\ t \geqslant t_0$$

for any real constant α. The former property is called the *additivity*, and the latter is called the *homogeneity*. They can be combined as

$$\left.\begin{array}{r} \alpha_1 x_1(t_0) + \alpha_2 x_2(t_0) \\ \alpha_1 u_1(t) + \alpha_2 u_2(t), t \geqslant t_0 \end{array}\right\} \to \alpha_1 y_1(t) + \alpha_2 y_2(t),\ t \geqslant t_0$$

for any real constants α_1 and α_2, which is generally called the *superposition* property. A system is called a *nonlinear system* if the superposition is not satisfied.

It should be noted that the additivity and the homogeneity cannot be substituted each other. A system with the homogeneity does not mean that it has the additivity, and vice versa. For example, consider a SISO system, at any time t, the relationship between the input and the output is given as

$$y(t) = \begin{cases} \dfrac{u^2(t)}{u(t-1)}, & u(t-1) \neq 0 \\ 0, & u(t-1) = 0 \end{cases}$$

It is easy to prove that this system satisfies the homogeneity but does not satisfy the additivity.

1.1.3 Time-Invariance

A system is called a *time-invariant system* if for every state-input-output pair

$$\left.\begin{array}{r} x(t_0) \\ u(t), t \geqslant t_0 \end{array}\right\} \to y(t), \ t \geqslant t_0$$

and any scalar T, it holds that

$$\left.\begin{array}{r} x(t_0 + T) \\ u(t-T), t \geqslant t_0 + T \end{array}\right\} \to y(t-T), \ t \geqslant t_0 + T$$

It indicates that if the initial state is shifted to time $t_0 + T$ and the same input waveform is applied from $t_0 + T$ instead of from t_0, then the output waveform will keep the same except that it starts to appear from $t_0 + T$. That is to say, for a time-invariant system, no matter at what time the input signal is applied, the output waveform is always the same. This fact is illustrated in Fig.1.3. Therefore, for a time-invariant system, one can assume that, without loss of generality, $t_0 = 0$. A system that does not satisfy the time-invariance is called a *time-varying system*.

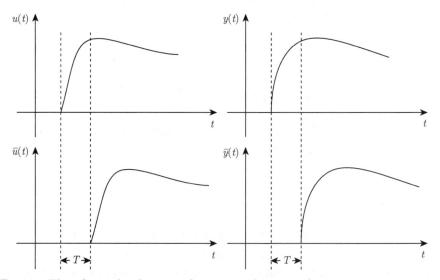

Fig.1.3 The relationship between the input and output of a time-invariant system

1.2 Input-Output Description of Linear Systems

For example, the circuit network in Fig.1.2 is time-invariant if R_1, R_2, C_1, C_2 and L_1 are all constants. Otherwise, it is time-varying.

It is noted that quantities of physical systems can be modeled as time-invariant systems over a limited time period. However, some physical systems must be modeled as time-varying systems. The burning rocket is a typical example since its mass decreases rapidly with time.

1.2 Input-Output Description of Linear Systems

The input-output description reveals the mathematical relationship between the input and the output of a system. When constructing the input-output description of a system, one can assume that the internal characteristics of the system is unknown, i.e., the system is viewed as a black box. Apply various input signals to the system, and then measure the corresponding output signals. According to the input-output data, the input-output description of a system can be determined.

When a linear system is represented by the input-output description, it is generally assumed that at time t_0, the output $y(t)$ for $t \geqslant t_0$ of the system only depends on the input $u(t)$ for $t \geqslant t_0$, and the system is called *relaxed* at t_0. In practical applications, it is always supposed that the system is relaxed at $t = -\infty$. In this way, if an input $u(t)$ for $t \in (-\infty, \infty)$ is applied, then the output $y(t)$ of the system only depends on $u(t)$ for $t \in (-\infty, \infty)$. As a matter of fact, the relaxed property of a system can be viewed as the zero initial condition.

1.2.1 Impulse Response of Linear Systems

The impulse response is an important class of input-output description of systems.

Let $\delta_\Delta(t - t_1)$ be the pulse shown in Fig.1.4, it can be described as

$$\delta_\Delta(t - t_1) = \begin{cases} 0, & t < t_1 \\ \dfrac{1}{\Delta}, & t_1 \leqslant t \leqslant t_1 + \Delta \\ 0, & t > t_1 + \Delta \end{cases}$$

For all Δ, the area of $\delta_\Delta(t - t_1)$ is always 1.

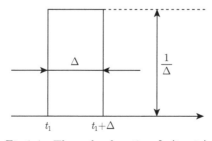

Fig.1.4 The pulse function $\delta_\Delta(t - t_1)$

Every piecewise-continuous input $u(t)$ can be appropriate by a sequence of pulses as shown in Fig.1.5. Since the value of $\delta_\Delta(t-t_i)\Delta$ is always 1, the input $u(t)$ can be described as

$$u(t) \approx \sum_i u(t_i)\delta_\Delta(t-t_i)\Delta \tag{1.2}$$

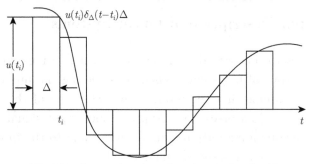

Fig.1.5 Approximation of an input signal $u(t)$

Let $g_\Delta(t,t_i)$ denote the output of a linear SISO at t excited by the pulse $\delta_\Delta(t-t_i)$. Owing to the homogeneity, the output of this system excited by $\delta_\Delta(t-t_i)u(t_i)\Delta$ is $g_\Delta(t,t_i)u(t_i)\Delta$. At the same time, owing to the additivity, the output excited by $\Sigma_i\delta_\Delta(t-t_i)u(t_i)\Delta$ is $\Sigma_i g_\Delta(t-t_i)u(t_i)\Delta$. Therefore, the output $y(t)$ excited by the input (1.2) can be appropriate by

$$y(t) \approx \sum_i g_\Delta(t,t_i)u(t_i)\Delta \tag{1.3}$$

If $\Delta \to 0$, the pulse $\delta_\Delta(t-t_i)$ becomes an *impulse* at t_i, denoted by $\delta(t-t_i)$, and the corresponding output is denoted by $g(t,t_i)$. As $\Delta \to 0$, the approximation in (1.3) becomes an equality, the summation becomes an integration, the discrete t_i becomes a continuum and can be replaced by τ, and Δ can be written as $\mathrm{d}\tau$. Consequently, (1.3) can be written as

$$y(t) = \int_{-\infty}^{\infty} g(t,\tau)u(\tau)\mathrm{d}\tau \tag{1.4}$$

It is noted that $g(t,\tau)$ is a function with two variables. The former variable denotes the time at which the output is observed, while the latter denotes the time at which the impulse input is applied. Since $g(t,\tau)$ is the response excited by an impulse, it is called the *impulse response*.

(1.4) can be further represented as

$$y(t) = \int_{-\infty}^{t_0} g(t,\tau)u(\tau)\mathrm{d}\tau + \int_{t_0}^{t} g(t,\tau)u(\tau)\mathrm{d}\tau + \int_{t}^{\infty} g(t,\tau)u(\tau)\mathrm{d}\tau$$

1.2 Input-Output Description of Linear Systems

If a system is causal, the output will not appear before an input is applied. That is to say, $g(t,\tau) = 0$ for $t < \tau$. Therefore, $\int_t^\infty g(t,\tau)u(\tau)\mathrm{d}\tau = 0$. Furthermore, if a system is relaxed at t_0, the output $y(t)$ for $t \geq t_0$ is excited exclusively by the input $u(t)$ for $t \geq t_0$. Therefore, $\int_{-\infty}^{t_0} g(t,\tau)u(\tau)\mathrm{d}\tau = 0$. In conclusion, a linear SISO system which is causal and relaxed at t_0 can be described as

$$y(t) = \int_{t_0}^t g(t,\tau)u(\tau)\mathrm{d}\tau \tag{1.5}$$

If a linear system which is causal and relaxed at t_0 has p inputs and q outputs, then (1.5) can be extended to

$$y(t) = \int_{t_0}^t G(t,\tau)u(\tau)\mathrm{d}\tau \tag{1.6}$$

where

$$G(t,\tau) = \begin{bmatrix} g_{11}(t,\tau) & \cdots & g_{1p}(t,\tau) \\ \vdots & & \vdots \\ g_{q1}(t,\tau) & \cdots & g_{qp}(t,\tau) \end{bmatrix}$$

which is called the *impulse response matrix* of the system. The physical meaning of the entry $g_{ij}(t,\tau)$ in $G(t,\tau)$ is the response of the ith output of the system at time t due to the jth impulse applied at time τ. That is, $g_{ij}(t,\tau)$ is the impulse response between the jth input and the ith output.

Consider a linear time-invariant SISO system which is causal and relaxed at t_0. Applying an impulse to the system at τ, the system output is $g(t,\tau)$ at the observing time t. If the applied time is $\tau + T$ (T is a real number), the system output at the observing time $t + T$ will equal to $g(t,\tau)$, that is

$$g(t+T, \tau+T) = g(t,\tau)$$

This equation holds for any t, τ and T. If $T = -\tau$, the above equation can be written as

$$g(t,\tau) = g(t-\tau, 0)$$

For convenience, denote $g(t-\tau, 0)$ as $g(t-\tau)$. This can be extended to the MIMO case. That is, for any t and τ, there exists

$$G(t,\tau) = G(t-\tau, 0) = G(t-\tau)$$

For the time-invariant case, one can always assume that the input signal is applied at $t_0 = 0$. Therefore, for a linear, time-invariant, causal system which is relaxed at $t_0 = 0$, its input-output description is

$$y(t) = \int_0^t G(t-\tau)u(\tau)\mathrm{d}\tau = \int_0^t G(\tau)u(t-\tau)\mathrm{d}\tau \tag{1.7}$$

1.2.2 Transfer Function Matrix of Linear Time-Invariant Systems

The transfer function is an important class mathematical description in the classical control theory. For a linear, time-invariant, causal, SISO system which is relaxed at time $t_0 = 0$, its input-output description is

$$y(t) = \int_0^t g(t-\tau)u(\tau)\mathrm{d}\tau$$

Taking the Laplace transform, one can obtain that[①]

$$y(s) = \mathcal{L}\left[y(t)\right] = \int_0^\infty y(t)\mathrm{e}^{-st}\mathrm{d}t$$

According to the convolution theorem of the Laplace transform, one can further obtain that

$$y(s) = g(s)u(s)$$

where $g(s)$ is called the transfer function of the system. It can be seen that the transfer function is the Laplace transform of the impulse response, and, conversely, the impulse response is the inverse Laplace transform of the transfer function.

The concept of transfer function in classical control theory can be extended to MIMO systems. For a linear, time-invariant, causal system which is relaxed at time $t_0 = 0$, the Laplace transform of its input-output description (1.7) is

$$y(s) = \mathcal{L}\left[y(t)\right] = \int_0^\infty y(t)\mathrm{e}^{-st}\mathrm{d}t = G(s)u(s)$$

where

$$G(s) = \int_0^\infty G(t)\mathrm{e}^{-st}\mathrm{d}t$$

is called the *transfer function matrix* of the system. It is the Laplace transform of the impulse response matrix.

Generally, $G(s)$ can be denoted as

$$G(s) = \begin{bmatrix} g_{11}(s) & \cdots & g_{1p}(s) \\ \vdots & & \vdots \\ g_{q1}(s) & \cdots & g_{qp}(s) \end{bmatrix}$$

---

① Note that $y(t)$ and $y(s)$ are different vector functions with different arguments. However, for convenience, the same symbol y is used.

1.2 Input-Output Description of Linear Systems

where $g_{ij}(s)$ in $G(s)$ is the transfer function from the jth input to the ith output, which equals to the ratio of the Laplace transform of the jth input to the ith output with zero initial conditions. The transfer function matrix is another form of the input-output description of systems.

1.2.3 Smith-McMillan Form of Transfer Function Matrices

The Smith-McMillan form is an important normal form of rational fraction matrix, which plays a significant role in defining and analyzing poles and zeros of transfer function matrices. The Smith-McMillan form is extended from the Smith form of polynomial matrices.

Let $N(s)$ be a real polynomial matrix whose elements are polynomials of s. For $N(s)$, the highest order of its nonzero minor is called the *rank* of $N(s)$ and can be written as rank $N(s)$. If $N(s)$ is a square matrix and $\det N(s) \neq 0$, then it is *regular* or *nonsingular* and the inverse matrix $N^{-1}(s)$ exists. The sufficient and necessary condition for $N^{-1}(s)$ to be a polynomial matrix is that $\det N(s)$ is a nonzero constant. A polynomial square matrix whose determinant is a nonzero constant is called a *unimodular matrix*, and the product of unimodular matrices is still a unimodular matrix.

The *determinant divisors* of $N(s)$ with rank $N(s) = r$ are polynomials $d_i(s)$, $i \in \{0, \cdots, r\}$, where $d_0(s) = 1$, and $d_i(s)$ is the monic greatest common divisor of all nonzero minors of $N(s)$ of order i.

Using the determinant divisors, the Smith form of polynomial matrices can be defined.

Definition 1.1 (Smith form of polynomial matrices) The *Smith form* of a $q \times p$ real polynomial matrix $N(s)$ with rank $N(s) = r$ is the diagonal real polynomial matrix defined by

$$N_S(s) = \begin{bmatrix} s_1(s) & & & & & \\ & s_2(s) & & & & 0 \\ & & \ddots & & & \\ & & & s_r(s) & & \\ \hdashline & & 0 & & & 0 \end{bmatrix} \quad (1.8)$$

where

$$s_i(s) = \frac{d_i(s)}{d_{i-1}(s)}, \quad i \in \{1, \cdots, r\}$$

which are called the *invariant factors* of $N(s)$.

Example 1.1 The real polynomial matrix

$$N(s) = \begin{bmatrix} s(s+3) & 0 \\ 0 & (s+2)^2 \\ s+3 & (s+2)(s+3) \\ 0 & s(s+2) \end{bmatrix}$$

has the nonzero minors, determinant divisors, and invariant factors as shown in Table 1.1.

Table 1.1 Nonzero minors, determinant divisors, and invariant factors of $N(s)$

Order	Nonzero minors	Determinant divisors	Invariant factors
$i = 0$	None	$d_0(s) = 1$	
$i = 1$	$s(s+3), (s+2)^2,$ $s+3, (s+2)(s+3),$ $s(s+2)$	$d_1(s) = 1$	$s_1(s) = 1$
$i = 2$	$s(s+2)^2(s+3),$ $s(s+2)(s+3)^2,$ $s^2(s+2)(s+3),$ $-(s+2)^2(s+3),$ $s(s+2)(s+3)$	$d_2(s) = (s+2)(s+3)$	$s_2(s) = (s+2)(s+3)$

Furthermore, the Smith form of $N(s)$ is

$$N_\mathrm{S}(s) = \begin{bmatrix} 1 & 0 \\ 0 & (s+2)(s+3) \\ 0 & 0 \\ 0 & 0 \end{bmatrix}$$

As a matter of fact, a real polynomial matrix can always transform into its Smith form by elementary transformations. The addition, subtraction and multiplication of a polynomial matrix are the same as those for a constant matrix. The elementary transformations include following three types.

(1) The row (column) interchange transform matrix is

$$U_1(s) = \begin{bmatrix} 1 & & & & & & \\ & \ddots & & & & & \\ & & 0 & \cdots & 1 & & \\ & & \vdots & & \vdots & & \\ & & 1 & \cdots & 0 & & \\ & & & & & \ddots & \\ & & & & & & 1 \end{bmatrix} \begin{array}{l} \\ \\ i\text{th row} \\ \\ j\text{th row} \\ \\ \end{array}$$

$$\phantom{U_1(s) = \begin{bmatrix}}i\text{th column}j\text{th column}$$

1.2 Input-Output Description of Linear Systems

Left-multiplying $N(s)$ by $U_1(s)$ means a row interchange, and right-multiplying $N(s)$ by $U_1(s)$ means a column interchange.

(2) Multiplying the jth row (ith column) of $N(s)$ by $f(s)$, and adding the results to the ith row (jth column), the corresponding transform matrix is

$$U_2(s) = \begin{bmatrix} 1 & & & & & & \\ & \ddots & & & & & \\ & & 1 & \cdots & f(s) & & \\ & & & \ddots & \vdots & & \\ & & & & 1 & & \\ & & & & & \ddots & \\ & & & & & & 1 \end{bmatrix} \begin{matrix} \\ \\ i\text{th row} \\ \\ j\text{th row} \\ \\ \end{matrix}$$

$$\qquad\qquad\quad i\text{th column} \quad j\text{th column}$$

Left-multiplying $N(s)$ by $U_2(s)$ means performing a row computation to $N(s)$, and right-multiplying $N(s)$ by $U_2(s)$ means performing a column computation.

(3) Multiplying the ith row (column) of $N(s)$ by a nonzero constant α, the corresponding transform matrix is

$$U_3(s) = \begin{bmatrix} 1 & & & & & & \\ & \ddots & & & & & \\ & & 1 & & & & \\ & & & \alpha & & & \\ & & & & 1 & & \\ & & & & & \ddots & \\ & & & & & & 1 \end{bmatrix} \begin{matrix} \\ \\ \\ i\text{th row} \\ \\ \\ \end{matrix}$$

$$\qquad\qquad\qquad i\text{th column}$$

Left-multiplying (right-multiplying) $N(s)$ by $U_3(s)$ means multiplying all the elements in the ith row (column) of $N(s)$ by α.

Definition 1.2 (Equivalence of polynomial matrices) For two polynomial matrices $N(s)$ and $M(s)$, if there is an unimodular matrix $U_L(s)$ satisfying

$$N(s) = U_L(s)M(s)$$

then $N(s)$ and $M(s)$ are *row equivalent*, if there is a unimodular matrix $U_R(s)$ satisfying

$$N(s) = M(s)U_R(s)$$

then $N(s)$ and $M(s)$ are *column equivalent*, if there are two unimodular matrices $U_L(s)$ and $U_R(s)$ satisfying

$$N(s) = U_L(s)M(s)U_R(s)$$

then $N(s)$ and $M(s)$ are *equivalent*.

Using elementary transformations, it is not difficult to transform a polynomial matrix $N(s)$ into the Smith form.

Theorem 1.1 *For a $q \times p$ real polynomial matrix $N(s)$ with $\operatorname{rank} N(s) = r$, there always exist $q \times q$ and $p \times p$ unimodular matrices $U(s)$ and $V(s)$ such that*

$$U(s)N(s)V(s) = N_S(s)$$

where $N_S(s)$ defined in (1.8) is the Smith form of $N(s)$.

The proof can be derived by elementary transformations, here we omit it.

Without loss of generality, suppose that a $q \times p$ transfer function matrix $G(s)$ can be described as

$$G(s) = \frac{1}{g(s)} N(s) \tag{1.9}$$

where $g(s)$ is the monic least common denominator of all entries of $G(s)$, and $N(s)$ is a $q \times p$ polynomial matrix and $\operatorname{rank} N(s) = r$.

Definition 1.3 (Smith-McMillan form of rational fraction matrices) The *Smith-McMillan form* of a $q \times p$ real rational fraction matrix $G(s)$ in (1.9) is the diagonal real rational fraction matrix defined by

$$G_M(s) = \frac{1}{g(s)} N_S(s) = \begin{bmatrix} \frac{\varepsilon_1(s)}{\psi_1(s)} & & & & \\ & \frac{\varepsilon_2(s)}{\psi_2(s)} & & & 0 \\ & & \ddots & & \\ & & & \frac{\varepsilon_r(s)}{\psi_r(s)} & \\ \hline & & 0 & & 0 \end{bmatrix} \tag{1.10}$$

where $N_S(s)$ denotes the Smith form of $N(s)$. All the common factors in the entries of (1.10) should be canceled, which indicates that the pairs of polynomials $\{\varepsilon_i(s), \psi_i(s)\}$ are all coprime, $i \in \{1, \cdots, r\}$.

According to the Smith form factorization in Theorem 1.1, one can conclude the Smith-McMillan factorization.

Theorem 1.2 *For a $q \times p$ real rational fraction matrix $G(s)$ in (1.9), there always exist $q \times q$ and $p \times p$ unimodular matrices $U(s)$ and $V(s)$ such that*

$$U(s)G(s)V(s) = G_M(s)$$

where $G_M(s)$ defined in (1.10) is the Smith-McMillan form of $G(s)$.

Example 1.2 Give out the Smith-McMillan form of the following transfer function

1.2 Input-Output Description of Linear Systems

matrix,

$$G(s) = \begin{bmatrix} \dfrac{s}{(s+1)^2(s+2)^2} & \dfrac{s}{(s+2)^2} \\ -\dfrac{s}{(s+2)^2} & -\dfrac{s}{(s+2)^2} \end{bmatrix}$$

The least common denominator of each entry of $G(s)$ is $g(s) = (s+1)^2(s+2)^2$. For $G(s) = N(s)/g(s)$, on can obtain that

$$N(s) = \begin{bmatrix} s & s(s+1)^2 \\ -s(s+1)^2 & -s(s+1)^2 \end{bmatrix}$$

The first-order minors of $N(s)$ are entries of $N(s)$, and then the determinant divisor $d_1(s) = s$. The second-order minor of $N(s)$ is $s^3(s+1)^2(s+2)$, and then the determinant divisor $d_2(s) = s^3(s+1)^2(s+2)$. Furthermore, the invariant factors of $N(s)$ are $s_1(s) = s$ and $s_2(s) = s^2(s+1)^2(s+2)$. Therefore, the Smith form of $N(s)$ is

$$\begin{bmatrix} s & 0 \\ 0 & s^2(s+1)^2(s+2) \end{bmatrix}$$

The Smith form of $N(s)$ can also be deduced by the elementary transformation. Add the $(s+1)^2$ times of the first row of $N(s)$ to the second row, one can get that

$$\begin{bmatrix} s & s(s+1)^2 \\ 0 & s(s+1)^4 - s(s+1)^2 \end{bmatrix}$$

Furthermore, add the $-(s+1)^2$ times of the first column of the above matrix to the second column, one can get that

$$\begin{bmatrix} s & 0 \\ 0 & s^2(s+1)^2(s+2) \end{bmatrix}$$

Divide each entry of the equation above by the least common denominator $g(s)$ to get the Smith-McMillan form of $G(s)$,

$$G_M(s) = U(s)G(s)V(s) = \begin{bmatrix} \dfrac{s}{(s+1)^2(s+2)^2} & \\ & \dfrac{s^2}{s+2} \end{bmatrix}$$

where

$$U(s) = \begin{bmatrix} 1 & 0 \\ (s+1)^2 & 1 \end{bmatrix}, \quad V(s) = \begin{bmatrix} 1 & -(s+1)^2 \\ 0 & 1 \end{bmatrix}$$

1.2.4 Poles and Zeros of Transfer Function Matrices

In linear time-invariant SISO systems, poles and zeros of a transfer function are very important, which determine the stability, the dynamical characteristics of a system and the frequency properties. The zeros are defined as the values of s which make the output y equal to 0 when the input u is limited. The poles are the values of s which make y equal to ∞. For convenience of computation, take the roots of the numerator polynomial of the transfer function as zeros, and use the roots of the denominator polynomial of the transfer function as poles. However, in linear time-invariant MIMO systems, the situation is more complicated. It should be understood from the concept, computation and physical meaning.

The Smith-McMillan form is especially useful to define poles and zeros for transfer function matrices.

Definition 1.4 (Poles and zeros of transfer function matrices) For a $q \times p$ real rational fraction transfer matrix $G(s)$, its regular rank is $\operatorname{rank} G(s) = r \leqslant \min\{q, p\}$. If its Smith-McMillan form is (1.10), the polynomial

$$p(s) = \prod_{i=1}^{r} \psi_i(s)$$

is called the *pole* (or *characteristic*) *polynomial* of $G(s)$, its degree is called the *McMillan degree* of $G(s)$, and its roots are called the *poles* of $G(s)$. The polynomial

$$z(s) = \prod_{i=1}^{r} \varepsilon_i(s)$$

is called the *zero polynomial* of $G(s)$, and its roots are called the *zeros* of $G(s)$.

Example 1.3 A $q \times p$ rational fraction transfer matrix is

$$G(s) = \begin{bmatrix} \dfrac{s}{(s+1)^2(s+2)^2} & \dfrac{s}{(s+2)^2} \\ -\dfrac{s}{(s+2)^2} & -\dfrac{s}{(s+2)^2} \end{bmatrix}$$

In Example 1.2, its Smith-McMillan form has been calculated,

$$G_M(s) = \begin{bmatrix} \dfrac{s}{(s+1)^2(s+2)^2} & \\ & \dfrac{s^2}{s+2} \end{bmatrix}$$

Based on Definition 1.4, $G(s)$ has 3 zeros at $s = 0$, 2 poles at $s = -1$, and 3 poles at $s = -2$. The zero polynomial of $G(s)$ is $z(s) = s^3$, and the pole polynomial of $G(s)$ is $p(s) = (s+1)^2(s+2)^3$.

It is worth mentioning that Definition 1.4 is only applicable in determining poles and zeros in finite complex plane. Poles and zeros of $G_M(s)$ in ∞ may not be poles and zeros of $G(s)$ in ∞.

Different from the scalar transfer function of a linear time-invariant SISO system, the transfer function matrix $G(s)$ may not be canceled when zeros and poles are at the same point in the complex plane. That is because for the Smith-McMillan form $G_M(s)$ in (1.10), there may be common factor between $\varepsilon_i(s)$ and $\psi_j(s)$, $i \in \{1,\cdots,r\}$, $j \in \{1,\cdots,r\}$, $i \neq j$.

The poles and zeros of a transfer function matrix $G(s)$ have other definitions. The following one is frequently used in engineering applications.

Definition 1.5 (Poles and zeros of transfer function matrices) For a $q \times p$ real rational fraction transfer matrix $G(s)$ with $\operatorname{rank} G(s) = r$, the least common denominator of its all nonzero minors is the pole polynomial $p(s)$ of $G(s)$. The roots of $p(s) = 0$ are the poles of $G(s)$. When all rth-order minors of $G(s)$ using $p(s)$ as their common denominator, the monic greatest common divisor of their numerators is the zero polynomial $z(s)$ of $G(s)$. The roots of $z(s) = 0$ are the zeros of $G(s)$.

Example 1.4 A given $q \times p$ rational fraction transfer function matrix $G(s)$ is

$$G(s) = \begin{bmatrix} \dfrac{1}{s+1} & 0 & \dfrac{s-1}{(s+1)(s+2)} \\ \dfrac{-1}{s-1} & \dfrac{1}{s+2} & \dfrac{1}{s+2} \end{bmatrix}$$

Based on Definition 1.5, $\operatorname{rank} G(s) = r = 2 \leqslant \min\{p,q\}$. The first-order minors of $G(s)$ are its entries. The second-order minors of $G(s)$ are

$$\frac{1}{(s+1)(s+2)},\ \frac{-(s-1)}{(s+1)(s+2)^2},\ \frac{2}{(s+1)(s+2)}$$

The least common denominator of these nonzero minors is $(s+1)(s+2)^2(s-1)$, which is the pole polynomial of $G(s)$. Furthermore, the poles of $G(s)$ are -1, -2, -2 and 1.

On this basis, when the above three second-order minors adopt the pole polynomial $p(s)$ as denominators, one can obtain that

$$\frac{(s+2)(s-1)}{(s+1)(s+2)^2(s-1)},\ \frac{-(s-1)^2}{(s+1)(s+2)^2(s-1)},\ \frac{2(s+2)(s-1)}{(s+1)(s+2)^2(s-1)}$$

The monic greatest common divisor of their numerators is $s - 1$. Therefore, the zero of $G(s)$ is 1.

1.3 State Space Description of Linear Systems

The input-output description of a system only fits the system that is relaxed and does not study the system internal information. At the beginning of 1960s, Kalman et al. introduced the concepts of the state space to the studying of systems and control theories. The studying results put forward the great development of modern control subjects.

1.3.1 Dynamical Equations of Linear Systems

Generally, for a continuous-time linear system with p inputs and q outputs, the state space description can be expressed as

$$\dot{x}(t) = A(t)x(t) + B(t)u(t) \tag{1.11}$$

$$y(t) = C(t)x(t) + D(t)u(t) \tag{1.12}$$

where $x(t) \in \mathbb{R}^n$ is the state vector, $u(t) \in \mathbb{R}^p$ is the input vector, $y(t) \in \mathbb{R}^q$ is the output vector, and $A(t) \in \mathbb{R}^{n \times n}$, $B(t) \in \mathbb{R}^{n \times p}$, $C(t) \in \mathbb{R}^{q \times n}$ and $D(t) \in \mathbb{R}^{q \times p}$ are continuous function matrices of time t defined in $(-\infty, \infty)$. They are called the *state coefficient matrix*, *control coefficient matrix*, *output coefficient matrix* and *feedforward coefficient matrix* of a system, respectively. The first-order differential equation (1.11) is called the *state equation* and the equation (1.12) is called the *output equation*. Both of them are collectively called the *dynamical equations* of a system.

Fig.1.6 shows the transfer relations described by dynamical equations between the internal structure of the system and signals.

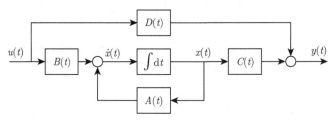

Fig.1.6 The linear system structure in the state space

In (1.11) and (1.12), coefficient matrices change with time. The system described by the above equations is a linear time-varying system. If all of the coefficient matrices are constant matrices, the dynamical equations can be written as

$$\dot{x}(t) = Ax(t) + Bu(t) \tag{1.13}$$

$$y(t) = Cx(t) + Du(t) \tag{1.14}$$

where A, B, C and D are constant coefficient matrices. The system described by (1.13) and (1.14) is a linear time-invariant system.

Example 1.5 Consider a circuit network shown in Fig.1.7. The input $u(t)$ is the voltage $u_s(t)$, which satisfies

$$LC\frac{\mathrm{d}^2 u_C(t)}{\mathrm{d}t^2} + RC\frac{\mathrm{d}u_C(t)}{\mathrm{d}t} + u_C(t) = u(t)$$

1.3 State Space Description of Linear Systems

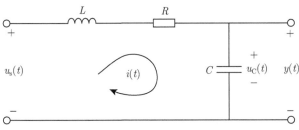

Fig.1.7 A circuit network

Denote $x_1(t) = u_C(t)$ and $x_2(t) = i(t)$, one can obtain that

$$\dot{x}_1(t) = \frac{1}{C}x_2(t)$$

$$\dot{x}_2(t) = -\frac{1}{L}x_1(t) - \frac{R}{L}x_2(t) + \frac{1}{L}u(t)$$

which can be written as the state equation(1.13) with $x(t) = [\ x_1(t)\ \ x_2(t)\ ]^T$, and

$$A = \begin{bmatrix} 0 & \frac{1}{C} \\ -\frac{1}{L} & -\frac{R}{L} \end{bmatrix},\ B = \begin{bmatrix} 0 \\ \frac{1}{L} \end{bmatrix}$$

Consider that the output is the voltage across the capacitor, i.e., $y(t) = u_C(t)$. It can be written as the output equation (1.14) with

$$C = \begin{bmatrix} 1 & 0 \end{bmatrix},\ D = 0$$

Example 1.6 Consider an armature controlling DC motor shown in Fig.1.8.

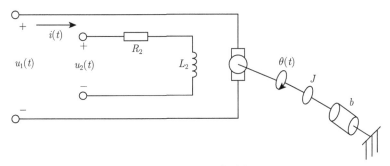

Fig.1.8 DC motor controlled by armature

Its excitation voltage is constant, the control voltage $u_1(t)$ is applied on the armature circuit, $\theta(t)$ is the angle of motor axis. The voltage equation of the armature circuit is

$$u_1(t) = Ri(t) + L\frac{di(t)}{dt} + \varepsilon(t)$$

where R is the equivalent resistance, L is an equivalent inductance, and $\varepsilon(t)$ is the counter-electromotive force of the motor. Based on the motor theory, one can deduce that

$$\varepsilon(t) = k_e \frac{\mathrm{d}\theta(t)}{\mathrm{d}t}$$

where k_e is the electrical potential coefficient of the motor. The moment equation of the motor is

$$J\frac{\mathrm{d}^2\theta(t)}{\mathrm{d}t^2} + b\frac{\mathrm{d}\theta(t)}{\mathrm{d}t} = M = k_m i(t)$$

where J is the equivalent inertia moment of the motor axis, b is the equivalent damping coefficient of the axis, and k_m is the moment coefficient of the motor. From the above equations, one can obtain that

$$\frac{\mathrm{d}i(t)}{\mathrm{d}t} = -\frac{R}{L}i(t) - \frac{k_e}{L}\frac{\mathrm{d}\theta(t)}{\mathrm{d}t} + \frac{1}{L}u_1(t)$$

$$\frac{\mathrm{d}^2\theta(t)}{\mathrm{d}t^2} = \frac{k_m}{J}i(t) - \frac{b}{J}\frac{\mathrm{d}\theta(t)}{\mathrm{d}t}$$

Denote $x_1(t) = \theta(t)$, $x_2(t) = \dot{\theta}(t)$ and $x_3(t) = i(t)$ as state variables, $u_1(t)$ is the control input, and $y_1(t) = x_1(t)$, $y_2(t) = x_3(t)$ are the output variables. Therefore, the state equation and the output equation of the armature control DC motor system can be written as

$$\dot{x}_1(t) = x_2(t)$$
$$\dot{x}_2(t) = -\frac{b}{J}x_2(t) + \frac{k_m}{J}x_3(t)$$
$$\dot{x}_3(t) = -\frac{k_e}{L}x_2(t) - \frac{R}{L}x_3(t) + \frac{1}{L}u_1(t)$$
$$y_1(t) = x_1(t)$$
$$y_2(t) = x_3(t)$$

They can also be described as dynamical equations (1.13) and (1.14), where $x(t) = [\ x_1(t)\ \ x_2(t)\ \ x_3(t)\ ]^\mathrm{T}$ is the state vector, $u(t) = u_1(t)$ is the control input, $y(t) = [\ y_1(t)\ \ y_2(t)\ ]^\mathrm{T}$ is the output vector, and the coefficient matrices are

$$A = \begin{bmatrix} 0 & 1 & 0 \\ 0 & -\dfrac{b}{J} & \dfrac{k_m}{J} \\ 0 & -\dfrac{k_e}{L} & -\dfrac{R}{L} \end{bmatrix},\ B = \begin{bmatrix} 0 \\ 0 \\ \dfrac{1}{L} \end{bmatrix},\ C = \begin{bmatrix} 1 & 0 & 0 \\ 0 & 0 & 1 \end{bmatrix},\ D = 0$$

It should be pointed out that the choice of state variable is not unique. For the same system, the system dynamical equations are different with different state variables.

1.3.2 Characteristic Structures of Linear Time-Invariant Systems

The characteristic structure of a linear time-invariant system is represented by eigenvalues and eigenvectors, which have great influence on motion characteristics of a system.

Consider a linear time-invariant system and its state equation is given by (1.13). $sI-A$ is called the *eigenmatrix*, where s denotes the complex variable and I is the identity matrix with the same dimension as A. $sI - A$ is a polynomial matrix which is nonsingular, and its inverse $(sI - A)^{-1}$ is called the *resolvent matrix*. $\det(sI - A)$ is called the *characteristic polynomial*, which is an nth-order polynomial,

$$\alpha(s) = \det(sI - A) = s^n + a_1 s^{n-1} + \cdots + a_{n-1} s + a_n \tag{1.15}$$

where $\alpha_1, \cdots, a_n$ are real constants determined by A. $\det(sI - A) = 0$ is called the *characteristic equation*. According to Cayley-Hamilton theorem, A is a "matrix root" of its characteristic equation. That is,

$$A^n + a_1 A^{n-1} + \cdots + a_{n-1} A + a_n I = 0 \tag{1.16}$$

which indicates that for $A \in \mathbb{R}^{n \times n}$, all of A^i with $i \geqslant n$ can be expressed as the linear combination of $\{I, A, \cdots, A^{n-1}\}$.

The resolvent matrix $(sI - A)^{-1}$ satisfies

$$(sI - A)^{-1} = \frac{\mathrm{adj}(sI - A)}{\alpha(s)} = \frac{P(s)}{\phi(s)} \tag{1.17}$$

where $\mathrm{adj}(sI - A)$ is the *adjoint matrix* of $sI - A$, which is a polynomial matrix. The order of s in the polynomial of $\mathrm{adj}(sI-A)$ is no greater than the order of s in $\alpha(s)$. Furthermore, canceling completely the common factor between $\alpha(s)$ and each entry of $\mathrm{adj}(sI - A)$, one can obtain the right expression in (1.17), where $P(s)$ is a polynomial matrix, and $\phi(s)$ is a polynomial whose order is no greater than the order of $\alpha(s)$. Moreover, $\phi(s)$ and each polynomial of $P(s)$ is coprime. On this basis, $\phi(s)$ is defined as the *minimal polynomial* of A. $\phi(s)$ also satisfies Cayley-Hamilton theorem, i.e., $\phi(A) = 0$. In this way, the minimal polynomial is also defined as the monic polynomial $\phi(s)$ with the least degree such that $\phi(A) = 0$. A matrix A is said to be *cyclic* if and only if its characteristic polynomial equals to its minimal polynomial.

The roots of the characteristic equation $\det(sI - A) = 0$ is called the eigenvalues of A. The $n \times n$ matrix A has n eigenvalues. Let λ_i be an eigenvalue of A and

$$\det(\lambda I - A) = (\lambda - \lambda_i)^{\sigma_i} \beta_i(\lambda), \ \beta_i(\lambda) \neq 0$$

then σ_i is called the *algebraic multiplicity* of λ_i. An eigenvalue with algebraic multiplicity 1 is called a *simple eigenvalue*, while an eigenvalue with algebraic multiplicity 2 or higher is called a *repeated eigenvalue*. Meanwhile,

$$\alpha_i = n - \mathrm{rank}\,(\lambda_i I - A)$$

is called the *geometric multiplicity* of λ_i. If λ_i is a simple eigenvalue, then $\sigma_i = \alpha_i = 1$. Otherwise, $\sigma_i \geqslant \alpha_i$.

A nonzero vector v_i satisfying $Av_i = \lambda_i v_i$ is called a (*right*) *eigenvector* of A associated with the eigenvalue λ_i. This definition is generally written as

$$(\lambda_i I - A) v_i = 0$$

The eigenvector is not unique. If eigenvalues of A are all distinct, then eigenvectors $\{v_1, \cdots, v_n\}$ associated with $\{\lambda_1, \cdots, \lambda_n\}$ are linearly independent.

A nonzero vector v_i is called a *generalized eigenvector* of grade k associated with the eigenvalue λ_i, if

$$(\lambda_i I - A)^k v_i = 0, \quad (\lambda_i I - A)^{k-1} v_i \neq 0$$

It reduces to the ordinary eigenvector when $k = 1$. If v_i is the generalized eigenvector of grade k associated with the eigenvalue λ_i, then the following k vectors are linearly independent,

$$v_i^{(k)} = v_i, \ v_i^{(k-1)} = (\lambda_i I - A) v_i, \ \cdots, \ v_i^{(1)} = (\lambda_i I - A)^{k-1} v_i$$

where the vector group $v_i^{(1)}, \cdots, v_i^{(k)}$ is called a chain of generalized eigenvectors. If λ_i is an eigenvalue with algebraic multiplicity σ_i, then its associated generalized eigenvector group contains σ_i linearly independent nonzero vectors. The generalized eigenvectors associated with different eigenvalues of A are linearly independent.

This subsection presents some definitions and properties about characteristic structures of linear time-invariant systems, which are important in the sequel.

1.3.3 Deriving Transfer Function Matrix from State Space Description

Consider a linear time-invariant system described by the state space model (1.13) and (1.14). Assume that the initial conditions satisfy $x(0) = 0$ and $u(0) = 0$. Taking the Laplace transform of both sides of (1.13) and (1.14), one can deduce that

$$sx(s) = Ax(s) + Bu(s)$$
$$y(s) = Cx(s) + Du(s)$$

Eliminating $x(s)$, one can get

$$G(s) = C(sI - A)^{-1}B + D \tag{1.18}$$

(1.18) indicates that $G(s)$ is a strictly proper rational fraction matrix when $D = 0$. When $D \neq 0$, $G(s)$ is a proper rational fraction matrix, and

$$\lim_{s \to \infty} G(s) = D$$

Let Λ_G denote the set consisting of all poles of $G(s)$, and Λ denote the set consisting of all eigenvalues of A. If and only if the system expressed by (1.13) and (1.14) is completely

1.3 State Space Description of Linear Systems

controllable and observable, $\Lambda_G = \Lambda$. Otherwise, $\Lambda_G \subset \Lambda$. The definitions of controllability and observability will be introduced in Chapter 3.

It is difficult to compute $G(s)$ by (1.18) since the inverse is involved. Faddeev-Leverrier algorithm is an efficient method to solve the resolvent matrix in (1.18).

Algorithm 1.1 (Faddeev-Leverrier algorithm) According to (1.17), $(sI - A)^{-1}$ can be written as

$$(sI - A)^{-1} = \frac{R_{n-1}s^{n-1} + R_{n-2}s^{n-2} + \cdots + R_1 s + R_0}{s^n + a_1 s^{n-1} + \cdots + a_{n-1}s + a_n} \tag{1.19}$$

where $R_0, \cdots, R_{n-1}$ are $n \times n$ constant matrices. Pre-multiply $\alpha(s)(sI - A)$ on both sides of (1.19), where $\alpha(s)$ is the characteristic polynomial of A, it holds that

$$\left(s^n + a_1 s^{n-1} + \cdots + a_{n-1}s + a_n\right) I = (sI - A)\left(R_{n-1}s^{n-1} + R_{n-2}s^{n-2} + \cdots + R_1 s + R_0\right)$$

Compare the coefficient matrices of the same order of s in the both sides of the above equation, then one can deduce that

$$\begin{cases} R_{n-1} = I \\ R_{n-2} = AR_{n-1} + a_1 I = A + a_1 I \\ R_{n-3} = AR_{n-2} + a_2 I = A^2 + a_1 A + a_2 I \\ \vdots \\ R_0 = AR_1 + a_{n-1} I = A^{n-1} + a_1 A^{n-2} + \cdots + a_{n-1} I \\ 0 = AR_0 + a_n I \end{cases} \tag{1.20}$$

The recursion formula for calculating the coefficients of the characteristic polynomial (1.15) is

$$\begin{cases} a_1 = -\operatorname{tr}(R_{n-1} A) \\ a_2 = -\frac{1}{2}\operatorname{tr}(R_{n-2} A) \\ \vdots \\ a_n = -\frac{1}{n}\operatorname{tr}(R_0 A) \end{cases} \tag{1.21}$$

where $\operatorname{tr}(X)$ denotes the trace of the square matrix X. Using (1.20) and (1.21) to compute R_{n-i} and a_i by turns, $i = 1, 2, \cdots, n$, and then the resolvent matrix can be obtained. The last equation in (1.20) can be used to validate the correctness of computation.

Example 1.7 Consider $A = \begin{bmatrix} 1 & 1 & 2 \\ 0 & 1 & 3 \\ 0 & 0 & 2 \end{bmatrix}$, compute $(sI - A)^{-1}$ by using (1.17).

The characteristic polynomial of matrix A is

$$\alpha(s) = \det(sI - A) = \det \begin{bmatrix} s-1 & -1 & -2 \\ 0 & s-1 & -3 \\ 0 & 0 & s-2 \end{bmatrix}$$

$$=(s-1)^2(s-2)=s^3-4s^2+5s-2$$

and thus, $a_1=-4$, $a_2=5$, $a_3=-2$. Using (1.20) to compute $R(s)=R_2s^2+R_1s+R_0$,

$$R_2=I,\ R_1=AR_2+a_1I=\begin{bmatrix}-3&1&2\\0&-3&3\\0&0&-2\end{bmatrix},\ R_0=AR_1+a_2I=\begin{bmatrix}2&-2&1\\0&2&-3\\0&0&1\end{bmatrix}$$

Based on (1.19), one can compute that

$$(sI-A)^{-1}=\frac{R(s)}{\alpha(s)}=\frac{1}{(s-1)^2(s-2)}\begin{bmatrix}s^2-3s+2&s-2&2s+1\\0&s^2-3s+2&3s-3\\0&0&s^2-2s+1\end{bmatrix}$$

$$=\begin{bmatrix}\dfrac{1}{s-1}&\dfrac{1}{(s-1)^2}&\dfrac{2s+1}{(s-1)^2(s-2)}\\0&\dfrac{1}{s-1}&\dfrac{3}{(s-1)(s-2)}\\0&0&\dfrac{1}{s-2}\end{bmatrix}$$

Based on Cayley-Hamilton theorem, an algorithm to compute the resolvent matrix can be derived.

Algorithm 1.2 (Cayley-Hamilton algorithm) *Express* $(sI-A)^{-1}$ *as a power series*

$$(sI-A)^{-1}=\sum_{k=0}^{\infty}\frac{1}{s^{k+1}}A^k$$

which can be written as the following form according to Cayley-Hamilton theorem,

$$(sI-A)^{-1}=q_0(s)I+q_1(s)A+\cdots+q_{n-1}(s)A^{n-1} \tag{1.22}$$

where $q_i(s)$, $i\in\{0,1,\cdots,n-1\}$ *are rational polynomials of s. Left-multiply* $sI-A$ *on both sides of* (1.22), *and then one can obtain that*

$$I=sq_0(s)I+sq_1(s)A+\cdots+sq_{n-1}(s)A^{n-1}-\left[q_0(s)A+q_1(s)A^2+\cdots+q_{n-1}(s)A^n\right]$$

Using Cayley-Hamilton theorem again, and rearrange the result, it holds that

$$I=[sq_0(s)+a_nq_{n-1}(s)]I+[sq_1(s)-q_0(s)+a_{n-1}q_{n-1}(s)]A+\cdots$$
$$+[sq_{n-1}(s)-q_{n-2}(s)+a_1q_{n-1}(s)]A^{n-1}$$

Comparing the coefficients of the same order of A in both sides of the equation above, it can be deduced that

$$\begin{cases}sq_0(s)+a_nq_{n-1}(s)=1\\sq_1(s)-q_0(s)+a_{n-1}q_{n-1}(s)=0\\\vdots\\(s+a_1)q_{n-1}(s)-q_{n-2}(s)=0\end{cases} \tag{1.23}$$

1.3 State Space Description of Linear Systems

Multiplying $1, s, s^2, \cdots, s^{n-1}$ *on both sides of the equations above in turn, and adding them, one can obtain that*

$$\left(s^n + a_1 s^{n-1} + \cdots + a_{n-1} s + a_n\right) q_{n-1}(s) = \alpha(s) q_{n-1}(s) = 1$$

Thus, it holds that

$$q_{n-1}(s) = \frac{1}{\alpha(s)}$$

where $\alpha(s)$ *is the characteristic polynomial of* A. *Substituting* $q_{n-1}(s)$ *into the last equation in* (1.23), *one can obtain that*

$$q_{n-2}(s) = \frac{1}{\alpha(s)}(s + a_1)$$

Substituting $q_{n-1}(s)$ *and* $q_{n-2}(s)$ *into the last second equation in* (1.23), *and repeat this process, it can be deduced that*

$$q_{n-1}(s) = \frac{1}{\alpha(s)} = \frac{1}{\alpha(s)} p_{n-1}(s)$$

$$q_{n-2}(s) = \frac{1}{\alpha(s)}(s + a_1) = \frac{1}{\alpha(s)} p_{n-2}(s)$$

$$\vdots$$

$$q_1(s) = \frac{1}{\alpha(s)}\left(s^{n-2} + a_1 s^{n-3} + \cdots + a_{n-3} s + a_{n-2}\right) = \frac{1}{\alpha(s)} p_1(s)$$

$$q_0(s) = \frac{1}{\alpha(s)}\left(s^{n-1} + a_1 s^{n-2} + \cdots + a_{n-2} s + a_{n-1}\right) = \frac{1}{\alpha(s)} p_0(s)$$

where

$$\begin{bmatrix} p_0(s) \\ p_1(s) \\ \vdots \\ p_{n-2}(s) \\ p_{n-1}(s) \end{bmatrix} = \begin{bmatrix} 1 & a_1 & a_2 & \cdots & a_{n-1} \\ & 1 & a_1 & \cdots & a_{n-2} \\ & & 1 & \ddots & \vdots \\ & & & \ddots & a_1 \\ & & & & 1 \end{bmatrix} \begin{bmatrix} s^{n-1} \\ s^{n-2} \\ \vdots \\ s \\ 1 \end{bmatrix}$$

Based on (1.22), *it can be further deduced that*

$$(sI - A)^{-1} = \sum_{k=0}^{n-1} q_k(s) A^k = \sum_{k=0}^{n-1} \frac{1}{\alpha(s)} p_k(s) A^k$$

1.3.4 Equivalence of Dynamical Systems

Example 1.8 In Example 1.5, if select $\bar{x}(t) = [\ u_C(t)\ \ \dot{u}_C(t)\ ]^{\mathrm{T}}$ as the state vector, the state equation is translated into

$$\dot{\bar{x}}(t) = \begin{bmatrix} 0 & 1 \\ -\dfrac{1}{LC} & -\dfrac{R}{L} \end{bmatrix} \bar{x}(t) + \begin{bmatrix} 0 \\ \dfrac{1}{LC} \end{bmatrix} u(t)$$

It can be seen the selection of state variables is not unique. As a result, there are many different dynamical equations for a given system. The relationships among these dynamical equations should be studied.

Definition 1.6 (Equivalence of linear time-invariant dynamical systems) Consider a nonsingular transformation $\bar{x}(t) = Tx(t)$, where T is a nonsingular constant matrix. The linear time-invariant dynamical system

$$\dot{\bar{x}}(t) = \bar{A}\bar{x}(t) + \bar{B}u(t) \tag{1.24}$$

$$y(t) = \bar{C}\bar{x}(t) + \bar{D}u(t) \tag{1.25}$$

with

$$\bar{A} = TAT^{-1},\ \bar{B} = TB,\ \bar{C} = CT^{-1},\ \bar{D} = D \tag{1.26}$$

is said to be (*algebraically*) *equivalent* to (1.13) and (1.14), and $\bar{x}(t) = Tx(t)$ is called an *equivalence transformation*.

From Definition 1.6, it holds that $x(t) = T^{-1}\bar{x}(t)$, and then one can deduce that

$$\dot{\bar{x}}(t) = T\dot{x}(t) = T(Ax(t) + Bu(t)) = TAT^{-1}\bar{x}(t) + TBu(t) = \bar{A}\bar{x}(t) + \bar{B}u(t)$$

$$y(t) = Cx(t) + Du(t) = CT^{-1}\bar{x}(t) + Du(t) = \bar{C}\bar{x}(t) + \bar{D}u(t)$$

thus, (1.26) can be obtained.

Let $\bar{G}(s)$ denote the transfer function matrix of the linear time-invariant system expressed by (1.24) and (1.25). Combining (1.18) and (1.26), we can derive that

$$\bar{G}(s) = \bar{C}(sI - \bar{A})^{-1}\bar{B} + \bar{D} = CT^{-1}\left(sI - TAT^{-1}\right)^{-1}TB + D$$

$$= CT^{-1}\left(TsT^{-1} - TAT^{-1}\right)^{-1}TB + D = CT^{-1}\left[T(sI - A)T^{-1}\right]^{-1}TB + D$$

$$= CT^{-1}\left[T(sI - A)^{-1}T^{-1}\right]TB + D = C(sI - A)^{-1}B + D = G(s)$$

It indicates that equivalent dynamical systems have the same transfer function matrices. In other words, taking the equivalence transformation do not change the transfer function matrix. As a matter of fact, all properties of (1.13) and (1.14) are preserved or invariant under any equivalence transformation. Let $\bar{\alpha}(s)$ denote the characteristic polynomial of $\bar{A}$. According to the fact $\det T \det T^{-1} = 1$, one can derive that

$$\bar{\alpha}(s) = \det(sI - \bar{A}) = \det\left(TsT^{-1} - TAT^{-1}\right) = \det\left[T(sI - A)T^{-1}\right]$$

1.3 State Space Description of Linear Systems

$$= \det T \det(sI - A) \det T^{-1} = \det(sI - A) = \alpha(s)$$

which indicates that equivalent dynamical systems have the same characteristic polynomials. Consequently, the eigenvalues keep unchanged under any equivalence transformation. Meanwhile, the poles of transfer function matrices are also invariant.

The equivalent transformation can be extended to linear time-varying systems. In this case, the transformation matrix should be time-varying and satisfying the requirement of differentiability.

Definition 1.7 (Equivalence of linear time-varying dynamical systems) Consider a nonsingular transformation $\bar{x}(t) = T(t)x(t)$, where $T(t)$ is a nonsingular and continuously differentiable matrix for all $t \in (-\infty, \infty)$. The linear time-varying dynamical system

$$\dot{\bar{x}}(t) = \bar{A}(t)\bar{x}(t) + \bar{B}(t)u(t) \tag{1.27}$$

$$y(t) = \bar{C}(t)\bar{x}(t) + \bar{D}(t)u(t) \tag{1.28}$$

with

$$\bar{A}(t) = T(t)A(t)T^{-1}(t) + \dot{T}(t)T^{-1}(t), \quad \bar{B}(t) = T(t)B(t)$$

$$\bar{C}(t) = C(t)T^{-1}(t), \quad \bar{D}(t) = D(t) \tag{1.29}$$

is said to be (*algebraically*) *equivalent* to (1.11) and (1.12), and $\bar{x}(t) = T(t)x(t)$ is called an *equivalence transformation*.

From Definition 1.7,

$$\dot{\bar{x}}(t) = T(t)\dot{x}(t) + \dot{T}(t)x(t) = T(t)[A(t)x(t) + B(t)u(t)] + \dot{T}(t)x(t)$$

$$= \left[T(t)A(t)T^{-1}(t) + \dot{T}(t)T^{-1}(t)\right]\bar{x}(t) + T(t)B(t)u(t) = \bar{A}(t)x(t) + \bar{B}(t)u(t)$$

$$y(t) = C(t)x(t) + D(t)u(t) = C(t)T^{-1}(t)\bar{x}(t) + D(t)u(t) = \bar{C}(t)\bar{x}(t) + \bar{D}(t)u(t)$$

and thus (1.29) holds.

According to the equivalent transformation, the state space description of a linear time-invariant system can be transformed into the Jordan normal form, which is diagonal or block diagonal.

Definition 1.8 (Jordan matrix) A matrix J is called a *Jordan matrix* if it is with the following form,

$$J = \begin{bmatrix} J_1 & & & \\ & \ddots & & \\ & & J_i & \\ & & & \ddots \\ & & & & J_l \end{bmatrix} \tag{1.30}$$

where

$$J_i = \begin{bmatrix} J_{i1} & & \\ & \ddots & \\ & & J_{i\alpha_i} \end{bmatrix} \quad (1.31)$$

$$J_{ik} = \begin{bmatrix} \lambda_i & 1 & & & \\ & \lambda_i & 1 & & \\ & & \ddots & \ddots & \\ & & & \lambda_i & 1 \\ & & & & \lambda_i \end{bmatrix} \quad (1.32)$$

λ_i is the eigenvalue of J with the algebraic multiplicity σ_i, $i \in \{1, \cdots, l\}$, J_i is called a *Jordan block matrix* associated with λ_i, and J_{ik}, $k \in \{1, \cdots, \alpha_i\}$ is called a *Jordan block*, α_i is the geometric multiplicity associated with λ_i.

Theorem 1.3 *Consider a linear time-invariant system* (1.13) *and* (1.14), λ_i *is the eigenvalue of A with the algebraic multiplicity* σ_i, $i \in \{1, \cdots, l\}$, *and* $\sigma_1 + \cdots + \sigma_i + \cdots + \sigma_l = n$. *There exists an equivalence transformation* $\bar{x}(t) = Tx(t)$ *such that A transforms into the Jordan normal form*

$$J = TAT^{-1} \quad (1.33)$$

where J is with the form of (1.30).

This theorem can be derived based on the construction of the Jordan normal form, which will be discussed in the sequel.

The definitions of the algebraic multiplicity and the geometric multiplicity are introduced in Subsection 1.3.2. The geometric multiplicity of λ_i equals to the number of Jordan blocks associated with λ_i, as well as the number of linearly independent eigenvectors associated with λ_i. That is to say, there only exists one linearly independent eigenvector for each Jordan block. The sum of orders of all Jordan blocks belong to λ_i equals to the algebraic multiplicity of λ_i. Furthermore, the number of Jordan blocks in (1.30) equals to the sum of geometric multiplicities of all eigenvalues, i.e., the number of linearly independent eigenvectors of A. For each eigenvalue of A, if its algebraic multiplicity is always equivalent to its geometric multiplicity, then the Jordan normal form of A is diagonal.

Let σ_{ik} denote the order of kth Jordan block associated with λ_i, $i \in \{1, \cdots, l\}$, $k \in \{1, \cdots, \alpha_i\}$, and $\sigma_{i\max} = \max_{1 \leqslant k \leqslant \alpha_i}\{\sigma_{ik}\}$. In this way, the minimal polynomial of A can also be defined as

$$\phi(s) = \prod_{i=1}^{l}(s - \lambda_i)^{\sigma_{i\max}}$$

To construct the Jordan normal form, one can start splitting in following cases.

1.3 State Space Description of Linear Systems

Firstly, consider that eigenvalues of A are all distinct. Let λ_i, $i \in \{1, \cdots, n\}$, be eigenvalues of A, and v_i be the corresponding eigenvector. In this way, the set of eigenvectors $\{v_1, \cdots, v_n\}$ is linearly independent and can be used as a basis. Therefore, $Q = [\ v_1 \ \cdots \ v_n\ ]$ is nonsingular. Let $T = Q^{-1}$. It is noted that $Av_i = \lambda_i v_i$. Taking the equivalence transformation $\bar{x}(t) = Tx(t)$, according to (1.26), one can deduce that

$$\begin{aligned}
\bar{A} &= TAT^{-1} = TA\begin{bmatrix} v_1 & \cdots & v_n \end{bmatrix} \\
&= T\begin{bmatrix} Av_1 & \cdots & Av_n \end{bmatrix} = T\begin{bmatrix} \lambda_1 v_1 & \cdots & \lambda_n v_n \end{bmatrix} \\
&= T\begin{bmatrix} v_1 & \cdots & v_n \end{bmatrix}\begin{bmatrix} \lambda_1 & & \\ & \ddots & \\ & & \lambda_n \end{bmatrix} = \begin{bmatrix} \lambda_1 & & \\ & \ddots & \\ & & \lambda_n \end{bmatrix}
\end{aligned} \qquad (1.34)$$

It can be seen that each square matrix with distinct eigenvalues can be translated into a diagonal matrix representation by using its eigenvalues as a basis. The transformation matrix T is selected as $[\ v_1 \ \cdots \ v_n\ ]^{-1}$. Different ordering of eigenvectors will lead to different diagonal matrices for the same A.

Example 1.9 Consider the matrix $A = \begin{bmatrix} 0 & 0 & 0 \\ 1 & 0 & 2 \\ 0 & 1 & 1 \end{bmatrix}$. Solve its Jordan normal form.

The characteristic polynomial of A is

$$\alpha(s) = \det(sI - A) = \det\begin{bmatrix} s & 0 & 0 \\ -1 & s & -2 \\ 0 & -1 & s-1 \end{bmatrix} = (s-2)(s+1)s$$

Therefore, eigenvalues of A are 2, -1 and 0. The eigenvector associated with $\lambda_1 = 2$ is any nonzero solution of

$$(\lambda_1 I - A)\, v_1 = \begin{bmatrix} 2 & 0 & 0 \\ -1 & 2 & -2 \\ 0 & -1 & 1 \end{bmatrix} v_1 = 0$$

and thus, $v_1 = [\ 0\ \ 1\ \ 1\ ]^T$ is an eigenvector associated with $\lambda_1 = 2$. Note that the eigenvector is not unique, $[\ 0\ \ \alpha\ \ \alpha\ ]^T$ with any nonzero real constant α can also be chosen as an eigenvector. The eigenvector associated with $\lambda_2 = -1$ is any nonzero solution of

$$(\lambda_2 I - A)\, v_2 = \begin{bmatrix} -1 & 0 & 0 \\ -1 & -1 & -2 \\ 0 & 1 & -2 \end{bmatrix} v_2 = 0$$

which yield $v_2 = [\ 0\ \ -2\ \ 1\ ]^T$. Similarly, the eigenvector associated with $\lambda_3 = 0$ can be computed as $v_3 = [\ 2\ \ 1\ \ -1\ ]^T$.

Therefore, the Jordan normal form of A is

$$\bar{A} = \begin{bmatrix} 2 & 0 & 0 \\ 0 & -1 & 0 \\ 0 & 0 & 0 \end{bmatrix}$$

It is a diagonal matrix with eigenvalues on the diagonal. This matrix can also be obtained by computing

$$\bar{A} = TAT^{-1}$$

with

$$T = \begin{bmatrix} v_1 & v_2 & v_3 \end{bmatrix}^{-1} = \begin{bmatrix} \frac{1}{6} & \frac{1}{3} & \frac{2}{3} \\ \frac{1}{3} & -\frac{1}{3} & \frac{1}{3} \\ \frac{1}{2} & 0 & 0 \end{bmatrix}$$

It is worth mentioning that there may arise complex numbers when computing eigenvalues and eigenvectors. In this situation, the diagonal Jordan normal form (1.30) will contain complex entries. Accordingly, the real linear space should be extended to the complex linear space, and the transpose should be replaced by complex-conjugate transpose. Although this form has no physical meaning in practice, it is effective to analyze the system structure. Incidentally, the diagonal matrix with complex eigenvalues can be transformed into a very useful real matrix. If there exists a pair of conjugate complex eigenvalues $\alpha \pm j\beta$, then (1.34) can be rewritten as

$$\bar{A} = \begin{bmatrix} \lambda_1 & & & & \\ & \ddots & & & \\ & & \lambda_{n-2} & & \\ & & & \alpha + j\beta & \\ & & & & \alpha - j\beta \end{bmatrix}$$

Taking the equivalence transformation, one can obtain that

$$\tilde{A} = \bar{T}\bar{A}\bar{T}^{-1}$$

$$= \begin{bmatrix} I & & \\ & 1 & 1 \\ & j & -j \end{bmatrix} \begin{bmatrix} \lambda_1 & & & & \\ & \ddots & & & \\ & & \lambda_{n-2} & & \\ & & & \alpha + j\beta & \\ & & & & \alpha - j\beta \end{bmatrix} \begin{bmatrix} I & & \\ & 0.5 & -0.5j \\ & 0.5 & 0.5j \end{bmatrix}$$

1.3 State Space Description of Linear Systems

$$= \begin{bmatrix} \lambda_1 & & & & \\ & \ddots & & & \\ & & \lambda_{n-2} & & \\ & & & \alpha & \beta \\ & & & -\beta & \alpha \end{bmatrix} \quad (1.35)$$

It can be seen that the above transformation transforms the complex eigenvalues on the diagonal into a block with the real part of the eigenvalues on the diagonal and the imaginary part on the off-diagonal. If there exist two or more pairs of conjugate complex eigenvalues, the similar transformation to (1.35) can be adopted. As a matter of fact, these two transformations (1.34) and (1.35) can be combined into one as

$$\tilde{T}^{-1} = T^{-1}\bar{T}^{-1}$$
$$= \begin{bmatrix} v_1 & \cdots & v_{n-2m} & \mathrm{Re}\,(v_{n-2m+1}) & \mathrm{Im}\,(v_{n-2m+1}) & \cdots & \mathrm{Re}\,(v_{n+m}) & \mathrm{Im}\,(v_{n+m}) \end{bmatrix}$$

where m denotes the number of pairs of conjugate complex eigenvalues. This form is useful in state space design.

In the next, consider that eigenvalues of A are not all distinct. In this case, if there still exist n linearly independent eigenvectors, these n eigenvectors can be selected as a basis. Similar to the case that all eigenvalues are distinct, the corresponding Jordan normal form is diagonal. Another case is that the number of linearly independent eigenvectors is less than n, then the corresponding Jordan normal form is block diagonal. Consider λ_i is an eigenvalue of A with the algebraic multiplicity σ_i, $i \in \{1, \cdots, l\}$. We subdivide into two cases to discuss this problem.

Firstly, consider that there exists only one linearly independent eigenvector associated with each eigenvalue. In this case, $\alpha_i = 1$ holds for all $i \in \{1, \cdots, l\}$. That is to say, there exists only one Jordan block for each eigenvalue, and the order of each Jordan block equals to the algebraic multiplicity of the corresponding eigenvalue. Under this circumstance, once all of eigenvalues are obtained, the Jordan normal form is determined,

$$\bar{A} = \begin{bmatrix} \bar{A}_1 & & \\ & \ddots & \\ & & \bar{A}_l \end{bmatrix}$$

where $\bar{A}_i$ is a $\sigma_i \times \sigma_i$ Jordan block associated with λ_i.

The transformation matrix $T = Q^{-1}$ can be obtained by solving the matrix equation

$$Q\bar{A} = AQ \quad (1.36)$$

Denoting $Q = [\ Q_1\ \cdots\ Q_l\ ]$, one can rewrite (1.36) as a matrix equation set

$$\begin{cases} Q_1 \bar{A}_1 = AQ_1 \\ \vdots \\ Q_l \bar{A}_l = AQ_l \end{cases} \quad (1.37)$$

Denoting $Q_i = [\ v_{i1}\ \cdots\ v_{i\sigma_i}\ ]$, and substituting it into (1.37), it holds that

$$[\ v_{i1}\ \cdots\ v_{i\sigma_i}\ ] \begin{bmatrix} \lambda_i & 1 & & & \\ & \lambda_i & 1 & & \\ & & \ddots & \ddots & \\ & & & \lambda_i & 1 \\ & & & & \lambda_i \end{bmatrix} = A [\ v_{i1}\ \cdots\ v_{i\sigma_i}\ ]$$

which is equivalent to

$$\begin{cases} \lambda_i v_{i1} = A v_{i1} \\ v_{i1} + \lambda_i v_{i2} = A v_{i2} \\ \vdots \\ v_{i(\sigma_i-1)} + \lambda_i v_{i\sigma_i} = A v_{i\sigma_i} \end{cases} \quad (1.38)$$

Solving (1.38), $Q_i = [\ v_{i1}\ \cdots\ v_{i\sigma_i}\ ]$ can be determined. (1.38) can be rewritten as

$$\begin{cases} (\lambda_i I - A)\, v_{i1} = 0 \\ (\lambda_i I - A)\, v_{i2} = -v_{i1} \\ \vdots \\ (\lambda_i I - A)\, v_{i\sigma_i} = -v_{i(\sigma_i-1)} \end{cases} \quad (1.39)$$

According to (1.39), it can be seen that v_{i1} is the eigenvector associated with λ_i, while $v_{i2}, \cdots, v_{i\sigma_i}$ are generalized eigenvectors associated with λ_i. In this way, the transformation matrix can be obtained, which satisfies

$$T = Q^{-1} = [\ Q_1\ \cdots\ Q_l\ ]^{-1} = [\ v_{11}\ \cdots\ v_{1\sigma_1}\ \cdots\ v_{l1}\ \cdots\ v_{l\sigma_l}\ ]^{-1}$$

Secondly, consider that there exist more than one linearly independent eigenvector for some eigenvalues. Here, assume that λ_i is a repeated eigenvalue with the corresponding geometric multiplicity $\alpha_i > 1$. In this case, there exist α_i linearly independent eigenvectors $v_{i1}, \cdots, v_{i\alpha_i}$, which can be determined by

$$(\lambda_i I - A)\, v_{ik} = 0$$

1.3 State Space Description of Linear Systems

for $k \in \{1, \cdots, \alpha_i\}$. Furthermore, one can use (1.39) to compute $\sigma_i - \alpha_i$ generalized eigenvectors $v_{i\alpha_i+1}, \cdots, v_{i\sigma_i}$ based on any one eigenvector in $\{v_{i1}, \cdots, v_{i\alpha_i}\}$. Let

$$Q_i = \begin{bmatrix} v_{i1} & \cdots & v_{i\alpha_i} & v_{i\alpha_i+1} & \cdots & v_{i\sigma_i} \end{bmatrix}$$

and the transformation matrix $T = Q^{-1} = [\begin{array}{ccc} Q_1 & \cdots & Q_l \end{array}]^{-1}$.

Example 1.10 Solve the Jordan normal form of the following dynamical equations,

$$\dot{x}(t) = \begin{bmatrix} 0 & 1 & 0 \\ 0 & 0 & 1 \\ 2 & 3 & 0 \end{bmatrix} x(t) + \begin{bmatrix} 0 \\ 0 \\ 1 \end{bmatrix} u(t), \quad y(t) = \begin{bmatrix} 1 & 0 & 0 \end{bmatrix} x(t)$$

Since $\det(sI - A) = s^3 - 3s - 2 = (s+1)^2(s-2) = 0$, eigenvalues of A are $\lambda_1 = \lambda_2 = -1$ and $\lambda_3 = 2$.

Next, calculate the transformation matrix T.

Firstly, calculate the rank of the eigenmatrix when $\lambda_1 = \lambda_2 = -1$,

$$\text{rank}\,(\lambda_1 I - A) = \text{rank} \begin{bmatrix} -1 & -1 & 0 \\ 0 & -1 & -1 \\ -2 & -3 & -1 \end{bmatrix} = 2 = 3 - 1 = n - \alpha_1$$

Since $\alpha_1 = 1$, there is only one independent eigenvector for $\lambda_1 = \lambda_2 = -1$, that is, there is only one Jordan block.

The eigenvector v_1 associated with $\lambda_1 = -1$ is obtained by

$$(\lambda_1 I - A)\, v_1 = \begin{bmatrix} -1 & -1 & 0 \\ 0 & -1 & -1 \\ -2 & -3 & -1 \end{bmatrix} v_1 = 0$$

and the solution of the above equation is $v_1 = [\begin{array}{ccc} 1 & -1 & 1 \end{array}]^\text{T}$.

Moreover, calculate another generalized eigenvector v_2 associated with $\lambda_1 = -1$, it can be obtained by the following equation

$$(\lambda_1 I - A)\, v_2 = -v_1$$

that is

$$\begin{bmatrix} -1 & -1 & 0 \\ 0 & -1 & -1 \\ -2 & -3 & -1 \end{bmatrix} v_2 = \begin{bmatrix} -1 \\ 1 \\ -1 \end{bmatrix}$$

and the solution of the above equation is $v_2 = [\begin{array}{ccc} 1 & 0 & -1 \end{array}]^\text{T}$.

Finally, the eigenvector v_3 associated with $\lambda_3 = 2$ is solved by the following equation

$$(\lambda_3 I - A)\, v_3 = \begin{bmatrix} 2 & -1 & 0 \\ 0 & 2 & -1 \\ -2 & -3 & 2 \end{bmatrix} v_3 = 0$$

and the solution of the above equation is $v_3 = \begin{bmatrix} 1 & 2 & 4 \end{bmatrix}^T$.

Therefore,

$$Q = \begin{bmatrix} v_1 & v_2 & v_3 \end{bmatrix} = \begin{bmatrix} 1 & 1 & 1 \\ -1 & 0 & 2 \\ 1 & -1 & 4 \end{bmatrix}$$

and

$$T = Q^{-1} = \frac{1}{9} \begin{bmatrix} 2 & -5 & 2 \\ 6 & 3 & -3 \\ 1 & 2 & 1 \end{bmatrix}$$

Furthermore,

$$\bar{A} = TAT^{-1} = \begin{bmatrix} -1 & 1 & 0 \\ 0 & -1 & 0 \\ 0 & 0 & 2 \end{bmatrix}, \quad \bar{B} = TB = \frac{1}{9}\begin{bmatrix} 2 \\ -3 \\ 1 \end{bmatrix}, \quad \bar{C} = CT^{-1} = \begin{bmatrix} 1 & 1 & 1 \end{bmatrix}$$

Assignments

1.1 For a matrix

$$A = \begin{bmatrix} 0 & 1 & 0 & \cdots & 0 \\ 0 & 0 & 1 & \cdots & 0 \\ \vdots & \vdots & \vdots & & \vdots \\ -a_n & -a_{n-1} & -a_{n-2} & \cdots & a_1 \end{bmatrix}$$

verify that the characteristic polynomial of A is $\alpha(s) = \det(sI - A) = s^n + a_1 s^{n-1} + a_2 s^{n-2} + \cdots + a_{n-1} s + a_n$.

Furthermore, if λ_i is an eigenvalue of A, try to verify that $\begin{bmatrix} 1 & \lambda_i & \lambda_i^2 & \cdots & \lambda_i^{n-1} \end{bmatrix}^T$ is the eigenvector associated with λ_i.

1.2 If λ_i is one of the eigenvalues of A, verify that $f(\lambda_i)$ is an eigenvalue of $f(A)$.

1.3 Try to get the characteristic polynomial and minimal polynomial of the following matrices,

$$\begin{bmatrix} \lambda & 1 & & \\ & \lambda & 1 & \\ & & \lambda & 1 \\ & & & \lambda \end{bmatrix}, \quad \begin{bmatrix} \lambda & 1 & & \\ & \lambda & & \\ & & \lambda & \\ & & & \lambda \end{bmatrix}, \quad \begin{bmatrix} \lambda & 1 & & \\ & \lambda & & \\ & & \lambda & 1 \\ & & & \lambda \end{bmatrix}$$

1.4 Consider $A = \begin{bmatrix} 1 & 1 & 0 \\ 0 & 0 & 1 \\ 0 & 0 & 1 \end{bmatrix}$, try to get A^{101} and $(sI - A)^{-1}$.

Assignments

1.5 If $A \in \mathbb{R}^{p \times q}$ and $B \in \mathbb{R}^{q \times p}$, try to verify that $\det(I_p + AB) = \det(I_q + BA)$.

1.6 Compute the transfer function matrix $G(s)$ of the following dynamical equations,

$$\dot{x}(t) = \begin{bmatrix} -2 & 1 & 0 \\ 0 & -3 & 0 \\ 0 & 1 & -4 \end{bmatrix} x(t) + \begin{bmatrix} 1 & -1 \\ 1 & 0 \\ 2 & 3 \end{bmatrix} u(t), \ y(t) = \begin{bmatrix} 1 & 2 & 1 \\ -3 & -1 & 0 \end{bmatrix} x(t)$$

1.7 Write out the zeros and poles of the following transfer function matrix,

$$G(s) = \begin{bmatrix} \dfrac{1}{s} & \dfrac{2s+1}{s(s+1)} \\ \dfrac{1}{s+1} & \dfrac{2s+1}{s(s+1)^2} \end{bmatrix}$$

1.8 Write out the Jordan normal form of the following state equations.

(1) $\dot{x}(t) = \begin{bmatrix} 4 & 1 & -2 \\ 1 & 0 & 2 \\ 1 & -1 & 3 \end{bmatrix} x(t) + \begin{bmatrix} 3 & 1 \\ 2 & 7 \\ 5 & 3 \end{bmatrix} u(t).$

(2) $\dot{x}(t) = \begin{bmatrix} 0 & 1 & 0 \\ 0 & 0 & 1 \\ -1 & -3 & -1 \end{bmatrix} x(t) + \begin{bmatrix} 1 \\ 1 \\ 3 \end{bmatrix} u(t).$

1.9 Given two square matrices

$$A = \begin{bmatrix} 0 & 1 & 0 & \cdots & 0 \\ 0 & 0 & 1 & \cdots & 0 \\ \vdots & \vdots & \vdots & & \vdots \\ 0 & 0 & 0 & \cdots & 1 \\ -a_n & -a_{n-1} & -a_{n-2} & \cdots & -a_1 \end{bmatrix}, \ \bar{A} = \begin{bmatrix} 0 & 0 & \cdots & 0 & -a_n \\ 1 & 0 & \cdots & 0 & -a_{n-1} \\ 0 & 1 & \cdots & 0 & -a_{n-2} \\ \vdots & \vdots & & \vdots & \vdots \\ 0 & 0 & \cdots & 1 & -a_1 \end{bmatrix}$$

determine the transformation matrix T such that $\bar{A} = TAT^{-1}$.

Chapter 2

Response of Linear Dynamical Systems

The state-space description provides the foundation for analyzing the motion of systems quantitatively. In quantitative analysis, it is important to understand how inputs and initial conditions affect responses of systems. It plays a significant role in control theory. For instance, it is significant to be able to select an input that will cause the system output to satisfy certain properties such as stability, controllability, observability and so on. Generally, responses of systems include state ones and output ones. Thereinto, responses are manifested as analytical solutions of state equations mathematically. This chapter is focused on studying responses in detail for linear systems.

2.1 Introduction

For continuous-time linear systems, the motion analysis can be come down to solve the state equation in certain initial states and external inputs. As introduced in Chapter 1, state equations are

$$\dot{x}(t) = A(t)x(t) + B(t)u(t), \; x(t_0) = x_0, \; t \in [t_0, \; t_f] \tag{2.1}$$

for a linear time-varying system, while

$$\dot{x}(t) = Ax(t) + Bu(t), \; x(0) = x_0, \; t \in [0, \infty) \tag{2.2}$$

for a linear time-invariant system. Solving vector differential equations (2.1) and (2.2), the analytical solution $x(t) \in \mathbb{R}^n$ can be achieved. Substituting it into output equations, the system output response $y(t)$ can be further obtained. Although responses of systems are excited by initial states and external inputs, the motion modality is mainly determined by the structure and related parameters of systems, i.e., parameter matrices $(A(t), B(t))$ or (A, B).

Obviously, it is meaningful to analyze the motion of systems only when solutions of state equations are existing and unique. Therefore, it is essential to import additional constraint conditions for system parameter matrices and inputs to ensure the existence and uniqueness of solutions.

For a linear time-varying system (2.1), if all entries of $A(t)$ and $B(t)$ are real continuous functions of t in $[t_0, t_f]$, and all entries of $u(t)$ are real continuous functions of t in

2.1 Introduction

$[t_0, t_f]$, then the solution of (2.1) is existing and unique. For physical systems in practice, these conditions are generally held. However, they are too strong from the viewpoint of mathematics and can be degraded into the following three conditions:

(1) each entry of the state coefficient matrix $A(t)$, $a_{ij}(t)$ is absolutely integrable in $[t_0, t_f]$, i.e.,

$$\int_{t_0}^{t_f} |a_{ij}(t)| \, dt < \infty, \quad i, j = 1, \cdots, n$$

(2) each entry of the control coefficient matrix $B(t)$, $b_{ik}(t)$ is square integrable in $[t_0, t_f]$, i.e.,

$$\int_{t_0}^{t_f} (b_{ik}(t))^2 \, dt < \infty, \quad i = 1, \cdots, n, \ k = 1, \cdots, p$$

(3) each entry of the input $u(t)$, $u_k(t)$ is square integrable in $[t_0, t_f]$, i.e.,

$$\int_{t_0}^{t_f} (u_k(t))^2 \, dt < \infty, \quad k = 1, \cdots, p$$

where n is the dimension of the state $x(t)$, and p is the dimension of the input $u(t)$. Using Schwarz inequality, it holds that

$$\sum_{k=1}^{p} \int_{t_0}^{t_f} |b_{ik}(t) u_k(t)| \, dt \leq \sum_{k=1}^{p} \left(\int_{t_0}^{t_f} (b_{ik}(t))^2 \, dt \int_{t_0}^{t_f} (u_k(t))^2 \, dt \right)^{\frac{1}{2}}$$

which indicates that conditions (2) and (3) are equivalent to the absolute integrability of each entry of $B(t)u(t)$ in $[t_0, t_f]$.

For a linear time-invariant system (2.2), since A and B are constant matrices and their entries are finite values, conditions (1) and (2) are always satisfied.

In subsequent sections, assume that systems always satisfy the above conditions.

It is noted that a linear system satisfies the superposition property. On this basis, the state response $x(t)$ excited by the initial state x_0 and input $u(t)$ simultaneously can be divided into the state response $x(t)$ individually excited by x_0 and $u(t)$, respectively.

The state response $x(t)$ only excited by the initial state x_0, i.e., $x_0 \neq 0$ and $u(t) = 0$, is called the *zero-input response*. It represents the free movement of states. Normally, a system with no inputs is called an *autonomous system* and can be expressed as

$$\dot{x}(t) = A(t)x(t), \ x(t_0) = x_0, \ t \in [t_0, t_f] \tag{2.3}$$

for the linear time-varying case, and

$$\dot{x}(t) = Ax(t), \ x(0) = x_0, \ t \in [0, \infty) \tag{2.4}$$

for the linear time-invariant case. The solution of (2.3) and (2.4) is the zero-input response of the corresponding system, which is generally denoted as $\phi(t; t_0, x_0, 0)$.

The state response $x(t)$ only excited by the input $u(t)$, i.e., $x_0 = 0$ and $u(t) \neq 0$ is called the *zero-state response*. It represents the forced movement of states. In the situation, the continuous-time system can be rewritten as

$$\dot{x}(t) = A(t)x(t) + B(t)u(t), \ x(t_0) = 0, \ t \in [t_0, t_f] \tag{2.5}$$

for the linear time-varying system, and

$$\dot{x}(t) = Ax(t) + Bu(t), \ x(t_0) = 0, \ t \in [0, \infty) \tag{2.6}$$

for the linear time-invariant system. The solution of (2.5) and (2.6) is the zero-state response of the corresponding system, which is generally denoted as $\phi(t; t_0, 0, u)$.

The state response $x(t)$ of a linear system excited by the initial state x_0 and input $u(t)$ simultaneously, which is denoted as $\phi(t; t_0, x_0, u)$, is the superposition of the zero-input response and the zero-state response. That is to say,

$$\phi(t; t_0, x_0, u) = \phi(t; t_0, x_0, 0) + \phi(t; t_0, 0, u) \tag{2.7}$$

2.2 Response of Linear Time-Varying Dynamical Systems

In this section, the properties of responses of linear time-varying dynamical systems are discussed.

2.2.1 State Transition Matrix of Linear Time-Varying Systems

First of all, introduce the solution space of a homogeneous state equation.

Consider a linear time-varying homogeneous state equation

$$\dot{x}(t) = A(t)x(t), \ x(t_0) = x_0, \ t \in [t_0, t_f] \tag{2.8}$$

where $x(t) \in \mathbb{R}^n$ is the state vector, $A(t)$ is a time-varying matrix function and its entries are absolutely integrable for $t \in [t_0, t_f]$. As discussed in Section 2.1, this equation has a unique solution.

Theorem 2.1 *The solution set of an n-dimensional linear time-varying homogeneous equation (2.8) forms an n-dimensional linear space in the real domain.*

Proof Firstly, prove that the solution set of the homogeneous equation (2.8) forms a linear space in the real domain.

If $\psi_1(t)$ and $\psi_2(t)$ are any two solutions of (2.8), then for any real numbers α_1 and α_2, $\alpha_1\psi_1(t) + \alpha_2\psi_2(t)$ is also a solution of (2.8). That is because

$$\frac{\mathrm{d}}{\mathrm{d}t}(\alpha_1\psi_1(t) + \alpha_2\psi_2(t)) = \alpha_1\frac{\mathrm{d}}{\mathrm{d}t}\psi_1(t) + \alpha_2\frac{\mathrm{d}}{\mathrm{d}t}\psi_2(t)$$
$$= \alpha_1 A(t)\psi_1(t) + \alpha_2 A(t)\psi_2(t)$$
$$= A(t)(\alpha_1\psi_1(t) + \alpha_2\psi_2(t))$$

2.2 Response of Linear Time-Varying Dynamical Systems

Therefore, the solution set forms a linear space in the real domain. It is called a solution space of (2.8).

Secondly, prove that this linear space is n-dimensional.

Suppose that $e_1, \cdots, e_n$ are n linear independent vectors, $\Psi_i(t)$ is the solution of (2.8) when the initial condition is $\Psi_i(t_0) = e_i$, $i \in \{1, \cdots, n\}$. By contradiction, suppose that $\Psi_1(t), \cdots, \Psi_n(t)$ are linearly dependent. Therefore, there must exist a real vector $\alpha \neq 0 \in \mathbb{R}^n$ such that

$$\begin{bmatrix} \psi_1(t) & \cdots & \psi_n(t) \end{bmatrix} \alpha = 0, \ \forall t \in [t_0, \infty)$$

Let $t = t_0$, the above equation still holds, i.e.,

$$\begin{bmatrix} \psi_1(t_0) & \cdots & \psi_n(t_0) \end{bmatrix} \alpha = \begin{bmatrix} e_1 & \cdots & e_n \end{bmatrix} \alpha = 0$$

which means that the vector group $e_1, \cdots, e_n$ are linearly dependent. It conflicts with the assumption. This inconsistency indicates that $\psi_1(t), \cdots, \psi_n(t)$ are linear independent for $t \in [t_0, \infty)$. Consequently, the dimension of the solution space is n. □

Suppose $\psi(t)$ is a solution of (2.8), and $\psi(t_0) = e$. Obviously, e can be denoted as

$$e = \alpha_1 e_1 + \alpha_2 e_2 + \cdots + \alpha_n e_n$$

where $\alpha_1, \cdots, \alpha_n$ are real numbers. Consider that $\Sigma_{i=1}^n \alpha_i \Psi_i(t)$ is also a solution of (2.8), which is satisfied with the initial condition $e = \sum_{i=1}^n \alpha_i e_i$. Based on the fact that a differential equation has a unique solution, one can obtain that

$$\psi(t) = \sum_{i=1}^n \alpha_i \psi_i(t) \tag{2.9}$$

which indicates that $\psi(t)$ can be described as a linear combination of $\psi_i(t)$, $i \in \{1, \cdots, n\}$.

According to Theorem 2.1, an n-dimensional linear time-varying homogeneous equation (2.8) has and only has n linearly independent solutions. According to this fact, introduce the following definition.

Definition 2.1 (Fundamental solution matrix of linear time-varying systems) Let $\psi_1(t), \cdots, \psi_n(t)$ be n linearly independent solutions of an n-dimensional linear time-varying homogeneous equation (2.8) for $t \in [t_0, t_f]$, and then the matrix

$$\Psi(t) = \begin{bmatrix} \psi_1(t) & \cdots & \psi_n(t) \end{bmatrix}$$

is called a *fundamental solution matrix* of the linear time-varying system.

Theorem 2.2 *The fundamental solution matrix $\Psi(t)$ of a linear time-varying system satisfies the matrix differential equation*

$$\dot{\Psi}(t) = A(t)\Psi(t), \ \Psi(t_0) = E, \ t \in [t_0, t_f] \tag{2.10}$$

where E is a nonsingular real constant matrix.

Proof It holds that

$$\dot{\Psi}(t) = \begin{bmatrix} \dot{\psi}_1(t) & \cdots & \dot{\psi}_n(t) \end{bmatrix} = \begin{bmatrix} A(t)\psi_1(t) & \cdots & A(t)\psi_n(t) \end{bmatrix} = A(t)\Psi(t)$$

On the other hand, according to the linear independence of $\psi_1(t), \cdots, \psi_n(t)$, one can easily deduce that E is nonsingular. $\square$

As a matter of fact, the fundamental solution matrix $\Psi(t)$ of a linear time-varying system can also be defined based on Theorem 2.2, i.e., for any nonsingular real constant matrix $E \in \mathbb{R}^{n \times n}$, a $n \times n$ matrix $\Psi(t)$ satisfying (2.10) is called a fundamental solution matrix of the linear time-varying system. It is obvious that $\Psi(t)$ is not unique for a linear time-varying system since the selection of matrix E is not unique.

Theorem 2.3 *If $\Psi_1(t)$ is a fundamental solution matrix of a linear time-varying system (2.8), then for any nonsingular matrix $C \in \mathbb{R}^{n \times n}$, $\Psi_2(t) = \Psi_1(t)C$ is also a fundamental solution matrix of this system. Moreover, if $\Psi_1(t)$ and $\Psi_2(t)$ are any two fundamental solution matrices of a linear time-varying system (2.8), then there exists a nonsingular matrix $C \in \mathbb{R}^{n \times n}$ such that $\Psi_2(t) = \Psi_1(t)C$.*

Proof It holds that

$$\dot{\Psi}_2(t) = \dot{\Psi}_1(t)C = A(t)\Psi_1(t)C = A(t)\Psi_2(t)$$

Meanwhile, $\Psi_2(t_0) = \Psi_1(t_0)C$ is a nonsingular real constant matrix since both $\Psi_1(t_0)$ and C are nonsingular real constant matrices. Therefore, $\Psi_2(t)$ is also a fundamental solution matrix.

Next, it is noticed that $\Psi_1^{-1}(t)$ exists since $\Psi_1(t)$ is nonsingular for $\forall t \in [t_0, t_f]$. Considering $\Psi_1(t)\Psi_1^{-1}(t) = I$ and differentiating its both sides, one can obtain that

$$\left(\frac{\mathrm{d}}{\mathrm{d}t}\Psi_1(t)\right)\Psi_1^{-1}(t) + \Psi_1(t)\frac{\mathrm{d}}{\mathrm{d}t}\Psi_1^{-1}(t) = 0$$

On this basis, one can further deduce that

$$\begin{aligned}\frac{\mathrm{d}}{\mathrm{d}t}\left(\Psi_1^{-1}(t)\Psi_2(t)\right) &= \left(\frac{\mathrm{d}}{\mathrm{d}t}\Psi_1^{-1}(t)\right)\Psi_2(t) + \Psi_1^{-1}(t)\frac{\mathrm{d}}{\mathrm{d}t}\Psi_2(t) \\ &= -\Psi_1^{-1}(t)\left(\frac{\mathrm{d}}{\mathrm{d}t}\Psi_1(t)\right)\Psi_1^{-1}(t)\Psi_2(t) + \Psi_1^{-1}(t)A(t)\Psi_2(t) \\ &= -\Psi_1^{-1}(t)A(t)\Psi_1(t)\Psi_1^{-1}(t)\Psi_2(t) + \Psi_1^{-1}(t)A(t)\Psi_2(t) \\ &= 0\end{aligned}$$

which indicates that $\Psi_1^{-1}(t)\Psi_2(t)$ equals to a constant matrix. $\square$

In the next, introduce the state transition matrix of the linear time-varying system. The homogeneous state equation (2.8) can be rewritten as

$$\mathrm{d}x(t) = A(t)x(t)\mathrm{d}t$$

2.2 Response of Linear Time-Varying Dynamical Systems

Integrating both sides of the above equation from t_0 to t, one can deduce that

$$x(t) - x(t_0) = \int_{t_0}^{t} A(\tau_1) x(\tau_1) \mathrm{d}\tau_1 \tag{2.11}$$

For the given initial condition $x(t_0) = x_0$, select the zero-order approximation $x(t) \approx x_0$. Substituting it into the right hand of (2.11), one can obtain that

$$x(t) = x_0 + \int_{t_0}^{t} A(\tau_1) x_0 \mathrm{d}\tau_1 = x_0 + \int_{t_0}^{t} A(\tau_1) \mathrm{d}\tau_1 x_0$$

Selecting the first-order approximation $x(t) \approx x_0 + \int_{t_0}^{t} A(\tau_1) \mathrm{d}\tau_1 x_0$, and substituting it into the right hand of (2.11), one can obtain that

$$x(t) = x_0 + \int_{t_0}^{t} A(\tau_1) \left(x_0 + \int_{t_0}^{\tau_1} A(\tau_2) \mathrm{d}\tau_2 x_0 \right) \mathrm{d}\tau_1$$

$$= x_0 + \int_{t_0}^{t} A(\tau_1) \mathrm{d}\tau_1 x_0 + \int_{t_0}^{t} A(\tau_1) \int_{t_0}^{\tau_1} A(\tau_2) \mathrm{d}\tau_2 \mathrm{d}\tau_1 x_0$$

$$= \left(I + \int_{t_0}^{t} A(\tau_1) \mathrm{d}\tau_1 + \int_{t_0}^{t} A(\tau_1) \int_{t_0}^{\tau_1} A(\tau_2) \mathrm{d}\tau_2 \mathrm{d}\tau_1 \right) x_0$$

Repeat this procedure, one can infinitely approach $x(t)$ as

$$x(t) = \left(I + \int_{t_0}^{t} A(\tau_1) \mathrm{d}\tau_1 + \int_{t_0}^{t} A(\tau_1) \int_{t_0}^{\tau_1} A(\tau_2) \mathrm{d}\tau_2 \mathrm{d}\tau_1 \right.$$

$$\left. + \int_{t_0}^{t} A(\tau_1) \int_{t_0}^{\tau_1} A(\tau_2) \int_{t_0}^{\tau_2} A(\tau_3) \mathrm{d}\tau_3 \mathrm{d}\tau_2 \mathrm{d}\tau_1 + \cdots \right) x_0$$

Define

$$\Phi(t, t_0) \triangleq I + \int_{t_0}^{t} A(\tau_1) \mathrm{d}\tau_1 + \int_{t_0}^{t} A(\tau_1) \int_{t_0}^{\tau_1} A(\tau_2) \mathrm{d}\tau_2 \mathrm{d}\tau_1$$

$$+ \int_{t_0}^{t} A(\tau_1) \int_{t_0}^{\tau_1} A(\tau_2) \int_{t_0}^{\tau_2} A(\tau_3) \mathrm{d}\tau_3 \mathrm{d}\tau_2 \mathrm{d}\tau_1 + \cdots \tag{2.12}$$

The series in (2.12) is called the *Peano-Baker series*. $\Phi(t, t_0)$ is generally called the *state transition matrix* of the linear time-varying system.

It is noted that

$$\frac{\partial}{\partial t} \Phi(t, t_0) = A(t) + A(t) \int_{t_0}^{t} A(\tau_1) \mathrm{d}\tau_1 + A(t) \int_{t_0}^{t} A(\tau_1) \int_{t_0}^{\tau_1} A(\tau_2) \mathrm{d}\tau_2 \mathrm{d}\tau_1 + \cdots$$

$$= A(t) \left(I + \int_{t_0}^{t} A(\tau_1) \mathrm{d}\tau_1 + \int_{t_0}^{t} A(\tau_1) \int_{t_0}^{\tau_1} A(\tau_2) \mathrm{d}\tau_2 \mathrm{d}\tau_1 + \cdots \right)$$

and

$$\Phi(t_0, t_0) = I + \int_{t_0}^{t_0} A(\tau_1) d\tau_1 + \int_{t_0}^{t_0} A(\tau_1) \int_{t_0}^{\tau_1} A(\tau_2) d\tau_2 d\tau_1 + \cdots = I$$

That is to say, it holds that

$$\frac{\partial}{\partial t} \Phi(t, t_0) = A(t) \Phi(t, t_0), \quad \Phi(t_0, t_0) = I, \quad t \in [t_0, t_\mathrm{f}] \tag{2.13}$$

Therefore, the state transition matrix of the linear time-varying system can also be defined as follows.

Definition 2.2 (State transition matrix of linear time-varying systems) For a linear time-varying system (2.8), the $n \times n$ matrix $\Phi(t, t_0)$ satisfying (2.13) is called a state transition matrix of the linear time-varying system.

Theorem 2.4 If $\Psi(t)$ is a fundamental solution matrix of a linear time-varying system (2.8), then the state transition matrix $\Phi(t, t_0)$ satisfies

$$\Phi(t, t_0) = \Psi(t) \Psi^{-1}(t_0), \quad t \in [t_0, t_\mathrm{f}] \tag{2.14}$$

Proof It holds that

$$\frac{\partial}{\partial t} \Phi(t, t_0) = \frac{\mathrm{d}}{\mathrm{d}t} \Psi(t) \Psi^{-1}(t_0) = A(t) \Psi(t) \Psi^{-1}(t_0) = A(t) \Phi(t, t_0)$$

Meanwhile, let $t = t_0$, then $\Phi(t_0, t_0) = \Psi(t_0) \Psi^{-1}(t_0) = I$. According to Definition 2.2, $\Phi(t, t_0)$ is the state transition matrix. □

Theorem 2.5 The state transition matrix $\Phi(t, t_0)$ of a linear time-varying system is unique.

Proof Consider $\Phi_1(t, t_0)$ and $\Phi_2(t, t_0)$ are two state transition matrices of (2.8) generated by two fundamental solution matrices $\Psi_1(t)$ and $\Psi_2(t)$, respectively. Combining Theorem 2.3 with Theorem 2.4, one can obtain that

$$\Phi_2(t, t_0) = \Psi_2(t) \Psi_2^{-1}(t_0) = \Psi_1(t) C C^{-1} \Psi_1^{-1}(t_0) = \Psi_1(t) \Psi_1^{-1}(t_0) = \Phi_1(t, t_0)$$

which indicates that $\Phi(t, t_0)$ is unique for a time-varying system and unrelated to the selection of $\Psi(t)$. □

In the next, the properties of $\Phi(t, t_0)$ is presented.
(1) $\Phi(t_0, t_0) = I$.
(2) $\Phi^{-1}(t, t_0) = \Phi(t_0, t)$.
(3) $\Phi(t_2, t_1) \Phi(t_1, t_0) = \Phi(t_2, t_0)$.
According to (2.14), one can easily deduce the above properties.
(4) $\dfrac{\partial}{\partial t} \Phi^{-1}(t, t_0) = \dfrac{\partial}{\partial t} \Phi(t_0, t) = -\Phi(t_0, t) A(t)$.

It is known that $\Phi(t_0, t_0) = \Phi(t_0, t) \Phi(t, t_0) = I$, then it can be deduced that

$$\left(\frac{\partial}{\partial t} \Phi(t_0, t) \right) \Phi(t, t_0) + \Phi(t_0, t) \frac{\partial}{\partial t} \Phi(t, t_0) = 0$$

2.2 Response of Linear Time-Varying Dynamical Systems

which means that

$$\frac{\partial}{\partial t}\Phi(t_0,t) = -\Phi(t_0,t)\left(\frac{\partial}{\partial t}\Phi(t,t_0)\right)\Phi^{-1}(t,t_0)$$

$$= -\Phi(t_0,t)A(t)\Phi(t,t_0)\Phi^{-1}(t,t_0)$$

$$= -\Phi(t_0,t)A(t)$$

2.2.2 Zero-Input Response of Linear Time-Varying Systems

In this subsection, the zero-input response of the linear time-varying system (2.8) is discussed.

Theorem 2.6 *The zero-input response of the linear time-varying system, i.e., the solution of the homogeneous state equation (2.8) is with the following form,*

$$x(t) = \phi(t;t_0,x_0,0) = \Phi(t,t_0)x_0, \quad t \in [t_0,t_f] \tag{2.15}$$

where $\Phi(t,t_0)$ is the state transition matrix of the linear time-varying system (2.8).

According to the discussion in previous subsection, this theorem can be easily deduced.

To solve the homogeneous state equation (2.8) in a closed form, the closed-form determination of $\Phi(t,t_0)$ is required. However, it is difficult for general time-varying systems. Specially, the following conclusion can be deduced.

Theorem 2.7 *For a linear time-varying system (2.8), if $A(t)$ and $\int_{t_0}^{t} A(\tau)d\tau$ are commutative, i.e.,*

$$A(t)\int_{t_0}^{t} A(\tau)d\tau = \int_{t_0}^{t} A(\tau)d\tau A(t) \tag{2.16}$$

then its state transition matrix is

$$\Phi(t,t_0) = \exp\left\{\int_{t_0}^{t} A(\tau)d\tau\right\} \tag{2.17}$$

Proof It is known that

$$\exp\left\{\int_{t_0}^{t} A(\tau)d\tau\right\} = I + \int_{t_0}^{t} A(\tau)d\tau + \frac{1}{2!}\left(\int_{t_0}^{t} A(\tau)d\tau\right)^2 + \cdots$$

On this basis, it holds that

$$\frac{\mathrm{d}}{\mathrm{d}t}\exp\left\{\int_{t_0}^{t} A(\tau)d\tau\right\} = A(t) + \frac{1}{2!}\left(A(t)\int_{t_0}^{t} A(\tau)d\tau + \int_{t_0}^{t} A(\tau)d\tau A(t)\right) + \cdots$$

$$= A(t) + A(t)\int_{t_0}^{t} A(\tau)d\tau + \cdots = A(t)\exp\left\{\int_{t_0}^{t} A(\tau)d\tau\right\}$$

and

$$\exp\left\{\int_{t_0}^{t} A(\tau)d\tau\right\}\bigg|_{t=t_0} = I$$

Therefore, it can be concluded that (2.17) holds according to Definition 2.2. □

If $A(t)$ is a diagonal matrix, (2.16) holds and (2.17) can be used to compute $\Phi(t, t_0)$. In general, (2.16) does not hold and $\Phi(t, t_0)$ cannot be expressed as a closed-form.

2.2.3 Zero-State Response of Linear Time-Varying Systems

In this subsection, the zero-state response of linear time-varying systems is discussed.

Consider a linear time-varying system

$$\dot{x}(t) = A(t)x(t) + B(t)u(t), x(t_0) = 0, \ t \in [t_0, t_{\mathrm{f}}] \qquad (2.18)$$

where $x(t) \in \mathbb{R}^n$ is the state vector, $u(t) \in \mathbb{R}^p$ is the input vector, $A(t)$ and $B(t)$ are time-varying matrix functions. Suppose that (2.18) satisfies the unique solution conditions. Its zero-state response satisfies the following conclusion.

Theorem 2.8 *The zero-state response of the linear time-varying system, i.e., the solution of the nonhomogeneous state equation* (2.18) *is with the following form,*

$$x(t) = \phi(t; t_0, 0, u) = \int_{t_0}^{t} \Phi(t, \tau) B(\tau) u(\tau) \mathrm{d}\tau, \ t \in [t_0, t_{\mathrm{f}}] \qquad (2.19)$$

Proof According to the properties of $\Phi(t, t_0)$, it holds that

$$\frac{\partial}{\partial t} \Phi(t_0, t) = -\Phi(t_0, t) A(t) \qquad (2.20)$$

Multiplying both sides of (2.18) by $\Phi(t_0, t)$, one can obtain that

$$\Phi(t_0, t) \dot{x}(t) - \Phi(t_0, t) A(t) x(t) = \Phi(t_0, t) B(t) u(t)$$

Along with (2.20), the above equation can be rewritten as

$$\frac{\partial}{\partial t} \left(\Phi(t_0, t) x(t) \right) = \Phi(t_0, t) B(t) u(t)$$

Integrating both sides of the above equation from t_0 to t, and under the zero initial condition, one can deduce that

$$\Phi(t_0, t) x(t) = \int_{t_0}^{t} \Phi(t_0, \tau) B(\tau) u(\tau) \mathrm{d}\tau$$

It is noted that $\Phi(t_0, t)\Phi(t, t_0) = I$ and $\Phi(t, t_0)\Phi(t_0, \tau) = \Phi(t, \tau)$, one can further deduce that

$$x(t) = \Phi(t, t_0) \int_{t_0}^{t} \Phi(t_0, \tau) B(\tau) u(\tau) \mathrm{d}\tau = \int_{t_0}^{t} \Phi(t, \tau) B(\tau) u(\tau) \mathrm{d}\tau, \ t \in [t_0, t_{\mathrm{f}}] \qquad □$$

2.2 Response of Linear Time-Varying Dynamical Systems

2.2.4 State Response and Output Response of Linear Time-Varying Systems

Combining (2.15) with (2.19), the state response of a linear time-varying system, i.e., the solution of the state equation excited by the initial state x_0 and input $u(t)$ simultaneously is with the following form,

$$x(t) = \phi(t; t_0, x_0, u) = \Phi(t, t_0) x_0 + \int_{t_0}^{t} \Phi(t, \tau) B(\tau) u(\tau) \mathrm{d}\tau, \ t \in [t_0, t_f] \quad (2.21)$$

The state response $x(t)$ of a linear system excited by the initial state x_0 and input $u(t)$ simultaneously, which is denoted as $\phi(t; t_0, x_0, u)$, is the superposition of the zero-input response and the zero-state response.

Intuitively, the state response consists of the initial state transition term related to x_0 and the forced term excited by the input $u(t)$. The existence of the forced term makes it possible to improve the system performance by importing appropriate input $u(t)$.

If the output equation is known, its output response can be deduced based on (2.21),

$$y(t) = C(t)\Phi(t, t_0) x_0 + \int_{t_0}^{t} C(t)\Phi(t, \tau) B(\tau) u(\tau) \mathrm{d}\tau + D(t) u(t), \ t \in [t_0, t_f] \quad (2.22)$$

In (2.22), consider $x_0 = 0$, and $u(t)$ can be represented as

$$u(t) = \int_{t_0}^{t} \delta(t - \tau) u(\tau) \mathrm{d}\tau$$

where $\delta(t)$ is the impulse function. In this way, one can deduce that

$$y(t) = \int_{t_0}^{t} (C(t)\Phi(t, \tau) B(\tau) + D(t)\delta(t - \tau)) u(\tau) \mathrm{d}\tau, \ t \in [t_0, t_f] \quad (2.23)$$

As discussed in Subsection 1.2.1, the linear time-varying system relaxed at t_0 can be described by

$$y(t) = \int_{t_0}^{t} G(t, \tau) u(\tau) \mathrm{d}\tau \quad (2.24)$$

where $G(t, \tau)$ is the impulse response matrix. Compared (2.23) with (2.24), it can be seen that the impulse response matrix of a linear time-varying system relaxed at t_0 can be obtained by

$$G(t, \tau) = C(t)\Phi(t, \tau) B(\tau) u(\tau) + D(t)\delta(t - \tau) \quad (2.25)$$

Example 2.1 Solve the output response and the impulse response matrix of the following dynamical equation,

$$\dot{x}(t) = \begin{bmatrix} 0 & \cos t \\ 0 & 0 \end{bmatrix} x(t) + \begin{bmatrix} 0 \\ 1 \end{bmatrix} u(t), \ y(t) = \begin{bmatrix} 1 & 0 \\ 0 & 1 \end{bmatrix} x(t) + \begin{bmatrix} 1 \\ 1 \end{bmatrix} u(t)$$

Firstly, find the fundamental solution matrix $\Psi(t)$. Because $\Psi(t)$ satisfies (2.9), here choose $E = I$, then one can obtain that

$$\dot{\psi}_{11}(t) = \cos t \psi_{21}(t), \ \psi_{11}(t_0) = 1, \ \dot{\psi}_{21}(t) = 0, \ \psi_{21}(t_0) = 0$$

$$\dot{\psi}_{12}(t) = \cos t \psi_{22}(t), \ \psi_{12}(t_0) = 0, \ \dot{\psi}_{22}(t) = 0, \ \psi_{22}(t_0) = 1$$

Solving the above equations, it holds that

$$\psi_1(t) = \begin{bmatrix} \psi_{11}(t) \\ \psi_{21}(t) \end{bmatrix} = \begin{bmatrix} 1 \\ 0 \end{bmatrix}, \ \psi_2(t) = \begin{bmatrix} \psi_{12}(t) \\ \psi_{22}(t) \end{bmatrix} = \begin{bmatrix} \sin t - \sin t_0 \\ 1 \end{bmatrix}$$

Therefore, the fundamental solution matrix $\Psi(t)$ is

$$\Psi(t) = \begin{bmatrix} \psi_1(t) & \psi_2(t) \end{bmatrix} = \begin{bmatrix} 1 & \sin t - \sin t_0 \\ 0 & 1 \end{bmatrix}$$

The state transition matrix $\Phi(t, t_0)$ is

$$\Phi(t, t_0) = \Psi(t)\Psi^{-1}(t_0) = \begin{bmatrix} 1 & \sin t - \sin t_0 \\ 0 & 1 \end{bmatrix}$$

Using (2.22), the output response is expressed as

$$y(t) = C(t)\Phi(t, t_0)x_0 + \int_{t_0}^{t} C(t)\Phi(t, \tau)B(\tau)u(\tau)\mathrm{d}\tau + D(t)u(t)$$

$$= \begin{bmatrix} 1 & \sin t - \sin t_0 \\ 0 & 1 \end{bmatrix} x_0 + \int_{t_0}^{t} \begin{bmatrix} \sin t - \sin \tau \\ 1 \end{bmatrix} u(\tau)\mathrm{d}\tau + \begin{bmatrix} 1 \\ 1 \end{bmatrix} u(t)$$

From (2.25), the impulse response of the system in this example is

$$G(t, \tau) = C(t)\Phi(t, \tau)B(\tau) + D(t)\delta(t - \tau) = \begin{bmatrix} \sin t - \sin \tau \\ 1 \end{bmatrix} + \begin{bmatrix} 1 \\ 1 \end{bmatrix} \delta(t - \tau)$$

The equivalence of linear dynamical systems has been introduced in Subsection 1.3.4. In the next, the properties of linear time-varying systems with equivalence is discussed.

If two linear time-varying systems $(A(t), B(t), C(t), D(t))$ and $(\bar{A}(t), \bar{B}(t), \bar{C}(t), \bar{D}(t))$ are equivalent[①], then there exists a nonsingular and continuously differentiable transformation matrix $T(t)$ such that $\bar{A}(t) = T(t)A(t)T^{-1}(t) + \dot{T}(t)T^{-1}(t)$, $\bar{B}(t) = T(t)B(t)$, $\bar{C}(t) = C(t)T^{-1}(t)$, $\bar{D}(t) = D(t)$, and $\bar{x}(t) = T(t)x(t)$. Let $\Psi(t)$ be a fundamental solution matrix of $(A(t), B(t), C(t), D(t))$, then, along with (2.10), one can deduce that

$$\frac{\mathrm{d}}{\mathrm{d}t}(T(t)\Psi(t)) = \frac{\mathrm{d}}{\mathrm{d}t}(T(t))\Psi(t) + T(t)\frac{\mathrm{d}}{\mathrm{d}t}\Psi(t) = \left(\frac{\mathrm{d}}{\mathrm{d}t}T(t) + T(t)A(t)\right)\Psi(t)$$

[①] For convenience, use $(A(t), B(t), C(t), D(t))$ to denote dynamical equations (1.11) and (1.12), while use $(\bar{A}(t), \bar{B}(t), \bar{C}(t), \bar{D}(t))$ to denote dynamical equations (1.24) and (1.25).

$$= \left(\frac{\mathrm{d}}{\mathrm{d}t}T(t) + T(t)A(t)\right) T^{-1}(t)T(t)\Psi(t) = \bar{A}(t)T(t)\Psi(t)$$

and $T(t_0)\Psi(t_0) = T(t_0)E$ is a nonsingular constant matrix. Therefore, $\bar{\Psi}(t) = T(t)\Psi(t)$ is a fundamental solution matrix of $(\bar{A}(t), \bar{B}(t), \bar{C}(t), \bar{D}(t))$. On this basis, Denote state transition matrices of these two systems as $\Phi(t, t_0)$ and $\bar{\Phi}(t, t_0)$, respectively. Along with (2.14), one can deduce that

$$\bar{\Phi}(t, t_0) = \bar{\Phi}(t)\bar{\Phi}^{-1}(t_0) = T(t)\Phi(t, t_0) T^{-1}(t_0) \tag{2.26}$$

Considering the zero-input case, the output of the system $(A(t), B(t), C(t), D(t))$ with the initial state x_0 is

$$y(t) = C(t)\Phi(t, t_0) x_0$$

while the output of the system $(\bar{A}(t), \bar{B}(t), \bar{C}(t), \bar{D}(t))$ with the initial state $\bar{x}_0 = T(t_0)x_0$ is

$$\bar{y}(t) = \bar{C}(t)\bar{\Phi}(t, t_0)\bar{x}_0 = C(t)T^{-1}(t)T(t)\Phi(t, t_0) T^{-1}(t_0) T(t_0) x_0 = C(t)\Phi(t, t_0) x_0$$

That is to say, for two linear time-varying systems with equivalence, when the output of one system with the initial state x_0 is $y(t)$, the other system always has a corresponding initial state so that they have the same outputs in the zero-input case.

Let $G(t, \tau)$ and $\bar{G}(t, \tau)$ denote impulse response matrices of $(A(t), B(t), C(t), D(t))$ and $(\bar{A}(t), \bar{B}(t), \bar{C}(t), \bar{D}(t))$, respectively. Along with (2.25) and (2.26), one can deduce that

$$\begin{aligned}\bar{G}(t, \tau) &= \bar{C}(t)\bar{\Phi}(t, \tau)\bar{B}(\tau) + \bar{D}(t)\delta(t - \tau) \\ &= C(t)T^{-1}(t)T(t)\Phi(t, \tau)T^{-1}(\tau)T(\tau)B(\tau) + D(t)\delta(t - \tau) \\ &= C(t)\Phi(t, \tau)B(\tau) + D\delta(t - \tau) = G(t, \tau)\end{aligned} \tag{2.27}$$

which indicates that two linear time-varying systems with equivalence have the same impulse response matrices.

Furthermore, considering the zero-state case, for the same input $u(t)$, outputs of two linear time-varying systems with equivalence are always equivalent. This fact can be easily deduced based on (2.24) and (2.27).

2.3 Response of Linear Time-Invariant Dynamical Systems

In this section, the properties of responses of linear time-invariant dynamical systems are discussed.

2.3.1 Matrix Exponential of Linear Time-Invariant Systems

Before discussing the response of linear time-invariant systems, firstly introduce the matrix exponential for the state coefficient matrix A.

Definition 2.3 (Matrix exponential) The *matrix exponential* of a given matrix $A \in \mathbb{R}^{n \times n}$ is defined by

$$e^{At} \triangleq I + At + \frac{1}{2!}A^2 t^2 + \cdots = \sum_{k=0}^{\infty} \frac{1}{k!} A^k t^k \tag{2.28}$$

The matrix exponential plays a significant role to analyze the response of linear time-invariant systems. In the next, the properties of e^{At} are discussed.

(1) The value of e^{At} at $t = 0$ is

$$e^{At}|_{t=0} = I \tag{2.29}$$

(2) Consider two independent time variables t and τ,

$$e^{A(t+\tau)} = e^{At} e^{A\tau} = e^{A\tau} e^{At} \tag{2.30}$$

(3) The transpose of e^{At} satisfies

$$\left(e^{At}\right)^{\mathrm{T}} = e^{A^{\mathrm{T}} t} \tag{2.31}$$

The above properties are easily deduced based on (2.28).

(4) The inverse of e^{At} satisfies

$$\left(e^{At}\right)^{-1} = e^{-At} \tag{2.32}$$

For any state coefficient matrix A, e^{At} is nonsingular. On this basis, selecting $\tau = -t$ in (2.30) and combining with (2.28), one can deduce that $e^{At} e^{-At} = I$. Therefore, (2.32) holds.

(5) For a nonnegative integer k,

$$\left(e^{At}\right)^k = e^{A(kt)} \tag{2.33}$$

It is easily deduced (2.33) according to (2.30).

(6) If matrices A and F are commutative, i.e., $AF = FA$, then

$$e^{(A+F)t} = e^{At} e^{Ft} = e^{Ft} e^{At} \tag{2.34}$$

According to (2.28), it holds that

$$e^{At} e^{Ft} = \left(I + At + \frac{1}{2!} A^2 t^2 + \frac{1}{3!} A^3 t^3 + \cdots\right)\left(I + Ft + \frac{1}{2!} F^2 t^2 + \frac{1}{3!} F^3 t^3 + \cdots\right)$$

$$= I + (A+F)t + \frac{1}{2!}(A^2 + 2AF + F^2)t^2 + \frac{1}{3!}(A^3 + 3A^2 F + 3F^2 A + F^3)t^3 + \cdots$$

2.3 Response of Linear Time-Invariant Dynamical Systems

and

$$e^{(A+F)t} = I + (A+F)t + \frac{1}{2!}(A+F)^2 t^2 + \frac{1}{3!}(A+F)^3 t^3 + \cdots$$

$$= I + (A+F)t + \frac{1}{2!}(A^2 + AF + FA + F^2)t^2$$

$$+ \frac{1}{3!}(A^3 + A^2F + AFA + AF^2 + FA^2 + FAF + F^2A + F^3)t^3 + \cdots$$

It can be concluded that $e^{(A+F)t} = e^{At}e^{Ft}$. If $AF = FA$, by the similar manipulation, $e^{(A+F)t} = e^{Ft}e^{At}$ can be proved.

(7) The derivative of e^{At} satisfies

$$\frac{\mathrm{d}}{\mathrm{d}t}e^{At} = Ae^{At} = e^{At}A \qquad (2.35)$$

According to (2.28), it holds that

$$\frac{\mathrm{d}}{\mathrm{d}t}e^{At} = A + A^2 t + \frac{1}{2!}A^3 t^2 + \cdots = A\sum_{k=0}^{\infty}\frac{1}{k!}A^k t^k = \left(\sum_{i=0}^{\infty}\frac{1}{k!}A^k t^k\right)A = Ae^{At} = e^{At}A$$

which indicates that (2.35) holds.

(8) The derivative of e^{-At} satisfies

$$\frac{\mathrm{d}}{\mathrm{d}t}e^{-At} = -Ae^{-At} = -e^{-At}A \qquad (2.36)$$

The derivation of (2.36) is similar to (2.35).

(9) If there exists a equivalent transformation matrix T such that $\bar{A} = TAT^{-1}$, then

$$e^{At} = T^{-1}e^{\bar{A}t}T \qquad (2.37)$$

According to (2.28), it holds that

$$e^{TAT^{-1}t} = I + TAT^{-1}t + \frac{1}{2!}(TAT^{-1})^2 t^2 + \frac{1}{3!}(TAT^{-1})^3 t^3 + \cdots$$

$$= T\left(I + At + \frac{1}{2!}A^2 t^2 + \frac{1}{3!}A^3 t^3 + \cdots\right)T^{-1}$$

which indicates that (2.37) holds.

(10) e^{At} is equivalent to the inverse Laplace transform of the resolvent matrix, i.e.,

$$e^{At} = \mathcal{L}^{-1}\left[(sI - A)^{-1}\right] \qquad (2.38)$$

This property will be proved in the next subsection.

In the next, discuss how to compute e^{At}.

(1) Use summation of the infinite series to compute e^{At}.

Utilizing the definition of matrix exponential (2.28) can compute e^{At} directly. However, this method is difficult to obtain the analytical solution in the closed form.

(2) Use the inverse Laplace transform to compute e^{At}.

The methods to solve the resolvent matrix $(sI - A)^{-1}$ have been introduced in Subsection 1.3.3. According to (2.38), one can take the inverse Laplace transform of $(sI - A)^{-1}$ to compute e^{At}.

(3) Use the Jordan normal form to compute e^{At}.

Before discussing this method, firstly introduce some typical e^{At} when A is with a special form.

If A is a diagonal matrix,

$$A = \begin{bmatrix} \lambda_1 & & \\ & \ddots & \\ & & \lambda_n \end{bmatrix}$$

then e^{At} is also a diagonal matrix,

$$e^{At} = \begin{bmatrix} e^{\lambda_1 t} & & \\ & \ddots & \\ & & e^{\lambda_n t} \end{bmatrix} \qquad (2.39)$$

It can be proved based on (2.28),

$$e^{At} = \begin{bmatrix} 1 & & \\ & \ddots & \\ & & 1 \end{bmatrix} + \begin{bmatrix} \lambda_1 & & \\ & \ddots & \\ & & \lambda_n \end{bmatrix} t + \frac{1}{2!} \begin{bmatrix} \lambda_1^2 & & \\ & \ddots & \\ & & \lambda_n^2 \end{bmatrix} t^2 + \cdots$$

$$= \begin{bmatrix} \sum_{k=0}^{\infty} \frac{1}{k!} \lambda_1^k t^k & & \\ & \ddots & \\ & & \sum_{k=0}^{\infty} \frac{1}{k!} \lambda_n^k t^k \end{bmatrix} = \begin{bmatrix} e^{\lambda_1 t} & & \\ & \ddots & \\ & & e^{\lambda_n t} \end{bmatrix}$$

If A is a block diagonal matrix,

$$A = \begin{bmatrix} A_1 & & \\ & \ddots & \\ & & A_l \end{bmatrix}$$

then e^{At} is also a block diagonal matrix,

2.3 Response of Linear Time-Invariant Dynamical Systems

$$e^{At} = \begin{bmatrix} e^{A_1 t} & & \\ & \ddots & \\ & & e^{A_l t} \end{bmatrix} \quad (2.40)$$

It can also be proved based on (2.28), here we omit it.

If a $n \times n$ matrix A is with the following form

$$A = \begin{bmatrix} 0 & 1 & & & \\ & 0 & 1 & & \\ & & \ddots & \ddots & \\ & & & 0 & 1 \\ & & & & 0 \end{bmatrix}$$

then it can be verified that $A^k = 0$ for $k = n, n+1, \cdots$. In this situation, using (2.28) can deduce that

$$e^{At} = \begin{bmatrix} 1 & t & \dfrac{t^2}{2!} & \dfrac{t^3}{3!} & \cdots & \dfrac{t^{n-1}}{(n-1)!} \\ 0 & 1 & t & \dfrac{t^2}{2!} & \cdots & \dfrac{t^{n-2}}{(n-2)!} \\ & 0 & 1 & t & \ddots & \vdots \\ \vdots & & \ddots & 1 & \ddots & \dfrac{t^2}{2!} \\ & & & & \ddots & t \\ 0 & & \cdots & & 0 & 1 \end{bmatrix} \quad (2.41)$$

If A is a Jordan block, it can be rewritten as

$$A = \begin{bmatrix} \lambda & 1 & & & \\ & \lambda & 1 & & \\ & & \ddots & \ddots & \\ & & & \lambda & 1 \\ & & & & \lambda \end{bmatrix} = \begin{bmatrix} \lambda & & & & \\ & \lambda & & & \\ & & \ddots & & \\ & & & \lambda & \\ & & & & \lambda \end{bmatrix} + \begin{bmatrix} 0 & 1 & & & \\ & 0 & 1 & & \\ & & \ddots & \ddots & \\ & & & 0 & 1 \\ & & & & 0 \end{bmatrix}$$

$\triangleq A_1 + A_2$

It can be seen that $A_1 A_2 = A_2 A_1$, and then one can derive that $e^{At} = e^{(A_1 + A_2)t} = e^{A_1 t} e^{A_2 t}$ according to (2.34). Combining (2.39) with (2.41), one can further obtain that

$$e^{At} = \begin{bmatrix} e^{\lambda t} & te^{\lambda t} & \dfrac{t^2 e^{\lambda t}}{2!} & \dfrac{t^3 e^{\lambda t}}{3!} & \cdots & \dfrac{t^{n-1} e^{\lambda t}}{(n-1)!} \\ 0 & e^{\lambda t} & te^{\lambda t} & \dfrac{t^2 e^{\lambda t}}{2!} & \cdots & \dfrac{t^{n-2} e^{\lambda t}}{(n-2)!} \\ 0 & & e^{\lambda t} & te^{\lambda t} & \ddots & \vdots \\ \vdots & & & e^{\lambda t} & \ddots & \dfrac{t^2 e^{\lambda t}}{2!} \\ & & & & \ddots & te^{\lambda t} \\ 0 & \cdots & & & 0 & e^{\lambda t} \end{bmatrix} \qquad (2.42)$$

If A is with the following form,

$$A = \begin{bmatrix} & \beta \\ -\beta & \end{bmatrix}$$

and then

$$\begin{aligned} e^{At} &= \begin{bmatrix} 1 & \\ & 1 \end{bmatrix} + \begin{bmatrix} & \beta \\ -\beta & \end{bmatrix} t + \frac{1}{2!} \begin{bmatrix} -\beta^2 & \\ & -\beta^2 \end{bmatrix} t^2 + \frac{1}{3!} \begin{bmatrix} & -\beta^3 \\ -\beta^3 & \end{bmatrix} t^3 + \cdots \\ &= \begin{bmatrix} \cos\beta t & \sin\beta t \\ -\sin\beta t & \cos\beta t \end{bmatrix} \end{aligned} \qquad (2.43)$$

If A is with the following form,

$$A = \begin{bmatrix} \alpha & \beta \\ -\beta & \alpha \end{bmatrix} = \begin{bmatrix} \alpha & \\ & \alpha \end{bmatrix} + \begin{bmatrix} & \beta \\ -\beta & \end{bmatrix}$$

and then using the similar method obtaining (2.42), one can deduce that

$$e^{At} = \begin{bmatrix} e^{\alpha t} \cos\beta t & e^{\alpha t} \sin\beta t \\ -e^{\alpha t} \sin\beta t & e^{\alpha t} \cos\beta t \end{bmatrix} \qquad (2.44)$$

In Subsection 1.3.4, it is discussed that the state coefficient matrix A can be transformed into the Jordan normal form $\bar{A} = TAT^{-1}$. The method to obtain the transformation matrix T can be find in Subsection 1.3.4. In the next, one can start splitting in following cases.

If eigenvalues of A are all distinct, then combining (2.37) with (2.39), one can derive that

$$e^{At} = T^{-1} e^{\bar{A}t} T = T^{-1} \begin{bmatrix} e^{\lambda_1 t} & & \\ & \ddots & \\ & & e^{\lambda_n t} \end{bmatrix} T \qquad (2.45)$$

If some eigenvalues in (2.45) are complex numbers, as discussed in Section 1.3.4, the Jordan normal form can be further transformed into a modal form. Combining (2.40) with

2.3 Response of Linear Time-Invariant Dynamical Systems

(2.44), and using the similar method obtaining (2.45), one can compute the corresponding e^{At}.

If there exist repeated eigenvalues of A, the corresponding Jordan normal form can be divided into some Jordan blocks. Combining (2.40) with (2.42), e^{At} can be obtained.

(4) Use the Cayley-Hamilton theorem to compute e^{At}.

As discussed in Subsection 1.3.3, the resolvent matrix can be solved by

$$(sI - A)^{-1} = \sum_{k=0}^{n-1} \frac{1}{\alpha(s)} p_k(s) A^k$$

where $\alpha(s) = s^n + a_1 s^{n-1} + \cdots + a_{n-1} s + a_n$ is the characteristic polynomial of A, and

$$\begin{bmatrix} p_0(s) \\ p_1(s) \\ \vdots \\ p_{n-2}(s) \\ p_{n-1}(s) \end{bmatrix} = \begin{bmatrix} 1 & a_1 & a_2 & \cdots & a_{n-1} \\ & 1 & a_1 & \cdots & a_{n-2} \\ & & 1 & \ddots & \vdots \\ & & & \ddots & a_1 \\ & & & & 1 \end{bmatrix} \begin{bmatrix} s^{n-1} \\ s^{n-2} \\ \vdots \\ s \\ 1 \end{bmatrix}$$

Denote

$$f_k(t) \triangleq \mathcal{L}^{-1}\left[\frac{p_k(s)}{\alpha(s)}\right], \quad k = 0, 1, \cdots, n-1$$

and then one can deduce that

$$e^{At} = \sum_{k=0}^{n-1} f_k(t) A^k$$

Example 2.2 Consider $A = \begin{bmatrix} -1 & 2 \\ 0 & -3 \end{bmatrix}$. Solve e^{At}.

Use summation of the infinite series to compute e^{At}. Basing on (2.28), it can be obtained that

$$e^{At} = I + At + \frac{1}{2!} A^2 t^2 + \cdots$$

$$= \begin{bmatrix} 1 & 0 \\ 0 & 1 \end{bmatrix} + \begin{bmatrix} -t & 2t \\ 0 & -3t \end{bmatrix} + \begin{bmatrix} \frac{1}{2} t^2 & -4t^2 \\ 0 & \frac{9}{2} t^2 \end{bmatrix} + \cdots$$

$$= \begin{bmatrix} 1 - t - \frac{1}{2} t^2 + \cdots & 2t - 4t^2 + \cdots \\ 0 & 1 - 3t + \frac{9}{2} t^2 + \cdots \end{bmatrix}$$

Use the inverse Laplace transform to compute e^{At}. The resolvent matrix of the state coefficient matrix A is obtained that

$$(sI - A)^{-1} = \begin{bmatrix} s+1 & -2 \\ 0 & s+3 \end{bmatrix}^{-1} = \begin{bmatrix} \dfrac{1}{s+1} & \dfrac{2}{(s+1)(s+3)} \\ 0 & \dfrac{1}{s+3} \end{bmatrix}$$

$$= \begin{bmatrix} \dfrac{1}{s+1} & \dfrac{1}{s+1} - \dfrac{1}{s+3} \\ 0 & \dfrac{1}{s+3} \end{bmatrix}$$

Taking the inverse Laplace transform, it can be obtained that

$$\mathrm{e}^{At} = \begin{bmatrix} \mathrm{e}^{-t} & \mathrm{e}^{-t} - \mathrm{e}^{-3t} \\ 0 & \mathrm{e}^{-3t} \end{bmatrix}$$

Use the Jordan normal form to compute e^{At}.

The eigenvalues of A are determined as $\lambda_1 = -1$ and $\lambda_2 = -3$. Furthermore, the transformation matrix T which transform the matrix A into a diagonal Jordan normal form are obtained that

$$T = \begin{bmatrix} 1 & 1 \\ 0 & -1 \end{bmatrix}, \quad T^{-1} = \begin{bmatrix} 1 & 1 \\ 0 & -1 \end{bmatrix}$$

On this basis, using (2.45) to obtain that

$$\mathrm{e}^{At} = T^{-1} \mathrm{e}^{\bar{A}t} T = \begin{bmatrix} 1 & 1 \\ 0 & -1 \end{bmatrix} \begin{bmatrix} \mathrm{e}^{\lambda_1 t} & \\ & \mathrm{e}^{\lambda_2 t} \end{bmatrix} \begin{bmatrix} 1 & 1 \\ 0 & -1 \end{bmatrix} = \begin{bmatrix} \mathrm{e}^{-t} & \mathrm{e}^{-t} - \mathrm{e}^{-3t} \\ 0 & \mathrm{e}^{-3t} \end{bmatrix}$$

Use the Cayley-Hamilton theorem to compute e^{At}. The characteristic polynomial of A is $\alpha(s) = s^2 + 4s + 3$, and then it holds that

$$\begin{bmatrix} p_0(s) \\ p_1(s) \end{bmatrix} = \begin{bmatrix} 1 & 4 \\ 0 & 1 \end{bmatrix} \begin{bmatrix} s \\ 1 \end{bmatrix} = \begin{bmatrix} s+4 \\ 1 \end{bmatrix}$$

Furthermore,

$$f_0(t) = \mathcal{L}^{-1}\left[\dfrac{p_0(s)}{\alpha(s)}\right] = \mathcal{L}^{-1}\left[\dfrac{s+4}{s^2+4s+3}\right] = \dfrac{3}{2}\mathrm{e}^{-t} - \dfrac{1}{2}\mathrm{e}^{-3t}$$

$$f_1(t) = \mathcal{L}^{-1}\left[\dfrac{p_1(s)}{\alpha(s)}\right] = \mathcal{L}^{-1}\left[\dfrac{1}{s^2+4s+3}\right] = \dfrac{1}{2}\mathrm{e}^{-t} - \dfrac{1}{2}\mathrm{e}^{-3t}$$

Therefore, it holds that

$$\mathrm{e}^{At} = f_0(t)I + f_1(t)A$$

$$= \begin{bmatrix} \dfrac{3}{2}e^{-t} - \dfrac{1}{2}e^{-3t} & 0 \\ 0 & \dfrac{3}{2}e^{-t} - \dfrac{1}{2}e^{-3t} \end{bmatrix} + \begin{bmatrix} -\dfrac{1}{2}e^{-t} + \dfrac{1}{2}e^{-3t} & e^{-t} - e^{-3t} \\ 0 & -\dfrac{3}{2}e^{-t} + \dfrac{3}{2}e^{-3t} \end{bmatrix}$$

$$= \begin{bmatrix} e^{-t} & e^{-t} - e^{-3t} \\ 0 & e^{-3t} \end{bmatrix}$$

2.3.2 Zero-Input Response of Linear Time-Invariant Systems

Consider a linear time-invariant autonomous system

$$\dot{x}(t) = Ax(t),\ x(0) = x_0,\ t \in [0, \infty) \tag{2.46}$$

where $x(t) \in \mathbb{R}^n$ is the state vector and A is a constant matrix.

Theorem 2.9 *The zero-input response of the linear time-invariant system, i.e., the solution of the homogeneous state equation* (2.46) *is with the following form,*

$$x(t) = \phi(t; 0, x_0, 0) = e^{At}x_0,\ t \geqslant 0 \tag{2.47}$$

Proof The solution of (2.46) can be represented as a power series

$$x(t) = b_0 + b_1 t + b_2 t^2 + \cdots = \sum_{k=0}^{\infty} b_k t^k,\ t \geqslant 0 \tag{2.48}$$

where $b_0, b_1, \cdots$ are coefficient vectors to be determined. Substituting (2.48) into (2.46), one can obtain that

$$b_1 + 2b_2 t + 3b_3 t^2 + \cdots = Ab_0 + Ab_1 t + Ab_2 t^2 + \cdots$$

which indicates that

$$\begin{cases} b_1 = Ab_0 \\ b_2 = \dfrac{1}{2}Ab_1 = \dfrac{1}{2!}A^2 b_0 \\ b_3 = \dfrac{1}{3}Ab_2 = \dfrac{1}{3!}A^3 b_0 \\ \vdots \\ b_k = \dfrac{1}{k}Ab_{k-1} = \dfrac{1}{k!}A^k b_0 \\ \vdots \end{cases} \tag{2.49}$$

Substituting (2.49) into (2.48), it holds that

$$x(t) = \left(I + At + \dfrac{1}{2!}A^2 t^2 + \cdots\right) b_0,\ t \geqslant 0 \tag{2.50}$$

Let $t = 0$ in (2.50) and consider the initial condition $x(0) = x_0$. One can further obtain that $b_0 = x_0$. According to the definition of the matrix exponential (2.28), (2.47) can be derived. □

It is worth mentioning that one can select the initial time $t_0 = 0$ for time-invariant systems. When $t_0 \neq 0$, (2.47) can be rewritten as

$$x(t) = \phi(t; t_0, x_0, 0) = e^{A(t-t_0)} x_0, \ t \geqslant t_0 \tag{2.51}$$

Based on Theorem 2.9, one can prove (2.38).

Taking the Laplace transform of both sides of the state equation (2.46), one can obtain that

$$sx(s) - x_0 = Ax(s)$$

which indicates that

$$x(s) = (sI - A)^{-1} x_0$$

Taking the inverse Laplace transform, it can be derived that

$$x(t) = \mathcal{L}^{-1}\left[(sI - A)^{-1}\right] x_0$$

Compared with (2.47), one can obtain (2.38).

In the next, define the fundamental solution matrix and the state transition matrix for linear time-invariant system (2.46).

Definition 2.4 (Fundamental solution matrix of linear time-invariant systems) Let $\psi_1(t), \cdots, \psi_n(t)$ be n linearly independent solutions of an n-dimensional linear time-invariant homogeneous equation (2.46) for $t \geqslant t_0$, and then the matrix

$$\Psi(t) = \begin{bmatrix} \psi_1(t) & \cdots & \psi_n(t) \end{bmatrix}$$

is called a *fundamental solution matrix* of the linear time-invariant system (2.46).

One can easily deduce that $\Psi(t)$ is nonsingular for $\forall t \geqslant t_0$. Similar to Theorem 2.2, $\Psi(t)$ also satisfies

$$\dot{\Psi}(t) = A\Psi(t), \ \Psi(t_0) = E, \ t \geqslant t_0 \tag{2.52}$$

where E is a nonsingular real constant matrix.

The fundamental solution matrix of a linear time-invariant system can also be defined based on (2.52). Meanwhile, if $\Psi_1(t)$ is a fundamental solution matrix of a linear time-invariant system, then for any nonsingular matrix $C \in \mathbb{R}^{n \times n}$, $\Psi_2(t) = \Psi_1(t)C$ is also a fundamental solution matrix of this system. Moreover, if $\Psi_1(t)$ and $\Psi_2(t)$ are any two fundamental solution matrices of a linear time-invariant system, then there exists a nonsingular matrix $C \in \mathbb{R}^{n \times n}$ such that $\Psi_2(t) = \Psi_1(t)C$.

It is worth mentioning that e^{At} is satisfied (2.52). Therefore, e^{At} is a fundamental solution matrix of the corresponding linear time-invariant system.

Definition 2.5 (State transition matrix of linear time-invariant systems) For a linear time-invariant system (2.46), the $n \times n$ matrix $\Phi(t - t_0)$ satisfying

$$\frac{\partial}{\partial t}\Phi(t - t_0) = A\Phi(t - t_0), \ \Phi(0) = I, \ t \geqslant t_0 \tag{2.53}$$

2.3 Response of Linear Time-Invariant Dynamical Systems

is called a *state transition matrix* of the linear time-invariant system.

Similar to Theorem 2.4, the state transition matrix of a linear time-invariant system satisfies

$$\Phi(t - t_0) = \Psi(t)\Psi^{-1}(t_0), \ t \geq t_0 \tag{2.54}$$

where $\Psi(t)$ is a fundamental solution matrix of the linear time-invariant system. Furthermore, the state transition matrix $\Phi(t - t_0)$ of a linear time-invariant system is unique. According to Theorem 2.9, it can be deduced that $\Phi(t - t_0) = e^{A(t-t_0)}$, $t \geq t_0$. Moreover, $\Phi(t) = e^{At}$, $t \geq 0$ when $t_0 = 0$. It indicates that the state transition matrix of a linear time-invariant system is uniquely determined by its state coefficient matrix A.

Combining with (2.51), the zero-input response of the linear time-invariant system can be expressed as

$$x(t) = \phi(t; t_0, x_0, 0) = \Phi(t - t_0)x_0, \ t \geq t_0$$

which coincides with the result for the linear time-varying system.

2.3.3 Zero-State Response of Linear Time-Invariant Systems

Consider a linear time-invariant system

$$\dot{x}(t) = Ax(t) + Bu(t), \ x(t_0) = 0, \ t \in [0, \infty) \tag{2.55}$$

where $x(t) \in \mathbb{R}^n$ is the state vector, $u(t) \in \mathbb{R}^p$ is the input vector, A and B are constant matrices. Suppose that (2.55) satisfies the unique solution conditions. Its zero-state response satisfies the following conclusion.

Theorem 2.10 *The zero-state response of the linear time-invariant system, i.e., the solution of the nonhomogeneous state equation (2.55) is with the following form,*

$$x(t) = \phi(t; 0, 0, u) = \int_0^t e^{A(t-\tau)}Bu(\tau)d\tau, \ t \geq 0 \tag{2.56}$$

Proof Pre-multiply both sides of (2.55) by e^{-At}, and then rewrite it as

$$e^{-At}(\dot{x}(t) - Ax(t)) = e^{-At}Bu(t) \tag{2.57}$$

Along with (2.36), it holds that

$$\frac{d}{dt}e^{-At}x(t) = -e^{-At}Ax(t) + e^{-At}\dot{x}(t) = e^{-At}(\dot{x}(t) - Ax(t)) \tag{2.58}$$

Combining (2.57) with (2.58), one can obtain that

$$\frac{d}{dt}e^{-At}x(t) = e^{-At}Bu(t) \tag{2.59}$$

Integrating both sides of (2.59) from 0 to t, one can deduce that

$$e^{-At}x(t) - x_0 = \int_0^t e^{-A\tau}Bu(\tau)d\tau \tag{2.60}$$

It is noticed that $x_0 = 0$. Meanwhile, pre-multiply both sides of (2.55) by e^{At}, one can obtain (2.56). □

In Theorem 2.10, it is supposed that $t_0 = 0$. When $t_0 \neq 0$, (2.56) can be rewritten as

$$x(t) = \phi(t; t_0, 0, u) = \int_{t_0}^{t} \mathrm{e}^{A(t-\tau)} Bu(\tau) \mathrm{d}\tau, \ t \geqslant 0 \tag{2.61}$$

Example 2.3 Consider the following linear time-invariant system,

$$\dot{x}(t) = \begin{bmatrix} -1 & 2 \\ 0 & -3 \end{bmatrix} x(t) + \begin{bmatrix} 0 \\ 1 \end{bmatrix} u(t), \ t \geqslant 0$$

The initial state $x_0 = [\ 0\ \ 0\ ]^{\mathrm{T}}$, and the input function $u(t) = 1(t)$, that is, it is the unit step function. Solve its zero-state response.

It has been exported in Example 2.1, the matrix exponential of A is

$$\mathrm{e}^{At} = \begin{bmatrix} \mathrm{e}^{-t} & \mathrm{e}^{-t} - \mathrm{e}^{-3t} \\ 0 & \mathrm{e}^{-3t} \end{bmatrix}$$

On this basis, using (2.61), the zero-state response of the system can be calculated,

$$x(t) = \int_0^t \mathrm{e}^{A(t-\tau)} Bu(\tau) \mathrm{d}\tau = \int_0^\tau \begin{bmatrix} \mathrm{e}^{-(t-\tau)} & \mathrm{e}^{-(t-\tau)} - \mathrm{e}^{-3(t-\tau)} \\ 0 & \mathrm{e}^{-3(t-\tau)} \end{bmatrix} \begin{bmatrix} 0 \\ 1 \end{bmatrix} \mathrm{d}\tau$$

$$= \int_0^t \begin{bmatrix} \mathrm{e}^{-(t-\tau)} - \mathrm{e}^{-3(t-\tau)} \\ \mathrm{e}^{-3(t-\tau)} \end{bmatrix} \mathrm{d}\tau = \begin{bmatrix} \dfrac{2}{3} - \mathrm{e}^{-t} + \dfrac{1}{3}\mathrm{e}^{-3t} \\ \dfrac{1}{3} - \dfrac{1}{3}\mathrm{e}^{-3t} \end{bmatrix}, \ t \geqslant 0$$

2.3.4 State Response and Output Response of Linear Time-Invariant Systems

As discussed in Section 2.1, the state response $x(t)$ of a linear system is the superposition of the zero-input response and the zero-state response. As a result, one can conclude that the state response of a linear time-invariant system, i.e., the solution of the state equation excited by the initial state x_0 and input $u(t)$ simultaneously is with the following form,

$$x(t) = \phi(t; 0, x_0, u) = \mathrm{e}^{At} x_0 + \int_0^t \mathrm{e}^{A(t-\tau)} Bu(\tau) \mathrm{d}\tau, \ t \geqslant 0 \tag{2.62}$$

When $t_0 \neq 0$, (2.62) can be rewritten as

$$x(t) = \phi(t; t_0, x_0, u) = \mathrm{e}^{A(t-t_0)} x_0 + \int_{t_0}^t \mathrm{e}^{A(t-\tau)} Bu(\tau) \mathrm{d}\tau, \ t \geqslant 0 \tag{2.63}$$

After solving the state response of a linear time-invariant system, its output response can be deduced based on its output equation,

$$y(t) = Cx(t) + Du(t) = C\mathrm{e}^{At} x_0 + \int_0^t C\mathrm{e}^{A(t-\tau)} Bu(\tau) \mathrm{d}\tau + Du(t), \ t \geqslant 0 \tag{2.64}$$

When $t_0 \neq 0$, (2.64) can be rewritten as

$$y(t) = Cx(t) + Du(t) = Ce^{A(t-t_0)}x_0 + \int_{t_0}^{t} Ce^{A(t-\tau)}Bu(\tau)d\tau + Du(t), \; t \geqslant 0 \quad (2.65)$$

Furthermore, the impulse response matrix of a linear time-invariant system relaxed at $t_0 = 0$ can be expressed as

$$G(t-\tau) = Ce^{A(t-\tau)}B + D\delta(t-\tau) \quad (2.66)$$

or

$$G(t) = Ce^{At}B + D\delta(t) \quad (2.67)$$

Since $\Phi(t - t_0) = e^{A(t-t_0)}$, $t \geqslant t_0$, (2.63), (2.65) and (2.67) can be expressed as

$$x(t) = \phi(t; t_0, x_0, u) = \Phi(t-t_0)x_0 + \int_{t_0}^{t} \Phi(t-\tau)Bu(\tau)d\tau, \; t \geqslant 0$$

$$y(t) = C\Phi(t-t_0)x_0 + \int_{t_0}^{t} C\Phi(t-\tau)Bu(\tau)d\tau + Du(t), \; t \geqslant 0$$

and

$$G(t) = C\Phi(t)B + D\delta(t)$$

which coincides with the results for linear time-varying systems.

As discussed in Subsection 1.3.4, if two linear time-invariant systems (A, B, C, D) and $(\bar{A}, \bar{B}, \bar{C}, \bar{D})$ are equivalent[①], then there exists an equivalent transformation matrix T such that $\bar{A} = TAT^{-1}$, $\bar{B} = TB$, $\bar{C} = CT^{-1}$, $\bar{D} = D$, and $\bar{x}(t) = Tx(t)$. Denote state transition matrices of these two systems as $\Phi(t-t_0)$ and $\bar{\Phi}(t-t_0)$, respectively. One can deduce that

$$\bar{\Phi}(t-t_0) = e^{\bar{A}(t-t_0)} = e^{TAT^{-1}(t-t_0)} = Te^{A(t-t_0)}T^{-1} = T\Phi(t-t_0)T^{-1}$$

It is not difficult to prove that for two linear time-invariant systems with equivalence, when the output of one system with the initial state x_0 is determined, the other system always has a corresponding initial state so that they have the same outputs in the zero-input case. Meanwhile, (A, B, C, D) and $(\bar{A}, \bar{B}, \bar{C}, \bar{D})$ have the same impulse response matrices and also transfer function matrices. In other words, in the zero-state case, for the same input $u(t)$, outputs of two linear time-invariant systems with equivalence are always equivalent.

Assignments

2.1 For the given matrix A, try to calculate e^{At}.

[①] For convenience, use (A, B, C, D) to denote dynamical equations (1.11) and (1.12), while use $(\bar{A}, \bar{B}, \bar{C}, \bar{D})$ to denote dynamical equations (1.24) and (1.25).

(1) $A = \begin{bmatrix} 1 & 2 \\ -2 & 1 \end{bmatrix}$.

(2) $A = \begin{bmatrix} -2 & 1 \\ 0 & -2 \end{bmatrix}$.

(3) $A = \begin{bmatrix} -1 & 0 & 0 \\ 0 & -3 & 1 \\ 0 & 0 & -3 \end{bmatrix}$.

(4) $A = \begin{bmatrix} -1 & 0 & 0 \\ 0 & 0 & 1 \\ 0 & -2 & 0 \end{bmatrix}$.

(5) $A = \begin{bmatrix} 2 & 1 & 1 \\ 1 & 2 & 1 \\ 1 & 1 & 2 \end{bmatrix}$.

2.2 The state equation and initial conditions are given by

$$\begin{bmatrix} \dot{x}_1(t) \\ \dot{x}_2(t) \\ \dot{x}_3(t) \end{bmatrix} = \begin{bmatrix} 1 & 0 & 0 \\ 0 & 1 & 0 \\ 0 & 1 & 2 \end{bmatrix} \begin{bmatrix} x_1(t) \\ x_2(t) \\ x_3(t) \end{bmatrix}, \quad x_0 = \begin{bmatrix} 1 \\ 0 \\ 1 \end{bmatrix}$$

(1) Solve the state transition matrix by using the inverse Laplace transform.
(2) Solve the state transition matrix based on the Jordan normal form.
(3) Solve the solution of this homogeneous state equation.

2.3 The state transition matrix $\Phi(t)$ of a system is

$$\Phi(t) = \begin{bmatrix} \frac{1}{2}(e^{-t} + e^{3t}) & \frac{1}{4}(-e^{-t} + e^{3t}) \\ (-e^{-t} + e^{3t}) & \frac{1}{2}(e^{-t} + e^{3t}) \end{bmatrix}$$

Try to get the coefficient matrix A of this system.

2.4 Consider the following linear time-invariant system

$$\dot{x}(t) = \begin{bmatrix} -3 & 1 \\ 1 & -3 \end{bmatrix} x(t) + \begin{bmatrix} 1 & 1 \\ 1 & 1 \end{bmatrix} u(t), \quad y(t) = \begin{bmatrix} 1 & 1 \\ 1 & -1 \end{bmatrix} x(t)$$

(1) Solve the state transition matrix of this system.
(2) Solve the zero-state response and output response under the unit step input $u(t) = \varepsilon(t)$.
(3) Solve the transfer function matrix of this system.

2.5 Solve the fundamental solution matrix and the state transition matrix of the following homogeneous equations.

(1) $\dot{x}(t) = \begin{bmatrix} 0 & t \\ 0 & 0 \end{bmatrix} x(t)$.

(2) $\dot{x}(t) = \begin{bmatrix} \sin t & 0 \\ 0 & 0 \end{bmatrix} x(t)$.

(3) $\dot{x}(t) = \begin{bmatrix} 0 & e^{-t} \\ e^{-t} & 0 \end{bmatrix} x(t)$.

(4) $\dot{x}(t) = \begin{bmatrix} 0 & \omega \\ -\omega & 0 \end{bmatrix} x(t)$.

2.6 Consider that A, B are constant square matrices, and $AB = BA$. Verify that the state transition matrix of $\dot{x}(t) = e^{-At} B e^{At} x(t)$ is $\Phi(t, t_0) = e^{-At} e^{(A+B)(t-t_0)} e^{At_0}$.

2.7 If $C_1 A_1^k B_1 = C_2 A_2^k B_2$, $k = 0, 1, 2, \cdots, D_1 = D_2$ hold between the linear time-invariant systems (A_1, B_1, C_1, D_1) and (A_2, B_2, C_2, D_2), try to verify that they have the same impulse response matrices.

2.8 For a linear system $\dot{x}(t) = A(t) x(t)$ with

$$A(t) = \begin{bmatrix} A_{11}(t) & A_{12}(t) \\ 0 & A_{22}(t) \end{bmatrix}$$

verify that the state transition matrix of the system is

$$\Phi(t, t_0) = \begin{bmatrix} \Phi_{11}(t, t_0) & \Phi_{12}(t, t_0) \\ 0 & \Phi_{22}(t, t_0) \end{bmatrix}$$

where $\dfrac{\partial}{\partial t} \Phi_{11}(t, t_0) = A_{11}(t) \Phi_{11}(t, t_0)$, $\dfrac{\partial}{\partial t} \Phi_{22}(t, t_0) = A_{22}(t) \Phi_{22}(t, t_0)$, $\dfrac{\partial}{\partial t} \Phi_{12}(t, t_0) = A_{11}(t) \Phi_{12}(t, t_0) + A_{12}(t) \Phi_{22}(t, t_0)$, and $\Phi_{12}(t_0, t_0)$ is a null matrix.

2.9 Use the conclusion in 2.8 to determine the state transition matrix $\Phi(t, 0)$ for

$$A(t) = \begin{bmatrix} -1 & e^{2t} \\ 0 & -1 \end{bmatrix}$$

Chapter 3

Controllability and Observability of Linear Systems

Controllability and observability are two basic structure properties of linear systems. In the early 1960s, Kalman studied them firstly. The development of control theories indicates that controllability and observability are very important for control theories studying and engineering design. The principal objective of this chapter is to investigate in depth the system properties of controllability and observability.

3.1 Controllability of Linear Systems

3.1.1 Basic Concepts

In the state space description of a system, inputs and outputs are external variables while state variables are internal ones. Controllability is the ability that whether or not internal states can be controlled and influenced by the input of the system. If each state variable can be controlled and influenced by the input of a system, and the state can be moved from any value to the origin, then the system is called state controllable. Otherwise, it is called incompletely state controllable.

Example 3.1 Consider a system with the following state equation,

$$\begin{bmatrix} \dot{x}_1(t) \\ \dot{x}_2(t) \end{bmatrix} = \begin{bmatrix} 1 & 0 \\ 0 & -3 \end{bmatrix} \begin{bmatrix} x_1(t) \\ x_2(t) \end{bmatrix} + \begin{bmatrix} 5 \\ 1 \end{bmatrix} u(t)$$

Apparently, this system is parallel connected by two scalar variables:

$$\dot{x}_1(t) = x_1(t) + 5u(t)$$
$$\dot{x}_2(t) = -3x_2(t) + u(t)$$

The above two equations indicates that the state variables $x_1(t)$ and $x_2(t)$ can be moved to the origin in the state space from any initial values by choosing a suitable input $u(t)$.

Example 3.2 Suppose that an operational amplifier circuit shown in Fig.3.1. The state variables of the system are the output of the operation amplifiers $x_1(t)$ and $x_2(t)$. The input of the operation amplifiers is $u(t)$.

Fig.3.1 shows that if $x_1(t_0)=x_2(t_0)$, the input voltage $u(t)$ can make $x_1(t)$ and $x_2(t)$ transfer to the same arbitrary target value, or $u(t)$ can make them return to $x_1(t) = x_2(t) =$

0 at the same time, $t \geqslant t_0$. However, $u(t)$ cannot make $x_1(t)$ and $x_2(t)$ transfer to different target values. If $x_1(t_0) \neq x_2(t_0)$, $u(t)$ cannot make them transfer to $x_1(t) = x_2(t) = 0$ at the same time, $t \geqslant t_0$.

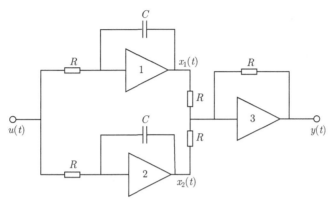

Fig.3.1 An incompletely controllable circuit

Example 3.3 Suppose that a circuit network shown in Fig.3.2. The state variable of the system is the voltage $x(t)$ across the capacitor, and the input is the voltage $u(t)$ from the electrical source.

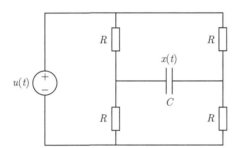

Fig.3.2 An uncontrollable bridge circuit

The circuit shows that for an initial state $x(t_0) = 0$, when $t \geqslant t_0$, whatever $u(t)$ is, $x(t) = 0$, i.e., $u(t)$ cannot influence and control $x(t)$.

The above three examples only are simple descriptions of controllability rather than the strict definition. For discussing the properties of controllability in depth, the strict definition of controllability is introduced subsequently.

Consider a linear time-varying system, its state equation is

$$\dot{x}(t) = A(t)x(t) + B(t)u(t), \ x(t_0) = x_0, \ t \in [t_0, t_f] \tag{3.1}$$

where $x(t) \in \mathbb{R}^n$ is the state vector, $u(t) \in \mathbb{R}^p$ is the input vector. $A(t)$ and $B(t)$ are time-varying matrices and their entries are absolutely integrable and square integrable in $[t_0, t_f]$, respectively. For convenience, denote the state equation (3.1) as a pair $(A(t), B(t))$.

For the time-invariant case, the state equation is

$$\dot{x}(t) = Ax(t) + Bu(t), \ x(t_0) = x_0, \ t \geqslant t_0 \tag{3.2}$$

where A and B are constant matrices. Simply denote (3.2) as (A, B).

Definition 3.1 (Controllability of a state) A state x_0 is *controllable* at time t_0 if for some finite $t_f > t_0$, there exists an input $u(t)$, $t \in [t_0, t_f]$ that transfers the state $x(t)$ from x_0 at t_0 to the origin at time t_f, i.e., from $x(t_0) = x_0$ to $x(t_f) = 0$.

Definition 3.2 (Controllability of a system) A system $(A(t), B(t))$ (or (A, B)) is *(completely state) controllable* at t_0 if all nonzero states x_0 in its state-space are controllable at t_0. A system $(A(t), B(t))$ (or (A, B)) is *incompletely (state) controllable* at t_0 if there exists one or more nonzero states in the state space that are not controllable at t_0. A system $(A(t), B(t))$ (or (A, B)) is *(completely state) uncontrollable* at t_0 if all nonzero states in its state-space are not controllable at t_0.

There are some remarks about the above definitions.

In Definition 3.1, all entries of $u(t)$ are required to be square integrable in $[t_0, t_f]$, and their amplitudes are unconstraint, but $u(t)$ may not be physical realizable. On the other hand, $u(t)$ is required that can transfer the nonzero state x_0 at t_0 to zero in a finite interval. However, the transfer trace of the state is arbitrary. Namely, controllability shows the qualitative property of the state movement of a system.

Controllability is defined with respect to given time t_0, which is necessary for time-varying systems. If a time-varying system is always controllable for any time in $[t_0, t_f]$, the system is said to be *uniformly (completely state) controllable*. For time-invariant systems, controllability is unrelated to the selection of t_0, i.e., (A, B) is controllable indicates that it is uniformly controllable.

Controllability is the structure property that is only determined by the structure and parameters of a system, and is independent of the input and the initial state of the system.

In practice, incomplete controllability is a kind of singular instance. When the parameter values of the system are changed a little, the system can become completely controllable. For example, if the value of one of the resistance R in the circuit shown in Fig.3.2 changes a little, the symmetry of the circuit will be destroyed and the voltage across the capacitor, i.e., the state $x(t)$ will change from uncontrollable to controllable.

Controllability is based on the transfer from nonzero states to zero states. If the transfer is from zero states to nonzero states, the property is called reachability.

Definition 3.3 (Reachability of a state) A state x_f is *reachable* at time t_f if for some finite $t_0 < t_f$, there exists an input $u(t)$, $t \in [t_0, t_f]$ that transfers the state $x(t)$ from the origin at t_0 to x_f at time t_f, i.e., from $x(t_0) = 0$ to $x(t_f) = x_f$.

Definition 3.4 (Reachability of a system) A system $(A(t), B(t))$ (or (A, B)) is *(completely state) reachable* at t_f if all nonzero states x_f in its state-space are reachable at t_f. A system $(A(t), B(t))$ (or (A, B)) is *incompletely (state) reachable* at t_f if there exists one or more nonzero states in the state space that are not reachable at t_f. A system $(A(t), B(t))$ (or (A, B)) is *(completely state) unreachable* at t_f if all nonzero states in its state-space are not reachable at t_f.

Reachability and controllability are equivalent for continuous linear time-invariant system. But they are not equivalent for time-varying systems. A system is incompletely

3.1 Controllability of Linear Systems

controllable, but it may be completely reachable.

3.1.2 Controllability Criteria of Linear Time-Varying Systems

In this subsection, some controllability criteria for a linear time-varying system $(A(t), B(t))$ is introduced.

Theorem 3.1 (Gramian criterion) *A linear time-varying system $(A(t), B(t))$ is completely controllable at t_0 if and only if there exists $t_{\mathrm{f}} > t_0$ such that*

$$W_{\mathrm{c}}(t_0, t_{\mathrm{f}}) \triangleq \int_{t_0}^{t_{\mathrm{f}}} \Phi(t_0, \tau) B(\tau) B^{\mathrm{T}}(\tau) \Phi^{\mathrm{T}}(t_0, \tau) \mathrm{d}\tau \tag{3.3}$$

is nonsingular, where $\Phi(t_0, \tau)$ is the state transition matrix of $(A(t), B(t))$.

$W_{\mathrm{c}}(t_0, t_{\mathrm{f}})$ is called the *controllability Gramian* of the linear time-varying system, which is an $n \times n$ matrix.

Proof First, prove the sufficiency. $W_{\mathrm{c}}(t_0, t_{\mathrm{f}})$ is nonsingular, then prove that $(A(t), B(t))$ is completely controllable at t_0.

Since $W_{\mathrm{c}}(t_0, t_{\mathrm{f}})$ is nonsingular, $W_{\mathrm{c}}^{-1}(t_0, t_{\mathrm{f}})$ is existing. On this basis, construct the following input for any $x(t_0) = x_0$,

$$u(t) = -B^{\mathrm{T}}(t) \Phi^{\mathrm{T}}(t_0, t) W_{\mathrm{c}}^{-1}(t_0, t_{\mathrm{f}}) x_0, \ t \in [t_0, t_{\mathrm{f}}]$$

Excited by the above $u(t)$, at t_{f},

$$\begin{aligned}
x(t_{\mathrm{f}}) =& \Phi(t_{\mathrm{f}}, t_0) x_0 + \int_{t_0}^{t_{\mathrm{f}}} \Phi(t_{\mathrm{f}}, \tau) B(\tau) u(\tau) \mathrm{d}\tau \\
=& \Phi(t_{\mathrm{f}}, t_0) x_0 - \int_{t_0}^{t_{\mathrm{f}}} \Phi(t_{\mathrm{f}}, t_0) \Phi(t_0, \tau) B(\tau) B^{\mathrm{T}}(\tau) \Phi^{\mathrm{T}}(t_0, \tau) W_{\mathrm{c}}^{-1}(t_0, t_{\mathrm{f}}) x_0 \mathrm{d}\tau \\
=& \Phi(t_{\mathrm{f}}, t_0) \left[x_0 - \int_{t_0}^{t_{\mathrm{f}}} \Phi(t_0, \tau) B(\tau) B^{\mathrm{T}}(\tau) \Phi^{\mathrm{T}}(t_0, \tau) \mathrm{d}\tau W_{\mathrm{c}}^{-1}(t_0, t_{\mathrm{f}}) x_0 \right] \\
=& \Phi(t_{\mathrm{f}}, t_0) \left[x_0 - W_{\mathrm{c}}(t_0, t_{\mathrm{f}}) W_{\mathrm{c}}^{-1}(t_0, t_{\mathrm{f}}) x_0 \right] = 0
\end{aligned}$$

which indicates that there exists a finite $t_{\mathrm{f}} > 0$ and a control input $u(t)$ such that the state x_0 transfers to $x(t_{\mathrm{f}}) = 0$ at t_{f}. According to Definition 3.2, $(A(t), B(t))$ is completely controllable at t_0.

Next, prove the necessity. Use reduction to absurdity. Suppose that $(A(t), B(t))$ is completely controllable at t_0, $W_{\mathrm{c}}(t_0, t_{\mathrm{f}})$ is singular. In this case, there must be a nonzero vector $\alpha \in \mathbb{R}^n$ such that

$$\alpha^{\mathrm{T}} W_{\mathrm{c}}(t_0, t_{\mathrm{f}}) \alpha = 0$$

Along with (3.3), it holds that

$$0 = \int_{t_0}^{t_{\mathrm{f}}} \alpha^{\mathrm{T}} \Phi(t_0, \tau) B(\tau) B^{\mathrm{T}}(\tau) \Phi^{\mathrm{T}}(t_0, \tau) \alpha \mathrm{d}\tau$$

$$= \int_{t_0}^{t_f} \alpha^T \Phi(t_0, \tau) B(\tau) \left(\alpha^T \Phi(t_0, \tau) B(\tau)\right)^T d\tau$$

It holds only when

$$\alpha^T \Phi(t_0, \tau) B(\tau) = 0, \ \forall t \in [t_0, t_f] \tag{3.4}$$

Let $x(t_0) = \alpha$ be an initial state of $(A(t), B(t))$. Since $(A(t), B(t))$ is completely controllable at t_0, there must exist $t_f > t_0$ and an input $u(t)$ such that $x(t_f) = 0$. That is to say,

$$0 = x(t_f) = \Phi(t_f, t_0)\alpha + \int_{t_0}^{t_f} \Phi(t_f, \tau) B(\tau) u(\tau) d\tau$$

which indicates that

$$\alpha = -\Phi^{-1}(t_f, t_0) \int_{t_0}^{t_f} \Phi(t_f, \tau) B(\tau) u(\tau) d\tau$$

$$= -\Phi(t_0, t_f) \int_{t_0}^{t_f} \Phi(t_f, \tau) B(\tau) u(\tau) d\tau$$

$$= -\int_{t_0}^{t_f} \Phi(t_0, \tau) B(\tau) u(\tau) d\tau$$

Furthermore,

$$\alpha^T \alpha = -\int_{t_0}^{t_f} \alpha^T \Phi(t_0, \tau) B(\tau) u(\tau) d\tau.$$

Combing with (3.4), one can deduce that $\alpha^T \alpha = 0$, and then $\alpha = 0$, which contradicts with the assumption. Therefore, if $(A(t), B(t))$ is completely controllable at t_0, then $W_c(t_0, t_f)$ is nonsingular. □

It can be verified that if $(A(t), B(t))$ is completely controllable at t_0, then $W_c(t_0, t_f)$ has the following properties.

(1) $W_c(t_0, t_f)$ is a positive definite constant matrix.
(2) $W_c(t_0, t_f) = W_c(t_0, t) + \Phi(t_0, t) W_c(t, t_f) \Phi^T(t_0, t)$.
(3) $W_c(t_f, t_f) = 0$.
(4) $\dfrac{\partial}{\partial t} W_c(t, t_f) = A(t) W_c(t, t_f) + W_c(t, t_f) A^T(t) - B(t) B^T(t)$.

Theorem 3.2 *A linear time-varying system $(A(t), B(t))$ is completely controllable at t_0 if and only if there exists $t_f > t_0$ such that rows of $\Phi(t_0, t) B(t)$ are linearly independent for $t \in [t_0, t_f]$.*

Theorem 3.2 can be deduced according to Theorem 3.1. Here, omit the proof.

It is worth mentioning that for the time-function vector $f_i(t)$, $i = 1, 2, \cdots, n$, whose elements are continuous real function defined in $[t_0, t_f]$, if there are some constants c_i, $i = 1, 2, \cdots, n$, that are not all zeros and can make

$$c_1 f_1(t) + c_2 f_2(t) + \cdots + c_n f_n(t) = 0$$

3.1 Controllability of Linear Systems

for any $t \in [t_0, t_f]$, then $f_1(t), f_2(t), \cdots, f_n(t)$ are linearly dependent. Otherwise, they are linearly independent. When the entries of vectors and constants are in different numerical domains, the property of the linear independence cannot be checked with the general method.

Theorem 3.3 (Rank criterion) *Suppose that $A(t)$ is $n-2$ continuously differentiable with respect to t, and $B(t)$ is $n-1$ continuously differentiable with respect to t. A linear time-varying system $(A(t), B(t))$ is completely controllable at t_0 if there exists $t_f > t_0$ such that*

$$\text{rank}\begin{bmatrix} M_0(t_f) & M_1(t_f) & \cdots & M_{n-1}(t_f) \end{bmatrix} = n \qquad (3.5)$$

where

$$M_0(t) = B(t)$$

$$M_1(t) = -A(t)M_0(t) + \frac{\mathrm{d}}{\mathrm{d}t}M_0(t)$$

$$\vdots$$

$$M_{n-1}(t) = -A(t)M_{n-2}(t) + \frac{\mathrm{d}}{\mathrm{d}t}M_{n-2}(t)$$

Proof Use reduction to absurdity. Supposing that $(A(t), B(t))$ is incompletely controllable at t_0, then according to Theorem 3.2, for any $t_f > t_0$, rows of $\Phi(t_0, t)B(t)$ is linearly dependent for $t \in [t_0, t_f]$. That is to say, there always exists a nonzero vector α such that

$$\alpha^\mathrm{T} \Phi(t_0, t) B(t) = 0, \ \forall t \in [t_0, t_f]$$

Taking partial derivative with respect to t, one can deduce that

$$\frac{\partial}{\partial t}\left(\alpha^\mathrm{T} \Phi(t_0, t) M_0(t)\right) = \alpha^\mathrm{T}\left(\frac{\partial}{\partial t}\Phi(t_0, t)\right)M_0(t) + \alpha^\mathrm{T}\Phi(t_0, t)\frac{\mathrm{d}}{\mathrm{d}t}M_0(t)$$

$$= -\alpha^\mathrm{T}\Phi(t_0, t)A(t)M_0(t) + \alpha^\mathrm{T}\Phi(t_0, t)\frac{\mathrm{d}}{\mathrm{d}t}M_0(t)$$

$$= \alpha^\mathrm{T}\Phi(t_0, t)M_1(t) = 0$$

Continue to take partial derivative to $\alpha^\mathrm{T}\Phi(t_0, t)M_1(t) = 0$ and repeat this procedure. Ultimately, one can obtain that for $\forall t \in [t_0, t_f]$,

$$\alpha^\mathrm{T}\Phi(t_0, t)\begin{bmatrix} M_0(t) & M_1(t) & \cdots & M_{n-1}(t) \end{bmatrix} = 0$$

Let $t = t_f$. Since $\Phi(t_0, t_f)$ is nonsingular and α is nonzero, it holds that

$$\text{rank}\begin{bmatrix} M_0(t_f) & M_1(t_f) & \cdots & M_{n-1}(t_f) \end{bmatrix} < n$$

which contradicts with (3.5). Therefore, $(A(t), B(t))$ is completely controllable at t_0. □

It is worth mentioning that the controllable criterion in Theorem 3.3 is sufficient but not necessary. That is to say, one cannot conclude that $(A(t), B(t))$ is incompletely controllable at t_0 when (3.5) is not satisfied.

Example 3.4 Consider the system

$$\begin{bmatrix} \dot{x}_1(t) \\ \dot{x}_2(t) \end{bmatrix} = \begin{bmatrix} t & 1 \\ 0 & t \end{bmatrix} \begin{bmatrix} x_1(t) \\ x_2(t) \end{bmatrix} + \begin{bmatrix} 0 \\ 1 \end{bmatrix} u(t)$$

One can compute that

$$M_0(t) = \begin{bmatrix} 0 \\ 1 \end{bmatrix}, \quad M_1(t) = -A(t)M_0(t) + \frac{\mathrm{d}}{\mathrm{d}t}M_0(t) = \begin{bmatrix} -1 \\ -t \end{bmatrix}$$

Furthermore, one can obtain that

$$\begin{bmatrix} M_0(t) & M_1(t) \end{bmatrix} = \begin{bmatrix} 0 & -1 \\ 1 & -t \end{bmatrix}$$

which is full rank for any t. Therefore, the system is completely controllable at any time.

When considering the reachability of linear time-varying system $(A(t), B(t))$, introduce the reachability Gramian

$$W_{\mathrm{r}}(t_0, t_{\mathrm{f}}) \triangleq \int_{t_0}^{t_{\mathrm{f}}} \Phi(t_{\mathrm{f}}, \tau) B(\tau) B^{\mathrm{T}}(\tau) \Phi^{\mathrm{T}}(t_{\mathrm{f}}, \tau) \mathrm{d}\tau \tag{3.6}$$

$(A(t), B(t))$ is completely reachable at t_{f} if and only if there exists $t_0 < t_{\mathrm{f}}$ such that $W_{\mathrm{r}}(t_0, t_{\mathrm{f}})$ is nonsingular. Referring to Theorem 3.1, this conclusion can be deduced. Meanwhile, combining (3.6) with (3.3), one can deduce that

$$W_{\mathrm{r}}(t_0, t_{\mathrm{f}}) = \Phi(t_{\mathrm{f}}, t_0) W_{\mathrm{c}}(t_0, t_{\mathrm{f}}) \Phi^{\mathrm{T}}(t_{\mathrm{f}}, t_0)$$

Since $\Phi(t_{\mathrm{f}}, t_0)$ is nonsingular for any t_{f} and t_0, rank $W_{\mathrm{r}}(t_0, t_{\mathrm{f}}) = $ rank $W_{\mathrm{c}}(t_0, t_{\mathrm{f}})$. That is to say, $(A(t), B(t))$ is completely reachable at t_{f} indicates that it is completely controllable at t_0, and vice versa. What is worth being highlighted is that $(A(t), B(t))$ is completely reachable at t_{f} does not mean that it is completely controllable at t_{f}, and vice versa.

3.1.3 Controllability Criteria of Linear Time-Invariant Systems

For a linear time-invariant system, controllability is independent of the initial time t_0. Without loss of generality, consider that $t_0 = 0$ in the sequel.

Consider a linear time-invariant system (A, B). In the next, introduce some frequently used controllability criteria.

Theorem 3.4 (Gramian criterion) *A linear time-invariant system (A, B) is completely controllable if and only if there exists $t_{\mathrm{f}} > 0$ such that*

$$W_{\mathrm{c}}(0, t_{\mathrm{f}}) \triangleq \int_0^{t_{\mathrm{f}}} \mathrm{e}^{-A\tau} B B^{\mathrm{T}} \mathrm{e}^{-A^{\mathrm{T}}\tau} \mathrm{d}\tau \tag{3.7}$$

is nonsingular.

3.1 Controllability of Linear Systems

$W_c(0, t_f)$ is called the *controllability Gramian* of the linear time-invariant system, which is an $n \times n$ matrix.

Proof First, prove the sufficiency. $W_c(0, t_f)$ is nonsingular, then prove that (A, B) is completely controllable.

Since $W_c(0, t_f)$ is nonsingular, $W_c^{-1}(0, t_f)$ is existing. On this basis, construct the following input for any $x(t_0) = x_0$,

$$u(t) = -B^{\mathrm{T}} e^{-A^{\mathrm{T}} t} W_c^{-1}(0, t_f) x_0, \ t \in [0, t_f]$$

Excited by the above $u(t)$, at t_f,

$$\begin{aligned} x(t_f) &= e^{At_f} x_0 + \int_0^{t_f} e^{A(t_f - \tau)} B u(\tau) \mathrm{d}\tau \\ &= e^{At_f} x_0 - \int_0^{t_f} e^{At_f} e^{-A\tau} B B^{\mathrm{T}} e^{-A^{\mathrm{T}}\tau} W_c^{-1}(0, t_f) x_0 \mathrm{d}\tau \\ &= e^{At_f} x_0 - e^{At_f} \int_0^{t_f} e^{-A\tau} B B^{\mathrm{T}} e^{-A^{\mathrm{T}}\tau} \mathrm{d}\tau \ W_c^{-1}(0, t_f) x_0 \\ &= e^{At_f} x_0 - e^{At_f} x_0 = 0 \end{aligned}$$

which indicates that there exists a finite $t_f > 0$ and a control input $u(t)$ such that the state x_0 transfers to $x(t_f) = 0$ at t_f. According to Definition 3.2, (A, B) is completely controllable.

Next, prove the necessity. Use reduction to absurdity. Suppose that that (A, B) is completely controllable, $W_c(0, t_f)$ is singular. In this case, there must be a nonzero vector $\alpha \in \mathbb{R}^n$ such that

$$\alpha^{\mathrm{T}} W_c(0, t_f) \alpha = 0$$

Along with (3.7), it holds that

$$0 = \int_0^{t_f} \alpha^{\mathrm{T}} e^{-A\tau} B B^{\mathrm{T}} e^{-A^{\mathrm{T}}\tau} \alpha \mathrm{d}\tau = \int_0^{t_f} \alpha^{\mathrm{T}} e^{-A\tau} B \left(\alpha^{\mathrm{T}} e^{-A\tau} B\right)^{\mathrm{T}} \mathrm{d}\tau$$

It holds only when

$$\alpha^{\mathrm{T}} e^{-At} B = 0, \ \forall t \in [0, t_f] \tag{3.8}$$

Let $x(t_0) = \alpha$ be an initial state of (A, B). Since (A, B) is completely controllable, there must exist $t_f > 0$ and an input $u(t)$ such that $x(t_f) = 0$. That is to say,

$$0 = x(t_f) = e^{At_f} \alpha + \int_0^{t_f} e^{A(t_f - \tau)} B u(\tau) \mathrm{d}\tau$$

which indicates that

$$\alpha = -\int_0^{t_f} e^{-A\tau} B u(\tau) \mathrm{d}\tau$$

Furthermore,

$$\alpha^T \alpha = -\int_0^{t_f} \alpha^T e^{-A\tau} B u(\tau) d\tau$$

Combing with (3.8), one can deduce that $\alpha^T \alpha = 0$, and then $\alpha = 0$, which contradicts with the assumption. Therefore, if (A, B) is completely controllable, then $W_c(0, t_f)$ is nonsingular. □

As a matter of fact, the controllability Gramian of the linear time-invariant system (3.7) is a special form of the one of the linear time-varying system (3.6). Therefore, this criterion can be deduced directly according to Theorem 3.1.

Theorem 3.5 (Rank criterion) *A linear time-invariant system (A, B) is completely controllable if and only if* $\operatorname{rank} U = n$ *with*

$$U \triangleq \begin{bmatrix} B & AB & A^2 B & \cdots & A^{n-1} B \end{bmatrix} \tag{3.9}$$

U is called the *controllability matrix* of the linear time-invariant system.

Proof Firstly, prove the sufficiency. Use reduction to absurdity. Suppose that (A, B) is incompletely controllable, $\operatorname{rank} U = n$. In this situation, $W_c(0, t_f)$ is singular for $\forall t_f > 0$ according to Theorem 3.4. That is to say, there exists a nonzero vector $\alpha \in \mathbb{R}^n$ such that

$$\alpha^T W_c(0, t_f) \alpha = \int_0^{t_f} \alpha^T e^{-A\tau} B \left(\alpha^T e^{-A\tau} B \right)^T d\tau = 0$$

which indicates that

$$\alpha^T e^{-At} B = 0, \ \forall t \in [0, t_f]$$

Taking derivatives with respect to t, and then setting $t = 0$, one can deduce that

$$\alpha^T B = 0, \ \alpha^T AB = 0, \ \alpha^T A^2 B = 0, \ \cdots, \ \alpha^T A^{n-1} B = 0$$

which means that

$$\alpha^T \begin{bmatrix} B & AB & A^2 B & \cdots & A^{n-1} B \end{bmatrix} = \alpha^T U = 0$$

Since $\alpha \neq 0$, then rows of U are linearly dependent, i.e., $\operatorname{rank} U < n$, which contradicts with $\operatorname{rank} U = n$. Consequently, the sufficiency is proved.

Next, prove the necessity. Still use reduction to absurdity. Suppose that $\operatorname{rank} U < n$, then there exists a nonzero vector $\alpha \in \mathbb{R}^n$ such that

$$\alpha^T U = \alpha^T \begin{bmatrix} B & AB & A^2 B & \cdots & A^{n-1} B \end{bmatrix} = 0$$

One can further deduce that

$$\alpha^T A^i B = 0, \ i = 0, 1, 2, \cdots, n-1$$

According to Cayley-Hamilton theorem, A^k, $k = n, n+1, \cdots$, can be expressed as a linear combination of $I, A, \cdots, A^{n-1}$. Furthermore, one can deduce that

$$\alpha^T A^i B = 0, \ i = 0, 1, 2, \cdots$$

3.1 Controllability of Linear Systems

According to the definition of the matrix exponential,

$$\alpha^{\mathrm{T}} e^{-At} B = \alpha^{\mathrm{T}} \left(I - At + \frac{1}{2!} A^2 t^2 - \cdots \right) B = 0, \ \forall t \in [0, t_f]$$

On this foundation, one can deduce that

$$\alpha^{\mathrm{T}} W_c(0, t_f) \alpha = \int_0^{t_f} \alpha^{\mathrm{T}} e^{-A\tau} B B^{\mathrm{T}} e^{-A^{\mathrm{T}} \tau} \alpha \mathrm{d}\tau = 0$$

Since $\alpha \neq 0$, $W_c(0, t_f)$ is singular. According to Theorem 3.4, (A, B) is incompletely controllable. There exists a contradiction. Therefore, the necessity is proved. □

Example 3.5 The state equation of the given system is

$$\dot{x}(t) = \begin{bmatrix} 1 & 2 & 3 \\ 0 & 4 & 0 \\ 5 & 0 & 6 \end{bmatrix} x(t) + \begin{bmatrix} 1 \\ 0 \\ 0 \end{bmatrix} u(t)$$

Since

$$\mathrm{rank} \begin{bmatrix} B & AB & A^2 B \end{bmatrix} = \mathrm{rank} \begin{bmatrix} 1 & 3 & 26 \\ 1 & 4 & 16 \\ 0 & 5 & 45 \end{bmatrix} = 3$$

the system is completely controllable.

In the next, introduce Popov-Belevitch-Hautus criteria, which are generally termed as PBH criteria. PBH criteria contain PBH eigenvector criterion and PBH rank criterion.

Theorem 3.6 (PBH eigenvector criterion) *A linear time-invariant system (A, B) is completely controllable if and only if A has no nonzero left eigenvector which is orthogonal to all columns of B. In other words, for any eigenvalue λ_i of A, $i \in \{1, \cdots, n\}$, only $\alpha = 0$ can satisfy*

$$\alpha^{\mathrm{T}} A = \lambda_i \alpha^{\mathrm{T}}, \ \alpha^{\mathrm{T}} B = 0 \tag{3.10}$$

Proof First, prove the sufficiency. There exists no $\alpha \neq 0$ such that (3.10) holds, then (A, B) is completely controllable.

Use reduction to absurdity. Suppose that (A, B) is incompletely controllable, then according to Theorem 3.5, $\mathrm{rank}\, U < n$. Therefore, there exists a nonzero $\alpha \in \mathbb{R}^n$ such that

$$\alpha^{\mathrm{T}} \begin{bmatrix} B & AB & A^2 B & \cdots & A^{n-1} B \end{bmatrix} = 0$$

which indicates that $\alpha^{\mathrm{T}} B = 0$, $\alpha^{\mathrm{T}} AB = 0$. According to these two equations, one can further deduce that there exists at least one eigenvalue λ_i of A such that $\alpha^{\mathrm{T}} A = \lambda_i \alpha^{\mathrm{T}}$. That is to say, there exists a nonzero α such that (3.10) holds, which contradicts with the assumption.

Next, prove the necessity. Still use reduction to absurdity. Suppose that there exists a nonzero $\alpha \in \mathbb{R}^n$, such that (3.10) holds, and (A, B) is controllable. One can deduce that

$$\alpha^T B = 0, \ \alpha^T AB = \lambda_i \alpha^T B = 0, \ \cdots, \ \alpha^T A^{n-1} B = \lambda_i^{n-1} \alpha^T B = 0$$

which indicates that

$$\alpha^T \begin{bmatrix} B & AB & A^2 B & \cdots & A^{n-1} B \end{bmatrix} = 0$$

In this way, rank $U < n$. According to Theorem 3.5, (A, B) is incompletely controllable, which contradicts with the assumption. $\square$

Theorem 3.7 (PBH rank criterion) *A linear time-invariant system (A, B) is completely controllable if and only if*

$$\operatorname{rank} \begin{bmatrix} \lambda_i I - A & B \end{bmatrix} = n \tag{3.11}$$

holds for all eigenvalues λ_i of A, $i \in \{1, \cdots, n\}$. Equivalently, (3.11) can be expressed as in the complex domain $\mathbb{C}$,

$$\operatorname{rank} \begin{bmatrix} sI - A & B \end{bmatrix} = n, \ \forall s \in \mathbb{C} \tag{3.12}$$

which indicates that $sI - A$ and B must be left coprime.

Proof In (3.12), if s is not the eigenvalue of A, then the characteristic polynomial $\det(sI - A) \neq 0$. Therefore, $\operatorname{rank}(sI - A) = n$. Furthermore, $\operatorname{rank}[\ sI - A \ \ B \ ] = n$. As a result, only need to prove (3.11).

Firstly, prove the sufficiency. Since (3.11) holds, rows of $[\ \lambda_i I - A \ \ B \ ]$ are linearly independent. In this situation, if $\alpha^T[\ \lambda_i I - A \ \ B \ ] = 0$, $\alpha = 0$ must hold. In other words, only $\alpha = 0$ can guarantee $\alpha^T A = \lambda_i \alpha^T$ and $\alpha^T B = 0$ simultaneously. According to Theorem 3.6, (A, B) is completely controllable.

Next, prove the necessity. Use reduction to absurdity. Suppose that there exists an eigenvalue λ_i such that $\operatorname{rank}[\ \lambda_i I - A \ \ B \ ] < n$, which indicates that rows of $[\ \lambda_i I - A \ \ B \ ]$ are linearly dependent. Therefore, there exists a nonzero $\alpha \in \mathbb{R}^n$ such that $\alpha^T[\ \lambda_i I - A \ \ B \ ] = 0$. That is to say, there exists $\alpha \neq 0$ such that $\alpha^T A = \lambda_i \alpha^T$ and $\alpha^T B = 0$. According to Theorem 3.6, (A, B) is incompletely controllable, which contradicts with the assumption. $\square$

Example 3.6 The state equation of a linear time-invariant system is

$$\dot{x}(t) = \begin{bmatrix} 0 & 1 & 0 \\ 0 & 0 & -1 \\ 0 & 0 & 0 \end{bmatrix} x(t) + \begin{bmatrix} 0 & 1 \\ 1 & 0 \\ 0 & 1 \end{bmatrix} u(t)$$

It is noted that

$$\begin{bmatrix} sI - A & B \end{bmatrix} = \begin{bmatrix} s & -1 & 0 & 0 & 1 \\ 0 & s & 1 & 1 & 0 \\ 0 & 0 & s & 0 & 1 \end{bmatrix}$$

3.1 Controllability of Linear Systems

The eigenvalues of matrix A are $\lambda_1 = \lambda_2 = \lambda_3 = 0$. Only need to check the rank of above matrix when $s = \lambda_1 = \lambda_2 = \lambda_3 = 0$.

$$\operatorname{rank}\begin{bmatrix} sI - A & B \end{bmatrix} = \operatorname{rank}\begin{bmatrix} 0 & -1 & 0 & 0 & 1 \\ 0 & 0 & 1 & 1 & 0 \\ 0 & 0 & 0 & 0 & 1 \end{bmatrix} = 3$$

which indicates that the system is completely controllable.

Finally, introduce Jordan normal form criterion.

Theorem 3.8 (**Jordan normal form criterion**) *Consider a linear time-invariant system (A, B), λ_i is the eigenvalue of A with multiplicity σ_i, $i \in \{1, \cdots, l\}$, and $\sigma_1 + \cdots + \sigma_i + \cdots + \sigma_l = n$. There exists an equivalence transformation $\bar{x}(t) = Tx(t)$ such that (A, B) transforms into the Jordan normal form $(\bar{A}, \bar{B})$, where*

$$\bar{A} = \begin{bmatrix} J_1 & & & \\ & J_2 & & \\ & & \ddots & \\ & & & J_l \end{bmatrix}, \quad \bar{B} = \begin{bmatrix} \bar{B}_1 \\ \bar{B}_2 \\ \vdots \\ \bar{B}_l \end{bmatrix}$$

$$J_i = \begin{bmatrix} J_{i1} & & & \\ & J_{i2} & & \\ & & \ddots & \\ & & & J_{i\alpha_i} \end{bmatrix}, \quad \bar{B}_i = \begin{bmatrix} \bar{B}_{i1} \\ \bar{B}_{i2} \\ \vdots \\ \bar{B}_{i\alpha_i} \end{bmatrix}$$

$$J_{ik} = \begin{bmatrix} \lambda_i & 1 & & \\ & \lambda_i & \ddots & \\ & & \ddots & 1 \\ & & & \lambda_i \end{bmatrix}, \quad \bar{B}_{ik} = \begin{bmatrix} \bar{b}_{ik1} \\ \bar{b}_{ik2} \\ \vdots \\ \bar{b}_{ik\sigma_{ik}} \end{bmatrix}$$

J_i is the $\sigma_i \times \sigma_i$ Jordan block matrix associated with λ_i, J_{ik}, $k \in \{1, \cdots, \alpha_i\}$ is the $\sigma_{ik} \times \sigma_{ik}$ Jordan block, α_i is the geometric multiplicity associated with λ_i, $\sigma_{i1} + \cdots + \sigma_{i\alpha_i} = \sigma_i$.

(A, B) is completely controllable if and only if rows of the matrix

$$\tilde{B}_i = \begin{bmatrix} \bar{b}_{i1\sigma_{i1}} \\ \bar{b}_{i2\sigma_{i2}} \\ \vdots \\ \bar{b}_{i\alpha_i\sigma_{i\alpha_i}} \end{bmatrix}$$

are linearly independent for $i \in \{1, \cdots, l\}$.

Firstly, consider that all eigenvalues of A are distinct. In this situation, $J_1 = \lambda_1$,

$J_2 = \lambda_2, \cdots, J_n = \lambda_n$, and $\bar{B}_1, \bar{B}_2, \cdots, \bar{B}_n$ have only one row. Construct

$$\begin{bmatrix} sI - \bar{A} & \bar{B} \end{bmatrix} = \begin{bmatrix} s-\lambda_1 & & & & \bar{B}_1 \\ & s-\lambda_2 & & & \bar{B}_2 \\ & & \ddots & & \vdots \\ & & & s-\lambda_n & \bar{B}_l \end{bmatrix}$$

It can be seen that if and only if $\bar{B}_i \neq 0$ holds for $i \in \{1, \cdots, n\}$, rank$[\ sI - \bar{A}\ \ \bar{B}\ ] = n$, and (A, B) is completely controllable according to Theorem 3.7.

Following the similar way, this theorem can be proved when there exist repeated eigenvalues. Here, omit it.

Example 3.7 Given the Jordan normal form of a linear time-invariant systems, judge the controllability of the system,

$$\dot{\bar{x}}(t) = \begin{bmatrix} -1 & 1 & & & & & & \\ 0 & -1 & & & & & & \\ & & -1 & & & & & \\ & & & -1 & & & & \\ & & & & 2 & 1 & & \\ & & & & 0 & 2 & & \\ & & & & & & 2 & \\ & & & & & & & 4 \end{bmatrix} \bar{x}(t) + \begin{bmatrix} 0 & 0 & 0 \\ 1 & 0 & 0 \\ 0 & 2 & 0 \\ 0 & 0 & 4 \\ 0 & 0 & 3 \\ 1 & 2 & 0 \\ 0 & 3 & 6 \\ 4 & 0 & 0 \end{bmatrix} u(t)$$

There are three eigenvalues for the state coefficient matrix of this system, $\lambda_1 = -1$, $\lambda_2 = 2$ and $\lambda_3 = 4$. The matrix formed by rows of matrix $\bar{B}_1$ corresponding to the last rows of three Jordan blocks associated with $\lambda_1 = -1$ is

$$\begin{bmatrix} \bar{b}_{112} \\ \bar{b}_{121} \\ \bar{b}_{131} \end{bmatrix} = \begin{bmatrix} 1 & 0 & 0 \\ 0 & 2 & 0 \\ 0 & 0 & 4 \end{bmatrix}$$

whose rows are linearly independent.

The matrix formed by rows of matrix $\bar{B}_2$ corresponding to the last rows of two Jordan blocks corresponding to $\lambda_2 = 2$ is

$$\begin{bmatrix} \bar{b}_{212} \\ \bar{b}_{221} \end{bmatrix} = \begin{bmatrix} 1 & 2 & 0 \\ 0 & 3 & 6 \end{bmatrix}$$

whose rows are linearly independent.

The row of matrix $\bar{B}_3$ corresponding to $\lambda_3 = 4$ is $[\ 4\ \ 0\ \ 0\ ]$, and its elements are not all zeros.

Therefore, the system is completely controllable.

3.1 Controllability of Linear Systems

When applying Jordan normal form criterion, if the state equation of the system does not match Jordan normal form, it can be transformed into Jordan normal form by an equivalent transformation, which does not change the inherent properties of the system.

When considering the reachability of a linear time-invariant system (A, B), introduce the reachability Gramian

$$W_{\mathrm{r}}(0, t_{\mathrm{f}}) \triangleq \int_0^{t_{\mathrm{f}}} \mathrm{e}^{(t_{\mathrm{f}}-\tau)A} B B^{\mathrm{T}} \mathrm{e}^{(t_{\mathrm{f}}-\tau)A^{\mathrm{T}}} \mathrm{d}\tau \tag{3.13}$$

(A, B) is completely reachable if and only if there exists $t_{\mathrm{f}} > 0$ such that $W_{\mathrm{r}}(0, t_{\mathrm{f}})$ is nonsingular. Referring to Theorem 3.4, this conclusion can be deduced. Meanwhile, combining (3.7) with (3.13), one can deduce that

$$W_{\mathrm{r}}(0, t_{\mathrm{f}}) = \mathrm{e}^{At_{\mathrm{f}}} W_{\mathrm{c}}(0, t_{\mathrm{f}}) \left(\mathrm{e}^{At_{\mathrm{f}}}\right)^{\mathrm{T}}$$

Since $\mathrm{e}^{At_{\mathrm{f}}}$ is nonsingular, it holds that $\operatorname{rank} W_{\mathrm{r}}(0, t_{\mathrm{f}}) = \operatorname{rank} W_{\mathrm{c}}(0, t_{\mathrm{f}})$. That is to say, (A, B) is completely reachable if and only if it is completely controllable.

3.1.4 Controllability Index of Linear Time-Invariant Systems

For a time-invariant system (A, B), define

$$U_\mu = \begin{bmatrix} B & AB & \cdots & A^{\mu-1}B \end{bmatrix} \tag{3.14}$$

where U_μ is an $n \times \mu p$ constant matrix, μ is a positive integer. When $\mu = n$, U_μ is the controllability matrix of the system and $\operatorname{rank} U_\mu = n$ if the system is completely controllable. It is noted that if $A^k b_i$ is linearly dependent on its left columns, $k = 1, \cdots, \mu - 1$, $i = 1, \cdots, p$, then all columns $A^l b_i$ with $l > k$ is linearly dependent on its all left columns, where b_i denotes the ith column of B. Owing to this fact, (3.14) indicates that when a subblock $A^k B$ is added into U_μ, the rank of U_μ increases at least 1 until $\operatorname{rank} U_\mu = n$.

Definition 3.5 (Controllability index of a linear time-invariant system) μ in (3.14) is called the *controllability index* of (A, B) when μ is the smallest positive integer which makes

$$\operatorname{rank} U_\mu = \operatorname{rank} U_{\mu+1} = n \tag{3.15}$$

Theorem 3.9 *The controllability index μ satisfies*

$$\frac{n}{p} \leqslant \mu \leqslant \min\{\bar{n}, n - r + 1\} \tag{3.16}$$

where r denotes the rank *of B, $\bar{n}$ denotes the order of the minimal polynomial of A.*

Proof Let $\phi(s) = s^{\bar{n}} + a_1 s^{\bar{n}-1} + \cdots + a_{\bar{n}-1} s + a_{\bar{n}}$ be the minimal polynomial of A, which satisfies $\phi(A) = 0$, i.e.,

$$A^{\bar{n}} = -a_1 A^{\bar{n}-1} - \cdots - a_{\bar{n}-1} A - a_{\bar{n}} I$$

Post-multiplying both sides of the above equation by B, one can deduce that

$$A^{\bar{n}}B = -a_1 A^{\bar{n}-1}B - \cdots - a_{\bar{n}-1}AB - a_{\bar{n}}B$$

which indicates that $A^{\bar{n}}B$ is linearly dependent on $B, AB, \cdots, A^{\bar{n}-1}B$. Furthermore, one can conclude that in (3.14), each column of $A^{\bar{n}}B$ is linearly dependent on its left columns. Therefore, it holds that $\mu \leqslant \bar{n}$.

On the other hand, the rank of U_μ increases at least 1 as μ increases 1 until rank $U_\mu = n$. In this case, $\mu - 1 \leqslant n - r$, i.e., $\mu \leqslant n - r + 1$.

There are μp columns in U_μ. In order to guarantee rank $U_\mu = n$, μp should be no less than n, i.e., $n \leqslant \mu p$.

To sum up, (3.16) holds. □

According to Theorem 3.9, for a single input system, $\mu = n$. Meanwhile, Theorem 3.5 can be revised as follows.

Corollary 3.1 *A linear time-invariant system (A, B) is completely controllable if and only if*

$$\operatorname{rank} U_{n-r+1} = \operatorname{rank} \begin{bmatrix} B & AB & A^2 B & \cdots & A^{n-r}B \end{bmatrix} = n \tag{3.17}$$

or

$$\operatorname{rank} U_{\bar{n}} = \operatorname{rank} \begin{bmatrix} B & AB & A^2 B & \cdots & A^{\bar{n}-1}B \end{bmatrix} = n \tag{3.18}$$

It is complicated to determine $\bar{n}$. Therefore, compared with (3.18), (3.17) is more frequently used.

Let $b_1, \cdots, b_p$ be columns of B. U_μ can be expressed as

$$U_\mu = \begin{bmatrix} b_1 & \cdots & b_p & Ab_1 & \cdots & Ab_p & \cdots & A^{\mu-1}b_1 & \cdots & A^{\mu-1}b_p \end{bmatrix} \tag{3.19}$$

Select n linearly independent column vectors in (3.19) from left to right. If a column cannot be expressed by the linear combination of the selected columns on the left, these columns are linearly independent and the column is one of the selected n linearly independent vectors. Otherwise, they are linearly dependent and another column will be selected. Suppose the rank of B is r, the n selected independent column vectors can be rearranged as

$$\begin{bmatrix} b_1 & Ab_1 & \cdots & A^{\mu_1-1}b_1 & b_2 & Ab_2 & \cdots & A^{\mu_2-1}b_2 & \cdots & b_r & Ab_r & \cdots & A^{\mu_r-1}b_r \end{bmatrix} \tag{3.20}$$

For a completely controllable system, one can deduce that

$$\mu_1 + \mu_2 + \cdots + \mu_r = n$$

The controllability index satisfies

$$\mu = \max\{\mu_1, \mu_2, \cdots, \mu_r\}$$

Generally, $\{\mu_1, \mu_2, \cdots, \mu_r\}$ is called the *controllability index set* or the *Kronecker invariable* of (A, B).

3.2 Observability of Linear Systems

Theorem 3.10 *Under equivalent transformations, μ and $\{\mu_1, \mu_2, \cdots, \mu_r\}$ of (A, B) keep invariant.*

Proof Let T is a nonsingular matrix and $\bar{A} = TAT^{-1}$, $\bar{B} = TB$, and then one can deduce that

$$\bar{U}_\mu = \begin{bmatrix} \bar{B} & \bar{A}\bar{B} & \cdots & \bar{A}^{\mu-1}\bar{B} \end{bmatrix} = \begin{bmatrix} TB & TAB & \cdots & TA^{\mu-1}B \end{bmatrix}.$$

Since $\operatorname{rank} T = n$, it holds that $\operatorname{rank} \bar{B} = \operatorname{rank} B, \cdots, \operatorname{rank} \bar{A}^{\mu-1}\bar{B} = \operatorname{rank} A^{\mu-1}B$. Therefore, μ and $\{\mu_1, \mu_2, \cdots, \mu_r\}$ are invariant. $\square$

Example 3.8 The state equation of a given linear time-invariant system is

$$\dot{x}(t) = \begin{bmatrix} 0 & 1 & 0 & 0 \\ 3 & 0 & 0 & 0 \\ 0 & 1 & 1 & 0 \\ 0 & 0 & 0 & 0 \end{bmatrix} x(t) + \begin{bmatrix} 0 & 0 \\ 1 & 0 \\ 0 & 0 \\ 0 & 1 \end{bmatrix} u(t)$$

By computing U_μ which satisfies $\operatorname{rank} U_\mu = \operatorname{rank} U_{\mu+1} = n$, one can obtain that

$$\operatorname{rank} \begin{bmatrix} B & AB & A^2B \end{bmatrix} = \operatorname{rank} \begin{bmatrix} 0 & 0 & 1 & 0 & 0 & 0 \\ 1 & 0 & 0 & 0 & 3 & 0 \\ 0 & 0 & 1 & 0 & 1 & 0 \\ 0 & 1 & 0 & 0 & 0 & 0 \end{bmatrix} = 4 = n$$

Therefore, the controllability index μ of this system is 3.

In above U_μ, select 4 linearly independent column vectors and rearrange them according to (3.20), one can obtain that

$$\begin{bmatrix} b_1 & Ab_1 & A^2b_1 & b_2 \end{bmatrix} = \begin{bmatrix} 0 & 1 & 0 & 0 \\ 1 & 0 & 3 & 0 \\ 0 & 1 & 1 & 0 \\ 0 & 0 & 0 & 1 \end{bmatrix}$$

Apparently, the controllability index set of this system is $\{\mu_1, \mu_2\} = \{3, 1\}$ and $\mu = \max\{\mu_1, \mu_2\} = \mu_1 = 3$.

3.2 Observability of Linear Systems

视频

3.2.1 Basic Concepts

In practice, states of a system are frequently required but not accessible. Under this circumstance, the question arises whether it is possible to determine the state by observing the output response of a system to some input over some period of time, which leads to the concept of observability of a system. Observability means that any movement of state variables of a system can be completely mapped by the output, i.e., the output of a system contains all the information about the internal state movements.

Example 3.9 Consider the system in Example 3.1, its output equation is

$$y(t) = \begin{bmatrix} 0 & -2 \end{bmatrix} \begin{bmatrix} x_1(t) \\ x_2(t) \end{bmatrix}$$

Apparently, the output $y(t)$ only contains the information of the state variable $x_2(t)$. The state variable $x_1(t)$ has no direct or indirect relation with $y(t)$, so the system is incompletely observable.

Example 3.10 Consider the system contains operational amplifiers and integrators shown in Fig.3.3. The state variables are $x_1(t)$ and $x_2(t)$ which are outputs of two integrators. The output of the system $y(t)$ is the sum of $x_1(t)$ and $x_2(t)$.

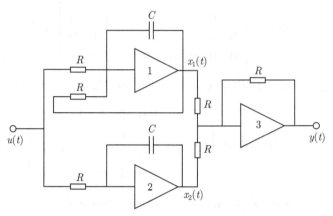

Fig.3.3 A completely observable operational circuit

Obviously, output $y(t)$ contains all the information of states $x_1(t)$ and $x_2(t)$, and one can obtain initial values $x_1(t_0)$ and $x_2(t_0)$ of $x_1(t)$ and $x_2(t)$ by measuring $y(t)$. Therefore, $x_1(t)$ and $x_2(t)$ are observable.

Example 3.11 Consider the bridge circuit in Fig.3.4. If $u(t) = 0$, then $y(t) = 0$ for all $t \geqslant t_0$ no matter what the initial voltage across the capacitor, i.e., the initial state $x(t_0)$ is. The information of $x(t)$ cannot be completely reflected into $y(t)$. It indicates that the state of this system is completely unobservable.

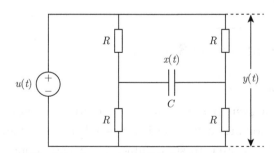

Fig.3.4 An unobservable bridge system

The above three examples are only the direct description of observability. In the next, the strict definition of observability is presented.

3.2 Observability of Linear Systems

The observability denotes the characteristic of states can be completely reflected by outputs. For discussing the observability of a system, we need to consider both the state equation and the output equation of a system,

$$\dot{x}(t) = A(t)x(t) + B(t)u(t), \ y(t) = C(t)x(t) + D(t)u(t), \ x(t_0) = x_0, \ t \in [t_0, t_f] \quad (3.21)$$

where $x(t) \in \mathbb{R}^n$ is the state vector, $u(t) \in \mathbb{R}^p$ is the input vector, $y(t) \in \mathbb{R}^q$ is the output vector. $A(t), B(t), C(t)$ and $D(t)$ are time-varying matrices and satisfy conditions such that there exists the unique solution.

As discussed in Subsection 2.2.4, solutions of (3.21) can be expressed as

$$x(t) = \Phi(t, t_0)x_0 + \int_{t_0}^{t} \Phi(t, \tau) B(\tau)u(\tau)d\tau$$

$$y(t) = C(t)\Phi(t, t_0)x_0 + \int_{t_0}^{t} C(t)\Phi(t, \tau) B(\tau)u(\tau)d\tau + D(t)u(t)$$

When considering the observability, $y(t)$ and $u(t)$ are supposed to be known and only the initial state x_0 is unknown. Denote

$$\bar{y}(t) \triangleq y(t) - \left[\int_{t_0}^{t} C(t)\Phi(t, \tau)B(\tau)u(\tau)d\tau + D(t)u(t) \right]$$

and then, it holds that

$$\bar{y}(t) = C(t)\Phi(t, t_0)x_0$$

which indicates that the observability is to study whether x_0 can be completely estimated by $\bar{y}(t)$. Because of the randomicity of $\bar{y}(t)$ and x_0, the study of the observability equals to the study of estimating x_0 by $y(t)$ when $u(t) = 0$, i.e., to study the observability of the zero-input dynamical system

$$\dot{x}(t) = A(t)x(t), \ y(t) = C(t)x(t), \ x(t_0) = x_0, \ t \in [t_0, t_f] \quad (3.22)$$

For convenience, denote the dynamical equation (3.22) as a pair $(A(t), C(t))$.

For the linear time-invariant case, we focus on the following system,

$$\dot{x}(t) = Ax(t), \ y(t) = Cx(t), \ x(t_0) = x_0, \ t \geqslant t_0 \quad (3.23)$$

and simply denote (3.23) as (A, C).

The definition of the observability of a system is introduced.

Definition 3.6 (Unobservability of a state) A state x_0 is *unobservable* at time t_0 if there exists finite $t_f > t_0$ such that the zero-input response $y(t) = 0$ for all $t \in [t_0, t_f]$.

Definition 3.7 (Observability of a system) A system $(A(t), C(t))$ (or (A, C)) is (*completely state*) *observable* at t_0 if the only state $x_0 \in \mathbb{R}^n$ that is unobservable at t_0 is the zero state, i.e., $x_0 = 0$. A system $(A(t), C(t))$ (or (A, C)) is *incompletely* (*state*) *observable* at t_0 if there exists one or more nonzero states $x_0 \in \mathbb{R}^n$ that are unobservable.

Observability is the structure property and is only determined by the structure and parameters of a system. It is unrelated to values of inputs and outputs of a system.

Observability is defined with respect to given time t_0, which is necessary for time-varying systems. If a linear time-varying system is always observable for any initial time in $[t_0, t_f]$, the system is said to be *uniformly (completely state) observable*. For time-invariant systems, observability is unrelated to the selection of t_0, i.e., if (A, C) is observable, then it is uniformly observable.

In practice, the case that a system is incomplete observable is a kind of singular instance, since a little change of the parameter values of a system can make the system completely observable.

Observability utilizes further output measurements to determine the present state. If past output measurements are used to accomplish the present state, constructibility is introduced.

Definition 3.8 (Unconstructibility of a state) A state $x(t_f)$ is *unconstructible* at time t_f if there exists finite $t_0 < t_f$ such that the zero-input response $y(t) = 0$ for all $t \in [t_0, t_f]$.

Definition 3.9 (Constructibility of a system) A system $(A(t), C(t))$ (or (A, C)) is *(completely state) constructible* at t_f if the only state $x(t_f) \in \mathbb{R}^n$ that is unconstructible at t_f is the zero state, i.e., $x(t_f) = 0$. A system $(A(t), C(t))$ (or (A, C)) is *incompletely (state) constructible* at t_f if there exists one or more nonzero states $x(t_f) \in \mathbb{R}^n$ that are unconstructible.

3.2.2 Observability Criteria of Linear Time-Varying Systems

Theorem 3.11 (Gramian criterion) *A linear time-varying system $(A(t), C(t))$ is completely observable at t_0 if and only if there exists $t_f > t_0$ such that*

$$W_o(t_0, t_f) \triangleq \int_{t_0}^{t_f} \Phi^T(t, t_0) C^T(t) C(t) \Phi(t, t_0) dt \tag{3.24}$$

is nonsingular, where $\Phi(t, t_0)$ is the state transition matrix of $(A(t), C(t))$.

$W_o(t_0, t_f)$ is called the *observability Gramian* of the linear time-varying system, which is an $n \times n$ matrix.

Proof Firstly, prove the sufficiency. It is noted that $y(t) = C(t)\Phi(t, t_0)x_0$ holds for $t \in [t_0, t_f]$. Multiplying its both sides with $\Phi^T(t, t_0)C^T(t)$, and integrating them from t_0 to t_f, one can deduce that

$$\int_{t_0}^{t_f} \Phi^T(t, t_0) C^T(t) y(t) dt = W_o(t_0, t_f) x_0$$

Since $W_o(t_0, t_f)$ is nonsingular, $W_o^{-1}(t_0, t_f)$ is existing. One can further deduce that

$$x_0 = W_o^{-1}(t_0, t_f) \int_{t_0}^{t_f} \Phi^T(t, t_0) C^T(t) y(t) dt$$

3.2 Observability of Linear Systems

which indicates that nonzero x_0 can be uniquely determined by $y(t)$ for $t \in [t_0, t_f]$. Thus, $(A(t), C(t))$ is completely observable at t_0.

Next, prove the necessity. Use reduction to absurdity. Suppose that $(A(t), C(t))$ is completely observable at t_0, $W_o(t_0, t_f)$ is singular. In this case, there must be a nonzero vector $\alpha \in \mathbb{R}^n$ such that

$$\alpha^T W_o(t_0, t_f) \alpha = 0$$

Along with (3.24), it holds that

$$0 = \int_{t_0}^{t_f} \alpha^T \Phi^T(t, t_0) C^T(t) C(t) \Phi(t, t_0) \alpha \, dt$$

$$= \int_{t_0}^{t_f} (C(t) \Phi(t, t_0) \alpha)^T C(t) \Phi(t, t_0) \alpha \, dt$$

It holds only when

$$C(t) \Phi(t, t_0) \alpha = 0, \ \forall t \in [t_0, t_f]$$

Let $x(t_0) = \alpha$ be an initial state of $(A(t), C(t))$, then the corresponding output $y(t) = C(t)\Phi(t, t_0)\alpha = 0$, which indicates that the nonzero state $x(t_0) = \alpha$ is unobservable at t_0. The necessity is proved. □

It can be verified that if $(A(t), C(t))$ is completely observable at t_0, then $W_o(t_0, t_f)$ has the following properties.

(1) $W_o(t_0, t_f)$ is a positive definite constant matrix.
(2) $W_o(t_0, t_f) = W_o(t_0, t) + \Phi(t_0, t) W_o(t, t_f) \Phi^T(t_0, t)$.
(3) $W_o(t_f, t_f) = 0$.
(4) $\dfrac{\partial}{\partial t} W_o(t, t_f) = A(t) W_o(t, t_f) + W_o(t, t_f) A^T(t) - C^T(t) C(t)$.

Theorem 3.12 *A linear time-varying system $(A(t), C(t))$ is completely observable at t_0 if and only if there exists $t_f > t_0$ such that columns of $C(t)\Phi(t, t_0)$ are linearly independent for $t \in [t_0, t_f]$.*

Theorem 3.13(Rank criterion) *Suppose that $A(t)$ is $n-2$ continuously differentiable with respect to t, and $C(t)$ is $n-1$ continuously differentiable with respect to t. A linear time-varying system $(A(t), C(t))$ is completely observable at t_0 if there exists $t_f > t_0$ such that*

$$\mathrm{rank} \begin{bmatrix} N_0(t_f) \\ N_1(t_f) \\ \vdots \\ N_{n-1}(t_f) \end{bmatrix} = n \tag{3.25}$$

where

$$N_0(t) = C(t)$$

$$N_1(t) = N_0(t)A(t) + \frac{\mathrm{d}}{\mathrm{d}t}N_0(t)$$

$$\vdots$$

$$N_{n-1}(t) = N_{n-2}(t)A(t) + \frac{\mathrm{d}}{\mathrm{d}t}N_{n-2}(t)$$

The proof of this criterion is similar to the proof of Theorem 3.3.

Example 3.12 Consider the system

$$\dot{x} = \begin{bmatrix} 0 & \mathrm{e}^{-t} \\ 6 & 0 \end{bmatrix} x(t), \ y(t) = \begin{bmatrix} 1 & 2 \end{bmatrix} x(t)$$

One can compute that

$$N_0(t) = \begin{bmatrix} 1 & 2 \end{bmatrix}, \ N_1(t) = N_0(t)A(t) + \frac{\mathrm{d}}{\mathrm{d}t}N_0(t) = \begin{bmatrix} 12 & \mathrm{e}^{-t} \end{bmatrix}$$

and then,

$$\begin{bmatrix} N_0(t) \\ N_1(t) \end{bmatrix} = \begin{bmatrix} 1 & 2 \\ 12 & \mathrm{e}^{-t} \end{bmatrix}$$

which is full rank for any t. Therefore, the system is completely observable at any time.

When considering the constructibility of linear time-varying system $(A(t), C(t))$, introduce the constructibility Gramian

$$W_{\mathrm{cn}}(t_0, t_\mathrm{f}) \triangleq \int_{t_0}^{t_\mathrm{f}} \Phi^{\mathrm{T}}(\tau, t_\mathrm{f}) C^{\mathrm{T}}(\tau) C(\tau) \Phi(\tau, t_\mathrm{f}) \mathrm{d}\tau \tag{3.26}$$

$(A(t), C(t))$ is completely constructible at t_f if and only if there exists $t_0 < t_\mathrm{f}$ such that $W_{\mathrm{cn}}(t_0, t_\mathrm{f})$ is nonsingular. Meanwhile, combining (3.7) with (3.26), one can deduce that

$$W_{\mathrm{o}}(t_0, t_\mathrm{f}) = \Phi^{\mathrm{T}}(t_\mathrm{f}, t_0) W_{\mathrm{cn}}(t_0, t_\mathrm{f}) \Phi(t_\mathrm{f}, t_0)$$

which indicates that rank $W_{\mathrm{o}}(t_0, t_\mathrm{f}) = $ rank $W_{\mathrm{cn}}(t_0, t_\mathrm{f})$.

3.2.3 Observability Criteria of Linear Time-Invariant Systems

Without loss of generality, consider that $t_0 = 0$ in this subsection.

Consider a linear time-invariant system (A, C). In the next, some frequently used observability criteria are introduced.

Theorem 3.14 (Gramian criterion) *A linear time-invariant system (A, C) is completely observable if and only if there exists $t_\mathrm{f} > 0$ such that*

$$W_{\mathrm{o}}(0, t_\mathrm{f}) \triangleq \int_0^{t_\mathrm{f}} \mathrm{e}^{A^{\mathrm{T}}\tau} C^{\mathrm{T}} C \mathrm{e}^{A\tau} \mathrm{d}\tau \tag{3.27}$$

is nonsingular.

$W_\text{o}(0, t_\text{f})$ is called the *observability Gramian* of the linear time-invariant system, which is an $n \times n$ matrix.

The proof is omitted since this theorem can be derived by Theorem 3.11 directly.

The following observability criteria are corresponding to controllability criteria, and their derivations follow the similar vein.

Theorem 3.15 (Rank criterion) *A linear time-invariant system (A, C) is completely observable if and only if* rank $V = n$ *with*

$$V \triangleq \begin{bmatrix} C \\ CA \\ \vdots \\ CA^{n-1} \end{bmatrix} \tag{3.28}$$

V is called the observability matrix of the linear time-invariant system.

Example 3.13 Consider a system given by

$$\dot{x} = \begin{bmatrix} 1 & 0 \\ 0 & -3 \end{bmatrix} x(t) + \begin{bmatrix} 5 \\ 1 \end{bmatrix} u(t), \ y(t) = \begin{bmatrix} 0 & -2 \end{bmatrix} x(t)$$

Since

$$\text{rank} \begin{bmatrix} C \\ CA \end{bmatrix} = \text{rank} \begin{bmatrix} 0 & -2 \\ 0 & 6 \end{bmatrix} < 2$$

the system is incompletely observable.

Theorem 3.16 (PBH eigenvector criterion) *A linear time-invariant system (A, C) is completely observable if and only if A has no nonzero right eigenvector which is orthogonal to all rows of C. In other words, for any eigenvalue λ_i of A, $i \in \{1, \cdots, n\}$, only $\alpha = 0$ can satisfy*

$$A\alpha = \lambda_i \alpha, \ C\alpha = 0 \tag{3.29}$$

Theorem 3.17 (PBH rank criterion) *A linear time-invariant system (A, C) is completely observable if and only if*

$$\text{rank} \begin{bmatrix} C \\ \lambda_i I - A \end{bmatrix} = n \tag{3.30}$$

holds for all eigenvalues λ_i of A, $i \in \{1, \cdots, n\}$. Equivalently, (3.30) can be expressed as in the complex domain $\mathbb{C}$,

$$\text{rank} \begin{bmatrix} C \\ sI - A \end{bmatrix} = n, \ \forall s \in \mathbb{C} \tag{3.31}$$

which indicates that $sI - A$ and C must be right coprime.

Theorem 3.18 (Jordan normal form criterion) Consider a linear time-invariant system (A,C), λ_i is the eigenvalue of A with multiplicity σ_i, $i \in \{1,\cdots,l\}$, and $\sigma_1 + \cdots + \sigma_i + \cdots + \sigma_l = n$. There exists an equivalence transformation $\bar{x}(t) = Tx(t)$ such that (A,C) transforms into the Jordan normal form $(\bar{A},\bar{C})$, where

$$\bar{A} = \begin{bmatrix} J_1 & & & \\ & J_2 & & \\ & & \ddots & \\ & & & J_l \end{bmatrix}, \quad \bar{C} = \begin{bmatrix} \bar{C}_1 & \bar{C}_2 & \cdots & \bar{C}_l \end{bmatrix}$$

$$J_i = \begin{bmatrix} J_{i1} & & & \\ & J_{i2} & & \\ & & \ddots & \\ & & & J_{i\alpha_i} \end{bmatrix}, \quad \bar{C}_i = \begin{bmatrix} \bar{C}_{i1} & \bar{C}_{i2} & \cdots & \bar{C}_{i\alpha_i} \end{bmatrix}$$

$$J_{ik} = \begin{bmatrix} \lambda_i & 1 & & \\ & \lambda_i & \ddots & \\ & & \ddots & 1 \\ & & & \lambda_i \end{bmatrix}, \quad \bar{C}_{ik} = \begin{bmatrix} \bar{C}_{ik1} & \bar{C}_{ik2} & \cdots & \bar{C}_{ik\sigma_{ik}} \end{bmatrix}$$

J_i is the $\sigma_i \times \sigma_i$ Jordan block matrix associated with λ_i, J_{ik}, $k \in \{1,\cdots,\alpha_i\}$, is the $\sigma_{ik} \times \sigma_{ik}$ Jordan block, α_i is the geometric multiplicity associated with λ_i, $\sigma_{i1} + \cdots + \sigma_{i\alpha_i} = \sigma_i$.

(A,C) is completely observable if and only if columns of the matrix

$$\tilde{C}_i = \begin{bmatrix} \bar{c}_{i11} & \bar{c}_{i21} & \cdots & \bar{c}_{i\alpha_i 1} \end{bmatrix}$$

is linearly independent for $i \in \{1,\cdots,l\}$.

Example 3.14 Given the Jordan normal form of a system, judge the observability of the system.

$$\dot{\bar{x}}(t) = \begin{bmatrix} -1 & 1 & & & & & & \\ 0 & -1 & & & & & & \\ & & -1 & & & & & \\ & & & -1 & & & & \\ & & & & 2 & 1 & & \\ & & & & 0 & 2 & & \\ & & & & & & 2 & \\ & & & & & & & 4 \end{bmatrix} \bar{x}(t)$$

$$y(t) = \begin{bmatrix} 2 & 0 & 0 & 0 & 1 & 0 & 0 & 0 \\ 0 & 0 & 1 & 0 & 2 & 4 & 0 & 6 \\ 0 & 0 & 0 & 3 & 3 & 0 & 1 & 0 \end{bmatrix} \bar{x}(t)$$

3.2 Observability of Linear Systems

Following the Jordan normal form criterion, list the columns of matrix $\bar{C}$ corresponding to the first columns of Jordan blocks associated with $\lambda_1 = -1$, $\lambda_2 = 2$ and $\lambda_3 = 4$ which form three matrices,

$$\begin{bmatrix} \bar{c}_{111} & \bar{c}_{121} & \bar{c}_{131} \end{bmatrix} = \begin{bmatrix} 2 & 0 & 0 \\ 0 & 1 & 0 \\ 0 & 0 & 3 \end{bmatrix}, \begin{bmatrix} \bar{c}_{211} & \bar{c}_{221} \end{bmatrix} = \begin{bmatrix} 1 & 0 \\ 2 & 0 \\ 3 & 1 \end{bmatrix}, \bar{c}_{311} = \begin{bmatrix} 0 \\ 6 \\ 0 \end{bmatrix}$$

Columns are linearly independent for all matrices. Therefore, the system is completely observable.

When considering the constructibility of linear time-invariant system (A, C), introduce the constructibility Gramian

$$W_{\text{cn}}(0, t_{\text{f}}) \triangleq \int_0^{t_{\text{f}}} e^{A^{\text{T}}(\tau - t_{\text{f}})} C^{\text{T}} C e^{A(\tau - t_{\text{f}})} d\tau \tag{3.32}$$

(A, C) is completely constructible if and only if there exists $t_{\text{f}} > 0$ such that $W_{\text{cn}}(0, t_{\text{f}})$ is nonsingular. Meanwhile, combining (3.27) with (3.32), one can deduce that

$$W_{\text{o}}(0, t_{\text{f}}) = \left(e^{At_{\text{f}}}\right)^{\text{T}} W_{\text{cn}}(0, t_{\text{f}}) e^{At_{\text{f}}}$$

which indicates that $\operatorname{rank} W_{\text{o}}(0, t_{\text{f}}) = \operatorname{rank} W_{\text{cn}}(0, t_{\text{f}})$. That is to say, (A, C) is completely constructible if and only if it is completely observable.

3.2.4 Observability Index of Linear Time-Invariant Systems

For a time-invariant system (A, C), define

$$V_\nu = \begin{bmatrix} C^{\text{T}} & A^{\text{T}} C^{\text{T}} & \cdots & \left(A^{\nu-1}\right)^{\text{T}} C^{\text{T}} \end{bmatrix}^{\text{T}} \tag{3.33}$$

where V_ν is a $\nu q \times n$ constant matrix, ν is a positive integer. When $\nu = n$, V_ν is the observability matrix of the system and $\operatorname{rank} V_\nu = n$ if the system is completely observable. It is noted that when a subblock CA^k is added into V_ν, $k = 1, \cdots, \nu - 1$, the rank of V_ν increases at least 1 until $\operatorname{rank} V_\nu = n$.

Definition 3.10 (Observability index of a linear time-invariant system) ν in (3.33) is called the *observability index* of (A, C) when ν is the smallest positive integer which makes

$$\operatorname{rank} V_\nu = \operatorname{rank} V_{\nu+1} = n \tag{3.34}$$

Theorem 3.19 *The observability index ν satisfies*

$$\frac{n}{q} \leqslant \nu \leqslant \max\{\bar{n}, n - m + 1\} \tag{3.35}$$

where m denotes the rank of C, $\bar{n}$ denotes the order of the minimal polynomial of A.

The proof is omitted since it is similar to the proof of Theorem 3.9.

According to Theorem 3.19, for a SISO system, $\nu = n$. Meanwhile, Theorem 3.15 can be revised as follows.

Corollary 3.2 *A linear time-invariant system (A, C) is completely observable if and only if*

$$\operatorname{rank} V_{n-m+1} = \operatorname{rank} \begin{bmatrix} C^{\mathrm{T}} & A^{\mathrm{T}}C^{\mathrm{T}} & \cdots & (A^{n-m})^{\mathrm{T}} C^{\mathrm{T}} \end{bmatrix}^{\mathrm{T}} = n \qquad (3.36)$$

or

$$\operatorname{rank} V_{\bar{n}} = \operatorname{rank} \begin{bmatrix} C^{\mathrm{T}} & A^{\mathrm{T}}C^{\mathrm{T}} & \cdots & (A^{\bar{n}-1})^{\mathrm{T}} C^{\mathrm{T}} \end{bmatrix}^{\mathrm{T}} = n \qquad (3.37)$$

Let $c_1, \cdots, c_q$ be rows of C. V_ν can be expressed as

$$V_\nu = \begin{bmatrix} c_1 \\ \vdots \\ c_q \\ \hline c_1 A \\ \vdots \\ c_q A \\ \hline \vdots \\ \hline c_1 A^{\nu-1} \\ \vdots \\ c_q A^{\nu-1} \end{bmatrix} \qquad (3.38)$$

Search for n linearly independent rows in V_ν in the order from top to bottom. The rank of matrix C is m, so the n linearly independent rows can be rearranged as

$$\begin{bmatrix} c_1 \\ c_1 A \\ \vdots \\ c_1 A^{\nu_1 - 1} \\ \hline c_2 \\ c_2 A \\ \vdots \\ c_2 A^{\nu_2 - 1} \\ \hline \vdots \\ \hline c_m \\ c_m A \\ \vdots \\ c_m A^{\nu_m - 1} \end{bmatrix}$$

For a completely observable system, one can deduce that

$$\nu_1 + \nu_2 + \cdots + \nu_m = n$$

The observability index satisfies

$$\nu = \max\{\nu_1, \nu_2, \cdots, \nu_m\}$$

Generally, $\{\nu_1, \nu_2, \cdots, \nu_m\}$ is called the *observability index set* or the *Kronecker invariable* of (A, C)

Theorem 3.20 *Under equivalent transformations, ν and $\{\nu_1, \nu_2, \cdots, \nu_m\}$ of (A, C) keep invariant.*

The proof is omitted.

Example 3.15 A given linear time-invariant system is

$$\dot{x}(t) = \begin{bmatrix} 0 & 1 & 0 & 0 \\ 3 & 0 & 0 & 2 \\ 0 & 0 & 0 & 1 \\ 0 & -2 & 0 & 0 \end{bmatrix} x(t), \quad y(t) = \begin{bmatrix} 1 & 0 & 0 & 0 \\ 0 & 0 & 1 & 0 \end{bmatrix} x(t)$$

Since

$$\operatorname{rank}\begin{bmatrix} C \\ CA \end{bmatrix} = \begin{bmatrix} 1 & 0 & 0 & 0 \\ 0 & 0 & 1 & 0 \\ 0 & 1 & 0 & 0 \\ 0 & 0 & 0 & 1 \end{bmatrix} = 4$$

the observability index of the system is $\nu = 2$.

Select 4 independent row vectors and rearrange them as

$$\begin{bmatrix} c_1 \\ c_1 A \\ c_2 \\ c_2 A \end{bmatrix} = \begin{bmatrix} 1 & 0 & 0 & 0 \\ 0 & 1 & 0 & 0 \\ 0 & 0 & 1 & 0 \\ 0 & 0 & 0 & 1 \end{bmatrix}$$

Apparently, the observability index set is $\{\nu_1, \nu_2\} = \{2, 2\}$, and $\nu = \max\{\nu_1, \nu_2\} = 2$.

3.3 Duality Principle of Linear Systems

The discussion above shows that the controllability and the observability have the duality property in both concepts and criteria. This internal duality relation reflects the duality property between the problems of system control and system estimation. The relationship which denotes the duality relation of a system is called duality principle.

3.3.1 Dual Systems

Consider a linear time-varying system

$$\dot{x} = A(t)x(t) + B(t)u(t), \ y(t) = C(t)x(t) + D(t)u(t) \tag{3.39}$$

where $x(t) \in \mathbb{R}^n$ is the state vector, $u(t) \in \mathbb{R}^p$ is the input vector, $y(t) \in \mathbb{R}^q$ is the output vector, $A(t)$, $B(t)$, $C(t)$ and $D(t)$ are time-varying matrices.

Definition 3.11 (Dual system) For a linear time-varying system (3.39), construct

$$\dot{z}(t) = -A^{\mathrm{T}}(t)z(t) + C^{\mathrm{T}}(t)v(t), \ w(t) = B^{\mathrm{T}}(t)z(t) + D^{\mathrm{T}}(t)v(t) \tag{3.40}$$

where $z(t) \in \mathbb{R}^n$ is the harmonious state vector, $v(t) \in \mathbb{R}^q$ is the input vector, and $w(t) \in \mathbb{R}^p$ is the output vector. Then the system (3.40) is called the *dual system* of (3.39).

The block diagram of a linear time-varying system and its dual system is shown in Fig.3.5.

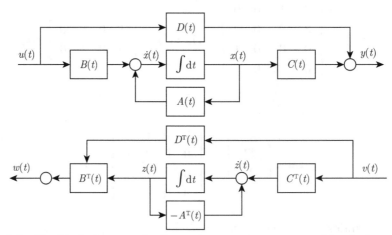

Fig.3.5 The block diagram of a linear time-varying system and its dual system

Suppose that $\Phi(t, t_0)$ is the state transition matrix of the system (3.39). According to properties of $\Phi(t, t_0)$, $\Phi(t, t_0)\Phi^{-1}(t, t_0) = I$ and $\Phi^{-1}(t, t_0) = \Phi(t_0, t)$. In this way, one can deduce that

$$0 = \frac{\partial}{\partial t}\left(\Phi(t, t_0)\Phi^{-1}(t, t_0)\right) = \frac{\partial}{\partial t}\left(\Phi(t, t_0)\right)\Phi^{-1}(t, t_0) + \Phi(t, t_0)\frac{\partial}{\partial t}\Phi^{-1}(t, t_0)$$

$$= A(t)\Phi(t, t_0)\Phi^{-1}(t, t_0) + \Phi(t, t_0)\frac{\partial}{\partial t}\Phi(t_0, t)$$

which indicates that

$$\frac{\partial}{\partial t}\Phi(t_0, t) = -\Phi^{-1}(t, t_0)A(t) = -\Phi(t_0, t)A(t)$$

Taking transpose and one can further obtain that

$$\frac{\partial}{\partial t}\Phi^{\mathrm{T}}(t_0, t) = -A^{\mathrm{T}}(t)\Phi^{\mathrm{T}}(t_0, t)$$

Meanwhile, $\Phi^{\mathrm{T}}(t_0, t_0) = I$. Therefore, $\Phi^{\mathrm{T}}(t_0, t)$ is the state transition matrix of the dual system (3.40). It can be seen that the state movement of the system (3.39) is the positive transition from t_0 to t, while the state movement of its dual system (3.40) is from t to t_0, inversely.

3.3.2 Duality Principle

Theorem 3.21 (Duality principle) *A linear time-varying system is completely controllable at t_0 if and only if its dual system is completely observable at t_0. A linear time-varying system is completely observable at t_0 if and only if its dual system is completely controllable at t_0.*

Proof Suppose that a linear time-varying system (3.39) is completely controllable at time t_0, which means that there exists $t_{\mathrm{f}} > t_0$ such that the controllability Gramian $W_{\mathrm{c}}(t_0, t_f)$ is nonsingular. It is noted that

$$W_{\mathrm{c}}(t_0, t_{\mathrm{f}}) \triangleq \int_{t_0}^{t_{\mathrm{f}}} \Phi(t_0, \tau) B(\tau) B^{\mathrm{T}}(\tau) \Phi^{\mathrm{T}}(t_0, \tau) \mathrm{d}\tau$$

$$= \int_{t_0}^{t_{\mathrm{f}}} \left[\Phi^{\mathrm{T}}(t_0, \tau)\right]^{\mathrm{T}} \left[B^{\mathrm{T}}(\tau)\right]^{\mathrm{T}} \left[B^{\mathrm{T}}(\tau)\right] \left[\Phi^{\mathrm{T}}(t_0, \tau)\right] \mathrm{d}\tau$$

which is equivalent to the observability Gramian matrix of the system (3.40). That is to say, the controllability of the system (3.39) is equivalent to the observability of the system (3.40).

Using the similar method, one can prove that the observability of system (3.39) is equivalent to the controllability of the system (3.40). □

Theorem 3.21 not only provides a method to evaluate the structure property of the original system based on the structure property of its dual system, but also establishes the corresponding relation between the control problem and the state estimation problem of a system. The duality principle is very important in subsequent studies.

If the system coefficient matrices in (3.39) and (3.40) are constant matrices, then the dual system of a linear time-invariant system is defined, which has the same duality property as the time-varying case. Let (A, B, C, D) denote a time-invariant system, and $(-A^{\mathrm{T}}, C^{\mathrm{T}}, B^{\mathrm{T}}, D^{\mathrm{T}})$ denote its dual system. As a matter of fact, there also exists the duality relationship between (A, B, C, D) and $(A^{\mathrm{T}}, C^{\mathrm{T}}, B^{\mathrm{T}}, D^{\mathrm{T}})$. The controllability matrix and the observability matrix of (A, B, C, D) are

$$U = \begin{bmatrix} B & AB & \cdots & A^{n-1}B \end{bmatrix}, \quad V = \begin{bmatrix} C^{\mathrm{T}} & A^{\mathrm{T}}C^{\mathrm{T}} & \cdots & (A^{n-1})^{\mathrm{T}}C^{\mathrm{T}} \end{bmatrix}^{\mathrm{T}}$$

respectively. Meanwhile, the controllability matrix and the observability matrix of $(A^{\mathrm{T}}, C^{\mathrm{T}}, B^{\mathrm{T}}, D^{\mathrm{T}})$ are

$$U_{\mathrm{d}} = \begin{bmatrix} C^{\mathrm{T}} & A^{\mathrm{T}}C^{\mathrm{T}} & \cdots & (A^{n-1})^{\mathrm{T}}C^{\mathrm{T}} \end{bmatrix}, \quad V_{\mathrm{d}} = \begin{bmatrix} B & AB & \cdots & A^{n-1}B \end{bmatrix}^{\mathrm{T}}$$

respectively. It can be seen that $U = V_{\mathrm{d}}^{\mathrm{T}}$ and $V = U_{\mathrm{d}}^{\mathrm{T}}$. According to the rank criterion, (A, B, C, D) is controllable if and only if $(A^{\mathrm{T}}, C^{\mathrm{T}}, B^{\mathrm{T}}, D^{\mathrm{T}})$ is observable, while

(A, B, C, D) is observable if and only if $(A^\mathrm{T}, C^\mathrm{T}, B^\mathrm{T}, D^\mathrm{T})$ is controllable. In subsequent sections, consider that the dual system of a linear time-invariant system (A, B, C, D) is $(A^\mathrm{T}, C^\mathrm{T}, B^\mathrm{T}, D^\mathrm{T})$.

3.4 Structure Decomposition of Linear Time-Invariant Systems

For incompletely controllable and observable systems, an important problem is how to find a method to decompose the structure according to controllability or observability. By decomposing, the structure of a system can be divided into the controllable and observable part, controllable and unobservable part, uncontrollable and observable part, and uncontrollable and unobservable part. With this decomposing one can profoundly show the structure property, understand the essential difference between state space description and the input-output description, and master the system analysis and design better.

3.4.1 Controllability and Observability of Linear Systems with Equivalent Transformation

Consider a linear time-varying system $(A(t), B(t), C(t), D(t))$. Taking the equivalent transformation $\bar{x}(t) = T(t)x(t)$, where $T(t)$ is a nonsingular and continuously differentiable transformation matrix, $(A(t), B(t), C(t), D(t))$ is transformed into $(\bar{A}(t), \bar{B}(t), \bar{C}(t), \bar{D}(t))$ with $\bar{A}(t) = T(t)A(t)T^{-1}(t) + \dot{T}(t)T^{-1}(t)$, $\bar{B}(t) = T(t)B(t)$, $\bar{C}(t) = C(t)T^{-1}(t)$, $\bar{D}(t) = D(t)$. Denote state transition matrices of these two systems as $\Phi(t, t_0)$ and $\bar{\Phi}(t, t_0)$, respectively. As discussed in Subsection 2.2.4,

$$\bar{\Phi}(t, t_0) = T(t)\Phi(t, t_0)T^{-1}(t_0)$$

Denote the controllability and observability Gramian of $(A(t), B(t), C(t), D(t))$ as $W_\mathrm{c}(t_0, t_\mathrm{f})$ and $W_\mathrm{o}(t_0, t_\mathrm{f})$, respectively, and denote the controllability and observability Gramian matrix of $(\bar{A}(t), \bar{B}(t), \bar{C}(t), \bar{D}(t))$ as $\bar{W}_\mathrm{c}(t_0, t_\mathrm{f})$ and $\bar{W}_\mathrm{o}(t_0, t_\mathrm{f})$, respectively. One can deduce that

$$\bar{W}_\mathrm{c}(t_0, t_\mathrm{f}) = \int_{t_0}^{t_\mathrm{f}} \bar{\Phi}(t_0, \tau)\bar{B}(\tau)\bar{B}^\mathrm{T}(\tau)\bar{\Phi}^\mathrm{T}(t_0, \tau)\mathrm{d}\tau$$

$$= \int_{t_0}^{t_\mathrm{f}} T(t_0)\Phi(t_0, \tau)B(\tau)B^\mathrm{T}(\tau)\Phi^\mathrm{T}(t_0, \tau)T^\mathrm{T}(t_0)\mathrm{d}\tau$$

$$= T(t_0)\int_{t_0}^{t_\mathrm{f}} \Phi(t_0, \tau)B(\tau)B^\mathrm{T}(\tau)\Phi^\mathrm{T}(t_0, \tau)\mathrm{d}\tau\, T^\mathrm{T}(t_0)$$

$$= T(t_0)W_\mathrm{c}(t_0, t_\mathrm{f})T^\mathrm{T}(t_0)$$

Since $T(t_0)$ is nonsingular, it can be proved that

$$\mathrm{rank}\,\bar{W}_\mathrm{c}(t_0, t_\mathrm{f}) = \mathrm{rank}\,W_\mathrm{c}(t_0, t_\mathrm{f}) \qquad (3.41)$$

Similarly, it holds that

$$\mathrm{rank}\,W_\mathrm{o}(t_0, t_\mathrm{f}) = \mathrm{rank}\,\bar{W}_\mathrm{o}(t_0, t_\mathrm{f}) \qquad (3.42)$$

3.4 Structure Decomposition of Linear Time-Invariant Systems

(3.41) and (3.42) indicate that taking the equivalent transformation for a linear time-varying system does not change the controllability and the observability of a system, and does not change the incomplete controllability and the incomplete observability, neither. This conclusion also holds for the linear time-invariant system. This fact provides the theory basis approach to realize the structure decomposition of the system with the equivalent transformation.

3.4.2 Controllability Structure Decomposition of Linear Time-Invariant Systems

Consider a linear time-invariant system

$$\dot{x} = Ax(t) + Bu(t), \ y(t) = Cx(t) \tag{3.43}$$

where $x(t) \in \mathbb{R}^n$ is the state vector, $u(t) \in \mathbb{R}^p$ is the input vector, $y(t) \in \mathbb{R}^q$ is the output vector, A, B and C are constant matrices with appropriate dimensions.

Assume that (3.43) is incompletely controllable. In this situation, its controllability matrix U satisfies $\operatorname{rank} U = k_1 < n$. Select k_1 linearly independent columns arbitrarily in U and denote them as $q_1, \cdots, q_{k_1}$, respectively. Furthermore, select another $n - k_1$ column vectors in $\mathbb{R}^n$ and denote them as $q_{k_1+1}, \cdots, q_n$, such that $q_1, \cdots, q_{k_1}, q_{k_1+1}, \cdots, q_n$ are linearly independent. On this basis, construct a transformation matrix T_1 satisfying

$$T_1^{-1} = \begin{bmatrix} q_1 & \cdots & q_{k_1} & q_{k_1+1} & \cdots & q_n \end{bmatrix} \tag{3.44}$$

Theorem 3.22 *For a linear time-invariant system* (3.43) *which is incompletely controllable, if* $\operatorname{rank} U = k_1 < n$, *an equivalent transformation matrix* T_1 *satisfying* (3.44) *can be obtained and* $\bar{x}(t) = T_1 x(t)$ *transforms the system* (3.43) *into*

$$\begin{cases} \begin{bmatrix} \dot{\bar{x}}_1(t) \\ \dot{\bar{x}}_2(t) \end{bmatrix} = \begin{bmatrix} \bar{A}_{11} & \bar{A}_{12} \\ 0 & \bar{A}_{22} \end{bmatrix} \begin{bmatrix} \bar{x}_1(t) \\ \bar{x}_2(t) \end{bmatrix} + \begin{bmatrix} \bar{B}_1 \\ 0 \end{bmatrix} u(t) \\ y(t) = \begin{bmatrix} \bar{C}_1 & \bar{C}_2 \end{bmatrix} \begin{bmatrix} \bar{x}_1(t) \\ \bar{x}_2(t) \end{bmatrix} \end{cases} \tag{3.45}$$

where $\bar{x}_1(t) \in \mathbb{R}^{k_1}$ is the controllable state, and $\bar{x}_2(t) \in \mathbb{R}^{n-k_1}$ is the uncontrollable state.

Proof Denote $T_1 = \begin{bmatrix} p_1 & \cdots & p_n \end{bmatrix}^{\mathrm{T}}$. Owing to $T_1 T_1^{-1} = I$, one can deduce that

$$p_i^{\mathrm{T}} q_j = \begin{cases} 1 & i = j \\ 0 & i \neq j \end{cases} \tag{3.46}$$

for $i \in \{1, \cdots, n\}$ and $j \in \{1, \cdots, n\}$.

Meanwhile, it is known that the controllable subspace is the invariant subspace of A. Since $q_1, \cdots, q_{k_1}$ are linearly independent, and then Aq_j is a linear combination of

$\{q_1, \cdots, q_{k_1}\}$ for $j = 1, \cdots, k_1$. Along with (3.46), one can deduce that $p_i^T A q_j = 0$ for $i = k_1 + 1, \cdots, n$ and $j = 1, \cdots, k_1$. Therefore, it holds that

$$\bar{A} = T_1 A T_1^{-1} = \begin{bmatrix} p_1^T A q_1 & \cdots & p_1^T A q_{k_1} & p_1^T A q_{k_1+1} & \cdots & p_1^T A q_n \\ \vdots & & \vdots & \vdots & & \vdots \\ p_{k_1}^T A q_1 & \cdots & p_{k_1}^T A q_{k_1} & p_{k_1}^T A q_{k_1+1} & \cdots & p_{k_1}^T A q_n \\ p_{k_1+1}^T A q_1 & \cdots & p_{k_1+1}^T A q_{k_1} & p_{k_1+1}^T A q_{k_1+1} & \cdots & p_{k_1+1}^T A q_n \\ \vdots & & \vdots & \vdots & & \vdots \\ p_n^T A q_1 & \cdots & p_n^T A q_{k_1} & p_n^T A q_{k_1+1} & \cdots & p_n^T A q_n \end{bmatrix}$$

$$= \begin{bmatrix} \bar{A}_{11} & \bar{A}_{12} \\ 0 & \bar{A}_{22} \end{bmatrix}$$

On the other hand, each column of B can be expressed as a linear combination of $\{q_1, \cdots, q_{k_1}\}$. In this way, $p_i^T B = 0$ for $i = k_1 + 1, \cdots, n$, and then,

$$\bar{B} = T_1 B = \begin{bmatrix} p_1^T B \\ \vdots \\ p_{k_1}^T B \\ p_{k_1+1}^T B \\ \vdots \\ p_n^T B \end{bmatrix}$$

$\begin{bmatrix} \bar{C}_1 & \bar{C}_2 \end{bmatrix}$ has no special form. It is viewed that $\bar{x}(t) = T_1 x(t)$ transforms the system (3.43) into (3.45). In the next, one can further deduce that

$$k_1 = \text{rank}\,\bar{U} = \text{rank}\begin{bmatrix} \bar{B} & \bar{A}\bar{B} & \cdots & \bar{A}^{n-1}\bar{B} \end{bmatrix}$$

$$= \text{rank}\begin{bmatrix} \bar{B}_1 & \bar{A}_{11}\bar{B}_1 & \cdots & \bar{A}_{11}^{n-1}\bar{B}_1 \\ 0 & 0 & \cdots & 0 \end{bmatrix}$$

$$= \text{rank}\begin{bmatrix} \bar{B}_1 & \bar{A}_{11}\bar{B} & \cdots & \bar{A}_{11}^{n-1}\bar{B}_1 \end{bmatrix}$$

$\bar{A}_{11}$ is a $k_1 \times k_1$ matrix. According to Cayley-Hamilton theorem, $\bar{A}_{11}^{k_1}\bar{B}_1, \cdots, \bar{A}_{11}^{n-1}\bar{B}_1$ can be expressed as a linear combination of $\bar{B}_1, \bar{A}_{11}\bar{B}, \cdots, \bar{A}_{11}^{k_1-1}\bar{B}_1$. Therefore, it holds that

$$\text{rank}\begin{bmatrix} \bar{B}_1 & \bar{A}_{11}\bar{B} & \cdots & \bar{A}_{11}^{n-1}\bar{B}_1 \end{bmatrix} = \text{rank}\begin{bmatrix} \bar{B}_1 & \bar{A}_{11}\bar{B} & \cdots & \bar{A}_{11}^{k_1-1}\bar{B}_1 \end{bmatrix} = k_1$$

which indicates that $(\bar{A}_{11}, \bar{B}_1)$ is completely controllable, and $\bar{x}_1(t)$ is the controllable state. □

3.4 Structure Decomposition of Linear Time-Invariant Systems

Example 3.16 Consider a linear time-invariant system given by

$$\dot{x}(t) = \begin{bmatrix} -1 & 0 & 0 \\ 1 & 0 & 5 \\ 0 & 1 & 3 \end{bmatrix} x(t) + \begin{bmatrix} 0 & 0 \\ 1 & 0 \\ 0 & 1 \end{bmatrix} u(t)$$

It can be computed that

$$\operatorname{rank} \begin{bmatrix} B & AB \end{bmatrix} = \operatorname{rank} \begin{bmatrix} 0 & 0 & 0 & 0 \\ 1 & 0 & 0 & 5 \\ 0 & 1 & 1 & 3 \end{bmatrix} = 2$$

$n = 3$, and rank $B = 2$. Thus, the system is incompletely controllable. Select the linearly independent columns $q_1 = [\ 0\ 1\ 0\ ]^T$ and $q_2 = [\ 0\ 0\ 1\ ]^T$ in the controllability matrix, and select $q_3 = [\ 1\ 0\ 0\ ]^T$, which can guarantee that q_1, q_2 and q_3 are linearly independent. Construct

$$T_1^{-1} = \begin{bmatrix} 0 & 0 & 1 \\ 1 & 0 & 0 \\ 0 & 1 & 0 \end{bmatrix},\ T_1 = \begin{bmatrix} 0 & 1 & 0 \\ 0 & 0 & 1 \\ 1 & 0 & 0 \end{bmatrix}$$

One can deduce that

$$\bar{A} = T_1 A T_1^{-1} = \begin{bmatrix} 0 & 5 & 1 \\ 1 & 3 & 0 \\ 0 & 0 & -1 \end{bmatrix},\ \bar{B} = T_1 B = \begin{bmatrix} 1 & 0 \\ 0 & 1 \\ 0 & 0 \end{bmatrix},\ \bar{C} = CT_1^{-1} = \begin{bmatrix} 0 & 1 & 1 \end{bmatrix}$$

According to Theorem 3.22, an incompletely controllable system (3.43) can be decomposed into two parts, while one part is a controllable subsystem

$$\dot{\bar{x}}_1(t) = \bar{A}_{11}\bar{x}_1(t) + \bar{A}_{12}\bar{x}_2(t) + \bar{B}_1 u(t),\ \bar{y}_1(t) = \bar{C}_1 \bar{x}_1(t)$$

the other part is an uncontrollable subsystem

$$\dot{\bar{x}}_2(t) = \bar{A}_{22}\bar{x}_2(t),\ \bar{y}_2(t) = \bar{C}_2 \bar{x}_2(t)$$

and $y(t) = \bar{y}_1(t) + \bar{y}_2(t)$. It is noted that

$$\det(sI - A) = \det(sI - \bar{A}) = \det \begin{bmatrix} sI - \bar{A}_{11} & -\bar{A}_{12} \\ 0 & sI - \bar{A}_{22} \end{bmatrix}$$

which indicates that eigenvalues of A consist of two parts. One part is determined by A_{11}, and the input $u(t)$ can influence and change these eigenvalues. These eigenvalues are generally called *controllable eigenvalues* of A. The other part is determined by A_{22}, and $u(t)$ cannot influence or change these eigenvalues. These eigenvalues are generally called *uncontrollable eigenvalues* of A.

Based on (3.45), the block diagram of the system after the controllability structure decomposition can be obtained and shown in Fig.3.6. It can be seen that the control input $u(t)$ cannot directly or indirectly influence the uncontrollable states.

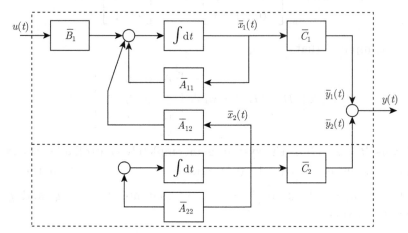

Fig.3.6 The block diagram of the system after the controllability structure decomposition

Since the equivalent transformation does not change the transfer function matrix of a system, the transfer function matrix of the system (3.45) is the transfer function matrix of the system (3.43). In this way,

$$G(s) = \begin{bmatrix} \bar{C}_1 & \bar{C}_2 \end{bmatrix} \begin{bmatrix} sI - \bar{A}_{11} & -\bar{A}_{12} \\ 0 & sI - \bar{A}_{22} \end{bmatrix}^{-1} \begin{bmatrix} \bar{B}_1 \\ 0 \end{bmatrix}$$

$$= \begin{bmatrix} \bar{C}_1 & \bar{C}_2 \end{bmatrix} \begin{bmatrix} (sI - \bar{A}_{11})^{-1} & (sI - \bar{A}_{11})^{-1}\bar{A}_{12}(sI - \bar{A}_{22})^{-1} \\ 0 & (sI - \bar{A}_{22})^{-1} \end{bmatrix} \begin{bmatrix} \bar{B}_1 \\ 0 \end{bmatrix}$$

$$= \bar{C}_1(sI - \bar{A}_{11})^{-1}\bar{B}_1$$

Since the choice of the transformation matrix T_1 is not unique, there exist different controllability structure decompositions of the system (3.43). These canonical forms have the same structure but different numerical values.

3.4.3 Observability Structure Decomposition of Linear Time-Invariant Systems

In this subsection, consider that the system (3.43) is incompletely observable. Assume that

$$\text{rank } V = \text{rank} \begin{bmatrix} C \\ CA \\ \vdots \\ CA^{n-1} \end{bmatrix} = k_2 < n$$

Select arbitrarily k_2 linearly independent rows in V and denote them as $h_1, \cdots, h_{k_2}$. Select another $n - k_2$ row vectors and denote them as $h_{k_2+1}, \cdots, h_n$ such that $h_1, \cdots, h_{k_2}$,

3.4 Structure Decomposition of Linear Time-Invariant Systems

$h_{k_2+1}, \cdots, h_n$ are linearly independent. On this basis, construct a transformation matrix T_2 satisfying

$$T_2 = \begin{bmatrix} h_1 \\ \vdots \\ h_{k_2} \\ h_{k_2+1} \\ \vdots \\ h_n \end{bmatrix} \tag{3.47}$$

Theorem 3.23 *For a linear time-invariant system (3.43) which is incompletely observable, if* $\operatorname{rank} V = k_2 < n$, *an equivalent transformation matrix* T_2 *satisfying (3.47) can be obtained and* $\hat{x}(t) = T_2 x(t)$ *transforms the system (3.43) into*

$$\begin{cases} \begin{bmatrix} \dot{\hat{x}}_1(t) \\ \dot{\hat{x}}_2(t) \end{bmatrix} = \begin{bmatrix} \hat{A}_{11} & 0 \\ \hat{A}_{21} & \hat{A}_{22} \end{bmatrix} \begin{bmatrix} \hat{x}_1(t) \\ \hat{x}_2(t) \end{bmatrix} + \begin{bmatrix} \hat{B}_1 \\ \hat{B}_2 \end{bmatrix} u(t) \\ y(t) = \begin{bmatrix} \hat{C}_1 & 0 \end{bmatrix} \begin{bmatrix} \hat{x}_1(t) \\ \hat{x}_2(t) \end{bmatrix} \end{cases} \tag{3.48}$$

where $\hat{x}_1(t) \in \mathbb{R}^{k_2}$ is the observable state, and $\hat{x}_2(t) \in \mathbb{R}^{n-k_2}$ is the unobservable state.

The proof is omitted.

According to Theorem 3.23, an incompletely observable system (3.43) can be decomposed into two parts, while one part is an observable subsystem

$$\dot{\hat{x}}_1(t) = \hat{A}_{11}\hat{x}_1(t) + \hat{B}_1 u(t), \quad \hat{y}_1(t) = \hat{C}_1 \hat{x}_1(t)$$

the other part is an unobservable subsystem

$$\dot{\hat{x}}_2(t) = \hat{A}_{21}\hat{x}_1(t) + \hat{A}_{22}\hat{x}_2(t) + \hat{B}_2 u(t), \quad \hat{y}_2(t) = 0$$

and $y(t) = \hat{y}_1(t) + \hat{y}_2(t)$. It is noted that

$$\det(sI - A) = \det(sI - \hat{A}) = \det \begin{bmatrix} sI - \hat{A}_{11} & 0 \\ -\hat{A}_{11} & sI - \hat{A}_{22} \end{bmatrix}$$

$$= \det(sI - \hat{A}_{11})\det(sI - \hat{A}_{22})$$

which indicates that eigenvalues of A consist of two parts. One part is determined by $\hat{A}_{11}$, and the output $y(t)$ can reflect these eigenvalues. These eigenvalues are generally called *observable eigenvalues* of A. The other part is determined by $\hat{A}_{22}$, and $y(t)$ cannot reflect these eigenvalues. These eigenvalues are generally called *unobservable eigenvalues* of A.

Based on (3.48), the block diagram of the system after the observability structure decomposition can be obtained and shown in Fig.3.7. It can be seen that the information of the unobservable part can not be reflected by the output $y(t)$.

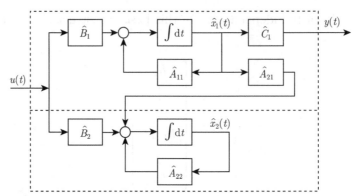

Fig.3.7 The block diagram of the system after the observability structure decomposition

The transfer function matrix of the system (3.48) is the transfer function matrix of the system (3.43), i.e.,

$$G(s) = \hat{C}_1(sI - \hat{A}_{11})^{-1}\hat{B}_1$$

There exist different observability structure decompositions of the system (3.43). These canonical forms have the same structure but different numerical values.

3.4.4 Kalman's Decomposition of Linear Time-Invariant Systems

In this subsection, consider that the time-invariant system (3.43) is not only incompletely controllable but also incompletely observable. In this situation, there exists the following conclusion, which is called the *canonical decomposition theorem* or *Kalman's decomposition theorem*.

Theorem 3.24 *For a linear time-invariant system (3.43) which is incompletely controllable and incompletely observable, there exists an equivalent transformation matrix T and $\tilde{x}(t) = Tx(t)$ transforms the system (3.43) into*

$$\begin{cases} \begin{bmatrix} \dot{\tilde{x}}_1(t) \\ \dot{\tilde{x}}_2(t) \\ \dot{\tilde{x}}_3(t) \\ \dot{\tilde{x}}_4(t) \end{bmatrix} = \begin{bmatrix} \tilde{A}_{11} & 0 & \tilde{A}_{13} & 0 \\ \tilde{A}_{21} & \tilde{A}_{22} & \tilde{A}_{23} & \tilde{A}_{24} \\ 0 & 0 & \tilde{A}_{33} & 0 \\ 0 & 0 & \tilde{A}_{43} & \tilde{A}_{44} \end{bmatrix} \begin{bmatrix} \tilde{x}_1(t) \\ \tilde{x}_2(t) \\ \tilde{x}_3(t) \\ \tilde{x}_4(t) \end{bmatrix} + \begin{bmatrix} \tilde{B}_1 \\ \tilde{B}_2 \\ 0 \\ 0 \end{bmatrix} u(t) \\ y(t) = \begin{bmatrix} \tilde{C}_1 & 0 & \tilde{C}_3 & 0 \end{bmatrix} \begin{bmatrix} \tilde{x}_1(t) \\ \tilde{x}_2(t) \\ \tilde{x}_3(t) \\ \tilde{x}_4(t) \end{bmatrix} \end{cases} \quad (3.49)$$

where $\tilde{x}_1(t)$ is the controllable and observable state, $\tilde{x}_2(t)$ is the controllable and unobservable state, $\tilde{x}_3(t)$ is the uncontrollable and observable state, $\tilde{x}_4(t)$ is the uncontrollable and unobservable state.

3.4 Structure Decomposition of Linear Time-Invariant Systems

Proof Use Theorem 3.22 to decompose the system (3.43) into a controllable part and an uncontrollable part. On this basis, using Theorem 3.23 to further decompose the controllable subsystem and the uncontrollable subsystem into observable subsystems and unobservable subsystems, respectively, and then the system (3.43) can be decomposed into (3.49). □

The block diagram of the structure canonical decomposition (3.49) is shown in Fig.3.8. It should be pointed out that this decomposition is not unique but the structure is unique.

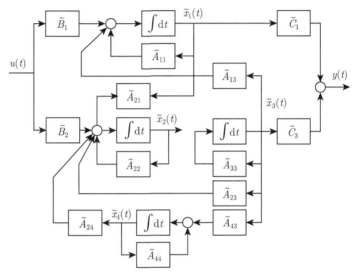

Fig.3.8 The block diagram of the system after the structure canonical decomposition

On the other hand, Theorem 3.24 is deduced by taking the controllability structure decomposition firstly. As a matter of fact, it can also take the observability structure decomposition firstly, and then take the controllability structure decomposition for the observable subsystem and the unobservable subsystem, respectively. In this way, the system (3.43) can also be decomposed into four parts.

The subsystem $(\tilde{A}_{11}, \tilde{B}_1, \tilde{C}_1)$ is both controllable and observable. The transfer function matrix of the system (3.43) equals to the one of $(\tilde{A}_{11}, \tilde{B}_1, \tilde{C}_1)$. That is to say, the transfer function matrix can only describe the controllable and observable subsystem, and thus

$$G(s) = \tilde{C}_1(sI - \tilde{A}_{11})^{-1}\tilde{B}_1$$

Example 3.17 Consider an incompletely controllable and incompletely observable linear time-invariant system

$$\dot{x}(t) = \begin{bmatrix} 0 & 0 & -1 \\ 1 & 0 & -3 \\ 0 & 1 & -3 \end{bmatrix} x(t) + \begin{bmatrix} 1 \\ 1 \\ 0 \end{bmatrix} u(t), \ y(t) = \begin{bmatrix} 0 & 1 & -2 \end{bmatrix} x(t)$$

Decompose the system into the canonical form.

Firstly, solve the controllability structure decomposition.

It can be computed that

$$\operatorname{rank} U = \operatorname{rank} \begin{bmatrix} b & Ab & A^2b \end{bmatrix} = \operatorname{rank} \begin{bmatrix} 1 & 0 & -1 \\ 1 & 1 & -3 \\ 0 & 1 & -2 \end{bmatrix} = 2 < n = 3$$

Select the linearly independent columns $q_1 = [\ 1\ \ 1\ \ 0\ ]^T$ and $q_2 = [\ 0\ \ 1\ \ 1\ ]^T$ in the controllability matrix, and select $q_3 = [\ 0\ \ 0\ \ 1\ ]^T$, which can guarantee that q_1, q_2 and q_3 are linearly independent. Construct

$$T_1^{-1} = \begin{bmatrix} q_1 & q_2 & q_3 \end{bmatrix} = \begin{bmatrix} 1 & 0 & 0 \\ 1 & 1 & 0 \\ 0 & 1 & 1 \end{bmatrix}$$

and it can be further computed that

$$T_1 = \begin{bmatrix} 1 & 0 & 0 \\ -1 & 1 & 0 \\ 1 & -1 & 1 \end{bmatrix}$$

One can obtain that

$$\bar{A} = T_1 A T_1^{-1} = \begin{bmatrix} 0 & -1 & -1 \\ 1 & -2 & -2 \\ 0 & 0 & -1 \end{bmatrix},\ \bar{B} = T_1 B = \begin{bmatrix} 1 \\ 0 \\ 0 \end{bmatrix},\ \bar{C} = C T_1^{-1} = \begin{bmatrix} 1 & -1 & -2 \end{bmatrix}$$

Secondly, solve the observability structure decomposition.

It can be computed that

$$\operatorname{rank} V = \operatorname{rank} \begin{bmatrix} C \\ CA \\ CA^2 \end{bmatrix} = \operatorname{rank} \begin{bmatrix} 0 & 1 & -2 \\ 1 & -2 & 3 \\ -2 & 3 & -4 \end{bmatrix} = 2 < n = 3$$

Select the linearly independent rows $h_1 = [\ 0\ \ 1\ \ -2\ ]$ and $h_2 = [\ 1\ \ -2\ \ 3\ ]$ in the observability matrix, and select $h_3 = [\ 0\ \ 0\ \ 1\ ]$, which can guarantee that h_1, h_2 and h_3 are linearly independent. Construct

$$T_2 = \begin{bmatrix} h_1 \\ h_2 \\ h_3 \end{bmatrix} = \begin{bmatrix} 0 & 1 & -2 \\ 1 & -2 & 3 \\ 0 & 0 & 1 \end{bmatrix}$$

and it can be further computed that

$$T_2^{-1} = \begin{bmatrix} 2 & 1 & 1 \\ 1 & 0 & 2 \\ 0 & 0 & 1 \end{bmatrix}$$

3.4 Structure Decomposition of Linear Time-Invariant Systems

One can obtain that

$$\hat{A} = T_2 A T_2^{-1} = \begin{bmatrix} 0 & 1 & 0 \\ -1 & -2 & 0 \\ 1 & 0 & -1 \end{bmatrix}, \hat{B} = T_2 B = \begin{bmatrix} 1 \\ -1 \\ 0 \end{bmatrix}, \hat{C} = CT_2^{-1} = \begin{bmatrix} 1 & 0 & 0 \end{bmatrix}$$

Thirdly, solve the controllability and observability structure decomposition. After the controllability structure decomposition, the controllable subsystem is

$$\dot{\bar{x}}(t) = \begin{bmatrix} 0 & -1 \\ 1 & -2 \end{bmatrix} \bar{x}_1(t) + \begin{bmatrix} -1 \\ -2 \end{bmatrix} \bar{x}_2(t) + \begin{bmatrix} 1 \\ 0 \end{bmatrix} u(t), \bar{y}_1(t) = \begin{bmatrix} 1 & -1 \end{bmatrix} \bar{x}_1(t)$$

Its observability matrix is

$$V_1 = \begin{bmatrix} 1 & -1 \\ -1 & 1 \end{bmatrix}$$

and rank $V_1 = 1 < 2$.

Hence the controllable subsystem is incompletely observable, take the observability structure decomposition. The transformation matrix is

$$T_{r1} = \begin{bmatrix} 1 & -1 \\ 0 & 1 \end{bmatrix}$$

It is obvious that the uncontrollable subsystem is observable, let $T_{r2} = 1$. T_{r1} and T_{r2} form a block diagonal matrix

$$T_r = \begin{bmatrix} T_{r1} & \\ & T_{r2} \end{bmatrix} = \begin{bmatrix} 1 & -1 & 0 \\ 0 & 1 & 0 \\ 0 & 0 & 1 \end{bmatrix}, T_r^{-1} = \begin{bmatrix} 1 & 1 & 0 \\ 0 & 1 & 0 \\ 0 & 0 & 1 \end{bmatrix}$$

Introduce an equivalent transformation $\tilde{x}(t) = T_r \bar{x}(t)$, and then one can obtain that

$$\tilde{A} = T_r \bar{A} T_r^{-1} = \begin{bmatrix} -1 & 0 & 1 \\ 1 & -1 & -2 \\ 0 & 0 & -1 \end{bmatrix}, \tilde{B} = T_r \bar{B} = \begin{bmatrix} 1 \\ 0 \\ 0 \end{bmatrix}, \tilde{C} = \bar{C} T_r^{-1} \begin{bmatrix} 1 & 0 & -2 \end{bmatrix}$$

It can be seen that the original system is ultimately decomposed into the controllable and observable part, the controllable but unobservable part, and the uncontrollable but observable part.

3.4.5 Relationships among Poles, Zeros and Eigenvalues

In Subsection 1.2.4, the poles and zeros of a transfer function matrix $G(s)$ are defined based on the Smith-McMillan form. In the next, introduce the poles and zeros of a system firstly.

Definition 3.12 (Rosenbrock's system matrix) For a linear time-invariant system (A, B, C, D), the matrix

$$S(s) \triangleq \begin{bmatrix} sI - A & B \\ -C & D \end{bmatrix} \qquad (3.50)$$

is called the *(Rosenbrock's) system matrix* of the system.

Consider the linear time-invariant system

$$\dot{x}(t) = Ax(t) + Bu(t), \ y(t) = Cx(t) + Du(t) \qquad (3.51)$$

where $x(t) \in \mathbb{R}^n$ is the state vector, $u(t) \in \mathbb{R}^p$ is the input vector, $y(t) \in \mathbb{R}^q$ is the output vector, A, B, C and D are constant matrices with appropriate dimensions. Combining (3.50) with (3.51), one can deduce that

$$S(s) \begin{bmatrix} -x(s) \\ u(s) \end{bmatrix} = \begin{bmatrix} 0 \\ y(s) \end{bmatrix}$$

where $x(s)$, $u(s)$ and $y(s)$ are the Laplace transform of $x(t)$, $u(t)$ and $y(t)$, respectively. In can be seen that the system matrix $S(s)$ can also be used to describe a linear time-invariant system.

Definition 3.13 (Poles of systems) For a linear time-invariant system (A, B, C, D), the polynomial

$$p_s(s) = \det(sI - A)$$

is called the *pole polynomial* of the system, and the roots of $p_s(s) = 0$ are called the *poles* of the system.

It can be seen that poles of a system (A, B, C, D) equal to eigenvalues of A.

Definition 3.14 (Zeros of systems) For a linear time-invariant system (A, B, C, D), consider that rank $S(s) = r$, and all those rth-order nonzero minors of $S(s)$ that are formed by taking the first n rows and n columns of $S(s)$, i.e., all rows and columns of $sI - A$, and then adding $r - n$ rows of $\begin{bmatrix} -C & D \end{bmatrix}$ and $r - n$ columns of $\begin{bmatrix} B^{\mathrm{T}} & D^{\mathrm{T}} \end{bmatrix}^{\mathrm{T}}$. The monic greatest common divisor of all these minors is called the *zero polynomial* of the system, and denoted as $z_s(s)$. The roots of $z_s(s) = 0$ are called the zeros of the system.

Definition 3.15 (Invariant zeros of systems) For a linear time-invariant system (A, B, C, D), consider that rank $S(s) = r$, and the invariant factors of $S(s)$ are $s_1(s)$, $\cdots$, $s_r(s)$. The polynomial

$$z_s^i(s) = \prod_{i=1}^{r} s_i(s)$$

is called the *invariant zero polynomial* of the system, and the roots of $z_s^i(s) = 0$ are called the *invariant zeros* of the system.

It can be deduced that the monic greatest common divisor of all the highest order nonzero minors of $S(s)$ equals to $z_s^i(s)$. In general,

$$\{\text{invariant zeros of the system}\} \subset \{\text{zeros of the system}\}.$$

Definition 3.16 (Input-decoupling zeros of systems) For a linear time-invariant system (A, B, C, D), the product of the invariant factors of $[\ sI - A\ \ B\ ]$ is a polynomial, and the roots of this polynomial are called the *input-decoupling zeros* of the system.

Definition 3.17 (Output-decoupling zeros of systems) For a linear time-invariant system (A, B, C, D), the product of the invariant factors of $[\ (sI - A)^{\mathrm{T}}\ \ -C^{\mathrm{T}}\ ]^{\mathrm{T}}$ is a polynomial, and the roots of this polynomial are called the *output-decoupling zeros* of the system.

According to the above definitions, it is not difficult to deduce that input-decoupling zeros and output-decoupling zeros are eigenvalues of A and also zeros of the system (A, B, C, D). Meanwhile, if λ is an input-decoupling zero, then rank$[\ \lambda I - A\ \ B\ ] < n$. According to PBH rank criterion, the system is incompletely controllable. As discussed in Subsection 3.4.2, an incompletely controllable linear time-invariant system can be divided into a controllable part and an uncontrollable part. As a matter of fact, the input-decoupling zeros of the system equal to the uncontrollable eigenvalues of A. Similarly, if there exists an output-decoupling zero, the system is incompletely observable, and the output-decoupling zeros equal to the unobservable eigenvalues of A.

As discussed in Subsection 3.4.4, there may exist eigenvalues of A that are both uncontrollable and unobservable. These eigenvalues are zeros of the system that are both input- and output-decoupling zeros and are called the *input-output-decoupling zeros* of the system.

If the zeros of a system are determined and the input- and output-decoupling zeros are removed, then the remaining zeros are the zeros of the transfer function matrix $G(s)$. In general, the zeros of the transfer function matrix $G(s)$ is called the *transmission zeros* of the system.

Summarizing the above discussion, one can obtain the following conclusions,

$$\{\text{zeros of the system}\} = \{\text{zeros of } G(s)\} \cup \{\text{input-decoupling zeros}\}$$
$$\cup \{\text{output-decoupling zeros}\} - \{\text{input-output-decoupling zeros}\}$$

It is noted that the invariant zeros of the system contain all zeros of $G(s)$ (transmission zeros), but not all decoupling zeros. When $S(s)$ is square and nonsingular, the zeros of the system are exactly the invariant zeros of the system. Furthermore, when (A, B, C) is completely controllable and observable, the zeros of the system, the invariant zeros and the zeros of $G(s)$ all coincide.

In addition, one can conclude that the set of eigenvalues of A satisfies

$$\{\text{eigenvalues of } A \text{ (poles of the system)}\} = \{\text{poles of } G(s)\} \cup \{\text{input-decoupling zeros}\}$$
$$\cup \{\text{output-decoupling zeros}\}$$

— {input-output-decoupling zeros}

As discussed in Subsection 3.4.4, a linear time-invariant system (A, B, C) can be divided into the controllable and observable part, the controllable and unobservable part, the uncontrollable and observable part, and the uncontrollable and unobservable part. The transfer function matrix of (A, B, C) equals to the one of the both controllable and observable subsystem $(\tilde{A}_{11}, \tilde{B}_1, \tilde{C}_1)$. On the other hand, the poles and zeros for a transfer function matrix $G(s)$ are defined based on the Smith-McMillan form in Subsection 1.2.4. As a matter of fact, for a linear time-invariant system (A, B, C), the poles of $G(s)$ are exactly the eigenvalues of $\tilde{A}_{11}$.

Example 3.18 Consider $A = \begin{bmatrix} 0 & -1 & 1 \\ 1 & -2 & 1 \\ 0 & 1 & -1 \end{bmatrix}$, $B = \begin{bmatrix} 1 & 0 \\ 1 & 1 \\ 1 & 2 \end{bmatrix}$ and $C = \begin{bmatrix} 0 & 1 & 0 \end{bmatrix}$.

The transfer function matrix is

$$G(s) = \begin{bmatrix} \dfrac{1}{s} & \dfrac{1}{s} \end{bmatrix}$$

The pole polynomial of $G(s)$ is $p(s) = s$, and thus $G(s)$ has only one pole $s_1 = 0$. $\lambda_1 = 0$ is the only controllable and observable eigenvalue of A. The other two eigenvalues of A, $\lambda_2 = -1$, $\lambda_3 = -2$, that are not completely controllable and observable, and do not appear as poles of $G(s)$.

The zero polynomial of $G(s)$ is $z(s) = 1$. $G(s)$ has no zero.

Let

$$S(s) = \begin{bmatrix} sI - A & B \\ -C & D \end{bmatrix} = \begin{bmatrix} s & 1 & -1 & 1 & 0 \\ -1 & s+2 & -1 & 1 & 1 \\ 0 & -1 & s+1 & 1 & 2 \\ 0 & -1 & 0 & 0 & 0 \end{bmatrix}$$

There are two fourth-order minors that include all rows and columns of $sI - A$ obtained by taking columns 1, 2, 3, 4 and columns 1, 2, 3, 5 of $S(s)$, that are $(s+1)(s+2)$ and $(s+1)(s+2)$. The zero polynomial of the system is $z_s(s) = (s+1)(s+2)$ and the zeros of the system are -1 and -2.

To determine the input-decoupling zeros, consider all the third-order nonzero minors of $[\,sI - A \quad B\,]$. The greatest common divisor is $s + 2$, which implies that the input-decoupling zero is -2. Similarly, consider $[\,(sI - A)^\mathrm{T} \quad -C^\mathrm{T}\,]^\mathrm{T}$ and it shows that $s + 1$ is the greatest common divisor of all third-order nonzero minors. Thus the output-decoupling zero is -1. It is noted that there are no input-output-decoupling zeros.

One could work with Smith form of the matrices of interest and the Smith-McMillan form of $G(s)$. In particular, it can be shown that the Smith form of $S(s)$ is

$$\begin{bmatrix} 1 & 0 & 0 & 0 & 0 \\ 0 & 1 & 0 & 0 & 0 \\ 0 & 0 & 1 & 0 & 0 \\ 0 & 0 & 0 & s+2 & 0 \end{bmatrix}$$

the Smith form of $[\ sI - A \quad B\ ]$ is

$$\begin{bmatrix} 1 & 0 & 0 & 0 & 0 \\ 0 & 1 & 0 & 0 & 0 \\ 0 & 0 & s+2 & 0 & 0 \end{bmatrix}$$

the Smith form of $[\ (sI - A)^\mathrm{T} \quad -C^\mathrm{T}\ ]^\mathrm{T}$ is

$$\begin{bmatrix} 1 & 0 & 0 \\ 0 & 1 & 0 \\ 0 & 0 & s+1 \\ 0 & 0 & 0 \end{bmatrix}$$

and the Smith form of $sI - A$ is

$$\begin{bmatrix} 1 & 0 & 0 \\ 0 & 1 & 0 \\ 0 & 0 & s(s+1)(s+2) \end{bmatrix}$$

Meanwhile, it can be shown that the Smith-McMillan form of $G(s)$ is

$$G_\mathrm{M}(s) = \begin{bmatrix} \dfrac{1}{s} & 0 \end{bmatrix}$$

It is straightforward to verify the above results. Note that in the present case the invariant zero polynomial is $z_\mathrm{s}^\mathrm{i}(s) = s + 2$ and there is only one invariant zero at -2.

Assignments

3.1 Determine the controllability and observability of the following systems.

(1) $\dot{x}(t) = \begin{bmatrix} 0 & 1 & 0 \\ 0 & 0 & 1 \\ -2 & -4 & -3 \end{bmatrix} x(t) + \begin{bmatrix} 1 & 0 \\ 0 & 1 \\ -1 & 1 \end{bmatrix} u(t),\ y(t) = \begin{bmatrix} 0 & 1 & -1 \\ 1 & 2 & 1 \end{bmatrix} x(t)$.

(2) $\dot{x}(t) = \begin{bmatrix} 1 & 1 & 0 \\ 0 & 1 & 0 \\ 0 & 0 & 1 \end{bmatrix} x(t) + \begin{bmatrix} 1 & 0 \\ 0 & 1 \\ 1 & 0 \end{bmatrix} u(t),\ y(t) = \begin{bmatrix} c_1 & c_2 & c_3 \end{bmatrix} x(t)$.

(3) $\dot{x}(t) = \begin{bmatrix} 1 & 0 & 0 & 0 \\ 0 & 2 & 0 & 0 \\ 0 & 0 & 3 & 1 \\ 0 & 0 & 0 & 3 \end{bmatrix} x(t) + \begin{bmatrix} 2 & 0 \\ 1 & 4 \\ 0 & 0 \\ 0 & 1 \end{bmatrix} u(t), \; y(t) = \begin{bmatrix} 1 & 2 & 0 & 1 \end{bmatrix} x(t).$

3.2 Determine the controllability and observability of the following systems.

(1) $\dot{x}(t) = \begin{bmatrix} 0 & 1 \\ 0 & t \end{bmatrix} x(t) + \begin{bmatrix} 0 \\ 1 \end{bmatrix} u(t), \; y(t) = \begin{bmatrix} 0 & 1 \end{bmatrix} x(t), \; t \geqslant 0.$

(2) $\dot{x}(t) = \begin{bmatrix} -1 & 0 \\ 0 & -2 \end{bmatrix} x(t) + \begin{bmatrix} e^{-t} \\ e^{-2t} \end{bmatrix} u(t), \; y(t) = \begin{bmatrix} 1 & e^{-t} \end{bmatrix} x(t), \; t \geqslant 0.$

3.3 Determine the range of undetermined coefficients when the following linear time-invariant systems are completely controllable and completely observable simultaneously.

(1) $\dot{x}(t) = \begin{bmatrix} -1 & 1 & a \\ 0 & -2 & 1 \\ 0 & 0 & -3 \end{bmatrix} x(t) + \begin{bmatrix} 0 \\ 0 \\ 1 \end{bmatrix} u(t), \; y(t) = \begin{bmatrix} 0 & 0 & 1 \end{bmatrix} x(t).$

(2) $\dot{x}(t) = \begin{bmatrix} 0 & 0 & 1 \\ 0 & 1 & 0 \\ -2 & -3 & -5 \end{bmatrix} x(t) + \begin{bmatrix} 0 \\ 1 \\ a \end{bmatrix} u(t), \; y(t) = \begin{bmatrix} 0 & 1 & b \end{bmatrix} x(t).$

3.4 Compute the controllability index and observability index of the following system

$$\dot{x}(t) = \begin{bmatrix} 0 & 1 & 0 \\ 0 & 0 & 1 \\ 0 & 3 & -1 \end{bmatrix} x(t) + \begin{bmatrix} 0 & 1 \\ 1 & 0 \\ 0 & 0 \end{bmatrix} u(t), \; y(t) = \begin{bmatrix} 1 & 0 & 1 \\ 0 & 1 & 0 \end{bmatrix} x(t)$$

3.5 Determine the controllability structure decomposition and observability structure decomposition for the following system

$$\dot{x}(t) = \begin{bmatrix} -1 & 1 \\ 0 & 0 \end{bmatrix} x(t) + \begin{bmatrix} 1 \\ 1 \end{bmatrix} u(t), \; y(t) = \begin{bmatrix} 0 & 1 \end{bmatrix} x(t)$$

3.6 Consider the system

$$\dot{x}(t) = \begin{bmatrix} -1 & 0 & 0 \\ 0 & -1 & 0 \\ 0 & 0 & 2 \end{bmatrix} x(t) + \begin{bmatrix} 1 & 0 \\ 0 & 1 \\ 0 & 0 \end{bmatrix} u(t), \; y(t) = \begin{bmatrix} 1 & 1 & 0 \\ 1 & 0 & 0 \end{bmatrix} x(t)$$

Determine the decoupling zeros of the system.

3.7 Give the dynamical equation of a system as

$$\dot{x}(t) = \begin{bmatrix} 1 & 0 & 0 & 0 \\ 0 & 1 & 0 & 0 \\ 0 & 1 & 5 & 1 \\ 0 & 0 & 0 & 6 \end{bmatrix} x(t) + \begin{bmatrix} 0 \\ 0 \\ 0 \\ 1 \end{bmatrix} u(t), \; y(t) = \begin{bmatrix} 1 & 0 & 0 & 1 \end{bmatrix} x(t)$$

(1) Canonically decompose the system.

(2) Determine the poles and zeros of the system.

3.8 Prove that $(A+bk, b)$ is controllable for any k if and only if (A, b) is controllable.

3.9 Consider a linear time-invariant SISO system (A, b, c). If (A, b) is controllable, determine whether there exists c such that (A, c) is always observable.

3.10 Prove that (A, b) is controllable if and only if only $X \equiv 0$ can guarantee $AX = XA$ and $Xb = 0$.

Chapter 4

Canonical Form and Realization of Linear Time-Invariant Systems

The discussion in previous chapters proves that an equivalent transformation does not change the structure and properties of a system. For further revealing the characteristics and the structure properties of a system clearly, analyzing and designing a system conveniently, an equivalent transformation is frequently used to transform a system into the form that can show the controllability and observability directly. Meanwhile, realization means to transform the transfer function matrices, which only show the external causal relation of a system, into the state space descriptions, which show the internal structure properties of a system. The research on realization is helpful to reveal the structure and properties of a system further and to discuss the canonical forms of the system under different descriptions.

4.1 Canonical Form of Linear Time-Invariant Systems

When taking an equivalent transformation for a linear time-invariant system, as discussed in Theorem 3.10 and Theorem 3.20, the controllability and the observability keep invariant. On this foundation, a completely controllable or observable linear time-invariant system can be transformed into the corresponding canonical forms based on an equivalent transformation. These canonical forms play significant roles in system synthesis.

4.1.1 Canonical Form of Linear Time-Invariant SISO Systems

Consider a linear time-invariant SISO system

$$\dot{x}(t) = Ax(t) + bu(t), \; y(t) = cx(t) \tag{4.1}$$

where $x(t) \in \mathbb{R}^n$ is the state vector, $u(t) \in \mathbb{R}^1$ is the input vector, $y(t) \in \mathbb{R}^1$ is the output vector, A, b and c are constant matrices with appropriate dimensions. Let

$$\alpha(s) = s^n + a_1 s^{n-1} + \cdots + a_{n-1} s + a_n \tag{4.2}$$

be the characteristic polynomial of (4.1).

Firstly, introduce the controllable canonical form of the system (4.1).

4.1 Canonical Form of Linear Time-Invariant Systems

Theorem 4.1 *If a linear time-invariant SISO system (4.1) is completely controllable, then there exists an equivalent transformation matrix T_c such that $\bar{x}(t) = T_c x(t)$ transforms the system (4.1) into the controllable canonical form,*

$$\dot{\bar{x}}(t) = \bar{A}_c \bar{x}(t) + \bar{b}_c u(t), \ y(t) = \bar{c}_c \bar{x}(t) \tag{4.3}$$

where

$$\bar{A}_c = \begin{bmatrix} 0 & 1 & 0 & \cdots & 0 \\ 0 & 0 & 1 & \cdots & 0 \\ \vdots & \vdots & \vdots & & \vdots \\ 0 & 0 & 0 & \cdots & 1 \\ -a_n & -a_{n-1} & -a_{n-2} & \cdots & -a_1 \end{bmatrix}, \ \bar{b}_c = \begin{bmatrix} 0 \\ 0 \\ \vdots \\ 0 \\ 1 \end{bmatrix}$$

$$\bar{c}_c = \begin{bmatrix} \beta_n & \beta_{n-1} & \beta_{n-2} & \cdots & \beta_1 \end{bmatrix}$$

$$= c \begin{bmatrix} b & Ab & \cdots & A^{n-1}b \end{bmatrix} \begin{bmatrix} a_{n-1} & a_{n-2} & \cdots & a_1 & 1 \\ a_{n-2} & \vdots & \cdot^{\cdot^{\cdot}} & 1 & \\ \vdots & a_1 & \cdot^{\cdot^{\cdot}} & & \\ a_1 & 1 & & & \\ 1 & & & & \end{bmatrix}$$

with $a_1, \cdots, a_n$ are coefficients in (4.2).

Proof Since the system (4.1) is controllable, then rank U = rank$\begin{bmatrix} b & Ab & \cdots & A^{n-1}b \end{bmatrix}$ = n. Construct

$$Q = \begin{bmatrix} q_1 & q_2 & \cdots & q_n \end{bmatrix}$$

$$= \begin{bmatrix} b & Ab & \cdots & A^{n-1}b \end{bmatrix} \begin{bmatrix} a_{n-1} & a_{n-2} & \cdots & a_1 & 1 \\ a_{n-2} & \vdots & \cdot^{\cdot^{\cdot}} & 1 & \\ \vdots & a_1 & \cdot^{\cdot^{\cdot}} & & \\ a_1 & 1 & & & \\ 1 & & & & \end{bmatrix} \tag{4.4}$$

It is obvious that Q is nonsingular. Select Q^{-1} as the transformation matrix T_c and introduce the transformation $\bar{x}(t) = T_c x(t)$, then $\bar{A}_c = T_c A T_c^{-1}$, i.e., $T_c^{-1} \bar{A}_c = A T_c^{-1}$, which can be rewritten as

$$\begin{bmatrix} q_1 & q_2 & \cdots & q_n \end{bmatrix} \bar{A}_c = A \begin{bmatrix} q_1 & q_2 & \cdots & q_n \end{bmatrix} \tag{4.5}$$

According to (4.4), it holds that

$$\begin{cases} q_1 = a_{n-1}b + a_{n-2}Ab + \cdots + A^{n-1}b \\ q_2 = a_{n-2}b + a_{n-3}Ab + \cdots + A^{n-2}b \\ \vdots \\ q_{n-1} = a_1 b + Ab \\ q_n = b \end{cases}$$

which indicates that

$$Aq_{i+1} = q_i - a_{n-i}q_n, \quad i = 1, \cdots, n-1 \tag{4.6}$$

According to Cayley-Hamilton theorem, $\alpha(A) = 0$. We can further deduce that

$$\begin{aligned} Aq_1 &= a_{n-1}Ab + a_{n-2}A^2b + \cdots + A^n b \\ &= (a_n I + a_{n-1}A + \cdots + A^n)b - a_n b = -a_n q_n \end{aligned} \tag{4.7}$$

Combining (4.6) with (4.7), one can deduce that

$$A \begin{bmatrix} q_1 & q_2 & \cdots & q_n \end{bmatrix} = \begin{bmatrix} q_1 & q_2 & \cdots & q_n \end{bmatrix} \begin{bmatrix} 0 & 1 & 0 & \cdots & 0 \\ 0 & 0 & 1 & \cdots & 0 \\ \vdots & \vdots & \vdots & & \vdots \\ 0 & 0 & 0 & \cdots & 1 \\ -a_n & -a_{n-1} & a_{n-2} & \cdots & -a_1 \end{bmatrix}$$

Compared with (4.5), one can conclude that $\bar{A}_c$ is satisfied with (4.3). Meanwhile, $\bar{b}_c = T_c b$, i.e., $T_c^{-1}\bar{b}_c = b$, which can be rewritten as

$$\begin{bmatrix} q_1 & q_2 & \cdots & q_n \end{bmatrix} \bar{b}_c = b$$

Since $q_n = b$, it can be deduced that $\bar{b}_c$ is satisfied with (4.3).

Finally, $\bar{c}_c = cT_c^{-1}$. Along with (4.1) it is obvious that $\bar{c}_c$ is satisfied with (4.3). $\square$

To achieve the controllable canonical form of a SISO system (4.1), the transformation matrix T_c is required. One can obtain T_c^{-1} according to (4.4), and then compute T_c by solving the inverse of T_c^{-1}. As a matter of fact, one can also construct T_c directly. Let the inverse of the controllability matrix be

$$U^{-1} = \begin{bmatrix} b & Ab & \cdots & A^{n-1}b \end{bmatrix}^{-1} = \begin{bmatrix} p_1^T \\ p_2^T \\ \vdots \\ p_n^T \end{bmatrix}$$

and then the transformation matrix is

$$T_c = \begin{bmatrix} p_n^T \\ p_n^T A \\ \vdots \\ p_n^T A^{n-1} \end{bmatrix} \tag{4.8}$$

Along with (4.4), (4.8) can be verified by $T_c T_c^{-1} = I$.

4.1 Canonical Form of Linear Time-Invariant Systems

Example 4.1 Consider the following linear time-invariant system,

$$\dot{x}(t) = \begin{bmatrix} 1 & 0 & 1 \\ 1 & 1 & 1 \\ 0 & 1 & 1 \end{bmatrix} x(t) + \begin{bmatrix} 1 \\ 0 \\ 1 \end{bmatrix} u(t), \; y(t) = \begin{bmatrix} 2 & 1 & 2 \end{bmatrix} x(t)$$

Obviously, the rank of its controllability matrix is

$$\operatorname{rank} \begin{bmatrix} b & Ab & A^2b \end{bmatrix} = \operatorname{rank} \begin{bmatrix} 1 & 2 & 3 \\ 0 & 2 & 5 \\ 1 & 1 & 3 \end{bmatrix} = 3$$

The system is completely controllable, so it can be transformed into the controllable canonical form. Here,

$$U^{-1} = \begin{bmatrix} b & Ab & A^2b \end{bmatrix}^{-1} = \begin{bmatrix} \dfrac{1}{5} & -\dfrac{3}{5} & \dfrac{4}{5} \\ 1 & 0 & -1 \\ -\dfrac{2}{5} & \dfrac{1}{5} & \dfrac{2}{5} \end{bmatrix}$$

Use the last row of U^{-1} to constitute the transformation matrix T_c,

$$T_c = \begin{bmatrix} p_3^T \\ p_3^T A \\ p_3^T A^2 \end{bmatrix} = \begin{bmatrix} -\dfrac{2}{5} & \dfrac{1}{5} & \dfrac{2}{5} \\ -\dfrac{1}{5} & \dfrac{3}{5} & \dfrac{1}{5} \\ \dfrac{2}{5} & \dfrac{4}{5} & \dfrac{3}{5} \end{bmatrix}, \; T_c^{-1} = \begin{bmatrix} -1 & -1 & 1 \\ -1 & 2 & 0 \\ 2 & -2 & 1 \end{bmatrix}$$

Furthermore,

$$\bar{A}_c = T_c A T_c^{-1} = \begin{bmatrix} 0 & 1 & 0 \\ 0 & 0 & 1 \\ 1 & -2 & 3 \end{bmatrix}, \; \bar{b}_c = T_c b = \begin{bmatrix} 0 \\ 0 \\ 1 \end{bmatrix}, \; \bar{c}_c = c T_c^{-1} = \begin{bmatrix} 1 & -4 & 4 \end{bmatrix}$$

Next, introduce the observable canonical form of the system (4.1).

Theorem 4.2 *If a linear time-invariant SISO system (4.1) is completely observable, then there exists an equivalent transformation matrix T_o such that $\hat{x}(t) = T_o x(t)$ transforms the system (4.1) into the observable canonical form,*

$$\dot{\hat{x}}(t) = \hat{A}_o \hat{x}(t) + \hat{b}_o u(t), y(t) = \hat{c}_o \hat{x}(t) \tag{4.9}$$

where

$$\hat{A}_\mathrm{o} = \begin{bmatrix} 0 & 0 & \cdots & 0 & -a_n \\ 1 & 0 & \cdots & 0 & -a_{n-1} \\ 0 & 1 & \cdots & 0 & -a_{n-2} \\ \vdots & \vdots & & \vdots & \vdots \\ 0 & 0 & \cdots & 1 & -a_1 \end{bmatrix},\ \hat{b}_\mathrm{o} = \begin{bmatrix} \beta_n \\ \beta_{n-1} \\ \beta_{n-2} \\ \vdots \\ \beta_1 \end{bmatrix},\ \hat{c}_\mathrm{o} = \begin{bmatrix} 0 & 0 & \cdots & 0 & 1 \end{bmatrix}$$

with $a_1, \cdots, a_n$ are coefficients in (4.2) and $\beta_1, \cdots, \beta_n$ are presented in Theorem 4.1.

Proof This theorem can be deduced based on Theorem 3.21 and Theorem 4.1. Since (A, b, c) is completely observable, then its dual system $(A^\mathrm{T}, c^\mathrm{T}, b^\mathrm{T})$ is completely controllable. In this way, the observable canonical form of the system (4.1) can be obtained with

$$\hat{A}_\mathrm{o} = \bar{A}_\mathrm{c}^\mathrm{T},\ \hat{b}_\mathrm{o} = \bar{c}_\mathrm{c}^\mathrm{T},\ \hat{c}_\mathrm{o} = \bar{b}_\mathrm{c}^\mathrm{T}$$

which is satisfied with (4.9). □

According to (4.4), it holds that

$$T_\mathrm{c}^{-1} = \begin{bmatrix} c^\mathrm{T} & A^\mathrm{T} c^\mathrm{T} & \cdots & (A^\mathrm{T})^{n-1} c^\mathrm{T} \end{bmatrix} \begin{bmatrix} a_{n-1} & a_{n-2} & \cdots & a_1 & 1 \\ a_{n-2} & \vdots & \iddots & 1 & \\ \vdots & a_1 & \iddots & & \\ a_1 & 1 & & & \\ 1 & & & & \end{bmatrix}$$

Since

$$\hat{A}_\mathrm{o} = T_\mathrm{o} A T_\mathrm{o}^{-1} = \bar{A}_\mathrm{c}^\mathrm{T} = (T_\mathrm{c} A^\mathrm{T} T_\mathrm{c}^{-1})^\mathrm{T} = (T_\mathrm{c}^{-1})^\mathrm{T} A T_\mathrm{c}^\mathrm{T}$$

then one can obtain that

$$T_\mathrm{o} = (T_\mathrm{c}^{-1})^\mathrm{T} = \begin{bmatrix} a_{n-1} & a_{n-2} & \cdots & a_1 & 1 \\ a_{n-2} & \vdots & \iddots & 1 & \\ \vdots & a_1 & \iddots & & \\ a_1 & 1 & & & \\ 1 & & & & \end{bmatrix} \begin{bmatrix} c \\ cA \\ \vdots \\ cA^{n-1} \end{bmatrix} \quad (4.10)$$

Example 4.2 Consider the system given in Example 4.1. The characteristic polynomial of this system is

$$\alpha(s) = s^3 + a_1 s^2 + a_2 s + a_3 = s^3 - 3s^2 + 2s - 1$$

The transformation matrix of the observable canonical form is

$$T_o = \begin{bmatrix} a_2 & a_1 & 1 \\ a_1 & 1 & 0 \\ 1 & 0 & 0 \end{bmatrix} \begin{bmatrix} c \\ cA \\ cA^2 \end{bmatrix} = \begin{bmatrix} 2 & -3 & 1 \\ -3 & 1 & 0 \\ 1 & 0 & 0 \end{bmatrix} \begin{bmatrix} 2 & 1 & 2 \\ 3 & 3 & 5 \\ 6 & 8 & 11 \end{bmatrix} = \begin{bmatrix} 1 & 1 & 0 \\ -3 & 0 & -1 \\ 2 & 1 & 2 \end{bmatrix}$$

and

$$\hat{A}_o = T_o A T_o^{-1} = \begin{bmatrix} 0 & 0 & 1 \\ 1 & 0 & -2 \\ 0 & 1 & 3 \end{bmatrix}, \quad \hat{b}_o = T_o b = \begin{bmatrix} 1 \\ -4 \\ 4 \end{bmatrix}^T, \quad \hat{c}_o = cT_o^{-1} = \begin{bmatrix} 0 & 0 & 1 \end{bmatrix}$$

There are other canonical forms of a linear time-invariant SISO system. Here, do not list them in detail.

4.1.2 Canonical Form of Linear Time-Invariant MIMO Systems

Consider a linear time-invariant MIMO system

$$\dot{x}(t) = Ax(t) + Bu(t), \quad y(t) = Cx(t) \tag{4.11}$$

where $x(t) \in \mathbb{R}^n$ is the state vector, $u(t) \in \mathbb{R}^p$ is the input vector, $y(t) \in \mathbb{R}^q$ is the output vector, A, B and C are constant matrices with appropriate dimensions.

Suppose that the system (4.11) is completely controllable and $\operatorname{rank} B = r \leqslant p$. The controllability matrix U is full rank. Write each column of U as

$$U = \begin{bmatrix} b_1 & \cdots & b_p & Ab_1 & \cdots & Ab_p & \cdots & A^{\mu-1}b_1 & \cdots & A^{\mu-1}b_p \end{bmatrix} \tag{4.12}$$

As discussed in Subsection 3.1.4, select n linearly independent column vectors in turn from left to right from the above matrix. If one column vector cannot be expressed as the linear combination of its left-hand selected columns, it is a linearly independent column vector and one can select it as one of the n linearly independent column vectors. Otherwise, it cannot be selected. It is noted that $\operatorname{rank} B = r$, one can rearrange the above n selected linearly independent column vectors to constitute an equivalent transformation matrix of the system (4.11), it satisfies

$$T_c^{-1} = \begin{bmatrix} b_1 & Ab_1 & \cdots & A^{\mu_1-1}b_1 & \cdots & b_r & Ab_r & \cdots & A^{\mu_r-1}b_r \end{bmatrix} \tag{4.13}$$

where $\{\mu_1, \mu_2, \cdots, \mu_r\}$ is the Kronecker invariable of the system (4.11), and $\mu_1 + \mu_2 + \cdots + \mu_r = n$.

In the sequel, "$*$" in coefficient matrices represents the entry that is possibly nonzero.

Theorem 4.3 *If a linear time-invariant system (4.11) is completely controllable, an equivalent transformation matrix T_c satisfying (4.13) can be obtained and $\bar{x}(t) = T_c x(t)$ transforms the system (4.11) into the Luenberger first controllable canonical form,*

$$\dot{\bar{x}}(t) = \bar{A}_c \bar{x}(t) + \bar{B}_c u(t), \quad y(t) = \bar{C}_c \bar{x}(t) \tag{4.14}$$

where

$$\bar{A}_c = \begin{bmatrix} \bar{A}_{c11} & \cdots & \bar{A}_{c1r} \\ \vdots & & \vdots \\ \bar{A}_{cr1} & \cdots & \bar{A}_{crr} \end{bmatrix}, \quad \bar{B}_c = \begin{bmatrix} \bar{B}_{c1} & * & \cdots & * \\ \vdots & \vdots & & \vdots \\ \bar{B}_{cr} & * & \cdots & * \end{bmatrix}$$

$\bar{C}_c = C T_c^{-1}$ has no special form, and $\bar{A}_{cii}$ is a $\mu_i \times \mu_i$ matrix, $\bar{A}_{cij}$ is a $\mu_i \times \mu_j$ matrix,

$$\bar{A}_{cii} = \begin{bmatrix} 0 & \cdots & 0 & * \\ 1 & \cdots & 0 & * \\ \vdots & & \vdots & \vdots \\ 0 & \cdots & 1 & * \end{bmatrix}, \quad \bar{A}_{cij} = \begin{bmatrix} 0 & \cdots & 0 & * \\ 0 & \cdots & 0 & * \\ \vdots & & \vdots & \vdots \\ 0 & \cdots & 0 & * \end{bmatrix}$$

for $i = 1, \cdots, r$, $j = 1, \cdots, r$, $i \neq j$, $\bar{B}_{ci}$ is a $\mu_i \times r$ matrix, and its entry in the first row and the ith column is 1, while the other entries are 0.

Proof It can be verified that

$$AT_c^{-1} = A \begin{bmatrix} b_1 & Ab_1 & \cdots & A^{\mu_1-1}b_1 & \cdots & b_r & Ab_r & \cdots & A^{\mu_r-1}b_r \end{bmatrix}$$
$$= \begin{bmatrix} Ab_1 & A^2b_1 & \cdots & A^{\mu_1}b_1 & \cdots & Ab_r & A^2b_r & \cdots & A^{\mu_r}b_r \end{bmatrix}$$
$$= \begin{bmatrix} b_1 & Ab_1 & \cdots & A^{\mu_1-1}b_1 & \cdots & b_r & Ab_r & \cdots & A^{\mu_r-1}b_r \end{bmatrix} \bar{A}_c$$

which indicates that $\bar{A}_c = T_c A T_c^{-1}$.

Meanwhile, it is noted that rank $B = r \leqslant p$. Without loss of generality, it is supposed that the first r columns of B are linear independent. In this way, it holds that

$$B = \begin{bmatrix} b_1 & \cdots & b_r & b_{r+1} & \cdots & b_p \end{bmatrix}$$
$$= \begin{bmatrix} b_1 & Ab_1 & \cdots & A^{\mu_1-1}b_1 & \cdots & b_r & Ab_r & \cdots & A^{\mu_r-1}b_r \end{bmatrix} \bar{B}_c$$
$$= T_c^{-1} \bar{B}_c$$

which indicates that $\bar{B}_c = T_c B$.

$\bar{C}_c = C T_c^{-1}$ has no special form. $\square$

More visually, in (4.14), it holds that

4.1 Canonical Form of Linear Time-Invariant Systems

$$\bar{A}_c = \begin{bmatrix} 0 & 0 & \cdots & 0 & * & & & & * & & & & & * \\ 1 & & & & * & & & & * & & & & & * \\ & 1 & & & * & & & & * & \cdots & & & & * \\ & & \ddots & & \vdots & & & & \vdots & & & & & \vdots \\ & & & 1 & * & & & & * & & & & & * \\ \hline & & & & * & 0 & 0 & \cdots & 0 & * & & & & * \\ & & & & * & 1 & & & & * & & & & * \\ & & & & * & & 1 & & & * & \cdots & & & * \\ & & & & \vdots & & & \ddots & & \vdots & & & & * \\ & & & & * & & & & 1 & * & & & & * \\ \hline & & & \vdots & & & & \vdots & & & \ddots & & \vdots & \\ \hline & & & & * & & & & * & 0 & 0 & \cdots & 0 & * \\ & & & & * & & & & * & 1 & & & & * \\ & & & & * & & & & * & & 1 & & & * \\ & & & & \vdots & & & & \vdots & & & \ddots & & * \\ & & & & * & & & & * & & & & 1 & * \end{bmatrix} \begin{matrix} \left.\vphantom{\begin{matrix}1\\1\\1\\1\\1\end{matrix}}\right\}\mu_1 \\ \\ \left.\vphantom{\begin{matrix}1\\1\\1\\1\\1\end{matrix}}\right\}\mu_2 \\ \\ \\ \left.\vphantom{\begin{matrix}1\\1\\1\\1\\1\end{matrix}}\right\}\mu_r \end{matrix}$$

$$\underbrace{}_{\mu_1} \underbrace{}_{\mu_2} \underbrace{}_{\mu_r}$$

$$\bar{B}_c = \begin{bmatrix} 1 & 0 & \cdots & 0 & * & \cdots & * \\ 0 & 0 & \cdots & 0 & * & \cdots & * \\ \vdots & \vdots & & \vdots & \vdots & & \vdots \\ 0 & 0 & \cdots & 0 & * & \cdots & * \\ \hline 0 & 1 & \cdots & 0 & * & \cdots & * \\ 0 & 0 & \cdots & 0 & * & \cdots & * \\ \vdots & \vdots & & \vdots & \vdots & & \vdots \\ 0 & 0 & \cdots & 0 & * & \cdots & * \\ \hline \vdots & \vdots & & \vdots & \vdots & & \vdots \\ \hline 0 & 0 & \cdots & 1 & * & \cdots & * \\ 0 & 0 & \cdots & 0 & * & \cdots & * \\ \vdots & \vdots & & \vdots & \vdots & & \vdots \\ 0 & 0 & \cdots & 0 & * & \cdots & * \end{bmatrix} \begin{matrix} \left.\vphantom{\begin{matrix}1\\1\\1\\1\end{matrix}}\right\}\mu_1 \\ \left.\vphantom{\begin{matrix}1\\1\\1\\1\end{matrix}}\right\}\mu_2 \\ \\ \left.\vphantom{\begin{matrix}1\\1\\1\\1\end{matrix}}\right\}\mu_r \end{matrix}$$

$$\underbrace{}_{r} \underbrace{}_{p-r}$$

Suppose that the system (4.11) is completely observable and rank $C = m \leqslant q$. As discussed in Subsection 3.2.4, one can construct

$$T_o = \begin{bmatrix} c_1 \\ c_1 A \\ \vdots \\ c_1 A^{\nu_1 - 1} \\ \vdots \\ c_m \\ c_m A \\ \vdots \\ c_m A^{\nu_m - 1} \end{bmatrix} \quad (4.15)$$

where $\{\nu_1, \nu_2, \cdots, \nu_m\}$ is the Kronecker invariable of the system (4.11), and $\nu_1 + \nu_2 + \cdots + \nu_m = n$.

Theorem 4.4 *If a linear time-invariant system (4.11) is completely observable, an equivalent transformation matrix T_o satisfying (4.15) can be obtained and $\hat{x}(t) = T_o x(t)$ transforms the system (4.11) into the Luenberger first observable canonical form,*

$$\dot{\hat{x}}(t) = \hat{A}_o \hat{x}(t) + \hat{B}_o u(t), \ y(t) = \hat{C}_o \hat{x}(t) \quad (4.16)$$

where

$$\hat{A}_o = \begin{bmatrix} \hat{A}_{o11} & \cdots & \hat{A}_{o1m} \\ \vdots & & \vdots \\ \hat{A}_{om1} & \cdots & \hat{A}_{omm} \end{bmatrix}, \ \hat{C}_o = \begin{bmatrix} \hat{C}_{o1} & \cdots & \hat{C}_{om} \\ * & \cdots & * \\ \vdots & & \vdots \\ * & \cdots & * \end{bmatrix}$$

$\hat{B}_o = T_o B$ *has no special form, and $\hat{A}_{oii}$ is a $\nu_i \times \nu_i$ matrix. $\hat{A}_{oij}$ is a $\nu_i \times \nu_j$ matrix, which has the following structure,*

$$\hat{A}_{oii} = \begin{bmatrix} 0 & 1 & \cdots & 0 \\ \vdots & \vdots & & \vdots \\ 0 & 0 & \cdots & 1 \\ * & * & * & * \end{bmatrix}, \ \hat{A}_{oij} = \begin{bmatrix} 0 & 0 & \cdots & 0 \\ \vdots & \vdots & & \vdots \\ 0 & 0 & \cdots & 0 \\ * & * & * & * \end{bmatrix}$$

for $i = 1, \cdots, m$, $j = 1, \cdots, m$, $i \neq j$, $\hat{C}_{oi}$ is an $m \times \nu_i$ matrix, and its entry in the ith row and the first column is 1, while the other entries are 0.

The proof is omitted.

Example 4.3 The parameters of a linear time-invariant system (4.11) are given as

4.1 Canonical Form of Linear Time-Invariant Systems

follows,

$$A = \begin{bmatrix} 0 & 0 & 0 & 1 \\ 1 & 0 & 0 & -2 \\ -22 & -11 & -4 & 0 \\ -23 & -6 & 0 & -6 \end{bmatrix}, \quad B = \begin{bmatrix} 0 & 0 \\ 0 & 0 \\ 0 & 1 \\ 1 & 3 \end{bmatrix}, \quad C = \begin{bmatrix} 0 & 0 & 0 & 1 \\ 0 & 0 & 1 & 0 \end{bmatrix}$$

(1) Solve the Luenberger first controllable canonical form.

The controllability matrix is

$$U = \begin{bmatrix} 0 & 0 & 1 & 3 & -6 & -18 & 25 & 75 \\ 0 & 0 & -2 & -6 & 13 & 39 & -56 & -168 \\ 0 & 1 & 0 & -4 & 0 & 16 & -11 & -97 \\ 1 & 3 & -6 & -18 & 25 & 75 & -90 & -270 \end{bmatrix}$$
$$\quad b_1 \quad b_2 \quad Ab_1 \quad Ab_2 \quad A^2b_1 \quad A^2b_2 \quad A^3b_1 \quad A^3b_2$$

Since rank $U = 4$, the system is completely controllable. Construct

$$T_c^{-1} = \begin{bmatrix} b_1 & Ab_1 & A^2b_1 & b_2 \end{bmatrix} = \begin{bmatrix} 0 & 1 & -6 & 0 \\ 0 & -2 & 13 & 0 \\ 0 & 0 & 0 & 1 \\ 1 & -6 & 25 & 3 \end{bmatrix}$$

and one can compute that

$$T_c = \begin{bmatrix} 28 & 11 & -3 & 1 \\ 13 & 6 & 0 & 0 \\ 2 & 1 & 0 & 0 \\ 0 & 0 & 1 & 0 \end{bmatrix}$$

Furthermore, one can obtain that

$$\bar{A}_c = T_c A T_c^{-1} = \begin{bmatrix} 0 & 0 & 27 & 12 \\ 1 & 0 & -11 & 13 \\ 0 & 1 & -6 & 0 \\ 0 & 0 & -11 & -4 \end{bmatrix}, \quad \bar{B}_c = T_c B = \begin{bmatrix} 1 & 0 \\ 0 & 0 \\ 0 & 0 \\ 0 & 1 \end{bmatrix}$$

$$\bar{C}_c = C T_c^{-1} = \begin{bmatrix} 1 & -6 & 25 & 3 \\ 0 & 0 & 0 & 1 \end{bmatrix}$$

(2) Solve the Luenberger first observable canonical form.

The observability matrix is

$$V = \begin{bmatrix} 0 & 0 & 0 & 1 \\ 0 & 0 & 1 & 0 \\ -23 & -6 & 0 & -6 \\ -22 & -11 & -4 & 0 \\ 132 & 36 & 0 & 25 \\ 77 & 44 & 16 & 0 \\ -539 & -150 & 0 & -90 \\ -308 & -176 & -64 & -11 \end{bmatrix} \begin{matrix} c_1 \\ c_2 \\ c_1 A \\ c_2 A \\ c_1 A^2 \\ c_2 A^2 \\ c_1 A^3 \\ c_2 A^3 \end{matrix}$$

Since rank $V = 4$, the system is completely observable. Construct

$$T_o = \begin{bmatrix} c_1 \\ c_1 A \\ c_2 \\ c_2 A \end{bmatrix} = \begin{bmatrix} 0 & 0 & 0 & 1 \\ -23 & -6 & 0 & -6 \\ 0 & 0 & 1 & 0 \\ -22 & -11 & -4 & 0 \end{bmatrix}$$

and then, it holds that

$$T_o^{-1} = \begin{bmatrix} -\dfrac{66}{121} & -\dfrac{11}{121} & \dfrac{24}{121} & \dfrac{6}{121} \\ \dfrac{132}{121} & \dfrac{22}{121} & -\dfrac{92}{121} & -\dfrac{23}{121} \\ 0 & 0 & 1 & 0 \\ 1 & 0 & 0 & 0 \end{bmatrix}$$

Therefore,

$$\hat{A}_o = T_o A T_o^{-1} = \begin{bmatrix} 0 & 1 & 0 & 0 \\ -\dfrac{935}{121} & -\dfrac{660}{121} & -\dfrac{144}{121} & -\dfrac{36}{121} \\ 0 & 0 & 0 & 1 \\ 6 & 1 & -\dfrac{24}{11} & -\dfrac{550}{121} \end{bmatrix}, \quad \hat{B}_o = T_o B = \begin{bmatrix} 1 & 3 \\ -6 & -18 \\ 0 & 1 \\ 0 & -4 \end{bmatrix}$$

$$\hat{C}_o = C T_o^{-1} = \begin{bmatrix} 1 & 0 & 0 & 0 \\ 0 & 0 & 1 & 0 \end{bmatrix}$$

Only the Luenberger first canonical form is introduced for a linear time-invariant MIMO system. There also exists the Luenberger second canonical form, and here do not elaborate it.

4.1 Canonical Form of Linear Time-Invariant Systems

In addition, one can construct the transformation matrix T_{wc} by the following way. Still select n linearly independent column vectors from U in (4.12). Start with b_1 and take $b_1, Ab_1, \cdots, A^{\bar{\mu}_1-1}b_1$ in turn until $A^{\bar{\mu}_1}b_1$ can be expressed by the linear combination of $b_1, Ab_1, \cdots, A^{\bar{\mu}_1-1}b_1$. If $\bar{\mu}_1 = n$, finish the selection of n linearly independent column vectors. It shows that only the first column of B can control the system. If $\bar{\mu}_1 < n$, then select $b_2, Ab_2, \cdots, A^{\bar{\mu}_2}b_2$ until $A^{\bar{\mu}_2}b_2$ can be expressed by the linear combination of all selected column vectors. If $\bar{\mu}_1 + \bar{\mu}_2 < n$, then repeat the above procedure until n linearly independent columns are all selected. Use the selected n linearly independent column vectors to constitute the equivalent transformation matrix which satisfies

$$T_{\text{wc}}^{-1} = \begin{bmatrix} b_1 & Ab_1 & \cdots & A^{\bar{\mu}_1-1}b_1 & \cdots & b_l & Ab_l & \cdots & A^{\bar{\mu}_l-1}b_l \end{bmatrix} \quad (4.17)$$

where $\bar{\mu}_1 + \bar{\mu}_2 + \cdots + \bar{\mu}_l = n$.

Theorem 4.5 *If a linear time-invariant system (4.11) is completely controllable, an equivalent transformation T_{wc} satisfying (4.17) can be obtained and $\bar{x}(t) = T_{\text{wc}}\bar{x}(t)$ transforms the system (4.11) into the Wonham first controllable canonical form,*

$$\dot{\bar{x}}(t) = \bar{A}_{\text{wc}}\bar{x}(t) + \bar{B}_{\text{wc}}u(t), \ y(t) = \bar{C}_{\text{wc}}\bar{x}(t) \quad (4.18)$$

where

$$\bar{A}_{\text{wc}} = \begin{bmatrix} \bar{A}_{\text{wc}11} & \cdots & \bar{A}_{\text{wc}1l} \\ & \ddots & \vdots \\ & & \bar{A}_{\text{wc}ll} \end{bmatrix}, \ \bar{B}_{\text{wc}} = \begin{bmatrix} \bar{B}_{\text{wc}1} & * & \cdots & * \\ \vdots & \vdots & & \vdots \\ \bar{B}_{\text{wc}l} & * & \cdots & * \end{bmatrix}$$

$\bar{C}_{\text{wc}} = CT_{\text{wc}}^{-1}$ *has no special form, and* $\bar{A}_{\text{wc}ii}$ *is a* $\bar{\mu}_i \times \bar{\mu}_i$ *matrix,* $\bar{A}_{\text{wc}ij}$ *is a* $\bar{\mu}_i \times \bar{\mu}_j$ *matrix,*

$$\bar{A}_{\text{wc}ii} = \begin{bmatrix} 0 & \cdots & 0 & * \\ 1 & \cdots & 0 & * \\ \vdots & & \vdots & \vdots \\ 0 & \cdots & 1 & * \end{bmatrix}, \ \bar{A}_{\text{wc}ij} = \begin{bmatrix} 0 & \cdots & 0 & * \\ 0 & \cdots & 0 & * \\ \vdots & & \vdots & \vdots \\ 0 & \cdots & 0 & * \end{bmatrix}$$

for $i = 1, \cdots, l, \ j = 1, \cdots, l, \ i \neq j$, $\bar{B}_{\text{wc}i}$ *is a* $\bar{\mu}_i \times l$ *matrix, and its entry in the first row and the ith column is 1, while the other entries are 0.*

The proof is omitted.

More visually, in (4.18),

$$\bar{A}_{\text{wc}} = \begin{bmatrix} 0 & 0 & \cdots & 0 & * & & & & * & & & & * \\ 1 & & & & * & & & & * & & & & * \\ & 1 & & & * & & & & * & \cdots & & & * \\ & & \ddots & & \vdots & & & & \vdots & & & & \vdots \\ & & & 1 & * & & & & * & & & & * \\ \hline & & & & 0 & 0 & \cdots & 0 & * & & & & * \\ & & & & & 1 & & & * & & & & * \\ & & & & & & 1 & & * & \cdots & & & * \\ & & & & & & & \ddots & \vdots & & & & * \\ & & & & & & & 1 & * & & & & * \\ \hline & & & & & & & & & \ddots & & & \vdots \\ \hline & & & & & & & & & 0 & 0 & \cdots & 0 & * \\ & & & & & & & & & & 1 & & & * \\ & & & & & & & & & & & 1 & & * \\ & & & & & & & & & & & & \ddots & * \\ & & & & & & & & & & & & 1 & * \end{bmatrix} \begin{matrix} \left.\vphantom{\begin{matrix}1\\1\\1\\1\\1\end{matrix}}\right\}\bar{\mu}_1 \\ \\ \left.\vphantom{\begin{matrix}1\\1\\1\\1\\1\end{matrix}}\right\}\bar{\mu}_2 \\ \\ \\ \left.\vphantom{\begin{matrix}1\\1\\1\\1\\1\end{matrix}}\right\}\bar{\mu}_l \end{matrix}$$

$$\underbrace{\qquad\qquad}_{\bar{\mu}_1} \underbrace{\qquad\qquad}_{\bar{\mu}_2} \underbrace{\qquad\qquad}_{\bar{\mu}_l}$$

$$\bar{B}_{\text{wc}} = \begin{bmatrix} 1 & 0 & \cdots & 0 & * & \cdots & * \\ 0 & 0 & \cdots & 0 & * & \cdots & * \\ \vdots & \vdots & & \vdots & \vdots & & \vdots \\ 0 & 0 & \cdots & 0 & * & \cdots & * \\ \hline 0 & 1 & \cdots & 0 & * & \cdots & * \\ 0 & 0 & \cdots & 0 & * & \cdots & * \\ \vdots & \vdots & & \vdots & \vdots & & \vdots \\ 0 & 0 & \cdots & 0 & * & \cdots & * \\ \hline \vdots & \vdots & & \vdots & \vdots & & \vdots \\ \hline 0 & 0 & \cdots & 1 & * & \cdots & * \\ 0 & 0 & \cdots & 0 & * & \cdots & * \\ \vdots & \vdots & & \vdots & \vdots & & \vdots \\ 0 & 0 & \cdots & 0 & * & \cdots & * \end{bmatrix} \begin{matrix} \left.\vphantom{\begin{matrix}1\\0\\ \vdots\\0\end{matrix}}\right\}\bar{\mu}_1 \\ \left.\vphantom{\begin{matrix}0\\0\\ \vdots\\0\end{matrix}}\right\}\bar{\mu}_2 \\ \\ \left.\vphantom{\begin{matrix}0\\0\\ \vdots\\0\end{matrix}}\right\}\bar{\mu}_l \end{matrix}$$

$$\underbrace{\qquad\qquad}_{l} \underbrace{\qquad\qquad}_{p-l}$$

It can be seen that the state coefficient matrix $\bar{A}_{\text{wc}}$ is an upper block-triangle matrix in the Wonham first controllable canonical form. There also exists the Wonham second controllable canonical form, in which the state coefficient matrix is a lower block-triangle matrix.

Following the similar vein to constitute T_{wc}^{-1}, we can construct

$$T_{\text{wo}} = \begin{bmatrix} c_1 \\ c_1 A \\ \vdots \\ c_1 A^{\bar{\nu}_1 - 1} \\ \vdots \\ c_k \\ c_k A \\ \vdots \\ c_k A^{\bar{\nu}_k - 1} \end{bmatrix} \tag{4.19}$$

where $\bar{\nu}_1 + \bar{\nu}_2 + \cdots + \bar{\nu}_k = n$.

Theorem 4.6 *If a linear time-invariant system(4.11) is completely observable, an equivalent transformation T_{wo} satisfying (4.19) can be obtained and $\hat{x}(t) = T_{\text{wo}} x(t)$ transforms the system (4.11) into the Wonham first observable canonical form,*

$$\dot{\hat{x}}(t) = \hat{A}_{\text{wo}} \hat{x}(t) + \hat{B}_{\text{wo}} u(t), \ y(t) = \hat{C}_{\text{wo}} \hat{x}(t) \tag{4.20}$$

where

$$\hat{A}_{\text{wo}} = \begin{bmatrix} \hat{A}_{\text{wo}11} & & \\ \vdots & \ddots & \\ \hat{A}_{\text{wo}k1} & \cdots & \hat{A}_{\text{wo}kk} \end{bmatrix}, \ \hat{C}_{\text{wo}} = \begin{bmatrix} \hat{C}_{\text{wo}1} & \cdots & \hat{C}_{\text{wo}k} \\ * & \cdots & * \\ \vdots & & \vdots \\ * & \cdots & * \end{bmatrix}$$

$\hat{B}_{\text{wo}} = T_{\text{wo}} B$ *has no special form, and $\hat{A}_{\text{wo}ii}$ is a $\bar{\nu}_i \times \bar{\nu}_i$ matrix, $\hat{A}_{\text{wo}ij}$ is a $\bar{\nu}_i \times \bar{\nu}_j$ matrix,*

$$\hat{A}_{\text{wo}ii} = \begin{bmatrix} 0 & 1 & \cdots & 0 \\ \vdots & \vdots & & \vdots \\ 0 & 0 & \cdots & 1 \\ * & * & * & * \end{bmatrix}, \ \hat{A}_{\text{wo}ij} = \begin{bmatrix} 0 & 0 & \cdots & 0 \\ \vdots & \vdots & & \vdots \\ 0 & 0 & \cdots & 0 \\ * & * & * & * \end{bmatrix}$$

for $i = 1, \cdots, k$, $j = 1, \cdots, k$, $i \neq j$, $\hat{C}_{\text{wo}i}$ is a $k \times \bar{\nu}_i$ matrix, and its entry in the ith row and the first column is 1, while the other entries are 0.

The proof is omitted.

Theorem 4.6 can be directly deduced based on the duality principle and Theorem 4.5. Similarly, there exists the Wonham second observable canonical form, which is a dual system of the system with the Wonham second controllable canonical form.

4.2 Controllability and Observability in Frequency Domain

In Chapter 3, the controllability and the observability of linear systems is discussed based on the state space description. As a matter of fact, the transfer function matrix $G(s)$ can also reflects these properties.

As discussed in Subsection 1.3.3, a transfer function matrix of a system (A, B, C, D) can be expressed as

$$G(s) = \frac{C \operatorname{adj}(sI - A) B}{\alpha(s)} + D \qquad (4.21)$$

where $\alpha(s) = \det(sI - A)$ is the characteristic polynomial of A.

Theorem 4.7 (A, B, C) *is both controllable and observable if $C \operatorname{adj}(sI - A) B$ and $\alpha(s)$ have no non-constant common factor.*

Proof Use reduction to absurdity. Suppose that $C \operatorname{adj}(sI - A) B$ and $\alpha(s)$ have no non-constant common factor, but (A, B, C) is incompletely controllable or incompletely observable. Based on Theorem 3.24, there must be a system (A_1, B_1, C_1) with dimension $n_1 < n$, and $n = \dim A$, such that (A_1, B_1, C_1) and (A, B, C) have the same transfer function matrix, i.e.,

$$\frac{C_1 \operatorname{adj}(sI - A_1) B_1}{\det(sI - A_1)} = \frac{C \operatorname{adj}(sI - A) B}{\alpha(s)}$$

Here, $\det(sI - A_1)$ is an n_1 degree polynomial of s. If the above equation holds, $C \operatorname{adj}(sI - A) B$ and $\alpha(s)$ must have a non-constant common factor, which contradicts with the assumption. □

Theorem 4.7 is not a necessary condition. In the next, a necessary condition is presented. Let $d(s)$ be the largest monic common factor of $\operatorname{adj}(sI - A)$, which satisfies $\operatorname{adj}(sI - A) = d(s) P(s)$ and $d(s) \phi(s) = \alpha(s)$. Thereinto, $\phi(s)$ is the minimal polynomial of A, and $P(s)$ is a polynomial matrix. In this way, (4.21) can be represented as

$$G(s) = \frac{C P(s) B}{\phi(s)} + D$$

Theorem 4.8 *If (A, B, C) is both controllable and observable, then $C P(s) B$ and $\phi(s)$ have no non-constant common factors.*

Proof Use reduction to absurdity. Suppose that (A, B, C) is both controllable and observable, but $C P(s) B$ and $\phi(s)$ have a non-constant common factor. In this situation, there exists $s = s_0$ such that $C P(s_0) B = 0$, $\phi(s_0) = 0$. On the other hand, it holds that $\phi(s) I = (sI - A) P(s)$, then one can further deduce that

$$A P(s_0) = \begin{cases} s_0 P(s_0), & s_0 \neq 0 \\ 0, & s_0 = 0 \end{cases}$$

Furthermore, one can obtain that

$$C P(s_0) B = 0, \ C A P(s_0) B = s_0 C P(s_0) B = 0, \ \cdots, \ C A^{n-1} P(s_0) B = 0$$

which indicates that

$$\begin{bmatrix} C \\ CA \\ \vdots \\ CA^{n-1} \end{bmatrix} P(s_0)B = 0$$

Since (A, C) is observable, one can deduce that $P(s_0)B = 0$.

On the other hand, as discussed in Subsection 1.3.3, $\mathrm{adj}(s_0 I - A)$ can be expressed as $\sum_{k=0}^{n-1} p_k(s_0) A^k$, then it holds that

$$d(s_0)P(s_0)B = \mathrm{adj}(s_0 I - A)B = \sum_{k=0}^{n-1} p_k(s_0) A^k B = 0$$

which indicates that

$$\begin{bmatrix} B & AB & \cdots & A^{n-1}B \end{bmatrix} \begin{bmatrix} p_0(s_0) \\ p_1(s_0) \\ \vdots \\ p_{n-1}(s_0) \end{bmatrix} = 0$$

Since $p_{n-1}(s_0) = 1$, then $\begin{bmatrix} B & AB & \cdots & A^{n-1}B \end{bmatrix}$ must be singular. That is to say, (A, B) is incompletely controllable, which contradicts with the assumption. □

Based on the above discussion, a sufficient and necessary condition can be deduced.

Theorem 4.9 *(A, B, C) is both controllable and observable if and only if $p(s) = \alpha(s)$, where $p(s)$ is the pole polynomial of $G(s)$, and $\alpha(s)$ is the characteristic polynomial of A.*

The proof is omitted.

4.3 Realization of Linear Time-Invariant Systems

视频

In this section, the problem that how to determine the state space description of a linear time-invariant system is discussed based on its transfer function (matrix).

4.3.1 Basic Concepts

Definition 4.1 (Realization) For a linear time-invariant system with transfer function matrix $G(s)$, if a state space description

$$\dot{x}(t) = Ax(t) + Bu(t), \ y(t) = Cx(t) + Du(t) \tag{4.22}$$

can be found such that its transfer function matrix satisfies

$$G(s) = C(sI - A)^{-1}B + D \tag{4.23}$$

then the state space description (4.22) is called a *realization* of the transfer function matrix $G(s)$.

In the following, discuss the properties of a realization.

(1) The dimension of the realization reflects the complexity of a system. It depends on the complexity of the structure of a transfer function matrix and the way of achieving the realization.

(2) There is more than one realization for a given transfer function matrix $G(s)$. For a given $G(s)$, based on the ways of achieving the realization, there are different realizations with different dimensions, or realizations with the same dimension but different parameters.

(3) Among all realizations of $G(s)$, there must be a realization, which has the minimal possible dimension, called a minimal realization. It is the simplest state space structure for the given $G(s)$. The minimal realization will be elaborated in Section 4.4.

(4) In general, there are no algebra equivalent relationships among all the realizations of $G(s)$ except the minimal realizations.

(5) The physical essence of the realization is to reflect the internal state structure of $G(s)$, and a minimal realization factually (in the sense of movement rule) reflects the internal state structure of $G(s)$.

Theorem 4.10 *The transfer function matrix $G(s)$ has a realization by a finite dimension dynamical system (4.22) if and only if $G(s)$ is a proper rational fraction matrix.*

Proof Firstly, prove the necessity. Suppose that (4.22) is a realization of $G(s)$, as discussed in Subsection 1.3.3, its transfer function matrix satisfies (4.23), which is a proper rational fraction matrix.

The sufficiency can be proved based on any methods to derive a realization of a transfer function matrix. Here, omit it. □

It can be further concluded that if $G(s)$ is proper, its realization has the form of (A, B, C, D) and

$$D = \lim_{s \to \infty} G(s) \qquad (4.24)$$

while if $G(s)$ is strictly proper, the realization of $G(s)$ must have the form of (A, B, C).

4.3.2 Realization of Linear Time-Invariant SISO Systems

Consider a proper rational transfer function

$$\bar{g}(s) = \frac{\beta_1 s^{n-1} + \beta_2 s^{n-2} + \cdots + \beta_n}{s^n + a_1 s^{n-1} + \cdots + a_n} + d \triangleq g(s) + d \qquad (4.25)$$

where $a_i, \beta_i, i = 1, \cdots, n$ and d are real constants. Since d is the direct transfer part of a realization, it only needs to consider the strictly proper part $g(s)$.

Theorem 4.11 *Consider a transfer function $g(s)$ of a linear time-invariant SISO*

4.3 Realization of Linear Time-Invariant Systems

system in (4.25), *its controllable realization* (A, b, c) *satisfies*

$$\begin{cases} A = \begin{bmatrix} 0 & 1 & 0 & \cdots & 0 \\ 0 & 0 & 1 & \cdots & 0 \\ \vdots & \vdots & \vdots & & \vdots \\ 0 & 0 & 0 & \cdots & 1 \\ -a_n & -a_{n-1} & -a_{n-2} & \cdots & -a_1 \end{bmatrix}, \quad b = \begin{bmatrix} 0 \\ 0 \\ \vdots \\ 0 \\ 1 \end{bmatrix} \\ c = \begin{bmatrix} \beta_n & \beta_{n-1} & \beta_{n-2} & \cdots & \beta_1 \end{bmatrix} \end{cases} \quad (4.26)$$

Proof Denote the numerator polynomial and denominator polynomial of $g(s)$ in (4.25) as $\beta(s)$ and $\alpha(s)$, respectively. It holds that

$$y(s) = g(s)u(s) = \beta(s)\alpha^{-1}(s)u(s)$$

where $y(s) = \mathcal{L}[y(t)]$, $u(s) = \mathcal{L}[u(t)]$.

Introduce $\bar{y}(s) = \alpha^{-1}(s)u(s)$, and then one can obtain that $y(s) = \beta(s)\bar{y}(s)$. Taking the inverse Laplace transform, it holds that

$$\frac{d^n}{dt^n}\bar{y}(t) + a_1 \frac{d^{n-1}}{dt^{n-1}}\bar{y}(t) + \cdots + a_{n-1}\frac{d}{dt}\bar{y}(t) + a_n\bar{y}(t) = u(t) \quad (4.27)$$

and

$$y(t) = \beta_1 \frac{d^{n-1}}{dt^{n-1}}\bar{y}(t) + \cdots + \beta_{n-1}\frac{d}{dt}\bar{y}(t) + \beta_n \bar{y}(t) \quad (4.28)$$

Select the following state variables,

$$x_1(t) = \bar{y}(t), \quad x_2(t) = \frac{d}{dt}\bar{y}(t), \quad \cdots, \quad x_n(t) = \frac{d^{n-1}}{dt^{n-1}}\bar{y}(t) \quad (4.29)$$

Combining (4.27) with (4.29), one can deduce that

$$\begin{cases} \dot{x}_1(t) = x_2(t) \\ \vdots \\ \dot{x}_{n-1}(t) = x_n(t) \\ \dot{x}_n(t) = -a_1 x_n(t) - \cdots - a_{n-1}x_2(t) - a_n x_1(t) + u(t) \end{cases}$$

which can be rewritten as $\dot{x}(t) = Ax(t) + bu(t)$ with A and b satisfying (4.26).

Combining (4.28) with (4.29), it holds that

$$y(t) = \beta_1 x_n(t) + \cdots + \beta_{n-1}x_2(t) + \beta_n x_1(t)$$

which can be rewritten as $y(t) = cx(t)$ with c satisfying (4.26). $\square$

It can be seen that the state space description (A, b, c) in Theorem 4.11 is the controllable canonical form. Similarly, $g(s)$ also has an observable realization.

Theorem 4.12 *Consider a transfer function $g(s)$ of a linear time-invariant SISO system in (4.25), its observable realization (A, b, c) satisfies*

$$A = \begin{bmatrix} 0 & 0 & \cdots & 0 & -a_n \\ 1 & 0 & \cdots & 0 & -a_{n-1} \\ 0 & 1 & \cdots & 0 & -a_{n-2} \\ \vdots & \vdots & & \vdots & \vdots \\ 0 & 0 & \cdots & 1 & -a_1 \end{bmatrix}, \quad b = \begin{bmatrix} \beta_n \\ \beta_{n-1} \\ \beta_{n-2} \\ \vdots \\ \beta_1 \end{bmatrix}, \quad c = \begin{bmatrix} 0 & 0 & \cdots & 0 & 1 \end{bmatrix} \quad (4.30)$$

The proof is omitted.

Example 4.4 A transfer function is given by

$$\bar{g}(s) = \frac{s^4 + 9s^3 + 30s^2 + 46s + 30}{s^4 + 9s^3 + 29s^2 + 39s + 18}$$

Using the polynomial division, one can obtain that

$$\bar{g}(s) = g(s) + d = \frac{s^2 + 7s + 12}{s^4 + 9s^3 + 29s^2 + 39s + 18} + 1$$

$d = 1$ is the direct transfer part of a realization. According to Theorem 4.11, (A_1, b_1, c_1, d) is a realization of $\bar{g}(s)$ with

$$A_1 = \begin{bmatrix} 0 & 1 & 0 & 0 \\ 0 & 0 & 1 & 0 \\ 0 & 0 & 0 & 1 \\ -18 & -39 & -29 & -9 \end{bmatrix}, \quad b_1 = \begin{bmatrix} 0 \\ 0 \\ 0 \\ 1 \end{bmatrix}, \quad c_1 = \begin{bmatrix} 12 & 7 & 1 & 0 \end{bmatrix}$$

According to Theorem 4.12, (A_2, b_2, c_2, d) is also a realization of $\bar{g}(s)$ with

$$A_2 = \begin{bmatrix} 0 & 0 & 0 & -18 \\ 1 & 0 & 0 & -39 \\ 0 & 1 & 0 & -29 \\ 0 & 0 & 1 & -9 \end{bmatrix}, \quad b_2 = \begin{bmatrix} 12 \\ 7 \\ 1 \\ 0 \end{bmatrix}, \quad c_2 = \begin{bmatrix} 0 & 0 & 0 & 1 \end{bmatrix}$$

Since the selection of state variables is different, the realization is not unique. It is worth mentioning that the controllable realization must be controllable while the observable realization must be observable.

4.3.3 Realization of Linear Time-Invariant MIMO Systems

Consider a $q \times p$ proper rational transfer function matrix $\bar{G}(s) = G(s) + D$, in which $D = \lim_{s \to \infty} \bar{G}(s)$ can be directly determined. In the following, only consider the realization

4.3 Realization of Linear Time-Invariant Systems

(A, B, C) of the strictly proper part $G(s)$. $G(s)$ can be represented as

$$G(s) = \begin{bmatrix} g_{11}(s) & \cdots & g_{1p}(s) \\ \vdots & & \vdots \\ g_{q1}(s) & \cdots & g_{qp}(s) \end{bmatrix} = \frac{1}{\alpha(s)} \begin{bmatrix} m_{11}(s) & \cdots & m_{1p}(s) \\ \vdots & & \vdots \\ m_{q1}(s) & \cdots & m_{qp}(s) \end{bmatrix}$$

$$= \frac{1}{\alpha(s)}(G_1 s^{r-1} + \cdots + G_{r-1} s + G_r) \tag{4.31}$$

where $\alpha(s) = s^r + a_1 s^{r-1} + \cdots + a_{r-1} s + a_r$ is the monic least common factor of the denominator of $g_{ij}(s)$, $i = 1, \cdots, q$, $j = 1, \cdots, p$, $\alpha(s) g_{ij}(s) = m_{ij}(s)$, G_k is a $q \times p$ constant matrix, $k = 1, \cdots, r$.

Theorem 4.13 *Consider a strictly proper rational transfer function matrix $G(s)$ of a linear time-invariant MIMO system in (4.31), its controllable realization (A, B, C) satisfies*

$$\begin{cases} A = \begin{bmatrix} O_p & I_p & O_p & \cdots & O_p \\ O_p & O_p & I_p & \cdots & O_p \\ \vdots & \vdots & \vdots & & \vdots \\ O_p & O_p & O_p & \cdots & I_p \\ -a_r I_p & -a_{r-1} I_p & -a_{r-2} I_p & \cdots & -a_1 I_p \end{bmatrix}, B = \begin{bmatrix} O_p \\ O_p \\ \vdots \\ O_p \\ I_p \end{bmatrix} \\ C = \begin{bmatrix} G_r & G_{r-1} & G_{r-2} & \cdots & G_1 \end{bmatrix} \end{cases} \tag{4.32}$$

where O_p denotes a $p \times p$ null matrix and I_p denotes a $p \times p$ identity matrix.

Proof Let

$$(sI - A)^{-1} = \begin{bmatrix} * & \begin{matrix} X_1 \\ X_2 \\ \vdots \\ X_r \end{matrix} \end{bmatrix}$$

where "$*$" represents the entry which does not need to know, and X_k is a $p \times p$ constant matrix, $k = 1, \cdots, r$. On this basis, one can obtain that

$$(sI - A) \begin{bmatrix} * & \begin{matrix} X_1 \\ X_2 \\ \vdots \\ X_r \end{matrix} \end{bmatrix} = \begin{bmatrix} I_{(r-1)p} & \begin{matrix} O_p \\ O_p \\ \vdots \\ I_p \end{matrix} \end{bmatrix}$$

Substituting A in (4.32) into the above equation, one can deduce that

$$\begin{cases} sX_1 = X_2 \\ \vdots \\ sX_{r-1} = X_r \\ sX_r = I_p - (a_r X_1 + a_{r-1} X_2 + \cdots + a_1 X_r) \end{cases} \tag{4.33}$$

which indicates that $X_k = sX_{k-1}$, $k = 2, \cdots, r$. Owing to this fact, it can be derived that

$$\begin{bmatrix} X_1 \\ X_2 \\ \vdots \\ X_r \end{bmatrix} = \frac{1}{\alpha(s)} \begin{bmatrix} I_p \\ sI_p \\ \vdots \\ s^{r-1}I_p \end{bmatrix}$$

Therefore, it holds that

$$C(sI - A)^{-1}B = \begin{bmatrix} G_r & G_{r-1} & G_{r-2} & \cdots & G_1 \end{bmatrix} * \begin{bmatrix} \frac{1}{\alpha(s)} \begin{bmatrix} I_p \\ sI_p \\ \vdots \\ s^{r-1}I_p \end{bmatrix} \end{bmatrix} \begin{bmatrix} O_p \\ O_p \\ \vdots \\ I_p \end{bmatrix}$$

$$= \frac{1}{\alpha(s)}(G_1 s^{r-1} + \cdots + G_{r-1}s + G_r) = G(s)$$

which indicates that (A, B, C) satisfying (4.32) is a realization of $G(s)$. □

It can be verified that $\text{rank} \begin{bmatrix} B & AB & A^2B & \cdots & A^{r-1}B \end{bmatrix} = rp$, and thus the controllable realization in Theorem 4.13 must be controllable, but it could be incompletely observable.

Similarly, we can deduce the observable realization of $G(s)$ in (4.31), which must be observable but could be incompletely controllable.

Theorem 4.14 *Consider a strictly proper rational transfer function matrix $G(s)$ of a linear time-invariant MIMO system in (4.31), its observable realization (A, B, C) satisfies*

$$\begin{cases} A = \begin{bmatrix} O_q & O_q & \cdots & O_q & -a_r I_q \\ I_q & O_q & \cdots & O_q & -a_{r-1} I_q \\ O_q & I_q & \cdots & O_q & -a_{r-2} I_q \\ \vdots & \vdots & & \vdots & \vdots \\ O_q & O_q & \cdots & I_q & -a_1 I_q \end{bmatrix}, \; B = \begin{bmatrix} G_r \\ G_{r-1} \\ G_{r-2} \\ \vdots \\ G_1 \end{bmatrix} \\ C = \begin{bmatrix} O_q & O_q & \cdots & O_q & I_q \end{bmatrix} \end{cases} \quad (4.34)$$

where O_q denotes a $q \times q$ null matrix and I_q denotes a $q \times q$ identity matrix.

The proof is omitted.

Example 4.5 A strictly proper transfer function matrix is given by

$$G(s) = \begin{bmatrix} \dfrac{2}{s+1} & \dfrac{1}{s+1} \\ \dfrac{1}{s+2} & \dfrac{1}{s+2} \end{bmatrix}$$

The least common denominator of each entry of $G(s)$ is $\alpha(s) = (s+1)(s+2) = s^2 + 3s + 2$. $G(s)$ can be expressed as

$$G(s) = \frac{1}{s^2+3s+2}\begin{bmatrix} 2(s+2) & s+2 \\ s+1 & s+1 \end{bmatrix} = \frac{1}{s^2+3s+2}\left(\begin{bmatrix} 2 & 1 \\ 1 & 1 \end{bmatrix}s + \begin{bmatrix} 4 & 2 \\ 1 & 1 \end{bmatrix}\right)$$

Based on Theorem 4.13, the controllable realization of $G(s)$ is

$$A_1 = \begin{bmatrix} O_2 & I_2 \\ -a_2 I_2 & -a_1 I_2 \end{bmatrix} = \begin{bmatrix} 0 & 0 & 1 & 0 \\ 0 & 0 & 0 & 1 \\ -2 & 0 & -3 & 0 \\ 0 & -2 & 0 & -3 \end{bmatrix}, \quad B_1 = \begin{bmatrix} O_2 \\ I_2 \end{bmatrix} = \begin{bmatrix} 0 & 0 \\ 0 & 0 \\ 1 & 0 \\ 0 & 1 \end{bmatrix}$$

$$C_1 = \begin{bmatrix} G_2 & G_1 \end{bmatrix} = \begin{bmatrix} 4 & 2 & 2 & 1 \\ 1 & 1 & 1 & 1 \end{bmatrix}$$

Based on Theorem 4.14, the observable realization of $G(s)$ is

$$A_2 = \begin{bmatrix} O_2 & -a_2 I_2 \\ I_2 & -a_1 I_2 \end{bmatrix} = \begin{bmatrix} 0 & 0 & -2 & 0 \\ 0 & 0 & 0 & -2 \\ 1 & 0 & -3 & 0 \\ 0 & 1 & 0 & -3 \end{bmatrix}, \quad B_2 = \begin{bmatrix} G_2 \\ G_1 \end{bmatrix} = \begin{bmatrix} 4 & 2 \\ 1 & 1 \\ 2 & 1 \\ 1 & 1 \end{bmatrix}$$

$$C_2 = \begin{bmatrix} O_2 & I_2 \end{bmatrix} = \begin{bmatrix} 0 & 0 & 1 & 0 \\ 0 & 0 & 0 & 1 \end{bmatrix}$$

4.4 Minimal Realization of Linear Time-Invariant Systems

 视频

In this section, properties of the minimal realization are discussed, and methods to obtain the minimal realization are introduced.

4.4.1 Basic Concepts

As discussed in Section 4.3, there are different realizations with different dimensions of the same transfer function matrix. Among these realizations, there must be a realization, which has the minimal possible dimension.

Since a proper rational fraction matrix $\bar{G}(s) = G(s) + D$, and $D = \lim_{s \to \infty} G(s)$ can be directly obtained, in the sequel, only discuss the minimal realization of a strictly proper rational fraction matrix $G(s)$.

Definition 4.2 (Minimal realization) Among all realizations of a transfer function matrix $G(s)$, a realization with minimal dimension is called a *minimal realization* or an *irreducible realization*.

Theorem 4.15 (A, B, C) *is a minimal realization of a strictly proper transfer function matrix $G(s)$ if and only if (A, B) is controllable and (A, C) is observable.*

Proof Firstly, prove the necessity.

Use the reduction to absurdity. Let (A, B, C) be the minimal realization of $G(s)$, but (A, B) is incompletely controllable or (A, C) is incompletely observable. In this situation, there exists a pair (A_1, B_1, C_1) which is the structure decomposition of (A, B, C) and is both controllable and observable. (A, B, C) and (A_1, B_1, C_1) have the same transfer function matrix $G(s)$, but $\dim A > \dim A_1$. It indicates that (A, B, C) is not the minimal realization of $G(s)$, which contradicts with the assumption. Therefore, the minimal realization (A, B, C) is both controllable and observable.

Next, prove the sufficiency.

Use the reduction to absurdity. Suppose that (A, B, C) is not a minimal realization of $G(s)$ but it is both controllable and observable. In this situation, there must be a minimal realization $(\bar{A}, \bar{B}, \bar{C})$ such that $n = \dim A > \dim \bar{A} = \bar{n}$.

Since (A, B, C) and $(\bar{A}, \bar{B}, \bar{C})$ has the same transfer function matrix, then any input $u(t)$ will excite the same output under zero initial conditions. That is to say,

$$\int_0^t Ce^{A(t-\tau)}Bu(\tau)d\tau = \int_0^t \bar{C}e^{\bar{A}(t-\tau)}\bar{B}u(\tau)d\tau$$

Owing to the randomicity of $u(\tau)$ and t in the above equation, one can obtain that

$$Ce^{A(t-\tau)}Bu(\tau)d\tau = \bar{C}e^{\bar{A}(t-\tau)}\bar{B}u(\tau)$$

holds for any t and τ.

Taking k-order derivative with respect to t on the both sides of this equation, one can deduce that

$$CA^k e^{A(t-\tau)}B = \bar{C}\bar{A}^k e^{\bar{A}(t-\tau)}\bar{B}, \ k = 0, 1, 2, \cdots$$

Let $t - \tau = 0$, we can further deduce that

$$CA^k B = \bar{C}\bar{A}^k \bar{B}, \ k = 0, 1, 2, \cdots \tag{4.35}$$

which can be expressed as

$$VU = \begin{bmatrix} C \\ CA \\ \vdots \\ CA^{n-1} \end{bmatrix} \begin{bmatrix} B & AB & \cdots & A^{n-1}B \end{bmatrix}$$

$$= \begin{bmatrix} \bar{C} \\ \bar{C}\bar{A} \\ \vdots \\ \bar{C}\bar{A}^{n-1} \end{bmatrix} \begin{bmatrix} \bar{B} & \bar{A}\bar{B} & \cdots & \bar{A}^{n-1}\bar{B} \end{bmatrix} = \bar{V}\bar{U} \tag{4.36}$$

Since (A, B, C) is both controllable and observable, then $\text{rank}\, V = n, \text{rank}\, U = n$, and $\text{rank}\, VU = n$. According to (4.36), it holds that $n = \text{rank}\, VU = \text{rank}\, \bar{V}\bar{U} \leqslant$

4.4 Minimal Realization of Linear Time-Invariant Systems

$\min\{\operatorname{rank}\bar{V}, \operatorname{rank}\bar{U}\}$. On the other hand, $\operatorname{rank}\bar{V} \leqslant \bar{n}$ and $\operatorname{rank}\bar{U} \leqslant \bar{n}$. As a result, $n \leqslant \bar{n}$, which contradicts with the assumption $n > \bar{n}$. This contradiction shows that (A, B, C) is a minimal realization of $G(s)$. □

Combining Theorem 4.9 with Theorem 4.15, one can obtain the following conclusion.

Corollary 4.1 *If (A, B, C) is a minimal realization of $G(s)$, then $\dim A$ equals to the degree of the pole polynomial $p(s)$ of $G(s)$.*

The minimal realization of $G(s)$ is not unique. Any two minimal realizations of $G(s)$ have the following relationship.

Theorem 4.16 *If (A, B, C) and $(\bar{A}, \bar{B}, \bar{C})$ are two minimal realizations of a transfer function matrix $G(s)$, then they must be algebraic equivalent, i.e., there must exist a nonsingular constant matrix T such that $\bar{A} = TAT^{-1}$, $\bar{B} = TB$ and $\bar{C} = CT^{-1}$. Moreover, $T = (\bar{V}^\mathrm{T}\bar{V})^{-1}\bar{V}^\mathrm{T}V$, $T^{-1} = U\bar{U}^\mathrm{T}(\bar{U}\bar{U}^\mathrm{T})^{-1}$, where U and V are the controllability matrix and the observability matrix of (A, B, C), respectively, and $\bar{U}$ and $\bar{V}$ are the controllability matrix and the observability matrix of $(\bar{A}, \bar{B}, \bar{C})$, respectively.*

Proof (A, B, C) and $(\bar{A}, \bar{B}, \bar{C})$ are minimal realizations. According to Theorem 4.15, they are both controllable and observable. Therefore, $\operatorname{rank} U = \operatorname{rank} V = \operatorname{rank} \bar{U} = \operatorname{rank} \bar{V} = n$. Moreover, according to the proof of Theorem 4.15, it holds that

$$VU = \bar{V}\bar{U} \tag{4.37}$$

which indicates that

$$\bar{U} = (\bar{V}^\mathrm{T}\bar{V})^{-1}\bar{V}^\mathrm{T}\bar{V}\bar{U} = (\bar{V}^\mathrm{T}\bar{V})^{-1}\bar{V}^\mathrm{T}VU = TU$$

and

$$\bar{V} = \bar{V}\bar{U}\bar{U}^\mathrm{T}(\bar{U}\bar{U}^\mathrm{T})^{-1} = VU\bar{U}^\mathrm{T}(\bar{U}\bar{U}^\mathrm{T})^{-1} = V\bar{T}$$

Along with (4.37), it can be verified that

$$T\bar{T} = (\bar{V}^\mathrm{T}\bar{V})^{-1}\bar{V}^\mathrm{T}VU\bar{U}^\mathrm{T}(\bar{U}\bar{U}^\mathrm{T})^{-1} = (\bar{V}^\mathrm{T}\bar{V})^{-1}\bar{V}^\mathrm{T}\bar{V}\bar{U}\bar{U}^\mathrm{T}(\bar{U}\bar{U}^\mathrm{T})^{-1} = I$$

That is to say, $T^{-1} = \bar{T} = U\bar{U}^\mathrm{T}(\bar{U}\bar{U}^\mathrm{T})^{-1}$.

Next, $\bar{U} = TU$ and $\bar{V} = VT^{-1}$ can be expressed as

$$\begin{bmatrix} \bar{B} & \bar{A}\bar{B} & \cdots & \bar{A}^{n-1}\bar{B} \end{bmatrix} = \begin{bmatrix} TB & TAB & \cdots & TA^{n-1}B \end{bmatrix}$$

and

$$\begin{bmatrix} \bar{C} \\ \bar{C}\bar{A} \\ \vdots \\ \bar{C}^{n-1}\bar{A} \end{bmatrix} = \begin{bmatrix} CT^{-1} \\ CAT^{-1} \\ \vdots \\ CA^{n-1}T^{-1} \end{bmatrix}$$

respectively, which indicate that $\bar{B} = TB$ and $\bar{C} = CT^{-1}$.

On the other hand, (4.35) holds when (A, B, C) and $(\bar{A}, \bar{B}, \bar{C})$ are both minimal realizations. Owing to this fact, it holds that

$$VAU = \begin{bmatrix} C \\ CA \\ \vdots \\ CA^{n-1} \end{bmatrix} A \begin{bmatrix} B & AB & \cdots & A^{n-1}B \end{bmatrix}$$

$$= \begin{bmatrix} \bar{C} \\ \bar{C}\bar{A} \\ \vdots \\ \bar{C}\bar{A}^{n-1} \end{bmatrix} \bar{A} \begin{bmatrix} \bar{B} & \bar{A}\bar{B} & \cdots & \bar{A}^{n-1}\bar{B} \end{bmatrix} = \bar{V}\bar{A}\bar{U}$$

Pre- and post-multiplying both sides of $VAU = \bar{V}\bar{A}\bar{U}$ by $\bar{V}^{\mathrm{T}}$ and $\bar{U}^{\mathrm{T}}$, respectively, it holds that

$$\bar{V}^{\mathrm{T}} V A U \bar{U}^{\mathrm{T}} = \bar{V}^{\mathrm{T}} \bar{V} \bar{A} \bar{U} \bar{U}^{\mathrm{T}}$$

which indicates that $\bar{A} = (\bar{V}^{\mathrm{T}}\bar{V})^{-1}\bar{V}^{\mathrm{T}} V A U \bar{U}^{\mathrm{T}} (\bar{U}\bar{U}^{\mathrm{T}})^{-1} = TAT^{-1}$. □

In fact, $\bar{V}^{+} = (\bar{V}^{\mathrm{T}}\bar{V})^{-1}\bar{V}^{\mathrm{T}}$ and $\bar{U}^{+} = \bar{U}^{\mathrm{T}}(\bar{U}\bar{U}^{\mathrm{T}})^{-1}$ are the Moore-Penrose inverse of $\bar{V}$ and $\bar{U}$, respectively. In this way, $T = \bar{V}^{+}V$ and $T^{-1} = U\bar{U}^{+}$.

A $q \times p$ proper rational transfer function matrix $\bar{G}(s)$ can be expressed as a Laurent series expansion,

$$\bar{G}(s) = \sum_{i=0}^{\infty} h_i s^{-i} = h_0 + h_1 s^{-1} + h_2 s^{-2} + h_3 s^{-3} + \cdots \qquad (4.38)$$

where $h_i \in \mathbb{R}^{q \times p}$ is called a *Markov parameter*, $i = 0, 1, 2, \cdots$, and can be determined by

$$h_0 = \lim_{s \to \infty} \bar{G}(s)$$

$$h_1 = \lim_{s \to \infty} s(\bar{G}(s) - h_0)$$

$$h_2 = \lim_{s \to \infty} s^2(\bar{G}(s) - h_0 - h_1 s^{-1})$$

$$\vdots$$

and so forth.

Definition 4.3 (Hankel matrix) The *Hankel matrix* $H(i,j)$ of order (i_q, j_p) corresponding to the Markov parameter sequence is defined as

$$H(i,j) \triangleq \begin{bmatrix} h_1 & h_2 & \cdots & h_j \\ h_2 & h_3 & \cdots & h_{j+1} \\ \vdots & \vdots & & \vdots \\ h_i & h_{i+1} & \cdots & h_{i+j+1} \end{bmatrix} \qquad (4.39)$$

4.4 Minimal Realization of Linear Time-Invariant Systems

Theorem 4.17 *Suppose that (A, B, C) is a minimal realization of $G(s)$ with the form of (4.38), then the dimension of the minimal realization n is*

$$\dim A = \operatorname{rank} H(n, n) \tag{4.40}$$

Proof Let (A, B, C) be a minimal realization of the given $G(s)$.
Substituting

$$(sI - A)^{-1} = \sum_{k=0}^{\infty} \frac{A^k}{s^{k+1}}$$

into $G(s) = C(sI - A)^{-1}B$, one can deduce that

$$G(s) = \sum_{k=0}^{\infty} CA^k B s^{-(k+1)} = \sum_{i=1}^{\infty} h_i s^{-i}$$

which indicates that

$$h_i = CA^{i-1}B, \ i = 1, 2, \cdots$$

On this basis, $H(n, n)$ can be expressed as

$$H(n,n) = \begin{bmatrix} C \\ CA \\ \vdots \\ CA^{n-1} \end{bmatrix} \begin{bmatrix} B & AB & \cdots & A^{n-1}B \end{bmatrix} = VU$$

Since (A, B, C) is the minimal realization of $G(s)$, (A, B, C) is both controllable and observable. Thus, $\operatorname{rank} V = \operatorname{rank} U = n$. Furthermore, it holds that

$$\operatorname{rank} V + \operatorname{rank} U - n \leqslant \operatorname{rank} H(n, n) \leqslant \min\{\operatorname{rank} V, \operatorname{rank} U\}$$

which indicates that $\operatorname{rank} H(n, n) = n$.
Using Cayley-Hamilton theorem, one can deduce that

$$\operatorname{rank} H(n + k, n + k) = \operatorname{rank} H(n, n) = n, \ k = 0, 1, 2, \cdots \qquad \square \tag{}$$

As a matter of fact, once the rank of the Hankel matrix keeps invariant as the dimension of Hankel matrix increases, the rank will no longer increase as the dimension of Hankel matrix continues to increase. (4.40) can be revised as

$$\dim A = \operatorname{rank} H(r, r) \tag{4.41}$$

where r is the degree of the least common denominator of all entries of $G(s)$. The proof is omitted. In general, (4.41) is more efficient to determine the dimension of the minimal realization.

4.4.2 Minimal Realization of Linear Time-Invariant SISO Systems

In Subsection 4.3.2, two realization methods are introduced for a linear time-invariant SISO system. On this basis, discuss its minimal realization.

If the given transfer function $\bar{g}(s)$ is proper but not strictly proper, $\bar{g}(s)$ can be expressed as $g(s) + d$ and d can be directly computed by the polynomial division. In the sequel, only discuss the minimal realization of the strictly proper part $g(s)$.

For a given $g(s)$, if its numerator polynomial and denominator polynomial are coprime, then realizations given by both Theorem 4.11 and Theorem 4.12 are minimal realizations of $g(s)$. That is because their characteristic polynomials are equals to the pole polynomial of $g(s)$, and then they are minimal realizations based on Corollary 4.1.

Next, consider the case that there exists a non-constant common factor in the numerator polynomial and denominator polynomial of $g(s)$. To obtain the minimal realization of $g(s)$, one can use Theorem 4.11 to get the controllable realization, on this basis, take the observability structure decomposition by Theorem 3.23, and then the observable subsystem is a minimal realization of $g(s)$ since it is both controllable and observable. Similarly, one can also use Theorem 4.12 to get the observable realization, and then take the controllability structure decomposition by Theorem 3.22 to obtain a minimal realization.

Example 4.6 For a transfer function

$$\bar{g}(s) = g(s) + d = \frac{s^2 + 7s + 12}{s^4 + 9s^3 + 29s^2 + 39s + 18} + 1$$

its controllable realization has been solved in Example 4.4,

$$A_1 = \begin{bmatrix} 0 & 1 & 0 & 0 \\ 0 & 0 & 1 & 0 \\ 0 & 0 & 0 & 1 \\ -18 & -39 & -29 & -9 \end{bmatrix}, \quad b_1 = \begin{bmatrix} 0 \\ 0 \\ 0 \\ 1 \end{bmatrix}, \quad c_1 = \begin{bmatrix} 12 & 7 & 1 & 0 \end{bmatrix}, \quad d = 1$$

The pair (A_1, c_1) is incompletely observable, that is because

$$\operatorname{rank} V_1 = \operatorname{rank} \begin{bmatrix} 12 & 7 & 1 & 0 \\ 0 & 12 & 7 & 1 \\ -18 & -39 & -7 & -2 \\ 36 & 60 & 19 & 1 \end{bmatrix} = 3$$

Select 3 linearly independent rows in V_1 and another row vector to construct an equivalent transformation matrix

4.4 Minimal Realization of Linear Time-Invariant Systems

$$T_1 = \begin{bmatrix} 12 & 7 & 1 & 0 \\ 0 & 12 & 7 & 1 \\ -18 & -39 & -17 & -2 \\ 0 & 1 & 0 & 0 \end{bmatrix}, \quad T_1^{-1} = \begin{bmatrix} \frac{1}{6} & \frac{1}{9} & \frac{1}{18} & -\frac{1}{3} \\ 0 & 0 & 0 & 1 \\ -1 & -\frac{4}{3} & -\frac{2}{3} & -3 \\ 7 & \frac{31}{3} & \frac{14}{3} & 9 \end{bmatrix}$$

It can be computed that

$$\hat{A}_1 = T_1 A_1 T_1^{-1} = \begin{bmatrix} 0 & 1 & 0 & 0 \\ 0 & 0 & 1 & 0 \\ -6 & -11 & -6 & 0 \\ -1 & -\frac{4}{3} & -\frac{2}{3} & -3 \end{bmatrix}, \quad \hat{b}_1 = T_1 b_1 = \begin{bmatrix} 0 \\ 1 \\ -2 \\ 0 \end{bmatrix}$$

$$\hat{c}_1 = c_1 T_1^{-1} = \begin{bmatrix} 1 & 0 & 0 & 0 \end{bmatrix}$$

The minimal realization of $\bar{g}(s)$ is

$$(A_{m1}, b_{m1}, C_{m1}, d) = \left(\begin{bmatrix} 0 & 1 & 0 \\ 0 & 0 & 1 \\ -6 & -11 & -6 \end{bmatrix}, \begin{bmatrix} 0 \\ 1 \\ -2 \end{bmatrix}, \begin{bmatrix} 1 & 0 & 0 \end{bmatrix}, 1 \right)$$

Another method is using the observable realization

$$A_2 = \begin{bmatrix} 0 & 0 & 0 & -18 \\ 1 & 0 & 0 & -39 \\ 0 & 1 & 0 & -29 \\ 0 & 0 & 1 & -9 \end{bmatrix}, \quad b_2 = \begin{bmatrix} 12 \\ 7 \\ 1 \\ 0 \end{bmatrix}, \quad c_2 = \begin{bmatrix} 0 & 0 & 0 & 1 \end{bmatrix}$$

(A_2, b_2) is incompletely controllable, that is because

$$\operatorname{rank} U_2 = \operatorname{rank} \begin{bmatrix} 12 & 0 & -18 & 36 \\ 7 & 12 & -39 & 60 \\ 1 & 7 & -17 & 19 \\ 0 & 1 & -2 & 1 \end{bmatrix} = 3$$

Select 3 linearly independent columns in U_2 and another column vector to construct an equivalent transformation matrix

$$T_2^{-1} = \begin{bmatrix} 12 & 0 & -18 & 0 \\ 7 & 12 & -39 & 1 \\ 1 & 7 & -17 & 0 \\ 0 & 1 & -2 & 0 \end{bmatrix}, \quad T_2 = \begin{bmatrix} \dfrac{1}{6} & 0 & -1 & 7 \\ \dfrac{1}{9} & 0 & -\dfrac{4}{3} & \dfrac{31}{3} \\ \dfrac{1}{18} & 0 & -\dfrac{2}{3} & \dfrac{14}{3} \\ -\dfrac{1}{3} & 1 & -3 & 9 \end{bmatrix}$$

It can be computed that

$$\bar{A}_2 = T_2 A_2 T_2^{-1} = \begin{bmatrix} 0 & 0 & -6 & -1 \\ 1 & 0 & -11 & -\dfrac{4}{3} \\ 0 & 1 & -6 & -\dfrac{2}{3} \\ 0 & 0 & 0 & -3 \end{bmatrix}, \quad \bar{b}_2 = T_2 b_2 = \begin{bmatrix} 1 \\ 0 \\ 0 \\ 0 \end{bmatrix}$$

$$\bar{c}_2 = c_2 T_2^{-1} = \begin{bmatrix} 0 & 1 & -2 & 0 \end{bmatrix}$$

The minimal realization of $\bar{g}(s)$ is

$$(A_{m2}, b_{m2}, c_{m2}, d) = \left(\begin{bmatrix} 0 & 0 & -6 \\ 1 & 0 & -11 \\ 0 & 1 & -6 \end{bmatrix}, \begin{bmatrix} 1 \\ 0 \\ 0 \end{bmatrix}, \begin{bmatrix} 0 & 1 & -2 \end{bmatrix}, 1 \right)$$

As a matter of fact, for a linear time-invariant SISO system (A, b, c), it is both controllable and observable if and only if its transfer function

$$g(s) = \frac{c\,\mathrm{adj}(sI - A)b}{\det(sI - A)}$$

has no zero-pole cancellation.

It is noted that

$$g(s) = \frac{s^2 + 7s + 12}{s^4 + 9s^3 + 29s^2 + 39s + 18} = \frac{s + 4}{s^3 + 6s^2 + 11s + 6} = \tilde{g}(s)$$

In $\tilde{g}(s)$, there is no zero-pole cancellation, and thus both the controllable canonical form of $\tilde{g}(s)$

$$(A_c, b_c, c_c, d) = \left(\begin{bmatrix} 0 & 1 & 0 \\ 0 & 0 & 1 \\ -6 & -11 & -6 \end{bmatrix}, \begin{bmatrix} 0 \\ 0 \\ 1 \end{bmatrix}, \begin{bmatrix} 4 & 1 & 0 \end{bmatrix}, 1 \right)$$

and the observable canonical form of $\tilde{g}(s)$

$$(A_o, b_o, c_o, d) = \left(\begin{bmatrix} 0 & 0 & -6 \\ 1 & 0 & -11 \\ 0 & 1 & -6 \end{bmatrix}, \begin{bmatrix} 4 \\ 1 \\ 0 \end{bmatrix}, \begin{bmatrix} 0 & 0 & 1 \end{bmatrix}, 1 \right)$$

are the minimal realization of $g(s)$.

4.4.3 Minimal Realization of Linear Time-Invariant MIMO Systems

Consider a $q \times p$ proper rational transfer function matrix $\bar{G}(s) = G(s) + D$, in which $D = \lim_{s \to \infty} \bar{G}(s)$ can be directly determined. As presented in Theorem 4.13, one can obtain its controllable realization (A, B, C) satisfies (4.32). If (A, C) is observable, then (A, B, C) is a minimal realization of $G(s)$ since it is both controllable and observable. Otherwise, one can take the observability structure decomposition by Theorem 3.23, and then the observable subsystem $(\hat{A}_{11}, \hat{B}_1, \hat{C}_1)$ is a minimal realization.

Similarly, one can obtain the observable realization by Theorem 4.14, and then take the controllability structure decomposition by Theorem 3.22 to obtain a minimal realization $(\bar{A}_{11}, \bar{B}_1, \bar{C}_1)$.

Example 4.7 For a transfer function matrix

$$G(s) = \begin{bmatrix} \dfrac{2}{s+1} & \dfrac{1}{s+1} \\ \dfrac{1}{s+2} & \dfrac{1}{s+2} \end{bmatrix}$$

its controllable realization has been solved in Example 4.5,

$$A_1 = \begin{bmatrix} 0 & 0 & 1 & 0 \\ 0 & 0 & 0 & 1 \\ -2 & 0 & -3 & 0 \\ 0 & -2 & 0 & -3 \end{bmatrix}, \quad B_1 = \begin{bmatrix} 0 & 0 \\ 0 & 0 \\ 1 & 0 \\ 0 & 1 \end{bmatrix}, \quad C_1 = \begin{bmatrix} 4 & 2 & 2 & 1 \\ 1 & 1 & 1 & 1 \end{bmatrix}$$

Since rank $V_1 = 2$, (A_1, C_1) is incompletely observable. It is noted that rank $C_1 = 2$. Use rows of C_1 to construct a transformation matrix

$$T_1 = \begin{bmatrix} 4 & 2 & 2 & 1 \\ 1 & 1 & 1 & 1 \\ 1 & 0 & 0 & 0 \\ 0 & 1 & 0 & 0 \end{bmatrix}, \quad T_1^{-1} = \begin{bmatrix} 0 & 0 & 1 & 0 \\ 0 & 0 & 0 & 1 \\ 1 & -1 & -3 & -1 \\ -1 & 2 & 2 & 0 \end{bmatrix}$$

One can further obtain that

$$\hat{A}_1 = T_1 A_1 T_1^{-1} = \begin{bmatrix} -1 & 0 & 0 & 0 \\ 0 & -2 & 0 & 0 \\ 1 & -1 & -3 & -1 \\ -1 & 2 & 2 & 0 \end{bmatrix}, \quad \hat{B}_1 = T_1 B_1 = \begin{bmatrix} 2 & 1 \\ 1 & 1 \\ 0 & 0 \\ 0 & 0 \end{bmatrix}$$

$$\hat{C}_1 = C_1 T_1^{-1} = \begin{bmatrix} 1 & 0 & 0 & 0 \\ 0 & 1 & 0 & 0 \end{bmatrix}$$

Therefore, the minimal realization of $G(s)$ is

$$(A_{m1}, B_{m1}, C_{m1}) = \left(\begin{bmatrix} -1 & 0 \\ 0 & -2 \end{bmatrix}, \begin{bmatrix} 2 & 1 \\ 1 & 1 \end{bmatrix}, \begin{bmatrix} 1 & 0 \\ 0 & 1 \end{bmatrix} \right)$$

Another method is using the observable realization of $G(s)$,

$$A_2 = \begin{bmatrix} 0 & 0 & -2 & 0 \\ 0 & 0 & 0 & -2 \\ 1 & 0 & -3 & 0 \\ 0 & 1 & 0 & -3 \end{bmatrix}, \quad B_2 = \begin{bmatrix} 4 & 2 \\ 1 & 1 \\ 2 & 1 \\ 1 & 1 \end{bmatrix}, \quad C_2 = \begin{bmatrix} 0 & 0 & 1 & 0 \\ 0 & 0 & 0 & 1 \end{bmatrix}$$

Since rank $U_2 = 2$, (A_2, B_2) is incompletely controllable. It is noted that rank $B_2 = 2$. Use columns of B_2 to construct a transformation matrix

$$T_2^{-1} = \begin{bmatrix} 4 & 2 & 0 & 0 \\ 1 & 1 & 0 & 0 \\ 2 & 1 & 1 & 0 \\ 1 & 1 & 0 & 1 \end{bmatrix}, \quad T_2 = \begin{bmatrix} \frac{1}{2} & -1 & 0 & 0 \\ -\frac{1}{2} & 2 & 0 & 0 \\ -\frac{1}{2} & 0 & 1 & 0 \\ 0 & -1 & 0 & 1 \end{bmatrix}$$

One can further obtain that

$$\bar{A}_2 = T_2 A_2 T_2^{-1} = \begin{bmatrix} 0 & 1 & -1 & 2 \\ -2 & -3 & 1 & -4 \\ 0 & 0 & -2 & 0 \\ 0 & 0 & 0 & -1 \end{bmatrix}, \quad \bar{B}_2 = T_2 B_2 = \begin{bmatrix} 1 & 0 \\ 0 & 1 \\ 0 & 0 \\ 0 & 0 \end{bmatrix}$$

$$\bar{C}_2 = C_2 T_2^{-1} = \begin{bmatrix} 2 & 1 & 1 & 0 \\ 1 & 1 & 0 & 1 \end{bmatrix}$$

Therefore, the minimal realization of $G(s)$ is

$$(A_{m2}, B_{m2}, C_{m2}) = \left(\begin{bmatrix} 0 & 1 \\ -2 & -3 \end{bmatrix}, \begin{bmatrix} 1 & 0 \\ 0 & 1 \end{bmatrix}, \begin{bmatrix} 2 & 1 \\ 1 & 1 \end{bmatrix} \right)$$

In the next, introduce the Hankel matrix method to obtain the minimal realization.

As discussed in Subsection 4.4.1, a $q \times p$ strictly proper rational transfer function matrix $G(s)$ can be expressed as a Laurent series expansion (4.38). On this basis, the Hankel matrix (4.39) is defined. Consider the rank of the Hankel matrix rank $H(n,n)$ is n, then construct the following matrices:

(1) $F \in \mathbb{R}^{n \times n}$, which is a nonsingular matrix constituted by the first n linearly independent rows and the first n linearly independent columns in $H(n,n)$;

(2) $F^* \in \mathbb{R}^{n \times n}$, which is constituted by the same columns with F but is q rows lower in $H(n,n)$;

(3) $F_1 \in \mathbb{R}^{q \times n}$, which is constituted by the first q row in $H(n,n)$ and has the same columns as F;

4.4 Minimal Realization of Linear Time-Invariant Systems

(4) $F_2 \in \mathbb{R}^{n \times p}$, which is constituted by the first n linearly independent rows in $H(n,n)$ with the first p columns.

Furthermore, let $A = F^*F^{-1}, B = F_2$ and $C = F_1 F^{-1}$, it can be verified (A,B,C) constructed by this way is both controllable and observable, then (A,B,C) is a minimal realization of $G(s)$.

Example 4.8 A given strictly proper rational transfer function matrix is

$$G(s) = \begin{bmatrix} \dfrac{1}{s(s+1)} & \dfrac{1}{s} \\ \dfrac{1}{s} & \dfrac{s-1}{s(s+1)} \end{bmatrix}$$

which can be expressed as a Laurent series expansion with

$$h_1 = \begin{bmatrix} 0 & 1 \\ 1 & 1 \end{bmatrix}, \; h_2 = \begin{bmatrix} 1 & 0 \\ 0 & -2 \end{bmatrix}, \; h_3 = -h_2, \; h_4 = h_2$$

Construct the Hankel matrix

$$H(3,2) = \begin{bmatrix} 0 & 1 & 1 & 0 \\ 1 & 1 & 0 & -2 \\ 1 & 0 & -1 & 0 \\ 0 & -2 & 0 & 2 \\ -1 & 0 & 1 & 0 \\ 0 & 2 & 0 & -2 \end{bmatrix}$$

It can be verified the rank of the Hankel matrix is 4. On this basis, construct

$$F = \begin{bmatrix} 0 & 1 & 1 & 0 \\ 1 & 1 & 0 & -2 \\ 1 & 0 & -1 & 0 \\ 0 & -2 & 0 & 2 \end{bmatrix}, \; F^* = \begin{bmatrix} 1 & 0 & -1 & 0 \\ 0 & -2 & 0 & 2 \\ -1 & 0 & 1 & 0 \\ 0 & 2 & 0 & -2 \end{bmatrix}$$

$$F_1 = \begin{bmatrix} 0 & 1 & 1 & 0 \\ 1 & 1 & 0 & -2 \end{bmatrix}, \; F_2 = \begin{bmatrix} 0 & 1 \\ 1 & 1 \\ 1 & 0 \\ 0 & -2 \end{bmatrix}$$

Furthermore,

$$A = F^*F^{-1} = \begin{bmatrix} 1 & 0 & -1 & 0 \\ 0 & -2 & 0 & 2 \\ -1 & 0 & 1 & 0 \\ 0 & 2 & 0 & -2 \end{bmatrix} \cdot \dfrac{1}{2} \begin{bmatrix} 1 & 1 & 1 & 1 \\ 1 & -1 & 1 & -1 \\ 1 & 1 & -1 & 1 \\ 1 & -1 & 1 & 0 \end{bmatrix} = \begin{bmatrix} 0 & 0 & 1 & 0 \\ 0 & 0 & 0 & 1 \\ 0 & 0 & -1 & 0 \\ 0 & 0 & 0 & -1 \end{bmatrix}$$

$$B = F_2 = \begin{bmatrix} 0 & 1 \\ 1 & 1 \\ 1 & 0 \\ 0 & -2 \end{bmatrix}, \quad C = F_1 F^{-1} = \begin{bmatrix} 1 & 0 & 0 & 0 \\ 0 & 1 & 0 & 0 \end{bmatrix}$$

It can be verified (A, B, C) is both controllable and observable, and $C(sI - A)^{-1}B = G(s)$. Therefore, (A, B, C) is a minimal realization of $G(s)$.

Assignments

4.1 Transform the following dynamical equations into the controllable canonical form and the observable canonical form, and figure out the corresponding transfer functions.

(1) $\dot{x}(t) = \begin{bmatrix} -1 & -2 & -2 \\ 0 & -1 & 1 \\ 0 & 0 & -1 \end{bmatrix} x(t) + \begin{bmatrix} 2 \\ 0 \\ 1 \end{bmatrix} u(t), \; y(t) = \begin{bmatrix} 1 & 1 & 0 \end{bmatrix} x(t).$

(2) $\dot{x}(t) = \begin{bmatrix} 0 & 0 & 1 \\ 1 & 0 & 0 \\ 1 & 1 & 1 \end{bmatrix} x(t) + \begin{bmatrix} 1 \\ 1 \\ 2 \end{bmatrix} u(t), \; y(t) = \begin{bmatrix} 1 & 0 & 1 \\ 1 & 1 & 1 \end{bmatrix} x(t).$

4.2 Transform the following dynamical equations into the Luenberger canonical form.

(1) $\dot{x}(t) = \begin{bmatrix} 1 & 0 & 1 & 0 \\ 2 & 1 & 0 & 1 \\ 1 & 0 & 2 & 0 \\ 0 & 2 & 0 & 3 \end{bmatrix} x(t) + \begin{bmatrix} 0 & 1 \\ 1 & 0 \\ 0 & 1 \\ 1 & 0 \end{bmatrix} u(t), \; y(t) = \begin{bmatrix} 1 & 0 & 0 & 1 \\ 0 & 1 & 1 & 0 \end{bmatrix} x(t).$

(2) $\dot{x}(t) = \begin{bmatrix} 1 & 0 & 1 & 2 \\ 3 & 2 & 1 & 0 \\ 0 & 1 & 1 & 0 \\ 1 & 0 & 0 & 1 \end{bmatrix} x(t) + \begin{bmatrix} 1 & 0 \\ 1 & 1 \\ 0 & 1 \\ 1 & 0 \end{bmatrix} u(t), \; y(t) = \begin{bmatrix} 1 & 0 & 2 & 2 \\ 0 & 1 & 1 & 0 \end{bmatrix} x(t).$

4.3 Transform the dynamical equations in 4.2 into the Wonham canonical form.

4.4 Calculate the minimal realization of the following transfer functions.

(1) $g(s) = \dfrac{s^3 + 1}{s^4 + 2s^3 + 3s + 1}.$

(2) $g(s) = \dfrac{s^2 - s + 1}{s^5 - s^4 + s^3 - s^2 + s - 1}.$

4.5 Use the different methods introduced in Subsection 4.4.3 to obtain the minimal realizations of the following transfer function matrices.

(1) $G(s) = \begin{bmatrix} \dfrac{s^2+1}{s^3} & \dfrac{2s+1}{s^2} \\ \dfrac{s+3}{s^2} & \dfrac{2}{s} \end{bmatrix}.$

(2) $G(s) = \begin{bmatrix} \dfrac{s+2}{s+1} & \dfrac{1}{s+3} \\ \dfrac{s}{s+1} & \dfrac{s+1}{s+2} \end{bmatrix}.$

(3) $G(s) = \begin{bmatrix} \dfrac{1}{s^2-1} & \dfrac{s^2}{s^2+5s+4} \\ \dfrac{1}{s+3} & \dfrac{s+4}{s^2+7s+12} \end{bmatrix}.$

(4) $G(s) = \begin{bmatrix} \dfrac{2s+1}{s^2-3} & \dfrac{s^2+4s+6}{s^2+s+1} \\ \dfrac{4s+5}{s+2} & \dfrac{3s^2+4s+1}{s^2+7s+12} \end{bmatrix}.$

4.6   For a given transfer function matrix

$$G(s) = \frac{1}{s^3+3s^2+2s}\begin{bmatrix} s+1 & 2s^2+s-1 & s^2-1 \\ -s^2-s & -s^2+s & s \end{bmatrix}$$

determine its minimal realization.

4.7   Consider the following system

$$\dot{x}(t) = \begin{bmatrix} 1 & 2 & 1 \\ 0 & 1 & 0 \\ 0 & 3 & 2 \end{bmatrix} x(t) + \begin{bmatrix} 0 & 1 \\ 1 & 0 \\ 1 & 1 \end{bmatrix} u(t), y(t) = \begin{bmatrix} 1 & 0 & 1 \\ 1 & 1 & 1 \end{bmatrix} x(t)$$

Determine which transfer function matrix is the realization and the minimal realization of this system.

(1) $G(s) = \begin{bmatrix} 1 & 0 \\ 1 & -2(s-1) \end{bmatrix} \begin{bmatrix} 0 & -2(s-1)^2 \\ \dfrac{1}{2}(s-2) & s^2+4s-4 \end{bmatrix}^{-1}.$

(2) $G(s) = \begin{bmatrix} 1 & 0 \\ 1 & -2(s-1) \end{bmatrix} \begin{bmatrix} 0 & (s-1)^2 \\ (s-2) & (s^2+2) \end{bmatrix}^{-1}.$

Chapter 5

Stability of Linear Systems

Stability is another important structure property of systems, and it is a qualitative property of systems as well as controllability and observability. The analysis and synthesis of stability is an important part of system control theories. A real system must be stable and only a stable system can be used in engineering practice. Although stability is not the unique requirement for a system to be successfully used in engineering applications, it is the premise and precondition for real system applications. In this chapter, the basic concepts of stability in the sense of Lyapunov stable are explained firstly. Based on these concepts, the internal stability, external stability, the relationships between different stabilities, and some criteria for stability are studied.

5.1 Lyapunov Stability Theory

5.1.1 Basic Concepts

The research of system stability is often limited to the systems with no inputs. Normally, these systems are called autonomous systems. This section is concerned with an autonomous system with the form of first-order ordinary differential equation,

$$\dot{x}(t) = f(x,t), \ x(t_0) = x_0, \ t \geqslant t_0 \tag{5.1}$$

where $x(t) \in \mathbb{R}^n$ is the state vector, $f(x,t)$ is an n-dimensional function vector.

Definition 5.1 (Equilibrium) A point $x_e \in \mathbb{R}^n$ is called an *equilibrium* of (5.1), if

$$f(x_e, t) = 0$$

holds for $\forall t \geqslant t_0$.

In most cases, $x_e = 0$, i.e., the origin of the state space, is an equilibrium of a system. Besides this state, a system can have several nonzero equilibriums. If equilibriums of a system present at several disjunctive isolated positions in the state space, they are called *isolated equilibriums*. For isolated equilibriums, they can always be converted to the origin of the state space by moving the coordinate. Therefore, in following discussion, it is supposed that the equilibrium x_e is at origin.

Suppose that the system (5.1) satisfies conditions of the unique solution, then the movement of $x(t)$ caused by its initial condition x_0 can be expressed as $x(t) = x(t; x_0, t_0)$,

5.1 Lyapunov Stability Theory

$t \geqslant t_0$, which is called the *disturbed movement* of a system. As a matter of fact, the disturbed movement equals to the zero-input response of states.

The stability of the system movement is to study the stability of its equilibrium, i.e., whether a disturbed movement, which departures from the equilibrium, can go back to the equilibrium, or it is restricted within a limited neighborhood around the equilibrium, only based on the inside structure property of the system.

Definition 5.2 (Stability) An equilibrium x_e is said to be *stable* if for any $\varepsilon > 0$, there exists a corresponding real number $\delta(t_0, \varepsilon) > 0$, such that whenever

$$\|x_0 - x_e\| \leqslant \delta(t_0, \varepsilon) \tag{5.2}$$

the disturbed movement satisfies

$$\|x(t; x_0, t_0) - x_e\| \leqslant \varepsilon, \ \forall t \geqslant t_0 \tag{5.3}$$

The geometrical meaning of this definition is that for any positive real number ε, it constructs a super-sphere in the state space with radius ε, and the center is x_e. Its globe field is marked as $S(\varepsilon)$. If there is a corresponding positive real number $\delta(t_0, \varepsilon)$, which depends on both ε and initial time t_0, then one can construct another super-sphere with radius $\delta(t_0, \varepsilon)$ and center at x_e. Its corresponding globe field is marked as $S(\delta)$. If any trajectory $x(t; x_0, t_0)$ of the movement which starts from any state within the field $S(\delta)$ does not break away from the field $S(\varepsilon)$ for all $t \geqslant t_0$, then the equilibrium x_e is said to be stable (in the sense of Lyapunov). For an example with $x(t) \in \mathbb{R}^2$ and $x_e = 0$, the above geometrical meaning is expressed in Fig.5.1.

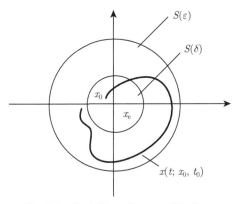

Fig.5.1 Stability of an equilibrium

Definition 5.3 (Uniform stability) An equilibrium x_e is said to be *uniformly stable* if for any $\varepsilon > 0$, there exists a corresponding real number $\delta(\varepsilon) > 0$, such that whenever

$$\|x_0 - x_e\| \leqslant \delta(\varepsilon) \tag{5.4}$$

the disturbed movement satisfies (5.3).

It can be seen that if in particular δ in Definition 5.2 is independent of t_0, the uniform stability is defined. For a time-invariant system, the stability equals to the uniform stability. However, this relationship does not hold for a time-varying system.

Definition 5.4 (Asymptotic stability) An equilibrium x_e is said to be *asymptotically stable* if

(1) it is stable,

(2) for $\delta(t_0, \varepsilon) > 0$ and $\varepsilon > 0$, there exists $T(\varepsilon, \delta, t_0) > 0$, such that whenever the initial state x_0 satisfies (5.2), the disturbed movement satisfies

$$\|x(t; x_0, t_0) - x_e\| \leqslant \varepsilon, \ \forall t \geqslant t_0 + T(\varepsilon, \delta, t_0) \tag{5.5}$$

Especially, if δ and T are both independent of the initial time t_0, then the equilibrium x_e is said to be *uniformly asymptotically stable*.

For $x(t) \in \mathbb{R}^2$, the geometrical meaning of the asymptotic stability is shown in Fig.5.2. Fig.5.2(a) indicates that the bounded property of movement, and Fig.5.2(b) explains the asymptotic property of the movement changing with time. Obviously, if $\varepsilon \to 0$, then $T \to \infty$. Therefore, when the origin equilibrium x_e is asymptotically stable, for any initial state satisfying (5.2), it holds that

$$\lim_{t \to \infty} x(t; x_0, t_0) = 0 \tag{5.6}$$

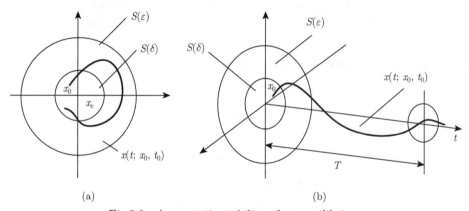

(a)　　　　　　　　　　　　　(b)

Fig.5.2 Asymptotic stability of an equilibrium

The set of all $x_0 \in \mathbb{R}^n$ such that $x(t; x_0, t_0) \to 0$ as $t \to \infty$ for $t_0 \geqslant 0$ is called the *domain of attraction* of the equilibrium $x_e = 0$.

Definition 5.5 (Exponential stability) An equilibrium x_e is said to be *exponential stable* if for $\delta(\varepsilon) > 0$ and $\varepsilon > 0$, there exists $\nu > 0$ such that for any initial state x_0 satisfying (5.4), it holds that

$$\|x(t; x_0, t_0) - x_e\| \leqslant \varepsilon e^{-\nu(t-t_0)}, \ \forall t \geqslant t_0 \tag{5.7}$$

Definition 5.5 gives out the quantitative conception for "asymptotic". Certainly, an equilibrium is exponential stable, then it must be uniformly asymptotically stable.

The aforementioned concepts pertain to local properties. In the next, consider global characterizations of an equilibrium.

5.1 Lyapunov Stability Theory

Definition 5.6 (Global asymptotic stability) The equilibrium $x_e = 0$ is said to be *globally asymptotically stable* if for any limited nonzero initial state x_0, the disturbed movement $x(t; x_0, t_0)$ is bounded, and (5.6) holds.

The equilibrium $x_e = 0$ is globally asymptotically stable, it is also said to be *asymptotically stable in the large*. If an equilibrium x_e is asymptotically stable but not globally asymptotically stable, it is said to be *locally asymptotically stable*. When the equilibrium $x_e = 0$ is globally asymptotically stable, its domain of attraction is $\mathbb{R}^n$. In this case, $x_e = 0$ is the only equilibrium of (5.1). That is to say, the necessary precondition for a system being globally asymptotically stable is that except the origin equilibrium, there is no any other isolated equilibrium for the system. For a linear system, it must be globally asymptotically stable when it is asymptotically stable. In general, it is expected that a system is globally asymptotically stable in engineering fields.

Definition 5.7 (Global exponential stability) An equilibrium $x_e = 0$ is said to be *global exponential stable* if for any $\delta > 0$, there exist $k(\delta) > 0$ and $\nu > 0$, such that whenever $x_0 < \delta$,

$$\|x(t; x_0, t_0) - x_e\| \leqslant k(\delta) \|x_0\| e^{-\nu(t-t_0)}, \quad \forall t \geqslant t_0 \tag{5.8}$$

Definition 5.8 (Instability) An equilibrium x_e is said to be *unstable* if no matter how large the bounded real number $\varepsilon > 0$ is, there is no a corresponding real number $\delta(t_0, \varepsilon) > 0$ such that for any initial state satisfying (5.2), the movement satisfies (5.3).

For $x(t) \in \mathbb{R}^2$, Fig.5.3 indicates the geometrical meaning of the instability. Obviously, if an equilibrium x_e is unstable, then no matter how large $S(\varepsilon)$ is and how small $S(\delta)$ is, there must exist a nonzero state $x_0 \in S(\delta)$ such that the movement starting form x_0 goes beyond $S(\varepsilon)$.

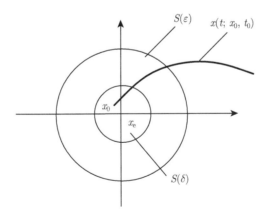

Fig.5.3 Instability of an equilibrium

Example 5.1 Consider a system with $\dot{x}_1(t) = -x_2(t)$ and $\dot{x}_2(t) = -x_1(t)$. The initial states are $x_1(t_0)$ and $x_2(t_0)$, respectively. Obviously, the origin is the unique equilibrium. It can be solved that

$$\begin{bmatrix} x_1(t) \\ x_2(t) \end{bmatrix} = \begin{bmatrix} \cos t & -\sin t \\ \sin t & \cos t \end{bmatrix} \begin{bmatrix} x_1(t_0) \\ x_2(t_0) \end{bmatrix}$$

On this basis, one can further deduce that

$$\|x(t)\| = \sqrt{x_1^2(t) + x_2^2(t)}$$
$$= \sqrt{(\cos t x_1(t_0) - \sin t x_2(t_0))^2 + (\sin t x_1(t_0) + \cos t x_2(t_0))^2}$$
$$= \sqrt{x_1^2(t_0) + x_2^2(t_0)} = \|x(t_0)\|$$

Obviously, for any $\varepsilon > 0$, there exists $\delta(\varepsilon) = \varepsilon$ such that $\|x(t_0)\| \leqslant \delta(\varepsilon)$ and $\|x(t)\| \leqslant \varepsilon$, $\forall t \geqslant t_0$. Therefore, the equilibrium $x_e = 0$ is uniformly stable.

On the other hand, since $\lim\limits_{t \to \infty} \|x(t)\| = \sqrt{x_1^2(t_0) + x_2^2(t_0)} \neq 0$, the equilibrium $x_e = 0$ is not asymptotically stable.

5.1.2 Main Theorems of Lyapunov's Second Method

In 1892, Lyapunov published his famous paper "Common Problem of Movement Stability". He put forward and established the base for the stability theory, and built up a new milestone for the stability research. Since that, Lyapunov's theory and methods have been infiltrating through mathematics, mechanics, control system theory, and many other fields. There are great developments in researches of both theories and applications.

The main methods of Lyapunov's research for stability are the first method and the second method, which are named after him. The first method, namely, the indirect method, is mainly used to analyze system stability by studying the solutions of system differential equations. The second method is called the direct method. It starts from system equations, constructs a Lyapunov function just like an energy function, and analyzes the function and its first-order differential coefficient sign to get the correlative information of the stability of a system. This method has a direct conception, clear physical meaning, and common applicability. Since it was introduced into system and control theory areas in 1960s, it has developed rapidly, applied broadly, and shown its importance in both theories and applications.

Consider a continuous-time nonlinear time-varying system

$$\dot{x}(t) = f(x, t), \ t \geqslant t_0 \tag{5.9}$$

with $f(0, t) = 0$ for any t, i.e., the origin of the state space is the equilibrium.

For the system (5.9), the global uniform asymptotic stability criterion is firstly introduced, which is generally termed as *Lyapunov's main stability theorem*.

Theorem 5.1 *For a system (5.9), if there exists a scalar function $V(x,t)$ such that*

(1) $V(x,t)$, $\dfrac{\partial}{\partial x}V(x,t)$ and $\dfrac{\partial}{\partial t}V(x,t)$ *are continuous for x and t, and $V(0,t) = 0$;*

(2) $V(x,t)$ *is positive-definite and bounded, namely, there are two continuous and unreduced scalar functions $\alpha(\|x\|)$ and $\beta(\|x\|)$ with $\alpha(0) = 0$ and $\beta(0) = 0$, for $\forall t \geqslant t_0$ and $x \neq 0$,*

$$0 < \alpha(\|x\|) \leqslant V(x,t) \leqslant \beta(\|x\|) \tag{5.10}$$

5.1 Lyapunov Stability Theory

(3) $\frac{\partial}{\partial t}V(x,t)$ is negative-definite and bounded, namely, there is a continuous and unreduced scalar function $\gamma(\|x\|)$ with $\gamma(0) = 0$, for $\forall t \geqslant t_0$ and $x \neq 0$,

$$\frac{\partial}{\partial t}V(x,t) \leqslant -\gamma(\|x\|) < 0 \tag{5.11}$$

(4) when $\|x\| \to \infty$, $\alpha(\|x\|) \to \infty$, namely, $V(x,t) \to \infty$,
then $x_e = 0$ is globally uniformly asymptotically stable.

Proof The proof is divided into three steps. Firstly, prove that the equilibrium $x_e = 0$ is uniformly stable. Secondly, prove that this stability is asymptotic. Thirdly, prove that the stability is global.

The first step is to prove that the origin equilibrium $x_e = 0$ is uniformly stable. The geometrical meaning of conditions (1) and (2) in Theorem 5.1 is displayed in Fig.5.4.

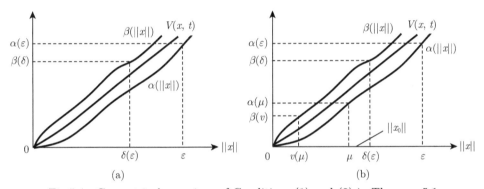

Fig.5.4 Geometrical meanings of Conditions (1) and (2) in Theorem 5.1

Because $\beta(\|x\|)$ is a continuous and unreduced function and $\beta(0) = 0$, there is a corresponding real number $\delta(\varepsilon) > 0$ for any real number $\varepsilon > 0$ such that $\beta(\delta) \leqslant \alpha(\varepsilon)$.

Since $\frac{\partial}{\partial t}V(x,t)$ is negative-definite, it holds that

$$V(x(t;x_0,t_0),t) - V(x_0,t_0) = \int_{t_0}^{t} \frac{\partial}{\partial \tau}V(x(\tau;x_0,t_0),\tau)\,\mathrm{d}\tau \leqslant 0, \quad \forall t \geqslant t_0$$

On this basis, for any t_0 and x_0 which satisfies $\|x_0\| \leqslant \delta(\varepsilon)$, and any $t \geqslant t_0$, one can deduce that

$$\alpha(\varepsilon) \geqslant \beta(\delta) \geqslant V(x_0,t_0) \geqslant V(x(t;x_0,t_0),t) \geqslant \alpha(\|x(t;x_0,t_0)\|)$$

Since $\alpha(\|x\|)$ is a continuous and unreduced function and $\alpha(0) = 0$, for any $t \geqslant t_0$ and any x_0 which satisfies $\|x_0\| \leqslant \delta(\varepsilon)$, it holds that

$$\|x(t;x_0,t_0)\| \leqslant \varepsilon, \quad \forall t \geqslant t_0$$

which indicates that for any real number $\varepsilon > 0$, there is a real number $\delta(\varepsilon) > 0$, such that all the disturbed movements starting from any initial state x_0, where $\|x_0\| \leqslant \delta(\varepsilon)$, satisfy the above inequality. Here, $\delta(\varepsilon)$ is unrelated to the initial time t_0. By Definition 5.3, the origin equilibrium $x_e = 0$ is uniformly stable.

The second step is to prove that $x_e = 0$ is uniformly asymptotically stable.

For $\mu > 0$ and $\delta(\varepsilon) > 0$, introduce $T(\mu, \delta) > 0$. Suppose that the initial time t_0 is arbitrary, and x_0 is an arbitrary initial state satisfying $\|x_0\| \leqslant \delta(\varepsilon)$. Without loss of generality, assume that $\mu \leqslant \|x_0\| \leqslant \delta(\varepsilon)$. Since $V(x,t)$ is bounded, a corresponding real number $0 < \nu(\mu) < \mu$ can be obtained for the given value $\mu > 0$ shown in Fig.5.4 such that $\beta(\nu) \leqslant \alpha(\mu)$. Meanwhile, $\gamma(\|x\|)$ is continuous and unreduced. Let $\rho(\mu, \delta)$ be the minimum value of $\gamma(\|x\|)$ for $0 \leqslant \|x\| \leqslant \varepsilon$, and then select

$$T(\mu, \delta) = \frac{\beta(\delta)}{\rho(\mu, \delta)}$$

In this way, for any $\mu > 0$, $T(\mu, \delta) > 0$ can be found which is independent of the initial time t_0. Furthermore, prove that there exists t_1 satisfying $t_0 \leqslant t_1 \leqslant t_2$ with $t_2 = t_0 + T(\mu, \delta)$ such that $\|x(t_1; x_0, t_0)\| = \nu(\mu)$. Use reduction to absurdity. Assume that for $\forall t \in [t_0, t_2]$, $\|x(t; x_0, t_0)\| > \nu(\mu)$. According to condition (3), one can deduce that

$$0 < \alpha(\nu) \leqslant \alpha(\|x(t_2; x_0, t_0)\|) \leqslant V(x(t_2; x_0, t_0), t_2)$$
$$\leqslant V(x_0, t_0) - (t_2 - t_0)\rho(\mu, \delta) \leqslant \beta(\delta) - T(\mu, \delta)\rho(\mu, \delta) = 0$$

which is contradictory. Therefore, there always exists t_1 in $[t_0, t_2]$ such that $x(t_1; x_0, t_0) = \nu(\mu)$. For $\forall t \geqslant t_1$, it holds that

$$\alpha(\|x(t; x_0, t_0)\|) \leqslant V(x(t; x_0, t_0), t) \leqslant V(x(t_1; x_0, t_0), t_1) \leqslant \beta(\nu) \leqslant \alpha(\mu)$$

which indicate that $\|x(t; x_0, t_0)\| \leqslant \mu$ holds for $\forall t \geqslant t_1$. Since $t_0 + T(\mu, \delta) \geqslant t_1$, then $\|x(t; x_0, t_0)\| \leqslant \mu$ also holds for $\forall t \geqslant t_0 + T(\mu, \delta)$. Meanwhile, as $\mu \to 0$, $T \to \infty$. Therefore, $x_e = 0$ is uniformly asymptotically stable.

The third step is to prove that $x_e = 0$ is globally uniformly asymptotically stable, i.e., to prove the disturbed movement caused by any nonzero initial state $x_0 \in \mathbb{R}^n$ is uniformly bounded.

If $\|x\| \to \infty$, then $\alpha(\|x\|) \to \infty$. For any arbitrarily large, limited real number $\delta > 0$, there must be a limited real number $\varepsilon(\delta) > 0$ such that $\beta(\delta) < \alpha(\varepsilon)$. Thus, for any nonzero x_0 and any $t \geqslant t_0$,

$$\alpha(\varepsilon) > \beta(\delta) \geqslant V(x_0, t_0) \geqslant V(x(t; x_0, t_0), t) \geqslant \alpha(\|x(t; x_0, t_0)\|)$$

Since $\alpha(\|x\|)$ is continuous and unreduced, $\|x(t; x_0, t_0)\| < \varepsilon(\delta)$ holds for $\forall t \geqslant t_0$ and $\forall x_0 \in \mathbb{R}^n$, where $\varepsilon(\delta)$ is independent of the initial time t_0. It indicates that $x(t; x_0, t_0)$ is uniformly bounded for any nonzero $x_0 \in \mathbb{R}^n$. □

It should be emphasized that Theorem 5.1 only gives out a sufficient condition to determine the system (5.9) is globally uniformly asymptotically stable. If a proper scalar function $V(x,t)$ cannot be found for the given system to make it satisfy the conditions in Theorem 5.1, this theorem cannot provide any information for the judgment of the stability.

If condition (4) in Theorem 5.1 is not satisfied and conditions (1)-(3) are only satisfied when $x \in \Omega$, where Ω is a neighborhood of the origin, then one can obtain that the local uniform asymptotic stability criterion.

If a system is defined in the time-invariant domain, i.e.,

$$\dot{x}(t) = f(x), \ t \geqslant 0 \tag{5.12}$$

where $f(0) = 0$ for any $t \geqslant 0$, then Theorem 5.1 can be revised as follow.

Theorem 5.2 *For a time-invariant system (5.12), if there exists a scalar function $V(x)$ such that*
   (1) *$V(x)$ and its first-order derivation are continuous;*
   (2) *$V(x) \geqslant 0$ and $V(x) = 0$ hold only for $x = 0$;*
   (3) *$\dfrac{\mathrm{d}}{\mathrm{d}t} V(x) < 0$ for $\forall x \neq 0$;*
   (4) *$V(x) \to \infty$, as $\|x\| \to \infty$,*
then, $x_{\mathrm{e}} = 0$ is globally uniformly asymptotically stable.

Example 5.2 Consider a continuous time-invariant system

$$\begin{cases} \dot{x}_1 = x_2 - x_1 \left(x_1^2 + x_2^2 \right) \\ \dot{x}_2 = -x_1 - x_2 \left(x_1^2 + x_2^2 \right) \end{cases}$$

Obviously, $x_1 = x_2 = 0$ is the unique equilibrium of the system. Let $V(x) = x_1^2 + x_2^2$, and then,

$$\begin{aligned}
\frac{\mathrm{d}}{\mathrm{d}t} V(x) &= \frac{\partial V(x)}{\partial x_1} \frac{\mathrm{d}x_1}{\mathrm{d}t} + \frac{\partial V(x)}{\partial x_2} \frac{\mathrm{d}x_2}{\mathrm{d}t} \\
&= \begin{bmatrix} \dfrac{\partial V(x)}{\partial x_1} & \dfrac{\partial V(x)}{\partial x_2} \end{bmatrix} \begin{bmatrix} \dot{x}_1 \\ \dot{x}_2 \end{bmatrix} \\
&= \begin{bmatrix} 2x_1 & 2x_2 \end{bmatrix} \begin{bmatrix} x_2 - x_1 \left(x_1^2 + x_2^2 \right) \\ -x_1 - x_2 \left(x_1^2 + x_2^2 \right) \end{bmatrix} \\
&= -2 \left(x_1^2 + x_2^2 \right)^2
\end{aligned}$$

The expressions of $V(x)$ and $\dfrac{\mathrm{d}}{\mathrm{d}t} V(x)$ show that they satisfy the conditions in Theorem 5.2, so $x_{\mathrm{e}} = 0$ is globally uniformly asymptotically stable.

Theorem 5.3 *For a time-invariant system (5.12), if there exists a scalar function $V(x)$ such that*
   (1) *$V(x)$ and its first-order derivation are continuous;*
   (2) *$V(x) \geqslant 0$ and $V(x) = 0$ hold only for $x = 0$;*

(3) $\dfrac{\mathrm{d}}{\mathrm{d}t}V(x) \leqslant 0$;

(4) $\dfrac{\mathrm{d}}{\mathrm{d}t}V(x(t;x_0,0)) \neq 0$ for any nonzero $x_0 \in \mathbb{R}^n$;

(5) $V(x) \to \infty$, as $\|x\| \to \infty$,

then, $x_\mathrm{e} = 0$ is globally uniformly asymptotically stable.

Example 5.3 Consider a continuous time-invariant system

$$\begin{cases} \dot{x}_1 = x_2 \\ \dot{x}_2 = -x_1 - (1+x_2)^2 x_2 \end{cases}$$

Obviously, $x_1 = x_2 = 0$ is the unique equilibrium of the system. Assume that $V(x) = x_1^2 + x_2^2$, and then, it holds that

$$\begin{aligned} \dfrac{\mathrm{d}}{\mathrm{d}t}V(x) &= \begin{bmatrix} \dfrac{\partial V(x)}{\partial x_1} & \dfrac{\partial V(x)}{\partial x_2} \end{bmatrix} \begin{bmatrix} \dot{x}_1 \\ \dot{x}_2 \end{bmatrix} \\ &= \begin{bmatrix} 2x_1 & 2x_2 \end{bmatrix} \begin{bmatrix} x_2 \\ -x_1 - (1+x_2)^2 x_2 \end{bmatrix} \\ &= -2x_2^2(1+x_2)^2 \end{aligned}$$

The expressions of $V(x)$ and $\dfrac{\mathrm{d}}{\mathrm{d}t}V(x)$ show that they satisfy conditions (1), (2), (3) and (5) of in Theorem 5.3. It should check the condition (4) for the expressions. $\dfrac{\mathrm{d}}{\mathrm{d}t}V(x) < 0$ holds except for two conditions: (a) x_1 is arbitrary and $x_2 = 0$; (b) x_1 is arbitrary and $x_2 = -1$. In these two cases, $\dfrac{\mathrm{d}}{\mathrm{d}t}V(x) = 0$.

In case (a), $x_2 = 0$. Therefore,

$$\begin{cases} \dot{x}_1 = x_2 = 0 \\ 0 = \dot{x}_2 = -x_1 - (1+x_2)^2 x_2 = -x_1 \end{cases}$$

which shows that except for $x_1 = x_2 = 0$, $\bar{x}(t;x_0,0) = \begin{bmatrix} x_1(t) & 0 \end{bmatrix}^\mathrm{T}$ is not the solution of the disturbed system.

In case (b), $x_2(t) = -1$. Therefore,

$$\begin{cases} \dot{x}_1 = x_2 = -1 \\ 0 = \dot{x}_2 = -x_1 - (1+x_2)^2 x_2 = -x_1 \end{cases}$$

which is an ambivalent result. It shows that $\tilde{x}(t;x_0,0) = \begin{bmatrix} x_1(t) & -1 \end{bmatrix}^\mathrm{T}$ is not the solution of the disturbed system neither. Thereby condition (4) in Theorem 5.3 holds. In conclusion, the origin equilibrium is globally uniformly asymptotically stable.

If it is difficult to determine the uniform asymptotic stability of a system, we can settle for the uniform stability of a system.

Theorem 5.4 *For a system (5.9), if there exists a continuous scalar function $V(x,t)$ with $V(0,t) = 0$, which has the continuous first-order partial derivation for x and t, and there exists a domain of attraction Ω which surrounds the origin, such that for any $x \in \Omega$ and $t \geqslant t_0$,*

(1) $V(x,t)$ *is positive-definite and bounded,*

(2) $\dfrac{\partial}{\partial t}V(x,t)$ *is semi-negative-definite and bounded,*

then, $x_e = 0$ is uniformly stable in the area Ω.

Theorem 5.5 *For a time-invariant system (5.12), if there exists a continuous scalar function $V(x)$ with $V(0) = 0$, which has the continuous first-order derivation for x, there exists a domain of attraction Ω which surrounds the origin, such that for any $x \in \Omega$ and $t \geqslant t_0$,*

(1) $V(x)$ *is positive-definite and bounded,*

(2) $\dfrac{\mathrm{d}}{\mathrm{d}t}V(x)$ *is semi-negative-definite,*

then, $x_e = 0$ is uniformly stable in the area Ω.

Finally, introduce the instability criteria.

Theorem 5.6 *For a system (5.9), if there exists a continuous scalar function $V(x,t)$ with $V(0,t) = 0$, which has the continuous first-order partial derivation for x and t, and there exists a domain of attraction Ω which surrounds the origin, such that for any $x \in \Omega$ and $t \geqslant t_0$,*

(1) $V(x,t)$ *is positive-definite and bounded,*

(2) $\dfrac{\partial}{\partial t}V(x,t)$ *is positive-definite and bounded,*

then, $x_e = 0$ is unstable.

Theorem 5.7 *For a time-invariant system (5.12), if there exists a continuous scalar function $V(x)$ with $V(0) = 0$, which has the continuous first-order partial derivation for x, there exists a domain of attraction Ω which surrounds the origin, such that for any $x \in \Omega$ and $t \geqslant t_0$,*

(1) $V(x)$ *is positive-definite,*

(2) $\dfrac{\mathrm{d}}{\mathrm{d}t}V(x)$ *is positive-definite,*

then, $x_e = 0$ is unstable.

5.2 Internal Stability of Linear Systems

In this section, the stability criteria are introduced for linear systems.

5.2.1 Internal Stability of Linear Time-Varying Systems

Consider the following linear time-invariant system,

$$\dot{x}(t) = A(t)x(t), \ x(0) = x_0, \ t \geqslant t_0 \quad (5.13)$$

where $x(t) \in \mathbb{R}^n$ is the state vector, $A(t)$ is a time-varying matrix. Assume that (5.13) has the unique solution. $x_e = 0$ is the equilibrium of the system (5.13).

Theorem 5.8 *For a linear time-varying system (5.13), the equilibrium $x_e = 0$ is stable if and only if*

$$\|\Phi(t, t_0)\| \leqslant k(t_0) < \infty$$

where $\Phi(t, t_0)$ is the state transition matrix of (5.13), and $k(t_0)$ is a constant that may depend on the choice of t_0.

Proof Firstly, prove the sufficiency.

It holds that

$$\|x(t; x_0, t_0)\| = \|\Phi(t, t_0)x_0\| \leqslant \|\Phi(t, t_0)\|\|x_0\|$$

Let $\delta(t_0, \varepsilon) = \varepsilon/k(t_0)$ for any real number $\varepsilon > 0$. If $\|x_0\| < \delta(t_0, \varepsilon)$, then one can deduce that

$$\|x(t; x_0, t_0)\| \leqslant \delta(t_0, \varepsilon)k(t_0) = \varepsilon, \ \forall t \geqslant t_0$$

which indicates that $x_e = 0$ is stable.

Next, prove the necessity.

Suppose that $x_e = 0$ is stable, but $\|\Phi(t, t_0)\|$ is unbounded. In this situation, there exists at least one entry in $\|\Phi(t, t_0)\|$ that is unbounded for $t \geqslant t_0$. Without loss of generality, assume that $\Phi_{ij}(t, t_0)$ satisfies

$$|\Phi_{ij}(t, t_0)| > k > 0$$

for $t \geqslant t_0$, where $\Phi_{ij}(t, t_0)$ denotes the entry of $\Phi(t, t_0)$ in the ith row and the jth column, $i \in \{1, \cdots, n\}$ and $j \in \{1, \cdots, n\}$. Select $x_0 = [\ 0 \ \cdots \ 0 \ \delta \ 0 \ \cdots \ 0\ ]^{\mathrm{T}}$, where the jth entry in x_0 is δ. One can deduce that

$$x(t; x_0, t_0) = \Phi_j(t, t_0)\delta$$

where $\Phi_j(t, t_0)$ is the jth column of $\Phi(t, t_0)$. It is noted that $x_e = 0$. Selecting $k = \varepsilon/\delta$, one can further obtain that

$$\|x(t; x_0, t_0) - x_e\| = \|\Phi_j(t, t_0)\delta\| > k\delta = \varepsilon, \ \forall t \geqslant t_0$$

which indicates that no matter how small δ is, $\|x(t; x_0, t_0) - x_e\| > \varepsilon$ always holds. It contradicts with the stability of the equilibrium. □

5.2 Internal Stability of Linear Systems

Theorem 5.9 *For a linear time-varying system* (5.13), *the equilibrium* $x_e = 0$ *is uniformly stable if and only if*

$$\|\Phi(t, t_0)\| \leqslant k < \infty$$

where $\Phi(t, t_0)$ *is the state transition matrix of* (5.13), *and* k *is a constant independent of* t_0.

The proof is similar to the proof of Theorem 5.8.

Theorem 5.10 *For a linear time-varying system* (5.13), *the following statements are equivalent*:

(1) *the equilibrium* $x_e = 0$ *is asymptotically stable*;
(2) *the equilibrium* $x_e = 0$ *is globally asymptotically stable*;
(3) $\lim\limits_{t \to \infty} \|\Phi(t, t_0)\| = 0$, *where* $\Phi(t, t_0)$ *is the state transition matrix of* (5.13).

Proof Assume that the statement (1) is true, then there exists a constant $\eta(t_0) > 0$ such that when $\|x_0\| \leqslant \eta(t_0)$, it holds that $\lim\limits_{t \to \infty} x(t; x_0, t_0) = 0$. As a matter of fact, for any $x_0 \neq 0$, owing to the homogeneity of linear system, one can deduce that

$$\lim_{t \to \infty} x(t; x_0, t_0) = \lim_{t \to \infty} x\left(t; \frac{\eta(t_0)}{\|x_0\|} x_0, t_0\right) \bigg/ \left(\frac{\eta(t_0)}{\|x_0\|}\right) = 0$$

It follows that the statement (2) is true.

Next, assume that the statement (2) is true. According to Definition 5.6, it holds that

$$\lim_{t \to \infty} x(t; x_0, t_0) = \lim_{t \to \infty} \Phi(t, t_0) x_0 = 0$$

for any $x_0 \neq 0$. Therefore, $\lim\limits_{t \to \infty} \Phi(t, t_0) = 0$, and the statement (3) can be further deduced.

Finally, assume that the statement (3) is true, then $\|\Phi(t, t_0)\|$ is bounded for all $t \geqslant t_0$. According to Theorem 5.8, $x_e = 0$ is stable. Furthermore, consider $t_0 \geqslant 0$, if there exists a constant $\eta(t_0) > 0$ such that $\|x_0\| \leqslant \eta(t_0)$, then $\|x(t; x_0, t_0)\| \leqslant \|x_0\| \|\Phi(t, t_0)\| \leqslant \eta(t_0) \|\Phi(t, t_0)\|$. In this way, $\lim\limits_{t \to \infty} \|x(t; x_0, t_0)\| = 0$. It can be further deduced that the statement (1) is true. □

Theorem 5.11 *For a linear time-varying system* (5.13), *the following statements are equivalent*:

(1) *the equilibrium* $x_e = 0$ *is uniformly asymptotically stable*;
(2) *the equilibrium* $x_e = 0$ *is exponentially stable*;
(3) *there exist* $\mu > 0$ *and* $\nu > 0$ *such that for any* t_0 *and* $t \geqslant t_0$,

$$\|\Phi(t, t_0)\| \leqslant \mu e^{-\nu(t-t_0)} \tag{5.14}$$

holds, where $\Phi(t, t_0)$ *is the state transition matrix of* (5.13).

Proof Assume that the statement (1) is true. In this situation, for any given $\varepsilon > 0$, there exist $\delta > 0$ and $T > 0$ such that for any initial state satisfying $\|x_0\| \leqslant \delta$, the disturbed movement stating from x_0 satisfies $\|x(t; x_0, t_0)\| \leqslant \varepsilon$ for $t \geqslant t_0 + T$. Without loss of generality, assume that when $t = t_0 + T$, $\|x_0\| = \delta$ and $\varepsilon = \delta/2$, then one can deduce that

$$\|x(t; x_0, t_0)\| = \|\Phi(t, t_0)x_0\| = \|\Phi(t_0 + T, t_0)x_0\| = \|\Phi(t_0 + T, t_0)\|\delta \leqslant \frac{\delta}{2}$$

holds for any t_0, which implies that

$$\|\Phi(t_0 + T, t_0)\| \leqslant \frac{1}{2}$$

Since $x_e = 0$ is uniformly stable, according to Theorem 5.9, there exists a constant k such that $\|\Phi(t, t_0)\| \leqslant k$ for $t \geqslant t_0$. Furthermore, it holds that for $\forall t \in [t_0 + T, t_0 + 2T]$,

$$\|\Phi(t, t_0)\| = \|\Phi(t, t_0 + T)\Phi(t_0 + T, t_0)\| \leqslant \|\Phi(t, t_0 + T)\|\|\Phi(t_0 + T, t_0)\| \leqslant \frac{k}{2}$$

On this basis, for $\forall t \in [t_0 + 2T, t_0 + 3T]$, one can deduce that

$$\|\Phi(t, t_0)\| = \|\Phi(t, t_0 + 2T)\Phi(t_0 + 2T, t_0 + T)\Phi(t_0 + T, t_0)\| \leqslant \frac{k}{2^2}$$

By induction, it holds that for $\forall t \in [t_0 + nT, t_0 + (n+1)T]$,

$$\|\Phi(t, t_0)\| \leqslant \frac{k}{2^n}$$

Construct a function $\mu e^{-\nu(t-t_0)}$ such that for $t = t_0 + nT$, $n = 1, 2, \cdots$, $\mu e^{-\nu(t-t_0)} = \frac{k}{2^{n-1}}$, then one can deduce that $\mu(e^{-\nu T})^n = 2k\left(\frac{1}{2}\right)^n$. Set $\nu = \frac{1}{T}\ln 2$ and $\mu = 2k$, one can obtain that (5.14) holds, which implies that the statement (3) is true.

It can be further deduced that

$$\|x(t; x_0, t_0)\| \leqslant \|\Phi(t, t_0)\|\|x_0\| \leqslant \mu\delta e^{-\nu(t-t_0)}$$

which implies that $x_e = 0$ is exponentially stable, i.e., the statement (2) is true.

In general, the statement (2) implies the statement (1), here do not elaborate it.

At last, assume that the statement (3) is true. (5.14) indicates that $\|\Phi(t, t_0)\| \leqslant \mu$. According to Theorem 5.9, $x_e = 0$ is uniformly stable. Meanwhile, it holds that

$$\|x(t; x_0, t_0)\| \leqslant \|\Phi(t, t_0)\|\|x_0\| \leqslant \mu e^{-\nu(t-t_0)}\|x_0\|$$

For any real number $\varepsilon > 0$ and $\delta > 0$, there always exists $T = -\frac{1}{\nu}\ln\frac{\varepsilon}{\mu\delta}$ which is independent with t_0. For $\|x_0\| \leqslant \delta$, and $t \geqslant t_0 + T$, it holds that

$$\|x(t; x_0, t_0)\| \leqslant \mu e^{-\nu T}\|x_0\| \leqslant \varepsilon$$

which indicates that $x_e = 0$ is uniformly asymptotically stable. That is to say, if the statement (3) is true, then the statement (1) is true. □

5.2 Internal Stability of Linear Systems

Example 5.4 Consider the following system,

$$\dot{x}(t) = -e^{2t}x(t), \ x(t_0) = x_0$$

The solution of this system is $x(t; x_0, t_0) = \Phi(t, t_0)x_0$, where

$$\Phi(t, t_0) = e^{\frac{1}{2}(e^{2t_0} - e^{2t})}$$

Since $\lim_{t \to \infty} \Phi(t, t_0) = 0$, it holds that the equilibrium $x_e = 0$ of the system is (globally) asymptotically stable.

$$\|x(t; x_0, t_0)\| = \left|x_0 e^{\frac{1}{2}e^{2t_0}} e^{-\frac{1}{2}e^{2t}}\right| \leqslant |x_0| e^{\frac{1}{2}e^{2t_0}} e^{-t}, \ t \geqslant t_0 \geqslant 0$$

holds since $e^{2t} > 2t$. Therefore, the equilibrium $x_e = 0$ of the system is globally uniformly asymptotically stable, and globally exponentially stable.

In the next, a stability criterion of linear time-varying systems is introduced based on the Lyapunov's Second method.

Consider a piecewise continuous matrix function $Q(t)$ for $t \in [t_0, \infty)$, which is symmetric, positive-definite and satisfied with

$$0 < c_1 I \leqslant Q(t) \leqslant c_2 I$$

where $c_2 > c_1 > 0$ are two positive real numbers, then $Q(t)$ is said to be *uniformly bounded and uniformly positive-definite*.

To deduce the stability criterion, the following lemma is presented.

Lemma 5.1 *Suppose that the equilibrium $x_e = 0$ of a linear time-varying system (5.13) is uniformly asymptotically stable, then for any uniformly bounded and uniformly positive-definite matrix function $Q(t)$, the integration*

$$0 \leqslant P(t) = \int_t^\infty \Phi^\mathrm{T}(\tau, t) Q(\tau) \Phi(\tau, t) \mathrm{d}\tau \tag{5.15}$$

is convergent for any t, and is the unique solution of

$$\dot{P}(t) + A^\mathrm{T}(t)P(t) + P(t)A(t) + Q(t) = 0, \ \forall t \geqslant t_0 \tag{5.16}$$

where $\Phi(\tau, t)$ is the state transition matrix of (5.13).

Proof Owing to the uniform asymptotic stability of the equilibrium, according to Theorem 5.11, there exist $\mu > 0$ and $\nu > 0$ such that for $\forall \tau \geqslant t$,

$$\|\Phi(\tau, t)\| \leqslant \mu e^{-\nu(\tau - t)}$$

Along with (5.15), it holds that

$$P(t) \leqslant \int_t^\infty c_2 \mu^2 e^{-2\nu(\tau - t)} I \mathrm{d}\tau = \frac{c_2 \mu^2}{2\nu} I$$

It is easy to verify that (5.15) is a solution of (5.16). In the next, prove that the solution is unique. Suppose that $P_1(t)$ and $P_2(t)$ are both solutions of (5.16). Combining (5.15) with (5.16), it holds that

$$P_2(t) = \int_t^\infty \Phi^T(\tau, t) Q(\tau) \Phi(\tau, t) d\tau$$

$$= -\int_t^\infty \Phi^T(\tau, t) \left(\dot{P}_1(t) + A^T(t) P_1(t) + P_1(t) A(t) \right) \Phi(\tau, t) d\tau$$

$$= -\int_t^\infty \frac{d}{d\tau} \left(\Phi^T(\tau, t) P_1(t) \Phi(\tau, t) \right) d\tau$$

$$= -\left. \Phi^T(\tau, t) P_1(t) \Phi(\tau, t) \right|_t^\infty = P_1(t)$$

which indicates that the solution of (5.16) is unique. □

Theorem 5.12 *For a linear time-varying system* (5.13), $x_e = 0$ *is uniformly asymptotically stable if and only if for any uniformly bounded and uniformly positive-definite $Q(t)$, the matrix differential function* (5.16) *has a unique, uniformly bounded and uniformly positive-definite solution $P(t)$.*

Proof The necessity can be proved by Lemma 5.1 directly. In the next, prove the sufficiency. Construct $V(x,t) = x^T P(t) x$. It can be proved that $V(x,t)$ is positive-definite and bounded. Meanwhile, it holds that

$$\frac{\partial}{\partial t} V(x,t) = x^T \left(\dot{P}(t) + A^T(t) P(t) + P(t) A(t) \right) x = -x^T Q(t) x$$

which indicates that $\frac{\partial}{\partial t} V(x,t)$ is negative-definite and bounded. In this way, it can be further deduced that $x_e = 0$ is uniformly asymptotically stable. □

5.2.2 Internal Stability of Linear Time-Invariant Systems

Consider the following linear time-invariant system,

$$\dot{x}(t) = Ax(t), \quad x(0) = x_0, \quad t \geqslant 0 \tag{5.17}$$

where $x(t) \in \mathbb{R}^n$ is the state vector, A is a constant matrix. $x_e = 0$ is the equilibrium of the system (5.17).

Theorem 5.13 *For a linear time-invariant system* (5.17), $x_e = 0$ *is stable, in fact uniformly stable, if and only if all eigenvalues of A have nonpositive real parts, and every eigenvalue with zero real part is the simple root of the minimal polynomial of A, i.e., every eigenvalue with zero real part has an associated Jordan block of order one.*

Proof Firstly, prove $\|e^{At}\| \leqslant k < \infty$ is the sufficient and necessary condition that guarantee $x_e = 0$ is stable.

It is known that

$$x(t; x_0, 0) = e^{At} x_0, \quad \forall t \geqslant 0$$

5.2 Internal Stability of Linear Systems

On the other hand, the equilibrium $x_e = 0$ satisfies

$$x_e = e^{At} x_e, \ \forall t \geq 0$$

Furthermore, it holds that

$$x(t; x_0, 0) - x_e = e^{At}(x_0 - x_e), \ \forall t \geq 0$$

For any given $\varepsilon > 0$, select $\delta(\varepsilon) = \varepsilon/k$, then for any initial state satisfying

$$\|x_0 - x_e\| \leq \delta(\varepsilon)$$

the solution starting from this initial state satisfies

$$\|x(t; x_0, 0) - x_e\| \leq \|e^{At}\| \|x_0 - x_e\| \leq \frac{\varepsilon}{k} \|e^{At}\|.$$

If $\|e^{At}\| \leq k < \infty$, then one can deduce that

$$\|x(t; x_0, 0) - x_e\| \leq \varepsilon$$

which indicates that $x_e = 0$ is stable.

Next, assume that $x_e = 0$ is stable but $\|e^{At}\|$ is unbounded. In this situation, there exists at least one entry in e^{At} that is unbounded. Without loss of generality, assume that $\varphi_{ij}(t)$ satisfies

$$|\varphi_{ij}(t)| > k > 0$$

for $t \geq 0$, where $\varphi_{ij}(t)$ denotes the entry of e^{At} in the ith row and the jth column, $i \in \{1, \cdots, n\}$ and $j \in \{1, \cdots, n\}$. Select $x_0 = [\ 0\ \cdots\ 0\ \delta\ 0\ \cdots\ 0\ ]^{\mathrm{T}}$, where the jth entry in x_0 is δ. One can deduce that

$$x(t; x_0, 0) = \varphi_j(t) \delta$$

where $\varphi_j(t)$ is the jth column of e^{At}. It is noted that $x_e = 0$. Selecting $k = \varepsilon/\delta$, one can further obtain that

$$\|x(t; x_0, 0) - x_e\| = \|\varphi_j(t) \delta\| > k\delta = \varepsilon, \ \forall t \geq 0$$

That is to say, no matter how small δ is, $\|x(t; x_0, 0) - x_e\| > \varepsilon$ always holds, which contradicts with the assumption that $x_e = 0$ is stable.

In conclusion, $x_e = 0$ is stable if and only if $\|e^{At}\| \leq k < \infty$.

Next, study the relationship between the boundedness of $\|e^{At}\|$ and eigenvalues of A. Let P be a nonsingular matrix such that $J = PAP^{-1}$ is the Jordan normal form. In this situation, $e^{Jt} = Pe^{At}P^{-1}$. It indicates that

$$\|e^{Jt}\| \leq \|P\| \|e^{At}\| \|P^{-1}\|, \ \|e^{At}\| \leq \|P^{-1}\| \|e^{Jt}\| \|P\|$$

which means that the boundedness of $\|e^{At}\|$ equals to the boundedness of $\|e^{Jt}\|$.

As discussed in Subsection 2.3.1, the entry of e^{Jt} is with the form of $ct^m\mathrm{e}^{(\alpha+j\beta)t}$, $0 \leqslant m \leqslant p-1$, where c is a constant, $\lambda = \alpha + j\beta$ is an eigenvalue of A, p is the order of the Jordan block associated with λ, p is the order of the Jordan block associated with λ. It can be concluded that e^{Jt} is bounded if and only if $\alpha < 0$ or $\alpha = 0$ and $p = 1$. e^{Jt} is bounded if and only if $\|\mathrm{e}^{Jt}\|$ is bounded, as well as $\|\mathrm{e}^{At}\|$ is bounded. Therefore, $\|\mathrm{e}^{At}\|$ is bounded if and only if all eigenvalues of A have nonpositive real parts, and every eigenvalue with zero real part is the simple root of the minimal polynomial of A. □

It is noted that Theorem 5.13 is a sufficient and necessary criterion to determine the stability of linear time-invariant systems. On this basis, one can conclude that $x_\mathrm{e} = 0$ is unstable, if and only if there exists at least one eigenvalue of A has positive real part, or the eigenvalue with zero real part is the repeated root of the minimal polynomial of A.

Theorem 5.14 *For a linear time-invariant system (5.17), $x_\mathrm{e} = 0$ is asymptotically stable if and only if all eigenvalues of A have negative real parts.*

Proof Based on the proof of Theorem 5.13,

$$\lim_{t\to\infty} x(t;x_0,0) = \lim_{t\to\infty} \mathrm{e}^{At} = 0$$

holds if and only if $\lim_{t\to\infty} \mathrm{e}^{Jt} = 0$, which equals to $\lim_{t\to\infty} ct^m \mathrm{e}^{(\alpha+j\beta)t} = 0$. It holds if and only if $\alpha < 0$, i.e., all eigenvalues of A have negative real parts. □

Next, a stability criterion of linear time-invariant systems is presented based on the Lyapunov's second method.

Theorem 5.15 *For a linear time-invariant system (5.17), $x_\mathrm{e} = 0$ is asymptotically stable if and only if for any symmetric and positive-definite matrix Q, there is a symmetric and positive-definite matrix P satisfying*

$$A^\mathrm{T} P + PA = -Q \tag{5.18}$$

and the solution P is unique.

(5.18) is called as the *matrix Lyapunov equation*.

Proof Firstly, prove the sufficiency.

Let $V(x) = x^\mathrm{T} P x$. Since $P > 0$, then $V(x) = x^\mathrm{T} P x \geqslant 0$ and $V(x) = 0$ only for $x = 0$. It holds that

$$\dot{V}(x) = \dot{x}^\mathrm{T} P x + x^\mathrm{T} P \dot{x} = x^\mathrm{T} A^\mathrm{T} P x + x^\mathrm{T} P A x$$
$$= x^\mathrm{T} \left(A^\mathrm{T} P + PA\right) x = -x^\mathrm{T} Q x < 0$$

Therefore, it can be concluded that the equilibrium $x_\mathrm{e} = 0$ is asymptotically stable. Next, prove the necessity.

Consider the matrix equation

$$\dot{X}(t) = A^\mathrm{T} X(t) + X(t)A, \ X(0) = Q \tag{5.19}$$

5.2 Internal Stability of Linear Systems

It can be deduced that the solution of this matrix equation is

$$X(t) = e^{A^T t} Q e^{At} \tag{5.20}$$

Performing integral operation on (5.19), one can obtain that

$$X(\infty) - X(0) = A^T \left(\int_0^\infty X(t) dt \right) + \left(\int_0^\infty X(t) dt \right) A$$

Since $x_e = 0$ is asymptotically stable, then $\lim_{t \to \infty} e^{At} = 0$. Along with (5.20), one can deduce that $X(\infty) = 0$. On this basis, one can further obtain that

$$A^T \left(\int_0^\infty e^{A^T t} Q e^{At} dt \right) + \left(\int_0^\infty e^{A^T t} Q e^{At} dt \right) A = -Q$$

which indicates that

$$P = \int_0^\infty e^{A^T t} Q e^{At} dt$$

is a solution of (5.18). Since the solution $X(t)$ of (5.19) is unique, and $X(\infty) = 0$, $P = \int_0^\infty X(t) dt$ is unique.

Furthermore, $P = P^T$ and $x^T P x = \int_0^\infty (e^{At} x)^T Q (e^{At} x) dt \geq 0$, and the equality holds only for $x = 0$. Therefore, P is a symmetric and positive-definite matrix. □

Example 5.5 Consider a linear time-invariant system

$$\dot{x} = Ax = \begin{bmatrix} a_{11} & a_{12} \\ a_{21} & a_{22} \end{bmatrix} x$$

Determine the conditions with which the system parameters should be satisfied if the system is asymptotically stable.

Let $Q = I$ in (5.18), and denote

$$P = \begin{bmatrix} p_{11} & p_{12} \\ p_{21} & p_{22} \end{bmatrix}$$

then we one can obtain that

$$\begin{bmatrix} 2a_{11} & 2a_{21} & 0 \\ a_{12} & a_{11} + a_{22} & a_{21} \\ 0 & 2a_{12} & 2a_{22} \end{bmatrix} \begin{bmatrix} p_{11} \\ p_{12} \\ p_{22} \end{bmatrix} \triangleq A_1 \begin{bmatrix} p_{11} \\ p_{12} \\ p_{22} \end{bmatrix} = \begin{bmatrix} -1 \\ 0 \\ -1 \end{bmatrix}$$

The determinant of the coefficient matrix A_1 of the above linear equation group is

$$\det A_1 = 4(a_{11} + a_{22})(a_{11} a_{22} - a_{12} a_{21})$$

If $\det A_1 \neq 0$, then the equation group has a unique solution

$$P = \frac{-2}{\det A_1} \begin{bmatrix} \det A + a_{21}^2 + a_{22}^2 & -(a_{12}a_{22} + a_{21}a_{11}) \\ -(a_{12}a_{22} + a_{21}a_{11}) & \det A + a_{21}^2 + a_{22}^2 \end{bmatrix}$$

Since P is positive-definite, one can obtain that

$$p_{11} = \frac{\det A + a_{21}^2 + a_{22}^2}{-2(a_{11}+a_{22})(a_{11}a_{22}-a_{12}a_{21})} > 0, \ \det P = \frac{(a_{11}+a_{22})^2 + (a_{12}-a_{21})^2}{4(a_{11}+a_{22})^2 \det A} > 0$$

Therefore, for the system to be asymptotically stable, the parameters of the system should satisfy

$$\det A = a_{11}a_{22} - a_{12}a_{21} > 0, \ a_{11} + a_{22} < 0$$

To simplify calculation, Q is tend to be selected as I when using (5.18).

Corollary 5.1 *For any symmetric and positive-definite matrix Q, the matrix equation (5.19) has the unique solution P, and P is symmetric and positive-definite, if and only if all eigenvalues of A have negative real parts.*

This corollary can be deduced based on Theorem 5.14 and Theorem 5.15. The proof is omitted.

Corollary 5.2 *For any symmetric and positive-definite matrix Q, and any real number $\sigma \geqslant 0$, the matrix equation*

$$2\sigma P + A^\mathrm{T} P + PA = -Q \tag{5.21}$$

has the unique solution P, and P is symmetric and positive-definite, if and only if

$$\mathrm{Re}\lambda_i < -\sigma, \ i = 1, \cdots, n$$

where λ_i denotes the eigenvalue of A.

Proof (5.21) can be rewritten as

$$(A + \sigma I)^\mathrm{T} P + P(A + \sigma I) = -Q$$

According to Corollary 5.1, (5.21) has a unique solution and this solution is symmetric and positive-definite, if and only if all eigenvalues of $A + \sigma I$ have negative real parts. It can be verified λ_i is the eigenvalue of A if and only if $\lambda_i + \sigma$ is the eigenvalue of $A + \sigma I$. $\mathrm{Re}(\lambda_i + \sigma) < 0$ equals that $\mathrm{Re}\lambda_i < -\sigma$. $\square$

5.3 External Stability of Linear Systems

In the preceding sections, the stability of equilibriums, namely, the stability of free movements of systems is mainly discussed. In general, the stability of equilibriums has no relation to the input of systems, and it is also called the *internal stability*. In

5.3 External Stability of Linear Systems

this section, the *external stability* is discussed, which requires that every bounded input of a system should produce a bounded output, i.e., the bounded-input, bounded-output (BIBO) stability. Furthermore, discuss whether the state is bounded under the bounded input, i.e., the bounded-input, bounded-state (BIBS) stability.

5.3.1 BIBO Stability of Linear Systems

For a linear system which is causal and relaxed at t_0, the relationship between its input $u(t) \in \mathbb{R}^p$ and output $y(t) \in \mathbb{R}^q$ is

$$y(t) = \int_{t_0}^{t} G(t,\tau)u(\tau)\mathrm{d}\tau, \ t \geqslant t_0 \qquad (5.22)$$

where $G(t,\tau)$ is the impulse response matrix of the system.

Furthermore, if the system is time-invariant, it can be considered that $t_0 = 0$, and the impulse response matrix can be denoted as $G(t-\tau)$. The relationship between its input $u(t)$ and output $y(t)$ is

$$y(t) = \int_{0}^{t} G(t-\tau)u(\tau)\mathrm{d}\tau, \ t \geqslant 0. \qquad (5.23)$$

Definition 5.9 (BIBO stability) Consider a causal linear system relaxed at t_0, for any bounded input $u(t)$, in other words, for any input $u(t)$ satisfying

$$\|u(t)\| \leqslant \bar{k} < \infty, \ \forall t \geqslant t_0$$

the corresponding output $y(t)$ satisfies

$$\|y(t)\| \leqslant c(t_0, u) < \infty, \ \forall t \geqslant t_0$$

where $\bar{k}$ and $c(t_0, u)$ are two constants, then the system is said to be *bounded-input, bounded-output (BIBO) stable*. Furthermore, if $c(t_0, u)$ is independent of t_0, which can be written as $c(u)$, then the system is staid to be *uniformly BIBO stable*.

It is worth mentioning that we restrict that the initial condition of the system is zero in Definition 5.9 since only in this way, the input-output description of a system is meaningful.

Firstly, a BIBO stability criterion is presented for the linear time-varying system (5.22).

Theorem 5.16 *A linear time-varying system (5.22) is BIBO stable if and only if there exists a constant k such that*

$$\int_{t_0}^{t} |g_{ij}(t,\tau)|\mathrm{d}\tau \leqslant k < \infty, \ i=1,2,\cdots,q, \ j=1,2,\cdots,p \qquad (5.24)$$

where $g_{ij}(t,\tau)$ is the entry of $G(t,\tau)$ in the ith row and the jth column.

Proof Firstly, prove the sufficiency.

Use $y_i(t)$ to denote the ith output and $u_j(t)$ to denote the jth input. Here consider $|u_j(t)| \leqslant k_j < \infty$. It holds that

$$|y_i(t)| = \left|\int_{t_0}^{t} \sum_{j=1}^{p} g_{ij}(t,\tau) u_j(\tau) d\tau\right| \leqslant \sum_{j=1}^{p} \int_{t_0}^{t} |g_{ij}(t,\tau)||u_j(\tau)| d\tau$$

$$\leqslant \sum_{j=1}^{p} k_j \int_{t_0}^{t} |g_{ij}(t,\tau)| d\tau \leqslant p k_j k$$

which indicates that the output is bounded. Therefore, the system is BIBO stable.

Next, prove the necessity.

Use reduction to absurdity. Suppose that the system (5.22) is BIBO stable, but there exists a certain $g_{ij}(t,\tau)$ and t_1 such that $\int_{0}^{t_1} |g_{ij}(t,\tau)| d\tau$ is unbounded.

Select the bounded input as $u_j(\tau) = \text{sign}[g_{ij}(t,\tau)]$, where $\text{sign}[g_{ij}(t,\tau)]$ is the sign of $g_{ij}(t,\tau)$. One can deduce that

$$y_i(t_1) = \int_{t_0}^{t_1} \sum_{j=1}^{p} g_{ij}(t,\tau) u_j(\tau) d\tau = \sum_{j=1}^{p} \int_{t_0}^{t_1} |g_{ij}(t,\tau)| d\tau$$

which is unbounded. It contradicts with the assumption that the system is BIBO stable. $\square$

Now turn attention to linear time-invariant systems.

Theorem 5.17 *A linear time-invariant system (5.23) is BIBO stable, if and only if there exists a constant k such that*

$$\int_{0}^{\infty} |g_{ij}(t-\tau)| d\tau \leqslant k < \infty, \ i=1,2,\cdots,q, \ j=1,2,\cdots,p \qquad (5.25)$$

where $g_{ij}(t-\tau)$ is the entry of $G(t-\tau)$ in the ith row and the jth column.

The proof is similar with the proof of Theorem 5.16. Here we omit it.

Theorem 5.18 *A linear time-invariant system (5.23) is BIBO stable if and only if its transfer function matrix $G(s)$ is a proper rational fraction matrix, and each pole of all entries of $G(s)$ has negative real part.*

Proof Let $g_{ij}(s)$ be the entry of $G(s)$ in the ith row and the jth column. When $g_{ij}(s)$ is a proper rational fraction, it can be expanded as the sum of finite terms, and each term can be expressed as $\beta_l/(s-\lambda_l)^{\alpha_l}$, where λ_l is the pole of $g_{ij}(s)$, α_l and β_l are constants. Since λ_l has negative real part, then the inverse Laplace transform of $g_{ij}(s)$ is the sum of $t^{\alpha_l-1}e^{\lambda_l t}$ and $\delta(t)$. Therefore, (5.25) is satisfied. According to Theorem 5.17, one can further deduce this theorem. $\square$

5.3.2 BIBS Stability of Linear Systems

Consider a linear time-varying system

$$\dot{x}(t) = A(t)x(t) + B(t)u(t), \ y(t) = C(t)x(t), \ x(t_0) = x_0, \ t \in [t_0, t_f] \tag{5.26}$$

where $x(t) \in \mathbb{R}^n$ is the state vector, $u(t) \in \mathbb{R}^p$ is the input vector, $y(t) \in \mathbb{R}^q$ is the output vector. $A(t)$, $B(t)$ and $C(t)$ are time-varying matrices. (5.26) omits the direct transfer part since it has no influence on the stability.

In the next, discuss whether a bounded input will lead to a bounded state response.

Definition 5.10 (BIBS stability) Consider a linear system (5.26). For any bounded input $u(t)$ and any initial nonzero initial state x_0, if the corresponding state $x(t)$ satisfies

$$\|x(t)\| \leqslant c(t_0, x_0, u) < \infty \tag{5.27}$$

where $c(t_0, x_0, u)$ is a constant, then the system is said to be *bounded-input, bounded-state* (*BIBS*) *stable*. Furthermore, if $c(t_0, x_0, u)$ is independent of t_0, which can be written as $c(x_0, u)$, then the system is said to be *uniformly BIBS stable*.

Theorem 5.19 *A linear time-varying system* (5.26) *is BIBS stable if and only if there exist two constants* $k_1(t_0)$ *and* $k_2(t_0)$ *such that*

$$\|\Phi(t, t_0)\| \leqslant k_1(t_0) \tag{5.28}$$

and

$$\int_{t_0}^{t} \|\Phi(t, \tau)B(\tau)\| d\tau \leqslant k_2(t_0) \tag{5.29}$$

where $\Phi(t, t_0)$ *is the state transition matrix of the system* (5.26).

Proof Firstly, prove the sufficiency.

Suppose that $\|u(t)\| \leqslant \bar{k} < \infty$, one can deduce that

$$\|x(\tau)\| = \left\|\Phi(t, t_0)x_0 + \int_{t_0}^{t} \Phi(t, \tau)B(\tau)u(\tau)d\tau\right\|$$

$$\leqslant \|\Phi(t, \tau_0)\| \| x_0\| + \int_{t_0}^{t} \|\Phi(t, \tau)B(\tau)\| \| u(\tau)\| d\tau$$

$$\leqslant k_1(t_0)\|x_0\| + k_2(t_0)\bar{k}$$

which indicates that (5.27) holds. Therefore, the system (5.26) is BIBS stable.

Next, prove the necessity.

Use reduction to absurdity. Suppose that the system (5.26) is BIBS stable, but there exists an initial state x_0 or a bounded input $u(t)$, such that $\|\Phi(t, t_0)\|$ or $\int_{t_0}^{t} \|\Phi(t, \tau)B(\tau)\| d\tau$ is unbounded. In this situation, $\|x(t)\|$ must be unbounded, which contradicts with the assumption. □

As a matter of fact, in Theorem 5.19, if $k_1(t_0)$ and $k_2(t_0)$ are independent of t_0, (5.28) and (5.29) are the sufficient and necessary conditions to guarantee that the system (5.26) is uniformly BIBS stable.

Theorem 5.19 also holds for linear time-invariant systems. Specifically, consider a linear time-invariant system (A, B, C), we can deduce the following conclusion.

Theorem 5.20 *A linear time-invariant system (A, B, C) is uniformly BIBS stable if and only if each pole of all entries of $(sI - A)^{-1}B$ has negative real part.*

The proof is omitted.

5.4 Relationships between External Stability and Internal Stability

If (A, B, C) is BIBS stable, it is obviously BIBO stable, but not vice versa. As discussed in Subsection 3.4.3, if (A, B, C) is incompletely observable, it can be decomposed into an observable subsystem and an unobservable subsystem. In fact, the output $y(t)$ cannot reflect the state information of the unobservable subsystem. Owing to this fact, we have the following conclusion.

Theorem 5.21 *If a linear time-invariant system (A, B, C) is completely observable, then its BIBS stability is equivalent to its BIBO stability.*

In the next, discuss relationships between the external stability and the internal stability for a linear time-invariant system (A, B, C).

Theorem 5.22 *If a linear time-invariant system (A, B, C) is asymptotically stable, then it is BIBO stable, and also BIBS stable.*

Proof The impulse response matrix of the system (A, B, C) is

$$G(t - \tau) = Ce^{A(t-\tau)}B$$

When the system is asymptotically stable, as discussed in Section 5.2.1, it holds that

$$\lim_{t \to \infty} e^{At} = 0$$

On this basis, it is not difficult to deduce that (5.24) is satisfied. Therefore, the system is BIBO stable.

On the other hand, it is known for a linear time-invariant system, its state transition matrix is e^{At}. In this way, it is not difficult to verify Theorem 5.19 holds. Therefore, the system is BIBS stable. □

It is worth mentioning that the BIBS stability cannot guarantee the asymptotic stability of a linear time-invariant system. As discussed in Subsection 3.4.2, if a linear system is incompletely controllable, it can de decomposed into a controllable subsystem and an uncontrollable subsystem. The input $u(t)$ can only affect the eigenvalues of the controllable part but cannot affect these of the uncontrollable part. Owing to this fact, we have the following conclusion.

5.4 Relationships between External Stability and Internal Stability

Theorem 5.23 *If a linear time-invariant system (A, B, C) is completely controllable, then its BIBS stability is equivalent to its internal stability.*

Obviously, the BIBO stability cannot guarantee the asymptotic stability of a linear time-invariant system. As discussed in Subsection 3.4.4, a linear time-invariant system can be divided into the controllable and observable part, the controllable but unobservable part, the uncontrollable but observable part, and the uncontrollable and unobservable part. The input-output property of a system can only implies the property of the controllable and observable part. As a result, the BIBO stability of a linear time-invariant system only indicates that its controllable and observable part is asymptotically stable. In this way, the following conclusion is deduced.

Theorem 5.24 *If a linear time-invariant system (A, B, C) is completely controllable and completely observable, then its BIBO stability is equivalent to its asymptotic stability.*

The relationship between the external stability and the internal stability of a linear time-invariant system is displayed in Fig.5.5.

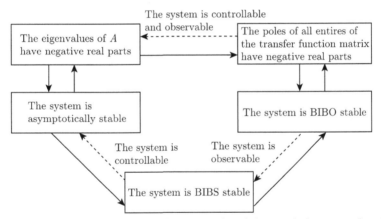

Fig.5.5 The relationship between the external stability and the internal stability

Example 5.6 Consider the following system

$$\dot{x}(t) = \begin{bmatrix} 1 & 0 \\ 1 & -1 \end{bmatrix} x(t) + \begin{bmatrix} 0 \\ 1 \end{bmatrix} u(t), \ y(t) = \begin{bmatrix} 1 & 1 \end{bmatrix} x(t)$$

The eigenvalues of the state coefficient matrix are 1 and -1, and thus the system is not asymptotically stable.

Meanwhile, it can be computed that

$$(sI - A)^{-1}b = \begin{bmatrix} s-1 & 0 \\ -1 & s+1 \end{bmatrix}^{-1} \begin{bmatrix} 0 \\ 1 \end{bmatrix} = \begin{bmatrix} 0 \\ \dfrac{1}{s+1} \end{bmatrix}$$

The pole of each entry of $(sI - A)^{-1}b$ is -1, and thus the system is BIBS stable.

Furthermore, one can obtain that

$$c(sI-A)^{-1}b = \begin{bmatrix} 1 & 1 \end{bmatrix} \begin{bmatrix} 0 \\ \dfrac{1}{s+1} \end{bmatrix} = \dfrac{1}{s+1}$$

and the pole is -1. Therefore, the system is BIBO stable.

It can be further computed that

$$U = \begin{bmatrix} b & Ab \end{bmatrix} = \begin{bmatrix} 0 & 0 \\ 1 & -1 \end{bmatrix}, \quad V = \begin{bmatrix} c \\ cA \end{bmatrix} = \begin{bmatrix} 1 & 1 \\ 2 & -1 \end{bmatrix}$$

Since $\operatorname{rank} U = 1$ and $\operatorname{rank} V = 2$. The system is completely observable but incompletely controllable. It can be seen that the external stability and the internal stability of this system satisfy the relationship shown in Fig.5.5.

Assignments

5.1 Consider the system $\dot{x} = x(x-1)$. Determine the equilibrium points and discuss the stability, uniform stability, asymptotic stability according to the definition.

5.2 Consider the system $\dot{x} = x\cos t$. Determine the equilibrium points and discuss the stability, uniform stability, asymptotic stability according to the definition.

5.3 Determine whether the equilibrium $x_e = 0$ of the given system is globally asymptotically stable,

$$\begin{cases} \dot{x}_1 = x_2 \\ \dot{x}_2 = -x_1 - x_1^2 x_2 \end{cases}$$

5.4 Determine whether the equilibrium $x_e = 0$ of the given system is globally asymptotically stable,

$$\begin{cases} \dot{x}_1 = x_2 \\ \dot{x}_2 = -x_1^3 - x_2 \end{cases}$$

5.5 Consider the system

$$\dot{x}(t) = \begin{bmatrix} e^{-t} & 0 \\ 0 & \dfrac{1}{t+1} \end{bmatrix} x(t), \quad t \geqslant 0$$

Discuss the stability, uniform stability, asymptotic stability of the equilibrium $x_e = 0$.

5.6 Consider the linear time-variant system

$$\dot{x}(t) = \begin{bmatrix} 0 & 1 \\ -\dfrac{1}{t+1} & -10 \end{bmatrix} x(t), \quad t \geqslant 0$$

Determine whether the equilibrium is globally asymptotically stable.

5.7 Consider the nonlinear time-invariant system $\dot{x} = f(x)$, $f(0) = 0$. Let $x = [\, x_1 \ \cdots \ x_n \,]^{\mathrm{T}}$ and $f(x) = [\, f(x_1) \ \cdots \ f(x_n) \,]^{\mathrm{T}}$. Define the Jacobi matrix of this system is

$$F(x) \triangleq \frac{\partial f(x)}{\partial x^{\mathrm{T}}} = \begin{bmatrix} \dfrac{\partial f_1(x)}{\partial x_1} & \cdots & \dfrac{\partial f_1(x)}{\partial x_n} \\ \vdots & & \vdots \\ \dfrac{\partial f_n(x)}{\partial x_1} & \cdots & \dfrac{\partial f_n(x)}{\partial x_n} \end{bmatrix}$$

Prove that if $F^{\mathrm{T}}(x) + F(x)$ is negative definite, the equilibrium $x_{\mathrm{e}} = 0$ is globally asymptotically stable.

5.8 Based on the conclusion in 5.6, determine whether the following system is globally asymptotically stable,

$$\begin{cases} \dot{x}_1 = -3x_1 + x_2 \\ \dot{x}_2 = x_1 - x_2 - x_2^3 \end{cases}$$

5.9 Determine whether these given systems are asymptotically stable, BIBS stable and BIBO stable.

(1) $\dot{x}(t) = \begin{bmatrix} -1 & -1 & 0 \\ 0 & 0 & 1 \\ 0 & -3 & -4 \end{bmatrix} x(t) + \begin{bmatrix} 1 & 0 \\ 0 & 1 \\ 1 & 1 \end{bmatrix} u(t)$, $y(t) = \begin{bmatrix} 0 & 1 & 2 \\ 0 & 0 & 1 \end{bmatrix} x(t)$

(2) $\dot{x}(t) = \begin{bmatrix} 0 & 1 & 0 \\ 0 & -2 & 0 \\ 0 & 0 & 0 \end{bmatrix} x(t) + \begin{bmatrix} 1 \\ 2 \\ 1 \end{bmatrix} u(t)$, $y(t) = \begin{bmatrix} 2 & 3 & 0 \end{bmatrix} x(t)$

(3) $\dot{x}(t) = \begin{bmatrix} -2 & 1 & 0 \\ 0 & 0 & 1 \\ 25 & 0 & -5 \end{bmatrix} x(t) + \begin{bmatrix} 0 \\ 0 \\ 1 \end{bmatrix} u(t)$, $y(t) = \begin{bmatrix} -5 & 1 & 0 \end{bmatrix} x(t)$

(4) $\dot{x}(t) = \begin{bmatrix} -1 & -1 & 0 & 0 \\ 1 & -0.1 & 0 & 0 \\ 0 & 0 & 0 & -1 \\ 0 & 0 & -1 & 0 \end{bmatrix} x(t) + \begin{bmatrix} 1 \\ 0 \\ 1 \\ 0 \end{bmatrix} u(t)$, $y(t) = \begin{bmatrix} -1 & 0 & -1 & 0 \end{bmatrix} x(t)$

(5) $G(s) = \begin{bmatrix} \dfrac{s+1}{s^2} & \dfrac{s+2}{s^2+1} \\ \dfrac{2}{s} & \dfrac{2s+3}{s^2+1} \end{bmatrix}$

(6) $G(s) = \begin{bmatrix} \dfrac{4}{s+1} & \dfrac{1}{(s+1)(s+2)} \\ \dfrac{1}{s+3} & \dfrac{1}{s(s+2)(s+3)} \end{bmatrix}$

5.10   Consider the system

$$\dot{x}(t) = \begin{bmatrix} a_{11} & a_{12} & a_{13} \\ 0 & a_{22} & a_{23} \\ 1 & 0 & a_{33} \end{bmatrix} x(t)$$

Using the Lyapunov equation to determine the condition which guarantees the global asymptotic stability of the equilibrium $x_e = 0$.

5.11   Suppose that the system

$$\dot{x}(t) = Ax(t) + bu(t), \quad y(t) = cx(t), \quad x(0) = x_0$$

is asymptotically stable when $u(t) = 0$, and P is the positive definite solution of

$$A^T P + PA = -c^T c$$

Prove that

$$\int_0^\infty y^2(t) \mathrm{d}t = x_0^T P x_0$$

5.12   Suppose that the system $\dot{x}(t) = Ax(t) + Bu(t)$ is completely controllable. Select $u(t) = -B^T e^{-A^T t} W^{-1}(0, t) x_0$, where

$$W(0, t) = \int_0^t e^{-A\tau} BB^T e^{-A^T \tau} \mathrm{d}\tau, \quad t > 0$$

Prove that the corresponding closed-loop system is asymptotically stable.

Chapter 6

Synthesis and Feedback Control of Linear Systems in Time Domain

Analysis and synthesis are two extremely important aspects for studying linear systems and they are related with each other tightly. Analysis aims at studying the qualitative actions (such as controllability, observability, stability and so on) and quantitative laws (such as dynamical response of linear systems). On the other hand, the aim of synthesis is to design a system to obtain the anticipant qualities. In this chapter, the synthesis problems are discussed for linear time-invariant systems.

6.1 State Feedback Control

Feedback is common in manufactured systems and is essential in automatic control of dynamical processes with uncertainties in their model descriptions and their interactions with the environment. In the classical control theory, the output of systems is selected as the feedback signal to stabilize the system or improve the system performance. When using the state space model to describe a system, state variables include all information of a system, and state responses imply all movement behaviors of a system. However, the output signals only reflect partial movement behaviors of a system which can be observed directly. Therefore, selecting state variables as feedback signals is more effective to improve the system performance. In this section, the state feedback is discussed in details.

6.1.1 Controllability and Observability of System with State Feedback

Consider a linear time-invariant system

$$\dot{x}(t) = Ax(t) + Bu(t), \quad y(t) = Cx(t) + Du(t) \tag{6.1}$$

where $x(t) \in \mathbb{R}^n$ is the state vector, $u(t) \in \mathbb{R}^p$ is the input vector and $y(t) \in \mathbb{R}^q$ is the output vector. A, B, C and D are real constant matrices with appropriate dimensions.

Introduce the state feedback control law

$$u(t) = Kx(t) + v(t) \tag{6.2}$$

where $v(t) \in \mathbb{R}^p$ is the reference input vector, K is a constant matrix, which is generally called the feedback gain. Substituting (6.2) into (6.1), the corresponding closed-loop system is

$$\dot{x}(t) = (A+BK)x(t) + Bv(t), \quad y(t) = (C+DK)x(t) + Dv(t) \tag{6.3}$$

and its block diagram is displayed in Fig.6.1 .

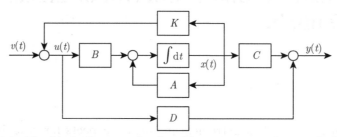

Fig.6.1   The block diagram of a closed-loop system with state feedback

Theorem 6.1   *For a linear time-invariant system (6.1), using the state feedback control law (6.2) cannot change the controllability, but may change the observability.*

Proof   The controllability matrix of the system (6.1) is $U = [\ B \quad AB \quad \cdots \quad A^{n-1}B\ ]$, while the controllability matrix of the corresponding closed-loop system (6.3) is $U_k = [\ B \quad (A+BK)B \quad \cdots \quad (A+BK)^{n-1}B\ ]$. Just need to prove that $\operatorname{rank} U = \operatorname{rank} U_k$, then it can be deduced that these two systems have the same controllability.

Let b_i denote the ith column vector of B, b_{j1} denote the jth entry of b_1, k_{ij} denote the entry of K in the ith row and the jth column, $i = 1, \cdots, p$, $j = 1, \cdots, n$. The first column of $(A+BK)B$ can be expressed as

$$(A+BK)b_1 = Ab_1 + \begin{bmatrix} b_1 & b_2 & \cdots & b_p \end{bmatrix} \begin{bmatrix} k_{11} & k_{12} & \cdots & k_{1n} \\ k_{21} & k_{22} & \cdots & k_{2n} \\ \vdots & \vdots & & \vdots \\ k_{p1} & k_{p2} & \cdots & k_{pn} \end{bmatrix} \begin{bmatrix} b_{11} \\ b_{21} \\ \vdots \\ b_{n1} \end{bmatrix}$$

$$= Ab_1 + \begin{bmatrix} b_1 & b_2 & \cdots & b_p \end{bmatrix} \begin{bmatrix} c_1 \\ c_2 \\ \vdots \\ c_p \end{bmatrix} = Ab_1 + c_1 b_1 + c_2 b_2 + \cdots + c_p b_p$$

where $c_l = \sum_{j=1}^{n} k_{lj} b_{j1}$, $l = 1, \cdots, p$. It indicates that $(A+BK)b_1$ is a linear combination of $Ab_1, b_1, \cdots, b_p$. Similarly, for the ith column of $(A+BK)B$, it can be deduced that $(A+BK)b_i$ is a linear combination of $Ab_i, b_1, \cdots, b_p$. As a result, each column of $(A+BK)B$ can be expressed as a linear combination of columns of $[\ B \quad AB\ ]$.

Following the similar vein, it can be concluded that each column of $(A+BK)^m B$ can be expressed as a linear combination of the columns of $[\ B \quad AB \quad \cdots \quad A^m B\ ]$, $m = 1, \cdots, n-1$. That is to say, each column of U_k can be expressed as a linear combination of the columns of U, which indicates that

$$\operatorname{rank} U \geqslant \operatorname{rank} U_k$$

6.1 State Feedback Control

On the other hand, it is noted that

$$\dot{x}(t) = Ax(t) + Bu(t) = [(A+BK) - BK]x(t) + Bu(t)$$

which can be viewed as a state feedback system of (6.3) with the feedback gain being $-K$. In this way, one can deduce that

$$\operatorname{rank} U \leqslant \operatorname{rank} U_k$$

In conclusion, $\operatorname{rank} U = \operatorname{rank} U_k$ and thus the controllability is invariant. □

The observability may be changed. Here is an example to illustrate this fact.

Example 6.1 A dynamical system is given by

$$\dot{x}(t) = \begin{bmatrix} 3 & 4 \\ 4 & 6 \end{bmatrix} x(t) + \begin{bmatrix} 0 \\ 1 \end{bmatrix} u(t), \quad y(t) = \begin{bmatrix} 3 & 4 \end{bmatrix} x(t)$$

It is easy to verify that the system is completely controllable and observable. Let

$$u(t) = \begin{bmatrix} -4 & -6 \end{bmatrix} x(t) + v(t)$$

The dynamical equations of the closed-loop system become

$$\dot{x}(t) = \begin{bmatrix} 3 & 4 \\ 0 & 0 \end{bmatrix} x(t) + \begin{bmatrix} 0 \\ 1 \end{bmatrix} v(t), \quad y(t) = \begin{bmatrix} 3 & 4 \end{bmatrix} x(t)$$

Obviously, the closed-loop system is still controllable but unobservable.

It is worth mentioning that the observability is possible to be unchanged after introducing the state feedback control law. Meanwhile, the controllability remaining unchanged indicates that if the original system (A, B) is completely controllable, then the closed-loop system $(A+BK, B)$ is also completely controllable. Accordingly, if (A, B) is incompletely controllable, then $(A+BK, B)$ is also incompletely controllable.

6.1.2 Eigenvalue Assignment with State Feedback for SISO Systems

As discussed in the previous subsection, the system coefficient matrix changed to be $A + BK$ after introducing the state feedback control law to the original system. This subsection is focused on designing the feedback gain K to assign the eigenvalues to desired locations. The eigenvalue assignment problem can be stated as: for a given (A, B), determine K to assign n eigenvalues of $A+BK$ to arbitrary real and/or complex conjugate locations. This problem is also termed as the pole assignment problem.

Firstly, discuss the eigenvalue assignment problem for SISO systems in this subsection. Consider a linear time-invariant SISO system

$$\dot{x}(t) = Ax(t) + bu(t), \quad y(t) = cx(t) + du(t) \tag{6.4}$$

where $x(t) \in \mathbb{R}^n$ is the state vector, $u(t) \in \mathbb{R}^1$ is the input and $y(t) \in \mathbb{R}^1$ is the output. A, b, c and d are real constant matrices with appropriate dimensions.

Select (6.2) as the state feedback control law, the corresponding closed-loop system is

$$\dot{x}(t) = (A + bK)x(t) + bv(t), \ y(t) = (c + dK)x(t) + dv(t) \tag{6.5}$$

Theorem 6.2 *For a linear time-invariant SISO system* (6.4), *its all eigenvalues can be assigned arbitrarily by a state feedback control law* (6.2) *if and only if the system is completely controllable.*

Proof Firstly, prove the necessity.

Use reduction to absurdity. Suppose that all eigenvalues of (6.4) can be assigned arbitrarily, but (A, b) is incompletely controllable. By the structure decomposition, one can obtain that

$$\bar{A} = TAT^{-1} = \begin{bmatrix} \bar{A}_{11} & \bar{A}_{12} \\ \bar{A}_{21} & \bar{A}_{22} \end{bmatrix}, \ \bar{b} = Tb = \begin{bmatrix} \bar{b}_1 \\ 0 \end{bmatrix}$$

where T is a nonsingular matrix.

Introduce a state feedback control law (6.2) with $K = \begin{bmatrix} K_1 & K_2 \end{bmatrix}$ and consider that $\bar{K} = KT^{-1} = \begin{bmatrix} \bar{K}_1 & \bar{K}_2 \end{bmatrix}$. One can deduce that

$$\det(sI - A - bK) = \det(sI - \bar{A} - \bar{b}\bar{K})$$

$$= \det \begin{bmatrix} sI - \bar{A}_{11} - \bar{b}_1\bar{K}_1 & -\bar{A}_{12} - \bar{b}_1\bar{K}_2 \\ 0 & sI - \bar{A}_{22} \end{bmatrix}$$

$$= \det(sI - \bar{A}_{11} - \bar{b}_1\bar{K}_1)\det(sI - \bar{A}_{22})$$

which indicates that the eigenvalues of the uncontrollable part cannot be changed. It contradicts with the assumption and thus the necessity holds.

Next, prove the sufficiency.

Since (A, b) is completely controllable, there exists a transformation matrix T which can transform (A, b) into a controllable canonical form $(\bar{A}, \bar{b})$ with

$$\bar{A} = TAT^{-1} = \begin{bmatrix} 0 & 1 & 0 & \cdots & 0 \\ 0 & 0 & 1 & \cdots & 0 \\ \vdots & \vdots & \vdots & & \vdots \\ 0 & 0 & 0 & \cdots & 1 \\ -a_n & -a_{n-1} & -a_{n-2} & \cdots & -a_1 \end{bmatrix}, \ \bar{b} = Tb = \begin{bmatrix} 0 \\ 0 \\ \vdots \\ 0 \\ 1 \end{bmatrix}$$

Choose a state feedback matrix

$$\bar{K} = KT^{-1} = \begin{bmatrix} a_n - \bar{a}_n & a_{n-1} - \bar{a}_{n-1} & \cdots & a_1 - \bar{a}_1 \end{bmatrix}$$

for $(\bar{A}, \bar{b})$, where the coefficients $\bar{a}_1, \cdots, \bar{a}_n$ can be selected arbitrarily. In this way, the

6.1 State Feedback Control

closed-loop system is

$$\bar{A}+\bar{b}\bar{K} = \begin{bmatrix} 0 & 1 & 0 & \cdots & 0 \\ 0 & 0 & 1 & \cdots & 0 \\ \vdots & \vdots & \vdots & & \vdots \\ 0 & 0 & 0 & \cdots & 1 \\ -\bar{a}_n & -\bar{a}_{n-1} & -\bar{a}_{n-2} & \cdots & -\bar{a}_1 \end{bmatrix}, \quad \bar{b} = \begin{bmatrix} 0 \\ \vdots \\ 0 \\ 1 \end{bmatrix}$$

The characteristic equation of the above closed-loop system is

$$\alpha_c(s) = s^n + \bar{a}_1 s^{n-1} + \cdots + \bar{a}_n = 0$$

Since the coefficients $\bar{a}_1, \cdots, \bar{a}_n$ can be selected arbitrarily, the eigenvalues can be assigned arbitrarily. □

The eigenvalue assignment algorithm is presented for SISO systems.

Algorithm 6.1 *For a completely controllable linear time-invariant SISO system (6.4), using the state feedback law (6.2), the eigenvalues of $A+bK$ can be assigned to $\lambda_1^*, \cdots, \lambda_n^*$. The state feedback gain K can be determined by following procedures.*

(1) *Calculate the characteristic polynomial of A,*

$$\alpha(s) = \det(sI - A) = s^n + a_1 s^{n-1} + \cdots + a_n$$

(2) *Calculate the desired characteristic polynomial based on desired eigenvalues,*

$$\alpha_c(s) = (s - \lambda_1^*)(s - \lambda_2^*) \cdots (s - \lambda_n^*) = s^n + \bar{a}_1 s^{n-1} + \cdots + \bar{a}_n$$

(3) *Construct the transformation matrix T,*

$$T = \left\{ \begin{bmatrix} b & Ab & \cdots & A^{n-1}b \end{bmatrix} \begin{bmatrix} a_{n-1} & a_{n-2} & \cdots & a_1 & 1 \\ a_{n-2} & \cdots & a_1 & 1 & \\ \vdots & \reflectbox{$\ddots$} & \reflectbox{$\ddots$} & & \\ a_1 & 1 & & & \\ 1 & & & & \end{bmatrix} \right\}^{-1}$$

(4) *Calculate the original state feedback matrix K of the system,*

$$K = \bar{K}T = \begin{bmatrix} a_n - \bar{a}_n & a_{n-1} - \bar{a}_{n-1} & \cdots & a_1 - \bar{a}_1 \end{bmatrix} T$$

This algorithm is derived based on the proof of Theorem 6.2. In the next, another eigenvalue assignment algorithm is presented for SISO systems.

Algorithm 6.2 *The problem is the same as the one in Algorithm 6.1. The state feedback gain $K = \begin{bmatrix} k_1 & k_2 & \cdots & k_n \end{bmatrix}$ can be determined by following procedures.*

(1) *Calculate the characteristic polynomial*

$$\alpha_k(s) = \det(sI - (A+bK)) = s^n + \tilde{a}_1 s^{n-1} + \cdots + \tilde{a}_n$$

where $\tilde{a}_1, \cdots, \tilde{a}_n$ are coefficients related to $k_1, \cdots, k_n$.

(2) Calculate the desired characteristic polynomial based on desired eigenvalues,

$$\alpha_c(s) = (s - \lambda_1^*)(s - \lambda_2^*) \cdots (s - \lambda_n^*) = s^n + \bar{a}_1 s^{n-1} + \cdots + \bar{a}_n$$

(3) Let $\tilde{a}_i = \bar{a}_i$, $i = 1, \cdots, n$, then n equations are obtained.
(4) Solve the simultaneous equations to obtain $K = [\ k_1 \ \ k_2 \ \ \cdots \ \ k_n\ ]$.

Example 6.2 Consider a SISO system

$$\dot{x}(t) = \begin{bmatrix} 1 & 0 & -1 \\ 1 & 2 & 1 \\ 2 & 2 & 3 \end{bmatrix} x(t) + \begin{bmatrix} 1 \\ 0 \\ 1 \end{bmatrix} u(t), \ y(t) = \begin{bmatrix} 1 & 0 & 0 \end{bmatrix} x(t)$$

Design a state feedback matrix such that the eigenvalues of the closed-loop system are $-1, -1 \pm 2j$.

It can be verified this system is completely controllable.

Firstly, use Algorithm 6.1 to design the control law.

Calculate the characteristic polynomial of A,

$$\det(sI - A) = s^3 - 6s^2 + 11s - 6$$

For the given desired eigenvalues, calculate the desired characteristic polynomial of the closed-loop system,

$$\alpha_c(s) = (s+1)(s+1+j2)(s+1-j2) = s^3 + 3s^2 + 7s + 5$$

Construct the transformation matrix T,

$$T = \left\{ \begin{bmatrix} 1 & 0 & -5 \\ 0 & 2 & 9 \\ 1 & 5 & 19 \end{bmatrix} \begin{bmatrix} 11 & -6 & 1 \\ -6 & 1 & 0 \\ 1 & 0 & 0 \end{bmatrix}^{-1} \right\}^{-1} = \begin{bmatrix} 6 & -6 & 1 \\ -3 & 2 & 0 \\ 0 & -1 & 1 \end{bmatrix}^{-1} = \begin{bmatrix} -\frac{2}{3} & -\frac{5}{3} & \frac{2}{3} \\ -1 & -2 & 1 \\ -1 & -2 & 2 \end{bmatrix}$$

Calculate the original state feedback matrix K of the system,

$$K = \bar{K}T = \begin{bmatrix} -6-5 & 11-7 & -6-3 \end{bmatrix} \begin{bmatrix} -\frac{2}{3} & -\frac{5}{3} & \frac{2}{3} \\ -1 & -2 & 1 \\ -1 & -2 & 2 \end{bmatrix} = \begin{bmatrix} \frac{37}{3} & \frac{85}{3} & -\frac{64}{3} \end{bmatrix}$$

Next, use Algorithm 6.2 to design the control law.

Consider $K = [\ k_1 \ \ k_2 \ \ k_3\ ]$, then one can obtain that

$$\det(sI - (A + bK)) = \det\left(\begin{bmatrix} s & & \\ & s & \\ & & s \end{bmatrix} - \begin{bmatrix} 1 & 0 & -1 \\ 1 & 2 & 1 \\ 2 & 2 & 3 \end{bmatrix} - \begin{bmatrix} k_1 & k_2 & k_3 \\ 0 & 0 & 0 \\ k_1 & k_2 & k_3 \end{bmatrix} \right)$$

6.1 State Feedback Control

$$= \det \left(\begin{bmatrix} s-1-k_1 & -k_2 & 1-k_3 \\ -1 & s-2 & -1 \\ -k_1-2 & -k_2-2 & s-k_3-3 \end{bmatrix} \right)$$

$$= s^3 - (k_1+k_3+6)s^2 + (11+6k_1-2k_2+k_3)s - 6 - 6k_1 + 3k_2$$

One can obtain the following equations,

$$\begin{cases} -(k_1+k_3+6) = 3 \\ 11+6k_1-2k_2+k_3 = 7 \\ -6-6k_1+3k_2 = 5 \end{cases}$$

and K can be obtained by solving the above equations, which is the same as the result computed by Algorithm 6.1.

6.1.3 Eigenvalue Assignment with State Feedback for MIMO Systems

In this subsection, discuss the eigenvalue assignment problem for MIMO systems.

Consider a linear time-invariant MIMO system (6.1), firstly introduce the following lemmas.

In Subsection 1.3.2, it is introduced that a matrix $A \in \mathbb{R}^{n \times n}$ is said to be cyclic if and only if its characteristic polynomial equals its minimal polynomial.

Lemma 6.1 *A is cyclic if and only if in the Jordan normal form of A, there only exists one Jordan block for each eigenvalue.*

Proof As discussed in Subsection 1.3.4, the minimal polynomial of A can be expressed as

$$\phi(s) = \prod_{i=1}^{l}(s-\lambda_i)^{\sigma_{i\max}}$$

where $\sigma_{i\max} = \max_{1 \leqslant k \leqslant \alpha_i} \sigma_{ik}$, σ_{ik} denotes the order of kth Jordan block associated with λ_i, α_i denotes the geometric multiplicity associated with λ_i, $i=1,2,\cdots,l$.

On the other hand, the characteristic polynomial of A can be expressed as

$$\alpha(s) = \prod_{i=1}^{l}(s-\lambda_i)^{\sigma_i}$$

where σ_i denotes the algebraic multiplicity of λ_i.

It is noted that $\sigma_i = \sum_{k=1}^{\alpha_i}\sigma_{ik}$. It can be seen that $\phi(s) = \alpha(s)$ if and only if $\alpha_i = 1$, which implies there only exists one Jordan block for each eigenvalue. $\square$

From the above discussion, Lemma 6.1 implies that A is cyclic if and only if the geometric multiplicity of each eigenvalue is 1. Specifically, if n eigenvalues of A are distinct from each other, A is cyclic.

Lemma 6.2 *if A is cyclic, there exists $b \in \mathbb{R}^n$ such that*

$$\text{rank} \begin{bmatrix} b & Ab & \cdots & A^{n-1}b \end{bmatrix} = n \tag{6.6}$$

i.e, (A, b) is completely controllable.

The proof is omitted.

As a matter of fact, Lemma 6.2 can be used to define the cyclicity of matrices. That is, a matrix $A \in \mathbb{R}^{n \times n}$ is said to be cyclic if there exists $b \in \mathbb{R}^n$ such that (6.6) holds.

Lemma 6.3 *For a linear time-invariant system (6.1), if A is cyclic, and (A, B) is completely controllable, then almost any $l \in \mathbb{R}^p$ will make (A, b) completely controllable, where $b = Bl$.*

Illustrate this lemma by an example.

Example 6.3 Consider a state equation with

$$A = \begin{bmatrix} 2 & 1 & 0 & 0 & 0 \\ 0 & 2 & 1 & 0 & 0 \\ 0 & 0 & 2 & 0 & 0 \\ 0 & 0 & 0 & 3 & 1 \\ 0 & 0 & 0 & 0 & 3 \end{bmatrix}, \quad B = \begin{bmatrix} 0 & 1 \\ 0 & 0 \\ 1 & 1 \\ 1 & 2 \\ 1 & 0 \end{bmatrix}$$

It can be seen that (A, B) is completely controllable, and A is cyclic. For any $l = \begin{bmatrix} l_1 & l_2 \end{bmatrix}^{\mathrm{T}}$,

$$Bl = \begin{bmatrix} l_2 \\ 0 \\ l_1 + l_2 \\ l_1 + 2l_2 \\ l_1 \end{bmatrix}$$

According to the Jordan normal form criterion, (A, Bl) is completely controllable except $l_1 + l_2 = 0$ or $l_1 = 0$.

Lemma 6.4 *For a linear time-invariant MIMO system (6.1) with $\text{rank} B = p$, if the system is completely controllable, then for any b_i, there exists a real constant matrix K_1 such that $(A + BK_1, b_i)$ is completely controllable, where b_i denotes the ith column vector of B, $i = 1, \cdots, p$.*

Proof Since (A, B) is controllable, one can select n linearly independent column vectors from the controllability matrix. Firstly, select $b_1, Ab_1, \cdots, A^{\mu_1-1}b_1$ until $A^{\mu_1}b_1$ can be expressed by a linear combination of $b_1, Ab_1, \cdots, A^{\mu_1-1}b_1$. Secondly, select $b_2, Ab_2, \cdots, A^{\mu_2-1}b_2$ until $A^{\mu_2}b_2$ can be expressed by a linear combination of the above column vectors that have been selected. Repeat this procedure until n independent column vectors are selected. $\mu_i \geqslant 0$, $i = 1, \cdots, p$. Use the n linearly independent column vectors to construct the following matrix

$$W = \begin{bmatrix} b_1 & Ab_1 & \cdots & A^{\mu_1-1}b_1 & \cdots & b_p & Ab_p & \cdots & A^{\mu_p-1}b_p \end{bmatrix} \tag{6.7}$$

6.1 State Feedback Control

which satisfies rank $W = n$.

Construct a $p \times n$ matrix S,

$$S = \begin{bmatrix} 0 & \cdots & 0 & \underset{\mu_1\text{th column}}{e_2} & 0 & \cdots & 0 & \cdots & \underset{\sum_{i=1}^{p-1}\mu_i\text{th column}}{e_p} & 0 & \cdots & 0 \end{bmatrix}$$

where e_i denotes the ith column vector of a $p \times p$ identity matrix. The μ_1th column of S is e_2, the $(\mu_1 + \mu_2)$th column is e_3, and so on, the $\sum_{i=1}^{p-1} \mu_i$th column is e_p, the other columns are zero vectors.

Select $K_1 = SW^{-1}$. One can deduce that

$$(A + BK_1)b_1 = Ab_1 + BK_1 b_1 = Ab_1$$
$$(A + BK_1)^2 b_1 = (A + BK_1)Ab_1 = A^2 b_1$$
$$\vdots$$
$$(A + BK_1)^{\mu_1 - 1} b_1 = A^{\mu_1 - 1} b_1$$
$$(A + BK_1)^{\mu_1} b_1 = (A + BK_1)(A + BK_1)^{\mu_1 - 1} b_1 = A^{\mu_1} b_1 + Be_2 = A^{\mu_1} b_1 + b_2$$

Since $A^{\mu_1} b_1$ is a linear combination of $b_1, Ab_1, \cdots, A^{\mu_1 - 1} b_1$, $A^{\mu_1} b_1$ can be expressed as

$$A^{\mu_1} b_1 = \alpha_0 b_1 + \alpha_1 A b_1 + \cdots + \alpha_{\mu_1 - 1} A^{\mu_1 - 1} b_1$$

where $\alpha_0, \alpha_1, \cdots, \alpha_{\mu_1 - 1}$ are constant coefficients. Furthermore, one can deduce that

$$(A + BK_1)^{\mu_1 + 1} b_1 = (A + BK_1)(A^{\mu_1} b_1 + b_2)$$
$$= Ab_2 + A^{\mu_1 + 1} b_1 + BK_1 A^{\mu_1} b_1 + BK_1 b_2$$
$$= Ab_2 + A^{\mu_1 + 1} b_1 + BK_1(\alpha_0 b_1 + \alpha_1 A b_1 + \cdots + \alpha_{\mu_1 - 1} A^{\mu_1 - 1} b_1)$$
$$= Ab_2 + A^{\mu_1 + 1} b_1 + \alpha_{\mu_1 - 1} b_2$$

Repeating this procedure, one can obtain that

$$U_k = \begin{bmatrix} b_1 & (A + BK_1)b_1 & \cdots & (A + BK_1)^{n-1} b_1 \end{bmatrix}$$

$$= W \begin{bmatrix} I_{\mu_1 \times \mu_1} & * & \cdots & * \\ & I_{\mu_2 \times \mu_2} & \ddots & \vdots \\ & & \ddots & * \\ & & & I_{\mu_p \times \mu_p} \end{bmatrix}$$

where "$*$" in coefficient matrices represents the entry that is possibly nonzero.

It can be seen that rank $U_k = n$ and thus $(A+BK_1, b_1)$ is completely controllable. By the similar method, one can construct K_{1i} such that $(A+BK_{1i}, b_i)$ is controllable for $i = 2, \cdots, p$. In conclusion, this lemma is proved. □

Lemma 6.4 indicates that if (A, B) is controllable and A is not cyclic, there exists a real constant matrix K_1 such that $A+BK_1$ is cyclic. As a matter of fact, if (A, B) is controllable and A is not cyclic, for almost any $K_1 \in \mathbb{R}^{p \times n}$, $A+BK_1$ is cyclic.

Theorem 6.3 *For a linear time-invariant MIMO system (6.1), its all eigenvalues can be assigned arbitrarily by a state feedback control law (6.2) if and only if the system is completely controllable.*

Proof The proof of the necessity is the same as that for Theorem 6.2. In the following, only prove the sufficiency.

According to Lemma 6.4, one can introduce a state feedback control law with the feedback gain K_1 such that $(A+BK_1, b_1)$ is controllable. On this basis, introduce another state feedback control law with the feedback gain K_2, which has the following form,

$$K_2 = \begin{bmatrix} k_2 \\ 0 \\ \vdots \\ 0 \end{bmatrix} \qquad (6.8)$$

where k_2 is the first row vector of K_2.

In this way, since $(A+BK_1, b_1)$ is controllable, according to the proof of the sufficiency for Theorem 6.2, one can select an appreciate row vector k_2 to assign the eigenvalues of $A+BK_1+b_1 k_2$ arbitrarily. According to the superposition principle, the feedback gain $K = K_1 + K_2$. Using such a state feedback control law, all eigenvalues of the system (6.1) can be assigned arbitrarily. □

The control scheme is displayed in Fig.6.2. Based on the above conclusions, one can conclude an algorithm to realize the eigenvalue assignment for the system (6.1).

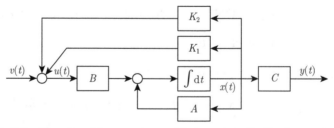

Fig.6.2 The closed-loop system with two state feedback matrices

Algorithm 6.3 *For a completely controllable MIMO system (6.1), using the state feedback law (6.2), the eigenvalues of $A+BK$ can be assigned to $\lambda_1^*, \cdots, \lambda_n^*$. The state feedback gain K can be determined by following procedures.*

(1) *Construct $K_1 = SW^{-1}$ such that $(A+BK_1, b_1)$ is controllable, where S and W are selected as the ones in Lemma 6.4, b_1 is the first column vector of B.*

6.1 State Feedback Control

(2) For the controllable pair $(A + BK_1, b_1)$, using Algorithm 6.1 or Algorithm 6.2 to determine k_2 such that the eigenvalues of $A + BK_1 + b_1 k_2$ are assigned to $\lambda_1^*, \cdots, \lambda_n^*$.

(3) Construct K_2 satisfying (6.8).

(4) Compute the feedback gain $K = K_1 + K_2$.

If there exists a column b of B, or a linear combination of B (also denote this linear combination as b), such that (A, b) is controllable, then the step (1) in Algorithm 6.3 can be omitted. As a matter of fact, according to Lemma 6.3, such a column vector b is frequently existed.

Example 6.4 Consider a completely controllable system

$$\dot{x}(t) = \begin{bmatrix} 1 & 1 & 0 & 0 \\ 0 & 2 & 0 & 0 \\ 1 & 0 & 0 & 0 \\ 0 & 1 & 0 & 0 \end{bmatrix} x(t) + \begin{bmatrix} 1 & 2 \\ 1 & 0 \\ 0 & 0 \\ 0 & 0 \end{bmatrix} u(t)$$

Design a state feedback gain matrix K such that the eigenvalues of the closed-loop system are assigned to $-1, -2, -3$ and -4.

Construct

$$W = \begin{bmatrix} b_1 & Ab_1 & b_2 & Ab_2 \end{bmatrix} = \begin{bmatrix} 1 & 2 & 2 & 2 \\ 1 & 2 & 0 & 0 \\ 0 & 1 & 0 & 2 \\ 0 & 1 & 0 & 0 \end{bmatrix}, \quad S = \begin{bmatrix} 0 & 0 & 0 & 0 \\ 0 & 1 & 0 & 0 \end{bmatrix}$$

Therefore, it can be computed that

$$K_1 = SW^{-1} = \begin{bmatrix} 0 & 0 & 0 & 0 \\ 0 & 0 & 0 & 1 \end{bmatrix}$$

The characteristic polynomial of $A + BK_1$ is

$$\det(sI - A - BK_1) = s^4 - 3s^3 + 2s^2$$

and the desired characteristic polynomial is

$$\alpha_c(s) = s^4 + 10s^3 + 35s^2 + 50s + 24$$

For the controllable pair $(A + BK_1, b_1)$, construct the transformation matrix T,

$$T = \left\{ \begin{bmatrix} 1 & 2 & 6 & 14 \\ 1 & 2 & 4 & 8 \\ 0 & 1 & 2 & 6 \\ 0 & 1 & 2 & 4 \end{bmatrix} \begin{bmatrix} 0 & 2 & -3 & 1 \\ 2 & -3 & 1 & \\ -3 & 1 & & \\ 1 & & & \end{bmatrix} \right\}^{-1} = \begin{bmatrix} 0 & 0 & 0.5 & -0.5 \\ 0.5 & -0.5 & 0 & 0 \\ 0.5 & -0.5 & 0 & 1 \\ 0.5 & -0.5 & 0 & 1 \end{bmatrix}$$

One can further calculate that

$$k_2 = \begin{bmatrix} 0-24 & 0-50 & 2-35 & -3-10 \end{bmatrix} \begin{bmatrix} 0 & 0 & 0.5 & -0.5 \\ 0.5 & -0.5 & 0 & 0 \\ 0.5 & -0.5 & 0 & 1 \\ 0.5 & -0.5 & 0 & 1 \end{bmatrix}$$

$$= \begin{bmatrix} -48 & -35 & -12 & -34 \end{bmatrix}$$

Finally, one can obtain that

$$K = K_1 + K_2 = \begin{bmatrix} -48 & -35 & -12 & -34 \\ 0 & 0 & 0 & 1 \end{bmatrix}$$

Based on the controllable canonical form of the linear time-invariant MIMO system, one can deduce some eigenvalue assignment algorithms. Here, only introduce an eigenvalue assignment algorithm based on the Luenberger second controllable canonical form.

Algorithm 6.4 *The problem is the same as the one in Algorithm 6.3. The state feedback gain K can be determined by following procedures.*

(1) *Transform the original system into the Luenberger second controllable canonical form with the transformation matrix T.*

(2) *Compute the desired characteristic polynomial $\alpha_c(s)$.*

(3) *Design $\bar{K}$ such that the characteristic polynomial of $\bar{A} + \bar{B}\bar{K}$ is $\alpha_c(s)$.*

(4) *Compute $K = \bar{K}T$.*

Example 6.5 Consider a system can be transformed into the Luenberger second controllable form,

$$\dot{\bar{x}}(t) = \begin{bmatrix} 0 & 1 & 0 & 0 & 0 & 0 \\ 0 & 0 & 1 & 0 & 0 & 0 \\ 3 & 2 & 0 & 4 & 7 & 9 \\ 0 & 0 & 0 & 0 & 1 & 0 \\ 0 & 0 & 0 & 0 & 0 & 1 \\ 2 & 3 & 4 & -5 & -3 & -4 \end{bmatrix} \bar{x}(t) + \begin{bmatrix} 0 & 0 \\ 0 & 0 \\ 1 & 1 \\ 0 & 0 \\ 0 & 0 \\ 0 & 1 \end{bmatrix} u(t)$$

The desired eigenvalues are $-1, -2, -3, -4, -5$ and -6. One can compute that

$$\alpha_{c1}(s) = (s+1)(s+2)(s+3) = s^3 + 6s^2 + 11s + 6$$

$$\alpha_{c2}(s) = (s+4)(s+5)(s+6) = s^3 + 15s^2 + 74s + 120$$

or

$$\alpha_c(s) = \alpha_{c1}(s)\alpha_{c2}(s) = s^6 + 21s^5 + 175s^4 + 735s^3 + 1624s^2 + 1764s + 720$$

Set the state feedback gain matrix $\bar{K}$ as

$$\bar{K} = \begin{bmatrix} k_{11} & k_{12} & k_{13} & k_{14} & k_{15} & k_{16} \\ k_{21} & k_{22} & k_{23} & k_{24} & k_{25} & k_{26} \end{bmatrix}$$

6.1 State Feedback Control

and then, it holds that

$$\bar{A} + \bar{B}\bar{K} = \left[\begin{array}{ccc|ccc} 0 & 1 & 0 & 0 & 0 & 0 \\ 0 & 0 & 1 & 0 & 0 & 0 \\ 3+k_{11}+k_{21} & 2+k_{12}+k_{22} & k_{13}+k_{23} & 4+k_{14}+k_{24} & 7+k_{15}+k_{25} & 9+k_{16}+k_{26} \\ \hline 0 & 0 & 0 & 0 & 0 & 0 \\ 0 & 0 & 0 & 0 & 0 & 0 \\ 2+k_{21} & 3+k_{22} & 4+k_{23} & -5+k_{24} & -3+k_{25} & -4+k_{26} \end{array}\right]$$

Meanwhile, according to the desired characteristic polynomial of $\bar{A} + \bar{B}\bar{K}$, one can obtain that

$$\bar{A} + \bar{B}\bar{K} = \left[\begin{array}{ccc|ccc} 0 & 1 & 0 & 0 & 0 & 0 \\ 0 & 0 & 1 & 0 & 0 & 0 \\ -6 & -11 & -6 & 0 & 0 & 0 \\ \hline 0 & 0 & 0 & 0 & 1 & 0 \\ 0 & 0 & 0 & 0 & 0 & 1 \\ 0 & 0 & 0 & -120 & -74 & -15 \end{array}\right]$$

or

$$\bar{A} + \bar{B}\bar{K} = \left[\begin{array}{cccccc} 0 & 1 & 0 & 0 & 0 & 0 \\ 0 & 0 & 1 & 0 & 0 & 0 \\ 0 & 0 & 0 & 1 & 0 & 0 \\ 0 & 0 & 0 & 0 & 1 & 0 \\ 0 & 0 & 0 & 0 & 0 & 1 \\ -720 & -1764 & -1264 & -735 & -175 & -21 \end{array}\right]$$

It can be further computed that

$$\bar{K} = \left[\begin{array}{cccccc} -7 & -10 & -2 & 111 & 64 & 2 \\ -2 & -3 & -4 & -115 & -71 & -11 \end{array}\right]$$

or

$$\bar{K} = \left[\begin{array}{cccccc} 719 & 1765 & 1268 & 727 & 165 & 8 \\ -722 & -1767 & -1268 & -730 & -172 & -17 \end{array}\right]$$

Finally, use $K = \bar{K}T$ to compute the feedback gain for the original system. It is noted that the designed state feedback gain matrix is not unique.

If all desired eigenvalues are distinct from all eigenvalues of the original system, the following algorithm can be used to realize the eigenvalue assignment.

Algorithm 6.5 *The problem is the same as the one in Algorithm 6.3, and it is additionally required that all desired eigenvalues are distinct from all eigenvalues of the original system. The state feedback gain K can be determined by following procedures.*

(1) *Select an $n \times n$ constant matrix F whose eigenvalues are equivalent to the desired eigenvalues.*

(2) *Select a $p \times n$ constant matrix $\bar{K}$ such that $(F, \bar{K})$ is completely observable.*

(3) *Solve the matrix equation $AT - TF = -B\bar{K}$.*

(4) *Judge whether T is nonsingular. If T is nonsingular, continue. Otherwise, return to the step (1) or (2) and select another F or $\bar{K}$.*

(5) *Compute $K = \bar{K}T^{-1}$.*

The matrix equation with the form of $XT + TY = Z$ is called the *Sylvester equation*, where $X \in \mathbb{R}^{n \times n}$, $Y \in \mathbb{R}^{m \times m}$ and $Z \in \mathbb{R}^{n \times m}$ are known, $T \in \mathbb{R}^{n \times m}$ is the solution of this equation. The solution T for any Z is unique if and only if there exists no common eigenvalue of X and $-Y$. Owing to this fact, it is required that all eigenvalues of A are distinct from ones of F, which can guarantee that the solution T of $AT - TF = -B\bar{K}$ is unique.

In the step (2), it is required that $(F, \bar{K})$ is completely observable. That is because the condition that (A, B) is completely controllable and $(F, \bar{K})$ is completely observable is the necessary condition to guarantee that T is nonsingular. Furthermore, if (A, B) is a SISO system, this condition is also sufficient. As a result, this manipulation is helpful to obtain a nonsingular T.

From $AT - TF = -B\bar{K}$ and $K = \bar{K}T^{-1}$, it holds that $TFT^{-1} = A + B\bar{K}T^{-1} = A + BK$. Therefore, the eigenvalues of $A + BK$ are the same as the ones of F.

If the n desired eigenvalues are distinct from each other. One can deduce another eigenvalue assignment algorithm.

Suppose that the feedback gain K in (6.2) can be expressed as the exterior product of two vectors, i.e. $K = lm$, where $l \in \mathbb{R}^p$ and $m^T \in \mathbb{R}^n$. On this basis, the state equation of the closed-loop system is

$$\dot{x}(t) = (A + BK)x(t) + Bv(t) = (A + Blm)x(t) + Bv(t) = (A + bm)x(t) + Bv(t)$$

where $b = Bl$ is a linear combination of the column vectors of B. If there exists a vector l, such that (A, b) is completely controllable, then the eigenvalue assignment problem for MIMO systems can be changed to the eigenvalue assignment problem for SISO systems.

For the closed-loop system, it holds that

$$\det(sI - A - bm) = \det\left((sI - A)(I - (sI - A)^{-1}bm)\right)$$
$$= \det(sI - A)\det\left(I - (sI - A)^{-1}bm\right)$$
$$= \det(sI - A)\left(1 - m(sI - A)^{-1}b\right)$$

6.1 State Feedback Control

$$= \det(sI - A) - m\,\text{adj}(sI - A)b$$

Denote $\alpha_c(s) = \det(sI - A - bm)$, $\alpha(s) = \det(sI - A)$ and $R(s) = \text{adj}(sI - A)$. It holds that

$$\alpha_c(s) = \alpha(s) - mR(s)b$$

Substitute the desired eigenvalues $\lambda_1^*, \cdots, \lambda_n^*$ into the above equation, one can obtain that

$$m \begin{bmatrix} R(\lambda_1^*)b & \cdots & R(\lambda_n^*)b \end{bmatrix} = \begin{bmatrix} \alpha(\lambda_1^*) & \cdots & \alpha(\lambda_n^*) \end{bmatrix} \tag{6.9}$$

To guarantee that $[\, R(\lambda_1^*)b \;\; \cdots \;\; R(\lambda_n^*)b \,]$ is nonsingular, $\lambda_1^*, \cdots, \lambda_n^*$ should be distinct from each other.

Algorithm 6.6 *The problem is the same as the one in Algorithm 6.3, and it is additionally required that all desired eigenvalues are distinct from each other. The feedback gain $K = lm$ can be determined by following procedures, where $l \in \mathbb{R}^p$ and $m^T \in \mathbb{R}^n$.*

(1) Select an arbitrary l such that (A, b) is completely controllable, where $b = Bl$.

(2) Use Algorithm 1.1 to determine $\alpha(s) = \det(sI - A)$ and $R(s) = \text{adj}(sI - A)$.

(3) Based on the desired eigenvalues, construct the equation (6.9), and then one can compute

$$m = \begin{bmatrix} \alpha(\lambda_1^*) & \cdots & \alpha(\lambda_n^*) \end{bmatrix} \begin{bmatrix} R(\lambda_1^*)b & \cdots & R(\lambda_n^*)b \end{bmatrix}^{-1}$$

(4) Compute $K = lm$.

Example 6.6 The longitudinal pose movement equation for a flight control system is

$$\dot{x}(t) = Ax(t) + Bu(t) = \begin{bmatrix} -0.605 & 0.023 & -5.816 \\ 9.29 & -0.343 & -33.6 \\ 0 & 0 & -20 \end{bmatrix} x(t) + \begin{bmatrix} 0 & 0 \\ 1 & 0 \\ 0 & 1 \end{bmatrix} u(t)$$

Design a state feedback gain matrix K such that the eigenvalues of the closed-loop system are assigned to -5, -10 and -1.

Select $l = [\, 33.6 \;\; 20 \,]^T$, and then $b = Bl = [\, 0 \;\; 33.6 \;\; 20 \,]^T$. It can be verified (A, b) is controllable.

Based on Algorithm 1.1, one can compute that

$R_2 = I$, $a_1 = -\text{tr}(R_2 A) = 20.94$

$$R_1 = AR_2 + a_1 I = \begin{bmatrix} 20.343 & 0.023 & -5.816 \\ 9.29 & 20.605 & -33.6 \\ 0 & 0 & 0.948 \end{bmatrix}, \quad a_2 = -\frac{1}{2}\text{tr}(R_1 A) = 18.954$$

$$R_0 = AR_1 + a_2 I = \begin{bmatrix} 6.86 & 0.46 & -2.768 \\ 185.8 & 12.1 & -74.359 \\ 0 & 0 & -0.006 \end{bmatrix}, \quad a_3 = -\frac{1}{3}\text{tr}(R_0 A) = -0.122$$

It is noted that $\alpha(s) = s^3 + a_1 s^2 + a_2 s + a_3$ and $R(s) = R_2 s^2 + R_1 s + R_0$. In this way, one can obtain that

$$m = \begin{bmatrix} \alpha(\lambda_1^*) & \alpha(\lambda_2^*) & \alpha(\lambda_3^*) \end{bmatrix} \begin{bmatrix} R(\lambda_1^*)b & R(\lambda_2^*)b & R(\lambda_3^*)b \end{bmatrix}^{-1}$$

$$= \begin{bmatrix} 303.808 & 905.138 & 1053.868 \end{bmatrix} \begin{bmatrix} 537.831 & 1115.566 & 1693.301 \\ -342.26 & 2076.1 & 6174.46 \\ 405.08 & 1810.28 & 4215.48 \end{bmatrix}^{-1}$$

$$= \begin{bmatrix} 2.080 & 0.6175 & -1.49 \end{bmatrix}$$

Finally, compute

$$K = lm = \begin{bmatrix} 33.6 \\ 20 \end{bmatrix} \begin{bmatrix} 2.080 & 0.618 & -1.49 \end{bmatrix} = \begin{bmatrix} 69.89 & 20.75 & -50.06 \\ 41.6 & 12.35 & -29.8 \end{bmatrix}$$

6.1.4 Stabilization with State Feedback

In this subsection, the concerned problem is to determine a state feedback control law (6.2) having the property that the corresponding closed-loop system is asymptotically stable when $v(t) = 0$. This is the stabilization problem.

For a linear time-invariant system (6.1), there exists a transformation matrix T which can decompose the system into

$$\begin{bmatrix} \dot{x}_c(t) \\ \dot{x}_{nc}(t) \end{bmatrix} = \begin{bmatrix} A_{11} & A_{12} \\ 0 & A_{22} \end{bmatrix} \begin{bmatrix} x_c(t) \\ x_{nc}(t) \end{bmatrix} + \begin{bmatrix} B_1 \\ 0 \end{bmatrix} u(t) \qquad (6.10)$$

where (A_{11}, B_1) is completely controllable. The eigenvalues of the system (6.10) are consisted of the eigenvalues of A_{11} and A_{22}, while the eigenvalues of A_{11} are controllable and the eigenvalues of A_{22} are uncontrollable.

Select

$$u(t) = \begin{bmatrix} K_c & K_{nc} \end{bmatrix} \begin{bmatrix} x_c(t) \\ x_{nc}(t) \end{bmatrix} \qquad (6.11)$$

Substituting (6.11) into (6.10), one can deduce that the closed-loop system is

$$\begin{bmatrix} \dot{x}_c(t) \\ \dot{x}_{nc}(t) \end{bmatrix} = \begin{bmatrix} A_{11} + B_1 K_c & A_{12} + B_1 K_{nc} \\ 0 & A_{22} \end{bmatrix} \begin{bmatrix} x_c(t) \\ x_{nc}(t) \end{bmatrix} \qquad (6.12)$$

The eigenvalues of the close-loop system (6.12) are the eigenvalues of $A_{11} + B_1 K_c$ and A_{22}. As discussed in previous subsection, the eigenvalues of $A_{11} + B_1 K_c$ can be arbitrarily assigned. Therefore, the following conclusion can be deduced.

Theorem 6.4 *A linear time-invariant system can be stabilized by a state feedback control law $u(t) = Kx(t)$ if and only if all uncontrollable eigenvalues of the system have negative real parts.*

6.1 State Feedback Control

The proof is omitted.

Obviously, if the system (6.1) is completely controllable, it can be stabilized by a linear state feedback control law. If it is incompletely controllable, the following algorithm can be used to deal with the stabilization problem.

Algorithm 6.7 *For an incompletely controllable system* (6.1), *using the state feedback law* (6.2), *the system can be stabilized. The state feedback gain* K *can be determined by following procedures.*

(1) *Taking the controllability structure decomposition, the transformation is* T *and the system is transformed into* (6.10). *The eigenvalues of* A_{22} *should have negative real part. Otherwise, the system* (6.1) *cannot be stabilized.*

(2) *Assign the eigenvalues of* A_{11} *to desired locations by the state feedback control law* (6.2) *with feedback gain* K_c, *the desired poles can be arbitrarily selected as long as they have negative real part.*

(3) *Compute* $K = [\ K_c\ \ K_{nc}\ ]T$, *where* K_{nc} *with appreciate dimensions can be arbitrarily selected.*

Example 6.7 Consider an incompletely controllable system

$$\dot{x}(t) = \begin{bmatrix} 1 & 0 & -1 \\ 0 & -2 & 0 \\ -1 & 0 & 2 \end{bmatrix} x(t) + \begin{bmatrix} 0 \\ 0 \\ 1 \end{bmatrix} u(t)$$

Using the transformation matrix

$$T = \begin{bmatrix} 0 & 0 & 1 \\ 1 & 0 & 0 \\ 0 & 1 & 0 \end{bmatrix}$$

the system can be decomposed into

$$\dot{\bar{x}}(t) = \begin{bmatrix} 2 & -1 & 0 \\ -1 & 1 & 0 \\ 0 & 0 & -2 \end{bmatrix} \bar{x}(t) + \begin{bmatrix} 1 \\ 0 \\ 0 \end{bmatrix} u(t)$$

Obviously, the system can be stabilized.

Select $\bar{K} = [\ -9\ \ 17\ \ 0\ ]$, the closed-loop system is

$$\dot{\bar{x}}(t) = \begin{bmatrix} -7 & 16 & 0 \\ -1 & 1 & 0 \\ 0 & 0 & -2 \end{bmatrix} \bar{x}(t)$$

which is asymptotically stable.

The feedback gain is

$$K = \bar{K}T = \begin{bmatrix} 17 & 0 & -9 \end{bmatrix}$$

Based on Theorem 5.15, the stabilization problem can also be solved. One can just solve the matrix equation

$$(A + BK)^T P + P(A + BK) = -Q \tag{6.13}$$

to obtain K, where P and Q are symmetric and positive definite matrix. For convenience, Q and P should be as simple as possible. For example, choose $P = Q = I$.

Example 6.8 Consider a completely controllable system

$$\dot{x}(t) = \begin{bmatrix} 0 & 1 \\ 2 & 2 \end{bmatrix} x(t) + \begin{bmatrix} 1 & 1 \\ -2 & 1 \end{bmatrix} u(t)$$

Let $u(t) = Kx(t)$ with

$$K = \begin{bmatrix} k_{11} & k_{12} \\ k_{21} & k_{22} \end{bmatrix}$$

One can compute that

$$A + BK = \begin{bmatrix} k_{11} + k_{21} & 1 + k_{12} + k_{22} \\ 2 - 2k_{11} + k_{21} & 2 - 2k_{12} + k_{22} \end{bmatrix}$$

Select $P = Q = I$. It holds that

$$\begin{bmatrix} k_{11} + k_{21} & 2 - 2k_{11} + k_{21} \\ 1 + k_{12} + k_{22} & 2 - 2k_{12} + k_{22} \end{bmatrix} + \begin{bmatrix} k_{11} + k_{21} & 1 + k_{12} + k_{22} \\ 2 - 2k_{11} + k_{21} & 2 - 2k_{12} + k_{22} \end{bmatrix} = -I$$

and then one can deduce the following equations,

$$\begin{cases} 2k_{11} + 2k_{21} = -1 \\ -2k_{11} + k_{12} + k_{21} + k_{22} = -3 \\ -4k_{12} + 2k_{22} = -5 \end{cases}$$

Select $k_{11} = 1$, and then one can obtain that $k_{12} = 1$, $k_{21} = -1.5$ and $k_{22} = -0.5$. It can be verified that eigenvalues of $A + BK$ have negative real parts. That is to say, the closed-loop system is asymptotically stable.

In addition, for a linear time-invariant system (6.1) the following quadratic index is frequently used in optimal regulation problem,

$$J = \int_0^\infty \left(x^T(t) Q x(t) + u^T(t) R u(t) \right) dt \tag{6.14}$$

where Q and R are two symmetric and positive definite matrices. This index corresponds to the energy consumed by the error and input signals. It is desired to design a state feedback control law $u(t) = Kx(t)$ to minimize the quadratic index J.

6.1 State Feedback Control

With the state feedback, (6.14) can be converted to

$$J = \int_0^\infty x^{\mathrm{T}}(t)(Q + K^{\mathrm{T}}RK)x(t)\mathrm{d}t$$

Let

$$x^{\mathrm{T}}(t)(Q + K^{\mathrm{T}}RK)x(t) = -\frac{\mathrm{d}}{\mathrm{d}t}x^{\mathrm{T}}(t)Px(t)$$

where P is a symmetric and positive definite matrix. It can be rewritten as

$$(A + BK)^{\mathrm{T}}P + P(A + BK) = -(Q + K^{\mathrm{T}}RK) \tag{6.15}$$

and then one can deduce that

$$J = \int_0^\infty -\frac{\mathrm{d}}{\mathrm{d}t}x^{\mathrm{T}}(t)Px(t)\mathrm{d}t = x^{\mathrm{T}}(0)Px(0) - x^{\mathrm{T}}(\infty)Px(\infty)$$

Since $A + BK$ should be asymptotically stable, then $x(\infty) = 0$ and the quadratic index J can be further represented as

$$J = x^{\mathrm{T}}(0)Px(0) \tag{6.16}$$

In this way, one can solve the matrix equation (6.15) to obtain the function related to entries of P and K. On this basis, substitute P into (6.16) to determine K by minimizing J.

Example 6.9 Consider the state equation and performance index are

$$\dot{x}(t) = \begin{bmatrix} 0 & 1 \\ 0 & 0 \end{bmatrix} x(t) + \begin{bmatrix} 0 \\ 1 \end{bmatrix} u(t)$$

$$J = \int_0^\infty (x^{\mathrm{T}}(t)x(t) + u^{\mathrm{T}}(t)u(t))\mathrm{d}t$$

Submit $u(t) = [\, k_1 \quad k_2 \,]x(t)$ into the state equation, and one can obtain that

$$\dot{x}(t) = \begin{bmatrix} 0 & 1 \\ k_1 & k_2 \end{bmatrix} x(t)$$

According to (6.15) and the performance index J, one can obtain that $Q = I$ and $R = I$. In this way, (6.15) can be written as

$$\begin{bmatrix} 0 & k_1 \\ 1 & k_2 \end{bmatrix} P + P \begin{bmatrix} 0 & 1 \\ k_1 & k_2 \end{bmatrix} = -\begin{bmatrix} 1+k_1^2 & k_1 k_2 \\ k_1 k_2 & 1+k_2^2 \end{bmatrix}$$

which indicates that the matrix P is the function of k_1 and k_2.

Combining with (6.16), the performance index J is the function of k_1 and k_2. Let $\dfrac{\partial J}{\partial k_1} = 0$ and $\dfrac{\partial J}{\partial k_2} = 0$ to minimize J for any $x(0)$. It can be solved that $k_1 = -1$ and $k_2 = -\sqrt{3}$. Therefore, one can obtain that

$$P = \begin{bmatrix} \sqrt{3} & 1 \\ 1 & \sqrt{3} \end{bmatrix}$$

and

$$J = x^{\mathrm{T}}(0) \begin{bmatrix} \sqrt{3} & 1 \\ 1 & \sqrt{3} \end{bmatrix} x(0)$$

6.2 Output Feedback Control

When considering the state feedback, it is supposed that values of states can be measured by using appropriate sensors. As a matter of fact, it may be either impossible or impractical to obtain measurements for all states. Compared with the state information, the output values are easier to be measured. In this section, the output feedback is discussed.

6.2.1 Controllability and Observability of System with Output Feedback

Consider a linear time-invariant system,

$$\dot{x}(t) = Ax(t) + Bu(t), \ y(t) = Cx(t) \tag{6.17}$$

where $x(t) \in \mathbb{R}^n$ is the state vector, $u(t) \in \mathbb{R}^p$ is the input vector, and $y(t) \in \mathbb{R}^q$ is the output vector. A, B and C are real constant matrices with appropriate dimensions.

Introduce the output feedback control law

$$u(t) = Ky(t) + v(t) \tag{6.18}$$

where $v(t) \in \mathbb{R}^p$ is the reference input vector, K is a constant matrix. In general, (6.18) is called a static output feedback control law.

Substituting (6.18) into (6.17), the corresponding closed-loop system is

$$\dot{x}(t) = (A + BKC)x(t) + Bv(t), \ y(t) = Cx(t) \tag{6.19}$$

and its block diagram is displayed in Fig.6.3.

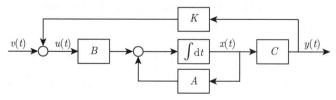

Fig.6.3 The block diagram of a closed-loop system with static output feedback

6.2 Output Feedback Control

Theorem 6.5 *For a linear time-invariant system* (6.17), *using the output feedback control law* (6.18) *cannot change the controllability and the observability.*

Proof For any output feedback control law (6.18), it can be viewed as a state feedback control law (6.2) with the state feedback gain be KC. According to Theorem 6.1, the controllability is unchanged.

Next, prove that the observability is also unchanged.

Observability matrices of (6.17) and (6.19) are $V = [\ C^{\mathrm{T}} \quad A^{\mathrm{T}}C^{\mathrm{T}} \quad \cdots \quad (A^{n-1})^{\mathrm{T}}C^{\mathrm{T}}\ ]^{\mathrm{T}}$ and $V_{\mathrm{ok}} = [\ C^{\mathrm{T}} \quad (A+BKC)^{\mathrm{T}}C^{\mathrm{T}} \quad \cdots \quad ((A+BKC)^{n-1})^{\mathrm{T}}C^{\mathrm{T}}\ ]^{\mathrm{T}}$, respectively. Next, prove $\operatorname{rank} V = \operatorname{rank} V_{\mathrm{ok}}$.

Similar to the proof of Theorem 6.1, it can be deduced that each row of $C(A+BKC)^m$ can be expressed as a linear combination of rows of $[\ C^{\mathrm{T}} \quad A^{\mathrm{T}}C^{\mathrm{T}} \quad \cdots \quad (A^m)^{\mathrm{T}}C^{\mathrm{T}}\ ]^{\mathrm{T}}$, $m = 1, \cdots, n-1$. Therefore, each row of V_{ok} can be expressed as a linear combination of the rows of V, which indicates that $\operatorname{rank} V \geqslant \operatorname{rank} V_{\mathrm{ok}}$.

On the other hand, it is noted that

$$\dot{x}(t) = Ax(t) + Bu(t) = [(A + BKC) - BKC]x(t) + Bu(t)$$

which can be viewed as an output feedback system of (6.17) with the feedback gain being $-K$. In this way, one can deduce that $\operatorname{rank} V \leqslant \operatorname{rank} V_{\mathrm{ok}}$.

In conclusion, $\operatorname{rank} V = \operatorname{rank} V_{\mathrm{ok}}$, and thus the controllability is invariant. □

6.2.2 Eigenvalue Assignment with Output Feedback

We have elaborated the cyclicity of matrices in Subsection 6.1.3. Here, introduce another two lemmas.

Lemma 6.5 *For a linear time-invariant system* (6.17), *if A is cyclic, and (A, C) is completely observable, then almost any $m^{\mathrm{T}} \in \mathbb{R}^q$ will make (A, c) completely observable, where $c = mC$.*

Lemma 6.6 *For a linear time-invariant system* (6.17), *if A is not cyclic, and (A, B, C) is completely controllable and observable, then there exists a constant matrix K such that $(A + BKC, B, C)$ is completely controllable and observable, and $A + BKC$ is cyclic.*

Here, omit the proof of the above two lemmas.

Theorem 6.6 *For a linear time-invariant system* (6.17) *with* $\operatorname{rank} B = p$ *and* $\operatorname{rank} C = q$, (A, B, C) *is completely controllable and observable, then there always exists a constant matrix K such that* $\max\{p, q\}$ *eigenvalues of $A + BKC$ can be assigned arbitrarily to approach the desired locations.*

Proof Without loss of generality, assume that A is cyclic. Otherwise, according to Lemma 6.6, a constant matrix K can be found such that $A + BKC$ is cyclic.

Consider $K = lm$, where $l \in \mathbb{R}^p$ and $m \in \mathbb{R}^q$. Substituting it into (6.19), one can obtain that

$$\dot{x}(t) = (A + BlmC)x(t) + Bv(t)$$

It is noted that

$$\det(sI - A - BlmC) = \det\left((sI - A)(I - (sI - A)^{-1}BlmC)\right)$$
$$= \det(sI - A)\det\left(I - (sI - A)^{-1}BlmC\right)$$
$$= \det(sI - A)(1 - mC(sI - A)^{-1}Bl)$$
$$= \det(sI - A) - mC\,\mathrm{adj}(sI - A)Bl$$

Denote $\alpha_c(s) = \det(sI - A - BlmC)$, $\alpha(s) = \det(sI - A)$ and $R(s) = \mathrm{adj}(sI - A)$. It holds that

$$\alpha_c(s) = \alpha(s) - mCR(s)Bl$$

Substitute the q desired eigenvalues $\lambda_1^*, \cdots, \lambda_q^*$ into the above equation, one can deduce that

$$m\begin{bmatrix} CR(\lambda_1^*)Bl & \cdots & CR(\lambda_q^*)Bl \end{bmatrix} = \begin{bmatrix} \alpha(\lambda_1^*) & \cdots & \alpha(\lambda_q^*) \end{bmatrix} \tag{6.20}$$

If $H \triangleq [\ CR(\lambda_1^*)Bl \ \cdots \ CR(\lambda_q^*)Bl\ ]$ is nonsingular, then for any pre-selected l which can guarantee (A, Bl) is controllable, there exists

$$m = \begin{bmatrix} \alpha(\lambda_1^*) & \cdots & \alpha(\lambda_q^*) \end{bmatrix} H^{-1} \tag{6.21}$$

and the corresponding $K = lm$ can assign q eigenvalues to desired locations exactly.

If H is singular, one can import small perturbations $\Delta\lambda_i^*$ to the desired eigenvalues, and replace λ_i^* by $\lambda_i^* + \Delta\lambda_i^*$, $i \in \{1, \cdots, q\}$ such that H is nonsingular. Since $\mathrm{rank}\,C = q$, this method is always effective. Furthermore, select m satisfying (6.21), $K = lm$ can assign q eigenvalues to desired locations approximately.

Consider the dual system (A^T, C^T, B^T) of (A, B, C), following the same way, we can find a constant matrix K such that $p = \mathrm{rank}\,B^T$ eigenvalues of A^T are assigned to desired locations exactly or approximately.

In conclusion, $\max\{p, q\}$ desired eigenvalues can be assigned exactly or approximately by the output feedback control law (6.18). □

Based on Theorem 6.6 and referred to Algorithm 6.6, one can conclude the algorithm to assign eigenvalues by the output feedback law (6.18). Here, do not elaborate it but illustrate it by an example.

Example 6.10 Consider a linear time-invariant system (6.17) with

$$A = \begin{bmatrix} 0 & 1 & 0 & 0 \\ 0 & 0 & 1 & 0 \\ 0 & 0 & 0 & 1 \\ -4 & 12 & -13 & 6 \end{bmatrix},\ B = \begin{bmatrix} 0 & 0 \\ 0 & 1 \\ 0 & 0 \\ 1 & 0 \end{bmatrix},\ C = \begin{bmatrix} 1 & 0 & 0 & 0 \\ 0 & 0 & 1 & 0 \\ 0 & 1 & 0 & 1 \end{bmatrix}$$

Try to assign the eigenvalues of the closed-loop system to $-1, -2, -3$ and -4.

6.2 Output Feedback Control

It can be verified that A is cyclic and (A, B, C) is controllable and observable. Based on Algorithm 1.1, one can obtain that

$$R_3 = I, \ a_1 = -\operatorname{tr}(R_3 A) = -6$$

$$R_2 = AR_3 + a_1 I = \begin{bmatrix} -6 & 1 & 0 & 0 \\ 0 & -6 & 1 & 0 \\ 0 & 0 & -6 & 1 \\ -4 & 12 & -13 & 0 \end{bmatrix}, \ a_2 = -\frac{1}{2}\operatorname{tr}(R_2 A) = 13$$

$$R_1 = AR_2 + a_2 I = \begin{bmatrix} 13 & -6 & 1 & 0 \\ 0 & 13 & -6 & 1 \\ -4 & 12 & 0 & 0 \\ 0 & -4 & 12 & 0 \end{bmatrix}, \ a_3 = -\frac{1}{3}\operatorname{tr}(R_1 A) = -12$$

$$R_0 = AR_1 + a_3 I = \begin{bmatrix} -12 & 13 & -6 & 1 \\ -4 & 0 & 0 & 0 \\ 0 & -4 & 0 & 0 \\ 0 & 0 & -4 & 0 \end{bmatrix}, \ a_4 = -\frac{1}{4}\operatorname{tr}(R_0 A) = 4.$$

Furthermore, one can deduce that $\alpha(s) = s^4 - 6s^3 + 13s^2 - 12s + 4$, $R(s) = R_3 s^3 + R_2 s^2 + R_1 s + R_0$.

Select $l = [\ 1 \ \ 0 \ ]^{\mathrm{T}}$, it can be verified that $(A, b = Bl)$ is controllable and

$$CR(s)b = \begin{bmatrix} 1 & 0 & 0 & 0 \\ 0 & 0 & 1 & 0 \\ 0 & 1 & 0 & 1 \end{bmatrix} (R_3 s^3 + R_2 s^2 + R_1 s + R_0) \begin{bmatrix} 0 \\ 0 \\ 0 \\ 1 \end{bmatrix}$$

$$= \begin{bmatrix} 0 \\ 0 \\ 1 \end{bmatrix} s^3 + \begin{bmatrix} 0 \\ 1 \\ 0 \end{bmatrix} s^2 + \begin{bmatrix} 0 \\ 0 \\ 1 \end{bmatrix} s + \begin{bmatrix} 1 \\ 0 \\ 0 \end{bmatrix}$$

Since $\operatorname{rank} C = 3$, select $\lambda_1^* = -1$, $\lambda_2^* = -2$, $\lambda_3^* = -3$. One can compute that

$$m = \begin{bmatrix} \alpha(\lambda_1^*) & \alpha(\lambda_2^*) & \alpha(\lambda_3^*) \end{bmatrix} \begin{bmatrix} CR(\lambda_1^*)b & CR(\lambda_2^*)b & CR(\lambda_3^*)b \end{bmatrix}^{-1}$$

$$= \begin{bmatrix} 36 & 144 & 400 \end{bmatrix} \begin{bmatrix} 1 & 1 & 1 \\ 1 & 4 & 9 \\ -2 & -10 & -30 \end{bmatrix}^{-1} = \begin{bmatrix} 7.6 & 5.6 & -11.4 \end{bmatrix}$$

Moreover, one can obtain that

$$K = lm = \begin{bmatrix} 1 \\ 0 \end{bmatrix} \begin{bmatrix} 7.6 & 5.6 & -11.4 \end{bmatrix} = \begin{bmatrix} 7.6 & 5.6 & -11.4 \\ 0 & 0 & 0 \end{bmatrix}$$

and
$$\alpha_c(s) = \det(sI - A - BKC) = s^4 + 5.4s^3 + 7.4s^2 - 0.6s - 3.6$$

It can be seen that only three eigenvalues can be assigned to desired locations. The fourth eigenvalue is 0.6, which moves to the right-half plane of s.

On the other hand, it holds that

$$A^{\mathrm{T}} = \begin{bmatrix} 0 & 0 & 0 & -4 \\ 1 & 0 & 0 & 12 \\ 0 & 1 & 0 & -13 \\ 0 & 0 & 1 & 6 \end{bmatrix}, \quad C^{\mathrm{T}} = \begin{bmatrix} 1 & 0 & 0 \\ 0 & 0 & 1 \\ 0 & 1 & 0 \\ 0 & 0 & 1 \end{bmatrix}, \quad B^{\mathrm{T}} = \begin{bmatrix} 0 & 0 & 0 & 1 \\ 0 & 1 & 0 & 0 \end{bmatrix}$$

Following the similar vein, one can compute that
$$R_3^{\mathrm{T}} = I, \quad a_1 = -\operatorname{tr}(R_3^{\mathrm{T}} A^{\mathrm{T}}) = -6$$

$$R_2^{\mathrm{T}} = A^{\mathrm{T}} R_3^{\mathrm{T}} + a_1 I = \begin{bmatrix} -6 & 0 & 0 & -4 \\ 1 & -6 & 0 & 12 \\ 0 & 1 & -6 & -13 \\ 0 & 0 & 1 & 0 \end{bmatrix}, \quad a_2 = -\frac{1}{2}\operatorname{tr}(R_2^{\mathrm{T}} A^{\mathrm{T}}) = 13$$

$$R_1^{\mathrm{T}} = A^{\mathrm{T}} R_2^{\mathrm{T}} + a_2 I = \begin{bmatrix} 13 & 0 & -4 & 0 \\ -6 & 13 & 12 & -4 \\ 1 & -6 & 0 & 12 \\ 0 & 1 & 0 & 0 \end{bmatrix}, \quad a_3 = -\frac{1}{3}\operatorname{tr}(R_1^{\mathrm{T}} A^{\mathrm{T}}) = -12$$

$$R_0^{\mathrm{T}} = A^{\mathrm{T}} R_1^{\mathrm{T}} + a_3 I = \begin{bmatrix} -12 & -4 & 0 & 0 \\ 13 & 0 & -4 & 0 \\ -6 & 0 & 0 & -4 \\ 1 & 0 & 0 & 0 \end{bmatrix}, \quad a_4 = -\frac{1}{4}\operatorname{tr}(R_0^{\mathrm{T}} A^{\mathrm{T}}) = 4$$

Furthermore, one can obtain that $\alpha(s) = s^4 - 6s^3 + 13s^2 - 12s + 4$, $R^{\mathrm{T}}(s) = R_3^{\mathrm{T}} s^3 + R_2^{\mathrm{T}} s^2 + R_1^{\mathrm{T}} s + R_0^{\mathrm{T}}$.

Select $l = \begin{bmatrix} 1 & 0 & 0 \end{bmatrix}^{\mathrm{T}}$, it can be verified that $(A^{\mathrm{T}}, C^{\mathrm{T}} l)$ is controllable and

$$B^{\mathrm{T}} R^{\mathrm{T}}(s) C^{\mathrm{T}} l = \begin{bmatrix} 0 & 0 & 0 & 1 \\ 0 & 1 & 0 & 0 \end{bmatrix} [R_3^{\mathrm{T}} s^3 + R_2^{\mathrm{T}} s^2 + R_1^{\mathrm{T}} s + R_0^{\mathrm{T}}] \begin{bmatrix} 1 & 0 & 0 \\ 0 & 0 & 1 \\ 0 & 1 & 0 \\ 0 & 0 & 1 \end{bmatrix} \begin{bmatrix} 1 \\ 0 \\ 0 \end{bmatrix}$$

$$= \begin{bmatrix} 0 \\ 1 \end{bmatrix} s^2 + \begin{bmatrix} 0 \\ -6 \end{bmatrix} s^1 + \begin{bmatrix} 1 \\ 13 \end{bmatrix}$$

Since $\operatorname{rank} B^{\mathrm{T}} = 2$, select $\lambda_1^* = -1$ and $\lambda_2^* = -2$. One can compute that that

$$m = \begin{bmatrix} 36 & 144 \end{bmatrix} \begin{bmatrix} 1 & 1 \\ 20 & 29 \end{bmatrix}^{-1} = \begin{bmatrix} -204 & 12 \end{bmatrix}$$

6.2 Output Feedback Control

and

$$\alpha_c(s) = \det(sI - A - BKC) = s^4 - 6s^3 + s^2 + 60s + 52$$

It can be seen that only two eigenvalues can be assigned to desired locations. The other two eigenvalues are $43.5 \pm j\sqrt{11.5}$, which move to the right-half plane of s.

In Theorem 6.6, it indicates that $\max\{p,q\}$ eigenvalues can be assigned to approach the desired locations, but the information about the other $n - \max\{p,q\}$ is not provided.

Theorem 6.7 *For a linear time-invariant system* (6.17) *with* $\text{rank } B = p$ *and* $\text{rank } C = q$, (A, B, C) *is completely controllable and observable, then to almost all* (A, B, C), *there exists a constant matrix* K *such that* $\min\{p + q - 1, n\}$ *eigenvalues of* $A + BKC$ *can be assigned arbitrarily to approach the desired locations. If* $p + q - 1 \geqslant n$, *then to almost all* (A, B, C), *all eigenvalues of* $A + BKC$ *can be assigned arbitrarily to approach the desired locations.*

The following algorithm can illustrate this theorem. Here, do not elaborate the proof.

Algorithm 6.8 *For a completely controllable and observable system* (6.17) *with* $\text{rank } B = p$ *and* $\text{rank } C = q$, *using the output feedback law* (6.18), $\min\{p + q - 1, n\}$ *eigenvalues of* $A + BK$ *can be assigned to approach desired locations. The output feedback gain* K *can be determined by following procedures.*

(1) *Consider* $K_1 = l_1 m_1$ *with* $l_1 = [\, l_{11} \;\; \cdots \;\; l_{1p} \,]^T$ *and* $m_1 = [\, m_{11} \;\; \cdots \;\; m_{1q} \,]$. *Denote* $W_1(s) = C \operatorname{adj}(sI - A)B$. *It can be deduced that* $\alpha_{c1}(s) = \alpha(s) - m_1 W_1(s) l_1$, *where* $\alpha_{c1}(s) = \det(sI - A - BK_1C)$, *and* $\alpha(s) = \det(sI - A)$. *Assign* $p - 1$ *eigenvalues to* $\lambda_1^*, \cdots, \lambda_{p-1}^*$. *It holds that*

$$\begin{cases} \alpha_{c1}(\lambda_1^*) = \alpha(\lambda_1^*) - m_1 W_1(\lambda_1^*) l_1 = 0 \\ \vdots \\ \alpha_{c1}(\lambda_{p-1}^*) = \alpha(\lambda_{p-1}^*) - m_1 W_1(\lambda_{p-1}^*) l_1 = 0 \end{cases}$$

Select m_1 *and* l_{11}, *and then solve the simultaneous equations to obtain* $l_{12}, \cdots, l_{1p}$. *Furthermore, solve* K_1.

(2) *Keep the* $p - 1$ *eigenvalues assigned in step* (1), *and then assign the left* q *eigenvalues to* $\lambda_p^*, \cdots, \lambda_{p+q-1}^*$. *After step* (1), *the system is changed to* $(A + BK_1C, B, C)$. *Consider* $K_2 = l_2 m_2$ *with* $l_2 = [\, l_{21} \;\; \cdots \;\; l_{2p} \,]^T$ *and* $m_2 = [\, m_{21} \;\; \cdots \;\; m_{2q} \,]$. *Denote* $W_2(s) = C \operatorname{adj}(sI - A - BK_1C)B$. *It can be deduced* $\alpha_{c2}(s) = \alpha_{c1}(s) - m_2 W_2(s) l_2$, *where* $\alpha_{c2}(s) = \det(sI - A - BK_1C - BK_2C)$.

To keep $\lambda_1^*, \cdots, \lambda_{p-1}^*$ *unchanged, no matter what* m_2 *is,*

$$\begin{cases} m_2 W_2(\lambda_1^*) l_2 = 0 \\ \vdots \\ m_2 W_2(\lambda_{p-1}^*) l_2 = 0 \end{cases}$$

should hold. Therefore,

$$\begin{cases} W_2(\lambda_1^*)l_2 = 0 \\ \vdots \\ W_2(\lambda_{p-1}^*)l_2 = 0 \end{cases}$$

Select l_2 satisfying the above simultaneous equations. Furthermore, assign q eigenvalues to $\lambda_p^*, \cdots, \lambda_{p+q-1}^*$, it holds that

$$\begin{cases} \alpha_{c1}(\lambda_p^*) - m_2 W_2(\lambda_p^*)l_2 = 0 \\ \vdots \\ \alpha_{c1}(\lambda_{p+q-1}^*) - m_2 W_2(\lambda_{p+q-1}^*)l_2 = 0 \end{cases}$$

Solve the above simultaneous equations to obtain m_2. Moreover, compute K_2.
(3) Compute $K = K_1 + K_2$.

In Algorithm 6.8, it is assumed that $p+q-1 \leqslant n$. If $p+q-1 > n$, just need to assign the remaining $n - p + 1$ eigenvalues in step (2).

Example 6.11 Consider a linear time-invariant system (6.17) with

$$A = \begin{bmatrix} 0 & -1 & 1 & 0 \\ 0 & 1 & 1 & 0 \\ 0 & 0 & 2 & 0 \\ 0 & 0 & 0 & -3 \end{bmatrix}, \quad B = \begin{bmatrix} 1 & 0 & 0 \\ 0 & 1 & 0 \\ 0 & 0 & 1 \\ 1 & 0 & 0 \end{bmatrix}, \quad C = \begin{bmatrix} 0 & 1 & 0 & 0 \\ 1 & 0 & 0 & 1 \end{bmatrix}$$

Try to assign the eigenvalues of closed-loop system to $-1, -2, -3$ and -4.

It can be verified that this system is completely controllable and observable. Since rank B + rank $C - 1 = 4$, all eigenvalues can be assigned by the output feedback law.

Based on Algorithm 1.1, one can obtain that $\alpha(s) = s^4 - 7s^2 + 6s$ and

$$W_1(s) = C \operatorname{adj}(sI - A)B = \begin{bmatrix} 0 & s(s-2)(s+3) & s(s+3) \\ (s-1)(s-2)(2s+3) & -(s-2)(s+3) & (s-2)(s+3) \end{bmatrix}$$

Design an output feedback matrix K_1 to assign $p - 1 = 2$ eigenvalues to -1 and -3. Select $m_1 = \begin{bmatrix} -9 & 1 \end{bmatrix}$ and $l_1 = \begin{bmatrix} 0 & l_{12} & l_{13} \end{bmatrix}^\mathrm{T}$. Solve

$$\begin{cases} \alpha(-1) - m_1 W_1(-1)l_1 = 0 \\ \alpha(-3) - m_1 W_1(-3)l_1 = 0 \end{cases}$$

$l_1 = \begin{bmatrix} 0 & 0 & 1 \end{bmatrix}^\mathrm{T}$ is a solution. Therefore,

$$K_1 = l_1 m_1 = \begin{bmatrix} 0 \\ 0 \\ -1 \end{bmatrix} \begin{bmatrix} -9 & 1 \end{bmatrix} = \begin{bmatrix} 0 & 0 \\ 0 & 0 \\ 9 & -1 \end{bmatrix}$$

6.2 Output Feedback Control

Next, one can compute that $\alpha_{c1}(s) = s^4 - 15s^2 - 20s - 6$, and

$$W_2(s) = C\,\mathrm{adj}(sI - A - BK_1C)$$
$$= \begin{bmatrix} -(2s+3) & (s+3)(s-1)^2 & s(s+3) \\ (2s+3)(s^2-3s-7) & -(s+3)(s-11) & (s+3)(s-2) \end{bmatrix}$$

To keep the desired eigenvalues -1 and -3 unchanged, it is required that

$$\begin{cases} W_2(-1)l_2 = \begin{bmatrix} -1 & 8 & -2 \end{bmatrix} l_2 = 0 \\ W_2(-3)l_2 = \begin{bmatrix} 3 & 0 & 0 \end{bmatrix} l_2 = 0 \end{cases}$$

$l_2 = \begin{bmatrix} 0 & 1 & 4 \end{bmatrix}^\mathrm{T}$ is a solution.

To assign the left two eigenvalues to -2 and -4, it is required that

$$\begin{cases} \alpha_{c1}(-2) - m_2 W_2(-2)l_2 = -10 - m_2 \begin{bmatrix} 1 & 9 & -2 \\ -3 & 13 & -4 \end{bmatrix} \begin{bmatrix} 0 \\ 1 \\ 4 \end{bmatrix} = 0 \\ \alpha_{c1}(-4) - m_2 W_2(-4)l_2 = 90 - m_2 \begin{bmatrix} 5 & -25 & 4 \\ -105 & -15 & 6 \end{bmatrix} \begin{bmatrix} 0 \\ 1 \\ 4 \end{bmatrix} = 0 \end{cases}$$

$m_2 = \begin{bmatrix} -10 & 0 \end{bmatrix}$ is the solution. Therefore, one can obtain that

$$K_2 = l_2 m_2 = \begin{bmatrix} 0 \\ 1 \\ 4 \end{bmatrix} \begin{bmatrix} -10 & 0 \end{bmatrix} = \begin{bmatrix} 0 & 0 \\ -10 & 0 \\ -40 & 0 \end{bmatrix}$$

Finally, it can be computed that

$$K = K_1 + K_2 = \begin{bmatrix} 0 & 0 \\ -10 & 0 \\ -31 & -1 \end{bmatrix}$$

Moreover,

$$A + B(K_1 + K_2)C = \begin{bmatrix} 0 & -1 & 1 & 0 \\ 0 & -9 & 1 & 0 \\ -1 & -31 & 2 & -1 \\ 0 & 0 & 0 & -3 \end{bmatrix}$$

whose eigenvalues are -1, -2, -3 and -4.

6.2.3 Dynamic Output Feedback Compensator

In this subsection, the dynamic output feedback control law is discussed for linear time-invariant systems.

Consider a linear time-invariant system (6.17). Introduce a dynamic feedback compensator

$$\dot{z}(t) = A_1 z(t) + B_1 y(t), \quad w(t) = C_1 z(t) + D_1 y(t) \tag{6.22}$$

where $z(t) \in \mathbb{R}^l$ and $w(t) \in \mathbb{R}^p$. A_1, B_1, C_1 and D_1 are constant matrices with appropriate dimensions.

Consider

$$u(t) = v(t) - w(t) = v(t) - C_1 z(t) - D_1 y(t) \tag{6.23}$$

where $v(t) \in \mathbb{R}^p$. The corresponding closed-loop system is

$$\begin{bmatrix} \dot{x}(t) \\ \dot{z}(t) \end{bmatrix} = \begin{bmatrix} A - BD_1 C & -BC_1 \\ B_1 C & A_1 \end{bmatrix} \begin{bmatrix} x(t) \\ z(t) \end{bmatrix} + \begin{bmatrix} B \\ 0 \end{bmatrix} v(t), \quad y(t) = Cx(t) \tag{6.24}$$

and its block diagram is displayed in Fig.6.4 .

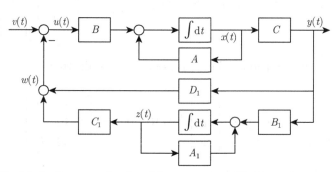

Fig.6.4 The block diagram of a closed-loop system with dynamic output feedback

Let

$$x_c(t) = \begin{bmatrix} x(t) \\ z(t) \end{bmatrix}, \quad v_c(t) = \begin{bmatrix} -v(t) \\ 0 \end{bmatrix}, \quad y_c(t) = \begin{bmatrix} y(t) \\ z(t) \end{bmatrix} = \begin{bmatrix} C & 0 \\ 0 & I \end{bmatrix} x_c(t)$$

the system (6.24) can be rewritten as

$$\dot{x}_c(t) = (A_e + B_e K_e C_e) x_c(t) + B_e v_c(t), \quad y_c(t) = C_e x_c(t) \tag{6.25}$$

where

$$A_e = \begin{bmatrix} A & 0 \\ 0 & 0 \end{bmatrix}, \quad B_e = \begin{bmatrix} -B & 0 \\ 0 & I \end{bmatrix}, \quad C_e = \begin{bmatrix} C & 0 \\ 0 & I \end{bmatrix}, \quad K_e = \begin{bmatrix} D_1 & C_1 \\ B_1 & A_1 \end{bmatrix}$$

6.2 Output Feedback Control

In this way, the closed-loop system (6.25) can be considered as an equivalent open-loop system (A_e, B_e, C_e) with a static output feedback control law, and the feedback gain is K_e. The block diagram is displayed in Fig.6.5.

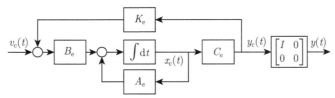

Fig.6.5 Equivalent closed-loop system with dynamic output feedback compensator

Theorem 6.8 The closed-loop system (6.25) is completely controllable and observable if and only if the system (6.17) is completely controllable and observable.

The proof is omitted.

Theorem 6.9 If the system (A_e, B_e, C_e) with rank $B_e = p + l$ and rank $C_e = q + l$ is completely controllable and observable, then to almost all (A_e, B_e, C_e), there exists a constant matrix K_e such that $\min\{p + q + 2l - 1, n + l\}$ eigenvalues of $A_e + B_e K_e C_e$ can be assigned arbitrarily to approach the desired locations. If $p + q + l - 1 \geqslant n$, then to almost all (A_e, B_e, C_e), all eigenvalues of $A_e + B_e K_e C_e$ can be assigned arbitrarily to approach the desired locations.

This theorem is an extension of Theorem 6.7. Here, the proof is omitted. Based on Theorem 6.9, one can deduce the following conclusion.

Corollary 6.1 If the system (6.17) is completely controllable and observable, to almost all (A, B, C), there exists a dynamic output feedback compensator (6.22) such that all eigenvalues of the closed-loop system (6.24) can be assigned arbitrarily to approach the desired locations.

The previous discussion shows that for any completely controllable and observable system, if a dynamic compensator is needed to form an output feedback closed-loop system, one can design a dynamic compensator with its equivalent static output feedback gain matrix, and the dimension l of the dynamic compensator can be selected by

$$l \geqslant n + 1 - p - q$$

If $l = 0$, the dynamic output feedback degrades into the static output feedback. The ability of dynamic compensator is superior to that of a pure static output feedback. For instance, a system (A, B, C) is completely controllable and observable, $\dim A = 5$, rank $B = 2$, rank $C = 2$ and all eigenvalues of A have positive real parts. The system cannot be stabilized with a static output feedback since $p + q - 1 = 3 < n = 5$. However, when using a dynamic output compensator, the system can be stabilized by choosing a dynamic compensator (A_1, B_1, C_1, D_1) with $\dim A_1 = l = n + 1 - p - q = 2$ to stabilize the system.

Example 6.12 Consider a linear time-invariant system (6.17) with

$$A = \begin{bmatrix} -2 & 1 \\ 0 & -1 \end{bmatrix}, \quad B = \begin{bmatrix} 0 \\ 1 \end{bmatrix}, \quad C = \begin{bmatrix} 1 & 0 \end{bmatrix}$$

Try to design a dynamic output feedback compensator such that all eigenvalues of the closed-loop system are at -3.

It can be verified (A, B, C) is completely controllable and observable. A dynamic output feedback compensator with dimension $l = n + 1 - p - q = 1$ is required.

Let the dynamic output feedback compensator be

$$\dot{z}(t) = a_1 z(t) + b_1 y(t), \quad w(t) = c_1 z(t) + d_1 y(t)$$

The equivalent open-loop system is

$$A_e = \begin{bmatrix} -2 & 1 & 0 \\ 0 & -1 & 0 \\ 0 & 0 & 0 \end{bmatrix}, \quad B_e = \begin{bmatrix} 0 & 0 \\ -1 & 0 \\ 0 & 1 \end{bmatrix}, \quad C_e = \begin{bmatrix} 1 & 0 & 0 \\ 0 & 0 & 1 \end{bmatrix}$$

and the equivalent static output feedback gain matrix is

$$K_e = \begin{bmatrix} d_1 & c_1 \\ b_1 & a_1 \end{bmatrix}$$

Consequently, the equivalent closed-loop system is

$$A_e + B_e K_e C_e = \begin{bmatrix} -2 & 1 & 0 \\ 0 & -1 & 0 \\ 0 & 0 & 0 \end{bmatrix} + \begin{bmatrix} 0 & 0 \\ -1 & 0 \\ 0 & 1 \end{bmatrix} \begin{bmatrix} d_1 & c_1 \\ b_1 & a_1 \end{bmatrix} \begin{bmatrix} 1 & 0 & 0 \\ 0 & 0 & 1 \end{bmatrix}$$

$$= \begin{bmatrix} -2 & 1 & 0 \\ -d_1 & -1 & -c_1 \\ b_1 & 0 & a_1 \end{bmatrix}$$

The corresponding characteristic polynomial is

$$\alpha_k(s) = \det(sI - A_e - B_e K_e C_e)$$
$$= s^3 + (3 - a_1)s^2 + (2 - 3a_1 + d_1)s - d_1 a_1 + b_1 c_1 - 2a_1$$

while the desired characteristic polynomial is

$$\alpha_c(s) = (s + 3)^3 = s^3 + 9s^2 + 27s + 27$$

It can be further computed that

$$K_e = \begin{bmatrix} d_1 & c_1 \\ b_1 & a_1 \end{bmatrix} = \begin{bmatrix} 7 & \dfrac{27}{b_1} \\ b_1 & -6 \end{bmatrix}$$

There exists a free parameter in K_e, which implies that K_e is not unique. That is to say, the dynamic compensator is not unique.

The transfer function of the dynamic compensator is

$$\frac{w(s)}{y(s)} = \frac{7s+15}{s+6}$$

which has no free parameter.

It is worth mentioning that the dynamic output feedback compensator should be stable. That is to say, it is required that all eigenvalues of the augmented system (6.24) have negative real parts.

6.3 Decoupling Control

In general, for a MIMO system, each input excites more than one output simultaneously, while each output is influenced by more than one input simultaneously. This phenomenon is called coupling, which is detrimental to the system control. If the number of inputs equals to that of outputs, one can try to design control laws such that each input only controls one output, and each output is only affected by one input. This is the decoupling control problem.

6.3.1 Basic Concepts

Consider a linear time-invariant system,

$$\dot{x}(t) = Ax(t) + Bu(t), \quad y(t) = Cx(t) \qquad (6.26)$$

where $x(t) \in \mathbb{R}^n$ is the state vector, $u(t) \in \mathbb{R}^p$ is the input vector, and $y(t) \in \mathbb{R}^p$ is the output vector. A, B and C are real constant matrices with appropriate dimensions. When considering zero initial conditions, the relationship between inputs and outputs can be described by the transfer function matrix,

$$y(s) = G(s)u(s) = C(sI-A)^{-1}Bu(s) = \begin{bmatrix} g_{11}(s) & \cdots & g_{1p}(s) \\ \vdots & & \vdots \\ g_{p1}(s) & \cdots & g_{pp}(s) \end{bmatrix} u(s)$$

Therefore, the ith output $y_i(s)$, $i = 1, \cdots, p$ can be expressed by

$$y_i(s) = \sum_{j=1}^{p} g_{ij}(s) u_j(s)$$

where $u_j(s)$ denotes the jth input, $j = 1, \cdots, p$. It can be seen that each output of the system is affected by more than one inputs if each row of $G(s)$ have more than one nonzero entry.

Definition 6.1 (Decoupled system) A system is said to be *decoupled* if its transfer function matrix is a diagonal nonsingular matrix. When a system is decoupled, the ith input only control the ith output, while the ith output only affected by the ith input, $i = 1, \cdots, p$.

The frequently used decoupling control law is

$$u(t) = Kx(t) + Hv(t) \tag{6.27}$$

where $v(t) \in \mathbb{R}^p$, K is a constant matrix, and H is a nonsingular constant matrix. It is a combination of the state feedback and the input transformation.

Substituting (6.27) into (6.26), one can obtain that the closed-loop system

$$\dot{x}(t) = (A + BK)x(t) + BHv(t), \quad y(t) = Cx(t) \tag{6.28}$$

and its transfer function matrix is

$$G_{\text{KH}}(s) = C(sI - A - BK)^{-1}BH$$

The schematic diagram of the system (6.28) is displayed in Fig.6.6. The decoupling control greatly simplifies the control progress since each variable can be controlled individually. This kind of decoupling problem is also called the *Morgan restrictive decoupling problem*.

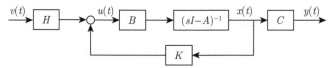

Fig.6.6 The block diagram of state feedback decoupling

Firstly, discuss the relationship between the open-loop transfer function matrix $G(s)$ and the closed-loop one $G_{\text{KH}}(s)$.

Lemma 6.7 *For a linear time-invariant system (6.26) with the control law (6.27), its open-loop transfer function matrix $G(s)$ and closed-loop one $G_{\text{KH}}(s)$ satisfy*

$$G_{\text{KH}}(s) = G(s)[I + K(sI - A - BK)^{-1}B]H = G(s)[I - K(sI - A)^{-1}B]^{-1}H \tag{6.29}$$

Proof It holds that

$$\begin{aligned}
G_{\text{KH}}(s) &= C(sI - A - BK)^{-1}BH \\
&= C(sI - A)^{-1}[(sI - A - BK) + BK](sI - A - BK)^{-1}BH \\
&= C(sI - A)^{-1}[B + BK(sI - A - BK)^{-1}B]H \\
&= C(sI - A)^{-1}B[I + K(sI - A - BK)^{-1}B]H \\
&= G(s)[I + K(sI - A - BK)^{-1}B]H
\end{aligned}$$

6.3 Decoupling Control

Meanwhile, it holds that

$$[I + K(sI - A - BK)^{-1}B][I - K(sI - A)^{-1}B]$$
$$= I + K(sI - A - BK)^{-1}B - K(sI - A)^{-1}B - K(sI - A - BK)^{-1}BK(sI - A)^{-1}B$$
$$= I + K(sI - A - BK)^{-1}B - K(sI - A - BK)^{-1}(sI - A - BK)(sI - A)^{-1}B$$
$$\quad - K(sI - A - BK)^{-1}BK(sI - A)^{-1}B$$
$$= I + K(sI - A - BK)^{-1}B - K(sI - A - BK)^{-1}(sI - A)(sI - A)^{-1}B$$
$$\quad + K(sI - A - BK)^{-1}BK(sI - A)^{-1}B - K(sI - A - BK)^{-1}BK(sI - A)^{-1}B$$
$$= I$$

As a result, it holds that

$$I + K(sI - A - BK)^{-1}B = [I - K(sI - A)^{-1}B]^{-1}$$

which indicates that (6.29) holds. □

Next, introduce two characteristic variables of a transfer function matrix.

Definition 6.2 (Characteristic variables of transfer function matrices) Let m_{ij} and n_{ij} denote the degree of the numerator polynomial and the denominator polynomial of $g_{ij}(s)$, respectively, and $\sigma_{ij} = n_{ij} - m_{ij}$, $i = 1, \cdots, p$, $j = 1, \cdots, p$. Define two characteristic variables of a transfer function matrix $G(s)$,

$$d_i \triangleq \min\{\sigma_{i1}, \sigma_{i2}, \cdots, \sigma_{ip}\} - 1$$
$$E_i \triangleq \lim_{s \to \infty} s^{d_i+1} G_i(s)$$

where $G_i(s)$ is the ith row of $G(s)$.

Example 6.13 For a given transfer function matrix

$$G(s) = \begin{bmatrix} \dfrac{s+2}{s^2 + 2s + 1} & \dfrac{1}{s^2 + s + 2} \\ \dfrac{1}{s^2 + 2s + 1} & \dfrac{3}{s^2 + s + 4} \end{bmatrix}$$

it can be computed that $d_1 = \min\{1, 2\} - 1 = 0$, $E_1 = \lim\limits_{s \to \infty} s^{0+1} G_1(s) = [\ 1\ \ 0\ ]$, $d_2 = \min\{2, 2\} - 1 = 1$, $E_2 = \lim\limits_{s \to \infty} s^{1+1} G_2(s) = [\ 1\ \ 3\ ]$.

As discussed in Subsection 1.3.3, the transfer function $G(s)$ can be expressed as

$$G(s) = \frac{C \operatorname{adj}(sI - A)B}{\det(sI - A)} = \frac{1}{\alpha(s)} C(R_{n-1}s^{n-1} + R_{n-2}s^{n-2} + \cdots + R_1 s + R_0)B$$

where $\alpha(s) = s^n + a_1 s^{n-1} + \cdots + a_{n-1}s + a_n$ is the characteristic polynomial of A, $R_0, \cdots, R_{n-1}$ are constant matrices, which can be computed by Algorithm 1.1.

On this basis, one can obtain that

$$G_i(s) = \frac{c_i \operatorname{adj}(sI-A)B}{\det(sI-A)} = \frac{1}{\alpha(s)} c_i (R_{n-1}s^{n-1} + R_{n-2}s^{n-2} + \cdots + R_1 s + R_0)B$$

where c_i is the ith row of C, $i = 1, \cdots, p$.

According to Definition 6.2, the minimal discrepancy of the denominator and the numerator of each entry of $G_i(s)$ is $d_i + 1$. Therefore, the coefficient of $s^{n-1}, \cdots, s^{n-d_i}$ should be zero. That is to say,

$$\begin{cases} c_i R_{n-1} B = c_i B = 0 \\ c_i R_{n-2} B = c_i (AR_{n-1} + a_1 I)B = c_i AB = 0 \\ \vdots \\ c_i R_{n-d_i} B = c_i A^{d_i-1} B = 0 \\ E_i = c_i R_{n-d_i-1} B = c_i A^{d_i} B \neq 0 \end{cases} \tag{6.30}$$

Denote two characteristic variables for the closed-loop transfer function matrix $G_{\mathrm{KH}}(s)$ as $\bar{d}_i$ and $\bar{E}_i$. Similarly, it holds that

$$\begin{cases} c_i BH = 0 \\ c_i (A+BK) BH = 0 \\ \vdots \\ c_i (A+BK)^{\bar{d}_i-1} BH = 0 \\ \bar{E}_i = c_i (A+BK)^{\bar{d}_i} BH \neq 0 \end{cases} \tag{6.31}$$

Lemma 6.8 *For a linear time-invariant system (6.26) with the control law (6.27), the characteristic variables of its open-loop transfer function matrix $G(s)$ and closed-loop one $G_{\mathrm{KH}}(s)$ satisfy*

$$\bar{d}_i = d_i, \quad \bar{E}_i = E_i H \tag{6.32}$$

Proof Firstly, prove that

$$c_i (A+BK)^k = c_i A^k, \quad k = 1, 2, \cdots, d_i \tag{6.33}$$

Use the mathematical induction to prove (6.33).

When $k = 1$, it holds that $c_i B = 0$ from (6.30). Therefore, $c_i (A+BK) = c_i A$.

Suppose that $c_i (A+BK)^L = c_i A^L$ for $0 < L \leqslant d_i - 1$. It is known that $c_i A^L B = 0$ from (6.30). Furthermore, one can deduce that

$$c_i (A+BK)^{L+1} = c_i A^L (A+BK) = c_i A^{L+1} + c_i A^L BK = c_i A^{L+1}$$

In conclusion, (6.33) holds.

Considering (6.30), (6.31) and (6.33), one can deduce that

$$c_i(A+BK)^k BH = c_i A^k BH = 0, \ k = 1, 2, \cdots, d_i - 1$$

Meanwhile, for $k = d_i$, it holds that

$$c_i(A+BK)^{d_i} BH = c_i A^{d_i} BH \neq 0$$

On this basis, one can further deduce that (6.32) holds. □

6.3.2 Decoupling with State Feedback

Theorem 6.10 *For a linear time-invariant system (6.26), it can be decoupled by the control law (6.27) if and only if*

$$E \triangleq \begin{bmatrix} E_1 \\ \vdots \\ E_p \end{bmatrix} \tag{6.34}$$

is nonsingular, where E_i is defined by Definition 6.2, $i = 1, \cdots, p$.

Proof Firstly, prove the necessity.

If $G(s)$ can be decoupled, then $G_{\mathrm{KH}}(s)$ is a nonsingular diagonal matrix. Thus,

$$\bar{E} \triangleq \begin{bmatrix} \bar{E}_1 \\ \vdots \\ \bar{E}_p \end{bmatrix}$$

must be a diagonal matrix. Meanwhile, $\bar{E}_i \neq 0$, $i = 1, \cdots, p$. Therefore, $\bar{E}$ is nonsingular. On the other hand, it is known $\bar{E}_i = E_i H$ and H is nonsingular. Therefore, E is nonsingular.

Next, prove the sufficiency.

Based on (6.30), $G_i(s)$ can be expressed as

$$G_i(s) = c_i(sI - A)^{-1} B = c_i \left(\sum_{k=0}^{\infty} \frac{A^k}{s^{k+1}} \right) B = \sum_{k=d_i}^{\infty} \frac{c_i A^k B}{s^{k+1}}$$

$$= \frac{1}{s^{d_i+1}} \left(c_i A^{d_i} B + c_i A^{d_i+1} \left(\frac{I}{s} + \frac{A}{s^2} + \frac{A^2}{s^3} + \cdots \right) B \right)$$

$$= \frac{1}{s^{d_i+1}} (E_i + F_i(sI - A)^{-1} B)$$

where $F_i \triangleq c_i A^{d_i+1}$, $i = 1, \cdots, p$.

In this way, $G(s)$ can be expressed as

$$G(s) = \begin{bmatrix} \frac{1}{s^{d_1+1}} & & & \\ & \frac{1}{s^{d_2+1}} & & \\ & & \ddots & \\ & & & \frac{1}{s^{d_p+1}} \end{bmatrix} \begin{bmatrix} E_1 + F_1(sI-A)^{-1}B \\ E_2 + F_2(sI-A)^{-1}B \\ \vdots \\ E_p + F_p(sI-A)^{-1}B) \end{bmatrix}$$

$$= \begin{bmatrix} \frac{1}{s^{d_1+1}} & & & \\ & \frac{1}{s^{d_2+1}} & & \\ & & \ddots & \\ & & & \frac{1}{s^{d_p+1}} \end{bmatrix} \begin{bmatrix} E + F(sI-A)^{-1}B \end{bmatrix}$$

where

$$F \triangleq \begin{bmatrix} F_1 \\ \vdots \\ F_p \end{bmatrix}$$

Select

$$u(t) = Kx(t) + Hv(t) = -E^{-1}Fx(t) + E^{-1}v(t) \tag{6.35}$$

According to Lemma 6.7, one can deduce that

$$\begin{aligned} G_{\text{KH}}(s) &= G(s)[I - K(sI-A)^{-1}B]^{-1}H \\ &= G(s)[I + E^{-1}F(sI-A)^{-1}B]^{-1}E^{-1} \\ &= G(s)[E + F(sI-A)^{-1}B]^{-1} \end{aligned}$$

Furthermore, we can deduce that

$$G_{\text{KH}}(s) = \begin{bmatrix} \frac{1}{s^{d_1+1}} & & & \\ & \frac{1}{s^{d_2+1}} & & \\ & & \ddots & \\ & & & \frac{1}{s^{d_p+1}} \end{bmatrix} \tag{6.36}$$

which indicates that the control law (6.35) can decouple $G(s)$. □

It can be seen that using (6.35), $G(s)$ can be decoupled as $G_{\text{KH}}(s)$ with the form of (6.36). In this way, all diagonal entries of $G_{\text{KH}}(s)$ are integral terms, and so this method is called the *integrating decoupling*.

6.3 Decoupling Control

Example 6.14 Consider a linear time-invariant system (6.26) with

$$A = \begin{bmatrix} 0 & 0 & 0 \\ 0 & 0 & 1 \\ -1 & -2 & -3 \end{bmatrix}, \quad B = \begin{bmatrix} 1 & 0 \\ 0 & 0 \\ 0 & 1 \end{bmatrix}, \quad C = \begin{bmatrix} 1 & 1 & 0 \\ 0 & 0 & 1 \end{bmatrix}$$

Try to design a control law (6.27) to decouple this system.
The transfer function is

$$G(s) = C(sI - A)^{-1}B = \begin{bmatrix} \dfrac{s^2 + 3s + 1}{s(s+1)(s+2)} & \dfrac{1}{(s+1)(s+2)} \\ \dfrac{-1}{(s+1)(s+2)} & \dfrac{s}{(s+1)(s+2)} \end{bmatrix}$$

The corresponding characteristic variables are $d_1 = 0$, $E_1 = \lim\limits_{s\to\infty} sG_1(s) = \begin{bmatrix} 1 & 0 \end{bmatrix}$, $d_2 = 0$, $E_2 = \lim\limits_{s\to\infty} sG_2(s) = \begin{bmatrix} 0 & 1 \end{bmatrix}$. $E = \begin{bmatrix} 1 & 0 \\ 0 & 1 \end{bmatrix}$ is nonsingular, and thus this system can be decoupled.

Calculate

$$F = \begin{bmatrix} c_1 A^{d_1+1} \\ c_2 A^{d_2+1} \end{bmatrix} = \begin{bmatrix} 0 & 0 & 1 \\ -1 & -2 & -3 \end{bmatrix}$$

which yields that

$$u(t) = Kx(t) + Hv(t) = -E^{-1}Fx(t) + E^{-1}v(t) = \begin{bmatrix} 0 & 0 & 1 \\ 1 & 2 & 3 \end{bmatrix} x(t) + \begin{bmatrix} 1 & 0 \\ 0 & 1 \end{bmatrix} v(t)$$

The coefficient matrices of the decoupled system is

$$\bar{A} = A + BK = \begin{bmatrix} 0 & 0 & -1 \\ 0 & 0 & 1 \\ 0 & 0 & 0 \end{bmatrix}, \quad \bar{B} = BH = \begin{bmatrix} 1 & 0 \\ 0 & 0 \\ 0 & 1 \end{bmatrix}, \quad \bar{C} = C = \begin{bmatrix} 1 & 1 & 0 \\ 0 & 0 & 1 \end{bmatrix}$$

and the corresponding transfer function is

$$G_{\mathrm{KH}}(s) = \begin{bmatrix} \dfrac{1}{s} & 0 \\ 0 & \dfrac{1}{s} \end{bmatrix}$$

By integrating decoupling, systems have higher order integration elements, which cannot work in practical projects. Therefore, pole assignment is needed to rearrange the poles.

Theorem 6.11 *For a linear time-invariant system (6.26), if the matrix E defined by (6.34) is nonsingular, then a control law (6.27) can be used such that*

$$G_{\mathrm{KH}}(s) = \begin{bmatrix} \dfrac{1}{s^{d_1+1} + a_{11}s^{d_1} + \cdots + a_{1d_1+1}} & & \\ & \ddots & \\ & & \dfrac{1}{s^{d_p+1} + a_{p1}s^{d_p} + \cdots + a_{pd_p+1}} \end{bmatrix}$$

where $a_{i1}, \cdots, a_{id_i+1}$ are coefficients of the desired pole polynomial of the ith diagonal entry of $G_{\mathrm{KH}}(s)$, $i = 1, \cdots, p$, $H = E^{-1}$, $K = -E^{-1}\tilde{F}$ with

$$\tilde{F} \triangleq \begin{bmatrix} \tilde{F}_1 \\ \vdots \\ \tilde{F}_p \end{bmatrix}$$

and $\tilde{F}_i \triangleq c_i A^{d_i+1} + a_{i1} c_i A^{d_i} + a_{i2} c_i A^{d_i-1} + \cdots + a_{id_i+1} c_i$.

Proof Suppose that the desired pole polynomial of the ith diagonal entry of $G_{\mathrm{KH}}(s)$ is

$$\alpha_{ci}(s) = s^{d_i+1} + a_{i1}s^{d_i} + a_{i2}s^{d_i-1} + \cdots + a_{id_i}s + a_{id_i+1}$$

Multiplying $\alpha_{ci}(s)$ by $G_i(s)$, which is the ith row of $G(s)$, one can deduce that

$$\begin{aligned}\alpha_{ci}(s)G_i(s) &= \alpha_{ci}(s)c_i(sI-A)^{-1}B = \alpha_{ci}(s)\sum_{k=0}^{\infty} \frac{c_i A^k B}{s^{k+1}} \\ &= c_i A^{d_i} B + (c_i A^{d_i+1} + a_{i1} c_i A^{d_i} + \cdots + a_{id_i+1} c_i)\left(\frac{I}{s} + \frac{A}{s^2} + \frac{A^2}{s^3} + \cdots\right)B \\ &= c_i A^{d_i} B + (c_i A^{d_i+1} + a_{i1} c_i A^{d_i} + \cdots + a_{id_i+1} c_i)(sI-A)^{-1}B \\ &= E_i + \tilde{F}_i(sI-A)^{-1}B \end{aligned}$$

Furthermore, it holds that

$$G_i(s) = \frac{1}{\alpha_{ci}(s)}[E_i + \tilde{F}_i(sI-A)^{-1}B]$$

which yields that

$$G(s) = \begin{bmatrix} \dfrac{1}{\alpha_{c1}(s)} & & & \\ & \dfrac{1}{\alpha_{c2}(s)} & & \\ & & \ddots & \\ & & & \dfrac{1}{\alpha_{cp}(s)} \end{bmatrix}[E + \tilde{F}(sI-A)^{-1}B]$$

6.3 Decoupling Control

Select the control law as

$$u(t) = Kx(t) + Hv(t) = -E^{-1}\tilde{F}x(t) + E^{-1}v(t) \qquad (6.37)$$

According to Lemma 6.7, one can further deduce that

$$G_{\text{KH}}(s) = G(s)[I - K(sI - A)^{-1}B]^{-1}H = G(s)[I + E^{-1}\tilde{F}(sI - A)^{-1}B]^{-1}E^{-1}$$

$$= G(s)[E + \tilde{F}(sI - A)^{-1}B]^{-1} = \begin{bmatrix} \dfrac{1}{\alpha_{c1}(s)} & & & \\ & \dfrac{1}{\alpha_{c2}(s)} & & \\ & & \ddots & \\ & & & \dfrac{1}{\alpha_{cp}(s)} \end{bmatrix}$$

which indicates that poles of $G_{\text{KH}}(s)$ can be assigned arbitrarily by (6.37). □

Example 6.15 Consider a linear time-invariant system (6.26) with

$$A = \begin{bmatrix} 0 & 0 & 0 \\ 0 & 0 & 1 \\ -1 & -2 & -3 \end{bmatrix}, \quad B = \begin{bmatrix} 0 & 0 \\ 1 & 0 \\ 0 & 1 \end{bmatrix}, \quad C = \begin{bmatrix} 1 & 1 & 0 \\ 0 & 0 & 1 \end{bmatrix}$$

Try to design a control law (6.27) to decouple this system and assign all poles to the left half of the s-plane.

The transfer function matrix is

$$G(s) = C(sI - A)^{-1}B = \begin{bmatrix} \dfrac{s+3}{(s+1)(s+2)} & \dfrac{1}{(s+1)(s+2)} \\ \dfrac{-2}{(s+1)(s+2)} & \dfrac{s}{(s+1)(s+2)} \end{bmatrix}$$

The corresponding characteristic variables are $d_1 = 0$, $E_1 = \lim\limits_{s \to \infty} sG_1(s) = [\, 1 \ \ 0 \,]$, $d_2 = 0$, $E_2 = \lim\limits_{s \to \infty} sG_2(s) = [\, 0 \ \ 1 \,]$. $E = \begin{bmatrix} 1 & 0 \\ 0 & 1 \end{bmatrix}$ is nonsingular, and so the system can be decoupled.

Let $\alpha_{c1}(s) = s + 4$, $\alpha_{c2}(s) = s + 5$. Calculate

$$\tilde{F}_1 = c_1 A + a_{11} c_1 = \begin{bmatrix} 1 & 1 & 0 \end{bmatrix} \begin{bmatrix} 0 & 0 & 0 \\ 0 & 0 & 1 \\ -1 & -2 & -3 \end{bmatrix} + 4 \begin{bmatrix} 1 & 1 & 0 \end{bmatrix} = \begin{bmatrix} 4 & 4 & 1 \end{bmatrix}$$

$$\tilde{F}_2 = c_2 A + a_{21} c_2 = \begin{bmatrix} 0 & 0 & 1 \end{bmatrix} \begin{bmatrix} 0 & 0 & 0 \\ 0 & 0 & 1 \\ -1 & -2 & -3 \end{bmatrix} + 5 \begin{bmatrix} 0 & 0 & 1 \end{bmatrix} = \begin{bmatrix} -1 & -2 & 2 \end{bmatrix}$$

and then one can obtain that

$$\tilde{F} = \begin{bmatrix} \tilde{F}_1 \\ \tilde{F}_2 \end{bmatrix} = \begin{bmatrix} 4 & 4 & 1 \\ -1 & -2 & 2 \end{bmatrix}$$

which yields that

$$\begin{aligned} u(t) &= Kx(t) + Hv(t) = -E^{-1}\tilde{F}x(t) + E^{-1}v(t) \\ &= \begin{bmatrix} -4 & -4 & -1 \\ 1 & 2 & -2 \end{bmatrix} x(t) + \begin{bmatrix} 1 & 0 \\ 0 & 1 \end{bmatrix} v(t) \end{aligned}$$

The coefficient matrices of the decoupled system is

$$\bar{A} = A + BK = \begin{bmatrix} 0 & 0 & 0 \\ -4 & -4 & 0 \\ 0 & 0 & -5 \end{bmatrix}, \quad \bar{B} = BH = \begin{bmatrix} 0 & 0 \\ 1 & 0 \\ 0 & 1 \end{bmatrix}, \quad \bar{C} = C = \begin{bmatrix} 1 & 1 & 0 \\ 0 & 0 & 1 \end{bmatrix}$$

and the corresponding transfer function is

$$G_{\text{KH}}(s) = \begin{bmatrix} \dfrac{1}{s+4} & 0 \\ 0 & \dfrac{1}{s+5} \end{bmatrix}$$

The above example shows that after decoupling and pole assignment, some zeros and poles are canceled. The system is converted from observable to incompletely observable. Meanwhile, the original system is incompletely controllable, but one can decouple it and assign its some poles, but not for all the poles or arbitrary poles.

6.3.3 Steady State Decoupling

Consider the system (6.26), if there exists a control law (6.27) such that the closed-loop system (6.28) satisfies the following properties,

(1) the closed-loop system (6.28) is asymptotically stable;

(2) $G_{\text{KH}}(s) = C(sI - A - BK)^{-1}BH$ is a diagonal and nonsingular matrix when $s \to 0$, then this method is called the *steady state decoupling*, while the decoupling method introducing in the previous subsection is called the *dynamic decoupling*.

In should be pointed out that when considering the steady state decoupling, $G_{\text{KH}}(s)$ is non-diagonal in general except for $s \to 0$. Meanwhile, the steady state decoupling only suitable for the case that the reference input is the step input, i.e.,

$$v(t) = \begin{bmatrix} b_1 & \cdots & b_p \end{bmatrix}^{\text{T}}$$

for $t \geqslant 0$, where $b_1, \cdots, b_p$ are constants.

6.3 Decoupling Control

In this way, one can deduce that

$$\lim_{t\to\infty} y(t) = \lim_{s\to 0} sG_{\mathrm{KH}}(s) \begin{bmatrix} b_1 \\ \vdots \\ b_p \end{bmatrix} \frac{1}{s} = \begin{bmatrix} g_{11}(0) & & \\ & \ddots & \\ & & g_{pp}(0) \end{bmatrix} \begin{bmatrix} b_1 \\ \vdots \\ b_p \end{bmatrix} = \begin{bmatrix} g_{11}(0)b_1 \\ \vdots \\ g_{pp}(0)b_p \end{bmatrix}$$

which indicates that

$$\lim_{t\to\infty} y_i(t) = g_{ii}(0)b_i \tag{6.38}$$

where $y_i(t)$ is the ith output, $g_{ii}(0)$ is the ith diagonal entry of $G_{\mathrm{KH}}(0)$.

(6.38) indicates that the steady state decoupling with the step input can make each output be controlled only by its corresponding input when the output is in steady-state. However, during the dynamical process, the coupling between inputs and outputs cannot be decoupled.

Theorem 6.12 *For a linear time-invariant system (6.26), it can be steady-state decoupled by a control law (6.27) if and only if this system can be stabilized by a state feedback control law and*

$$\operatorname{rank} \begin{bmatrix} A & B \\ C & 0 \end{bmatrix} = n + p \tag{6.39}$$

Proof Firstly, prove the sufficiency.

If a state feedback control law with the feedback gain K can stabilize the system, then all eigenvalues of $A + BK$ have negative real parts. Therefore, $(A+BK)^{-1}$ is existed.

Since (6.39) holds, one can deduce that

$$\begin{bmatrix} A+BK & B \\ C & 0 \end{bmatrix} = \begin{bmatrix} A & B \\ C & 0 \end{bmatrix} \begin{bmatrix} I & 0 \\ K & I \end{bmatrix} \tag{6.40}$$

is nonsingular. On this basis, one can further deduce that

$$\det \begin{bmatrix} A+BK & B \\ C & 0 \end{bmatrix} = \det(A+BK)\det(-C(A+BK)^{-1}B) \neq 0 \tag{6.41}$$

which indicates that $C(A+BK)^{-1}B$ is also nonsingular.

Let

$$H = (-C(A+BK)^{-1}B)^{-1}M \tag{6.42}$$

where M is a diagonal and nonsingular matrix to be designed. One can further obtain that

$$G_{\mathrm{KH}}(0) = C(sI - A - BK)^{-1}B(-C(A+BK)^{-1}B)^{-1}M\big|_{s=0} = M$$

which indicates that the system can be steady-state decoupled.

Next, prove the necessity.

If a system needs to be steady-state decoupled, $G_{KH}(s)$ is required to be capable to enter the steady state, and $G_{KH}(0) = -C(A+BK)^{-1}BH$ is required to be a diagonal and nonsingular matrix. The former condition implies that the system can be stabilized by a state feedback control law. The latter condition implies that $C(A+BK)^{-1}B$ and $A+BK$ are both nonsingular. Based on (6.40) and (6.41), one can deduce that (6.39) holds. $\square$

The proof of Theorem 6.12 provides a method to realize the steady state decoupling. Using the control law (6.27), we can select any K such that the closed-loop system is asymptotically stable, and then select H based on (6.42).

Example 6.16 Consider a linear time-invariant system (6.26) with

$$A = \begin{bmatrix} 0 & 0 & 1 \\ 1 & 0 & 0 \\ 1 & 1 & 1 \end{bmatrix}, \quad B = \begin{bmatrix} 1 & 1 \\ -1 & 1 \\ 0 & -1 \end{bmatrix}, \quad C = \begin{bmatrix} 0 & 1 & 1 \\ -1 & 1 & 0 \end{bmatrix}$$

The transfer function matrix can be obtained as

$$G(s) = \begin{bmatrix} \dfrac{-s^2 + 2s + 1}{s^3 - s^2 - s - 1} & \dfrac{2s - 2}{s^3 - s^2 - s - 1} \\ \dfrac{-2s^2 + 3s - 1}{s^3 - s^2 - s - 1} & \dfrac{-2}{s^3 - s^2 - s - 1} \end{bmatrix}$$

It can be computed that

$$E = \begin{bmatrix} -1 & 0 \\ -2 & 0 \end{bmatrix}$$

which is singular. Therefore, this system cannot be dynamic decoupled.

However, this system is completely controllable and satisfies (6.39), it can be steady state decoupled. One can select K such that all eigenvalues of $A + BK$ have negative real parts, and further select H based on (6.42) to realize the steady state decoupling.

6.4 State Observer

When considering the state feedback, it is supposed that states of a system can be measured by appropriately positioned sensors. Generally, however, it may be either impossible or simply impractical to obtain measurements for all states. In particular, some states may not be available for measurement at all. Therefore, it is necessary to be able to estimate the values of states of a system from available measurements, typically outputs and inputs.

6.4.1 Full-Order State Observer

Consider a linear time-invariant system,

$$\dot{x}(t) = Ax(t) + Bu(t), \quad y(t) = Cx(t), \quad x(0) = x_0 \qquad (6.43)$$

6.4 State Observer

where $x(t) \in \mathbb{R}^n$ is the state vector, $u(t) \in \mathbb{R}^p$ is the input vector, and $y(t) \in \mathbb{R}^p$ is the output vector. A, B and C are real constant matrices with appropriate dimensions.

A *full-order observer* is an n-dimensional model system constructed by $u(t)$ and $y(t)$ as inputs, which can guarantee that its output $\hat{x}(t) \in \mathbb{R}^n$ satisfies

$$\lim_{t \to \infty} \hat{x}(t) = \lim_{t \to \infty} x(t) \tag{6.44}$$

In the next, two methods are introduced to construct a full-order state observer for the system (6.43).

A full-order observer of $x(t)$ can be constructed into the following manner,

$$\dot{\hat{x}}(t) = A\hat{x}(t) + Bu(t) + G\left(y(t) - \hat{y}(t)\right), \ \hat{x}(0) = \hat{x}_0 \tag{6.45}$$

where $\hat{y}(t) \triangleq C\hat{x}(t)$, G is a constant matrix.

The observer (6.45) contains a feedback amendment part $G(y(t) - \hat{y}(t))$. When removing this part, it has the same structure as the original system (6.43). In this case, it is difficult to satisfy (6.44) since it is hard to keep the initial condition $\hat{x}_0$ identical to x_0. Specifically, if the original system is unstable, then any tiny difference between $\hat{x}_0$ and x_0 will result in the divergence of $\hat{x}(t) - x(t)$.

It is noted that (6.45) can be rewritten as

$$\dot{\hat{x}}(t) = (A - GC)\hat{x}(t) + Bu(t) + Gy(t)$$

its block diagram is show in Fig.6.7. Define the observing error $e(t) \triangleq x(t) - \hat{x}(t)$. Combining (6.43) with (6.45), one can obtain that

$$\dot{e}(t) = (A - GC)e(t), \ e(0) = e_0 = x_0 - \hat{x}_0 \tag{6.46}$$

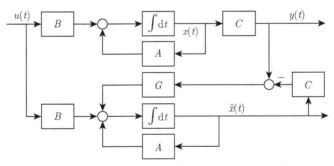

Fig.6.7 The block diagram of a full-order state observer

If eigenvalues of $A - GC$ have negative real parts, then $e(t) \to 0$ as $t \to \infty$, independently of the initial condition e_0. That is to say, (6.44) holds. This asymptotic state observer is known as the *Luenberger observer*. It can be seen that if one can select a matrix G such that the error system (6.46) is asymptotically stable, the Luenberger observer (6.45) is existed. Obviously, if the original system (6.46) is completely observable, the error system (6.46) can be stabilized. If (A, C) is incompletely observable, there exists a transformation matrix T which can decompose the system into

$$\begin{bmatrix} \dot{x}_o(t) \\ \dot{x}_{no}(t) \end{bmatrix} = \begin{bmatrix} A_{11} & 0 \\ A_{21} & A_{22} \end{bmatrix} \begin{bmatrix} x_o(t) \\ x_{no}(t) \end{bmatrix} + \begin{bmatrix} B_1 \\ B_2 \end{bmatrix} u(t), \ y(t) = [C_1 \ 0] \begin{bmatrix} x_o(t) \\ x_{no}(t) \end{bmatrix} \quad (6.47)$$

where (A_{11}, C_1) is completely observable. The eigenvalues of the system (6.47) are consisted of eigenvalues of A_{11} and A_{22}. If all eigenvalues of A_{22} have negative real parts, then (A, C) is said to be *detectable*.

Theorem 6.13 *For a linear time-invariant system (6.43), the full-order observer (6.45) is existed if and only if (A, C) is detectable.*

Proof For the system (6.47), adopt the Luenberger observer (6.45). Denote $G = [\ G_1^T \ G_2^T \ ]^T$, the corresponding error system can be expressed as

$$\dot{e}(t) = \begin{bmatrix} \dot{e}_1(t) \\ \dot{e}_2(t) \end{bmatrix} = \begin{bmatrix} A_{11} - G_1 C_1 & 0 \\ A_{21} - G_2 C_1 & A_{22} \end{bmatrix} \begin{bmatrix} e_1(t) \\ e_2(t) \end{bmatrix}$$

Since (A_{11}, C_1) is completely observable, one can select G_1 such that all eigenvalues of $A_{11} - G_1 C_1$ have negative real parts. Meanwhile, since (A, C) is detectable, all eigenvalues of A_{22} have negative real parts. Therefore, it holds that $\lim\limits_{t \to \infty} e(t) = 0$. That is to say, the Luenberger observer is existed.

The proof of the necessity is omitted. $\square$

Theorem 6.14 *For a linear time-invariant system (6.43), all eigenvalues of the full-order observer (6.45) can be assigned arbitrarily if and only if (A, C) is completely observable.*

The proof is omitted.

Algorithm 6.9 *For a detectable linear time-invariant system (6.43), the Luenberger observer (6.45) can be determined by following procedures.*

(1) Based on the dual system (A^T, C^T, B^T), calculate $A^T - C^T G^T$.

(2) Select a feedback gain matrix G^T to assign eigenvalues of $A^T - C^T G^T$ to desired locations.

(3) Write out the Luenberger observer equation (6.45).

In the next, introduce another way to design the full-order observer for the system (6.43). It is required that the system (6.43) is completely controllable and observable. The full-order state observer is with the following form,

$$\dot{z}(t) = Fz(t) + Gy(t) + Hu(t), \ z(0) = z_0, \ \hat{x}(t) = T^{-1} z(t) \quad (6.48)$$

where $z(t) \in \mathbb{R}^n$, F, G, H and T are constant matrices with appreciate dimensions, and T is nonsingular.

Theorem 6.15 *For a completely controllable and observable linear time-invariant system (6.43), (6.48) is a full-order observer of the system (6.43) if and only if all eigenvalues of F have negative real parts, $TA - FT = GC$, and $H = TB$.*

6.4 State Observer

Proof Firstly, prove the sufficiency.

The observing error is still defined as $e(t) \triangleq x(t) - \hat{x}(t)$. Combining (6.43) and (6.48), one can deduce that

$$\begin{aligned}
\dot{e}(t) =& \dot{x}(t) - \dot{\hat{x}}(t) = \dot{x}(t) - T^{-1}\dot{z}(t) \\
=& Ax(t) + Bu(t) - T^{-1}(Fz(t) + Hu(t) + Gy(t)) \\
=& Ax(t) + Bu(t) - T^{-1}FT\hat{x}(t) - T^{-1}Hu(t) - T^{-1}GCx(t) \\
=& T^{-1}FTe(t) + T^{-1}(TA - FT - GC)x(t) + T^{-1}(TB - H)u(t)
\end{aligned} \qquad (6.49)$$

If all eigenvalues of F have negative real parts, $TA - FT = GC$, and $H = TB$, it can be deduced that $\lim\limits_{t\to\infty} e(t) = 0$. That is to say, (6.44) is satisfied. Therefore, (6.48) is a full-order observer of (6.43).

Next, prove the necessity.

Use reduction to absurdity. Suppose $\lim\limits_{t\to\infty} e(t) = 0$, but the conditions in this theorem are not satisfied simultaneously. In this situation, from (6.49), as $t \to \infty$, $e(t) \to 0$ cannot be ensured, which contradicts with the assumption. $\square$

Algorithm 6.10 *For a completely controllable and observable linear time-invariant system (6.43), the full-order state observer (6.48) can be determined by following procedures.*

(1) Select a matrix F such that its eigenvalues have negative real parts and are distinct from the ones of A.

(2) Select a matrix G such that (F, G) is completely controllable.

(3) Solve the matrix equation $TA - FT = GC$ to determine the unique solution T.

(4) If T is nonsingular, solve $H = TB$, and the full-order state observer (6.48) is determined. Otherwise, reselect F or G and repeat the above procedures.

Example 6.17 Consider a linear time-invariant system

$$\dot{x}(t) = \begin{bmatrix} -1 & 2 \\ 0 & 1 \end{bmatrix} x(t) + \begin{bmatrix} 0 \\ 1 \end{bmatrix} u(t), \ y(t) = \begin{bmatrix} 1 & 0 \end{bmatrix} x(t)$$

Design a full-order state observer for this system.

Select

$$F = \begin{bmatrix} -2 & 1 \\ 0 & -3 \end{bmatrix}$$

It can be seen that both eigenvalues of F have negative real parts, and they distinct from the ones of A.

Select $G = \begin{bmatrix} 0 & 1 \end{bmatrix}^{\mathrm{T}}$, which can guarantee (F, G) is completely controllable.

By solving the matrix equation $TA - FT = GC$, on can get that

$$T = \begin{bmatrix} -\dfrac{1}{12} & -\dfrac{5}{12} \\ -\dfrac{1}{4} & -\dfrac{1}{4} \end{bmatrix}$$

which is nonsingular. Therefore,

$$H = TB = \begin{bmatrix} -\dfrac{5}{12} \\ -\dfrac{1}{4} \end{bmatrix}$$

Write out the equation of the full-dimensional observer,

$$\dot{z}(t) = \begin{bmatrix} -2 & 1 \\ 0 & -3 \end{bmatrix} z(t) + \begin{bmatrix} -\dfrac{5}{12} \\ -\dfrac{1}{4} \end{bmatrix} u(t) + \begin{bmatrix} 0 \\ 1 \end{bmatrix} y(t)$$

$$\hat{x}(t) = T^{-1} z(t) = \begin{bmatrix} 3 & -5 \\ -3 & 1 \end{bmatrix} z(t)$$

6.4.2 Reduced-Order State Observer

视频

Since the output $y(t)$ includes part of information about the system, a lower order state observer can be designed, which is called a *reduced-order state observer*.

Consider the linear time-invariant system (6.43) with rank $C = q$. The minimal dimension of the reduced-order state observer is $n - q$.

Consider $C = [\ C_1 \ \ C_2\ ]$, where $C_1 \in \mathbb{R}^{q \times q}$, $C_2 \in \mathbb{R}^{q \times (n-q)}$ and rank $C_1 = q$. If rank $C_1 = q$, we can always find a column transformation to ensure rank $C_1 = q$.

Select a transformation matrix

$$T = \begin{bmatrix} C_1 & C_2 \\ 0 & I \end{bmatrix}$$

and perform the equivalent transformation of the system (6.43) with $\bar{x}(t) = Tx(t)$. One can obtain that

$$\begin{bmatrix} \dot{\bar{x}}_1(t) \\ \dot{\bar{x}}_2(t) \end{bmatrix} = \begin{bmatrix} A_{11} & A_{12} \\ A_{21} & A_{22} \end{bmatrix} \begin{bmatrix} \bar{x}_1(t) \\ \bar{x}_2(t) \end{bmatrix} + \begin{bmatrix} B_1 \\ B_2 \end{bmatrix} u(t), \ y(t) = \begin{bmatrix} I & 0 \end{bmatrix} \begin{bmatrix} \dot{\bar{x}}_1(t) \\ \dot{\bar{x}}_2(t) \end{bmatrix} \quad (6.50)$$

It can be seen that $\bar{x}_1(t) \in \mathbb{R}^q$ is described by $y(t)$ directly. Therefore, only need to design a state observer to estimate $\bar{x}_2(t) \in \mathbb{R}^{n-q}$.

According to (6.50), one can obtain that

$$\dot{\bar{x}}_2(t) = A_{22}\bar{x}_2(t) + (A_{21}y(t) + B_2 u(t)), \ \dot{y}(t) - A_{11}y(t) - B_1 u(t) = A_{12}\bar{x}_2(t) \quad (6.51)$$

6.4 State Observer

Define $\bar{u}(t) \triangleq A_{21}y(t) + B_2 u(t)$ and $w(t) \triangleq \dot{y}(t) - A_{11}y(t) - B_1 u(t)$. Substituting them into (6.51), one can deduce that

$$\dot{\bar{x}}_2(t) = A_{22}\bar{x}_2(t) + \bar{u}(t), \ w(t) = A_{12}\bar{x}_2(t) \tag{6.52}$$

Lemma 6.9 (A_{22}, A_{12}) *is completely observable if and only if (A, C) is completely observable.*

Proof Since (A, C) is completely observable, then (6.50) is completely observable. The observability matrix of the system (6.50) satisfies

$$\operatorname{rank} V = \operatorname{rank} \begin{bmatrix} I & 0 \\ A_{11} & A_{12} \\ A_{11}^2 + A_{12}A_{21} & A_{11}A_{12} + A_{12}A_{22} \\ \vdots & \vdots \end{bmatrix}$$

$$= \operatorname{rank} \begin{bmatrix} I & 0 \\ 0 & A_{12} \\ 0 & A_{11}A_{12} + A_{12}A_{22} \\ \vdots & \vdots \end{bmatrix} = \operatorname{rank} \begin{bmatrix} I & 0 \\ 0 & A_{12} \\ 0 & A_{12}A_{22} \\ \vdots & \vdots \end{bmatrix} = n$$

which indicates that

$$\operatorname{rank} \begin{bmatrix} A_{12} \\ A_{12}A_{22} \\ \vdots \\ A_{12}A_{22}^{n-q-1} \end{bmatrix} = n - q$$

Therefore, (A_{22}, A_{12}) is completely observable.

The proof of the necessity is omitted. □

Next, if the original system is observable, then the system (6.52) is observable. One can design a Luenberger observer to estimate $\bar{x}_2(t)$,

$$\dot{\hat{\bar{x}}}_2(t) = (A_{22} - G_2 A_{12})\hat{\bar{x}}_2(t) + G_2 w(t) + \bar{u}(t) \tag{6.53}$$

where G_2 is a constant matrix to be determined. The eigenvalues of $A_{22} - G_2 A_{12}$ can be assigned arbitrarily by selecting G_2.

Substituting $\bar{u}(t)$ and $w(t)$ into (6.53), one can deduce that

$$\dot{\hat{\bar{x}}}_2(t) = (A_{22} - G_2 A_{12})\hat{\bar{x}}_2(t) + G_2\left(\dot{y}(t) - A_{11}y(t) - B_1 u(t)\right) + A_{21}y(t) + B_2 u(t) \tag{6.54}$$

which contains the derivative of $y(t)$. It is undesirable for the disturbance rejection.

Introduce a transformation

$$z(t) \triangleq \hat{\bar{x}}_2(t) - G_2 y(t)$$

and substitute it into (6.54). It can be derived that

$$\dot{z}(t) = (A_{22} - G_2 A_{12})z(t) + \bar{G}_2 y(t) + (B_2 - G_2 B_1)u(t) \tag{6.55}$$

where $\bar{G}_2 = (A_{22} - G_2 A_{12})G_2 + (A_{21} - G_2 A_{11})$.

Furthermore, the estimated state of $\bar{x}_2(t)$ can be expressed as

$$\hat{\bar{x}}_2(t) = z(t) + G_2 y(t) \tag{6.56}$$

Let $\hat{\bar{x}}_1(t)$ denote the observation of $\bar{x}_1(t)$. It holds that $\hat{\bar{x}}_1(t) = y(t)$. Meanwhile, let $\hat{x}(t)$ denote the observation of $x(t)$, it holds that

$$\begin{aligned}\hat{x}(t) &= T^{-1}\begin{bmatrix}\hat{\bar{x}}_1(t)\\ \hat{\bar{x}}_2(t)\end{bmatrix} = \begin{bmatrix}C_1^{-1} & -C_1^{-1}C_2\\ 0 & I_q\end{bmatrix}\begin{bmatrix}y(t)\\ z(t)+G_2 y(t)\end{bmatrix}\\ &= \begin{bmatrix}-C_1^{-1}C_2\\ I\end{bmatrix}z(t) + \begin{bmatrix}C_1^{-1}(I-C_2 G_2)\\ G_2\end{bmatrix}y(t)\end{aligned} \tag{6.57}$$

Theorem 6.16 *For a completely observable linear time-invariant system (6.43) with rank $C = q$, an $(n-q)$-dimensional state observer (6.55) and (6.57) of the system (6.43) can be constructed, and all eigenvalues of the observer can be assigned arbitrarily.*

The proof is omitted.

Based on the above discussion, one can conclude the way to obtain a reduced-order state observer.

Algorithm 6.11 *For a completely observable linear time-invariant system (6.43) with rank $C = q$, the $(n-q)$-dimensional state observer (6.55) and (6.57) can be determined by following procedures.*

(1) *Transform $C = [\ C_1\ \ C_2\ ]$, and select $T = \begin{bmatrix}C_1 & C_2\\ 0 & I\end{bmatrix}$. Transform the original system (6.43) into (6.50) by $\bar{x}(t) = Tx(t)$.*

(2) *Select $G_2 \in \mathbb{R}^{(n-q)\times q}$ to assign eigenvalues of $A_{22} - G_2 A_{12}$ to desired locations.*

(3) *Calculate parameters in (6.55) and (6.57). The reduced-order state observer is determined.*

Example 6.18 Consider a linear time-invariant system

$$\dot{x}(t) = \begin{bmatrix}-1 & 0 & 0\\ 0 & 1 & 1\\ 0 & 0 & 1\end{bmatrix}x(t) + \begin{bmatrix}1 & 0\\ 0 & 1\\ 0 & 1\end{bmatrix}u(t),\ y(t) = \begin{bmatrix}1 & 0 & 0\\ 0 & 1 & 1\end{bmatrix}x(t)$$

It can be verified this system is observable and rank $C = 2$. One can design a reduced-order state observer, and its dimension is $n - q = 1$.

Select

$$T = \begin{bmatrix}C_1 & C_2\\ 0 & I\end{bmatrix} = \begin{bmatrix}1 & 0 & 0\\ 0 & 1 & 1\\ 0 & 0 & 1\end{bmatrix},\ T^{-1} = \begin{bmatrix}1 & 0 & 0\\ 0 & 1 & -1\\ 0 & 0 & 1\end{bmatrix}$$

6.4 State Observer

The original system is transformed into

$$\begin{bmatrix} \dot{\bar{x}}_1(t) \\ \dot{\bar{x}}_2(t) \end{bmatrix} = \begin{bmatrix} -1 & 0 & 0 \\ 0 & 1 & 1 \\ 0 & 0 & 1 \end{bmatrix} \begin{bmatrix} \bar{x}_1(t) \\ \bar{x}_2(t) \end{bmatrix} + \begin{bmatrix} 1 & 0 \\ 0 & 2 \\ 0 & 1 \end{bmatrix} u(t)$$

$$y(t) = \begin{bmatrix} 1 & 0 & 0 \\ 0 & 1 & 0 \end{bmatrix} \begin{bmatrix} \bar{x}_1(t) \\ \bar{x}_2(t) \end{bmatrix}$$

Select $G_2 = [\ g_1 \ \ g_2\ ]$, the 1-dimensional state observer can be written as

$$\dot{z}(t) = (1 - g_2)z(t) + [-g_1 \quad 1 - 2g_2]\, u(t) + [2g_1 - g_1g_2 \quad -g_2^2]\, y(t)$$

and the reconstructed state is

$$\hat{x}(t) = \begin{bmatrix} 0 \\ -1 \\ 1 \end{bmatrix} z(t) + \begin{bmatrix} 1 & 0 \\ -g_1 & 1 - g_2 \\ g_1 & g_2 \end{bmatrix} y(t)$$

Here, select $G_2 = [\ 0\ \ 5\ ]^{\mathrm{T}}$ to assign both eigenvalues to -4. Hence, the reduced-order state observer is

$$\dot{z}(t) = -4z(t) + [0 \ -9]u(t) + [0 \ -25]y(t)$$

$$\hat{x}(t) = \begin{bmatrix} 0 \\ -1 \\ 1 \end{bmatrix} z(t) + \begin{bmatrix} 1 & 0 \\ 0 & -4 \\ 0 & 5 \end{bmatrix} y(t)$$

Next, introduce another method to design the reduced-order state observer for the system (6.43). It is required that the system (6.43) is completely controllable and observable and rank $C = q$. A $(n - q)$-dimensional state observer is constructed with the following form.

Construct a linear time-invariant system

$$\dot{z}(t) = Fz(t) + Gy(t) + Hu(t), \ z(0) = z_0 \tag{6.58}$$

where $z(t) \in \mathbb{R}^{n-q}$, F, G and H are constant matrices with appreciate dimensions.

Theorem 6.17 *For a completely controllable and observable linear time-invariant system (6.43) with* rank $C = q$, *the system (6.58) is a $(n - q)$-dimensional reduced-order state observer of the system (6.43) if and only if there exists a row full rank matrix $T \in \mathbb{R}^{(n-q) \times n}$ such that*

$$P \triangleq \begin{bmatrix} C \\ T \end{bmatrix}$$

is nonsingular, all eigenvalues of F have negative real parts, $TA - FT = GC$ and $H = TB$. Moreover, the estimated state $\hat{x}(t)$ is

$$\hat{x}(t) = P^{-1} \begin{bmatrix} y(t) \\ z(t) \end{bmatrix} = \begin{bmatrix} C \\ T \end{bmatrix}^{-1} \begin{bmatrix} y(t) \\ z(t) \end{bmatrix} = \begin{bmatrix} Q_1 & Q_2 \end{bmatrix} \begin{bmatrix} y(t) \\ z(t) \end{bmatrix} \qquad (6.59)$$

where $Q = \begin{bmatrix} Q_1 & Q_2 \end{bmatrix} \triangleq P^{-1}$.

Following the similar proof of Theorem 6.15, this theorem can be derived. Here do not elaborate it.

According to Theorem 6.17, the following algorithm can be concluded.

Algorithm 6.12 For a completely controllable and observable linear time-invariant system (6.43) with rank $C = q$, the $(n-q)$-dimensional state observer (6.58) and (6.59) can be determined by following procedures.

(1) Select a matrix $F \in \mathbb{R}^{(n-q) \times (n-q)}$ such that its eigenvalues have negative real parts, and are distinct from the ones of A.

(2) Select a matrix $G \in \mathbb{R}^{(n-q) \times q}$ such that (F, G) is completely controllable.

(3) Solve $TA - FT = GC$ to obtain the unique solution T.

(4) Construct the matrix P and determine whether P is nonsingular. If P is nonsingular, then calculate $H = TB$, and the reduced-order state observer (6.58) and (6.59) can be further determined. Otherwise, reselect F or G.

It should be point out that owing to the lower dimension of a reduced-order state observer, it has a simpler structure, lower price and error rate, and relatively more reliable. All these make the reduced-order state observer very effective in practice. It is different from a full-order state observer since the output of a reduced-order state observer reflects to $\hat{x}(t)$ through the constant matrix with $y(t)$ directly, while a full-order state observer gets it through an integrator. As a result, if there exists disturbance or noise in $y(t)$, the reduced-order state observer will be detrimental. Hence, it is better to analyze its characters before adopting a reduced-order observer based on the balance of its advantages and disadvantages.

6.4.3 Functional State Observer

In some applications, only a linear combination of states is concerned. The main objective of constructing state observers is to realize the state feedback. One can reconstruct state function Kx directly, which is helpful to further reduce the order of observers. In general, this kind of observer is called a *functional state observer*.

Consider a linear time-invariant system(6.43), which is completely controllable and observable. Construct a completely observable system

$$\dot{z}(t) = Fz(t) + Gy(t) + Hu(t), \quad w(t) = Ez(t) + My(t), \quad z(0) = z_0 \qquad (6.60)$$

where $z(t) \in \mathbb{R}^r$ is the state vector, $w(t) \in \mathbb{R}^p$ is the output vector of this system, F, G, H, E and M are constant matrices with appreciate dimensions. The objective is to determine

6.4 State Observer

proper matrix parameters in (6.60) such that

$$\lim_{t\to\infty} w(t) = \lim_{t\to\infty} Kx(t) \tag{6.61}$$

where K is the feedback gain. In this case, (6.61) is a functional state observer of the system (6.43).

Theorem 6.18 *For a completely controllable and observable linear time-invariant system (6.43) with* $\operatorname{rank} C = q$, *the system (6.60) is a functional state observer of the system (6.43) if and only if all eigenvalues of F have negative real parts, and there exists a full rank matrix $T \in \mathbb{R}^{r\times n}$ such that $TA - FT = GC$, $H = TB$ and $K = ET + MC$.*

Proof Firstly, prove the sufficiency.

It holds that

$$\begin{aligned}\dot{z}(t) - T\dot{x}(t) &= (Fz(t) + Gy(t) + Hu(t)) - T(Ax(t) + Bu(t))\\ &= (GC - TA)x(t) + (H - TB)u(t) + Fz(t)\\ &= Fz(t) - FTx(t) = F(z(t) - Tx(t))\end{aligned}$$

which implies that for any $u(t)$, x_0 and z_0, there exists

$$z(t) - Tx(t) = \mathrm{e}^{Ft}(z_0 - Tx_0)$$

Since all eigenvalues of F have negative real parts, $\mathrm{e}^{Ft} \to 0$ as $t \to \infty$. Therefore, it holds that

$$\lim_{t\to\infty} z(t) = \lim_{t\to\infty} Tx(t)$$

On this basis, it can be further derived that

$$\lim_{t\to\infty} w(t) = \lim_{t\to\infty}(Ez(t) + My(t)) = \lim_{t\to\infty}(ET + MC)x(t) = \lim_{t\to\infty} Kx(t)$$

which indicates that (6.60) is a functional state observer of the system (6.43).

Next, prove the necessity.

Assume that for any $u(t)$, x_0 and z_0, (6.61) holds. On this basis, one can obtain that

$$\lim_{t\to\infty}(Kx(t) - w(t)) = 0$$

$$\lim_{t\to\infty}(K\dot{x}(t) - \dot{w}(t)) = 0$$

$$\vdots$$

$$\lim_{t\to\infty}\left(Kx^{(r-1)}(t) - w^{(r-1)}(t)\right) = 0$$

Along with (6.60), it holds that

$$Kx(t) - w(t) = (K - MC)x(t) - Ez(t) \triangleq M_0 x(t) - Ez(t)$$

$$K\dot{x}(t) - \dot{w}(t) = [(K - MC)A - EGC]x(t) - EFz(t) + [(K - MC)B - EH]u(t)$$
$$\triangleq M_1 x(t) - EFz(t) + N_1 u(t)$$
$$K\ddot{x}(t) - \ddot{w}(t) \triangleq M_2 x(t) - EF^2 z(t) + N_1 \dot{u}(t) + N_2 u(t)$$
$$\vdots$$
$$Kx^{(r-1)}(t) - w^{(r-1)}(t) \triangleq M_{r-1} x(t) - EF^{r-1} z(t) + \sum_{j=1}^{r-1} N_1 u^{(r-1-j)}(t)$$

On this basis, one can deduce that
$$\lim_{t \to \infty} (M_0 x(t) - Ez(t)) = 0$$
$$\lim_{t \to \infty} (M_1 x(t) - EFz(t) + N_1 u(t)) = 0$$
$$\lim_{t \to \infty} (M_2 x(t) - EF^2 z(t) + N_1 \dot{u}(t) + N_2 u(t)) = 0$$
$$\vdots$$
$$\lim_{t \to \infty} \left(M_{r-1} x(t) - EF^{r-1} z(t) + \sum_{j=1}^{r-1} N_j u^{(r-1-j)}(t) \right) = 0$$

Owing to the randomicity of the control input, one can just set $u(t) = \dot{u}(t) = \cdots = u^{(r-1)}(t) = 0$. In this way, it holds that
$$\lim_{t \to \infty} (M_0 x(t) - Ez(t)) = 0$$
$$\lim_{t \to \infty} (M_1 x(t) - EFz(t)) = 0$$
$$\lim_{t \to \infty} (M_2 x(t) - EF^2 z(t)) = 0$$
$$\vdots$$
$$\lim_{t \to \infty} (M_{r-1} x(t) - EF^{r-1} z(t)) = 0$$

Define
$$R \triangleq \begin{bmatrix} M_0 \\ M_1 \\ \vdots \\ M_{r-1} \end{bmatrix}, \quad Q \triangleq \begin{bmatrix} E \\ EF \\ \vdots \\ EF^{r-1} \end{bmatrix}$$

and then one can obtain that
$$\lim_{t \to \infty} (Rx(t) - Qz(t)) = 0$$

Since (F, E) is completely observable and Q is column full rank, the generalized inverse Q^+ exists. Let $T = Q^+ R$, and then it holds that

$$\lim_{t \to \infty} (Q^+ R x(t) - z(t)) = \lim_{t \to \infty} (T x(t) - z(t)) = 0$$

In order to guarantee that the above equation holds, according to the derivation process of the sufficiency, it holds that $TA - FT = GC$, $H = TB$ and all eigenvalues of F have negative real parts. Moreover, for any $u(t)$, x_0 and z_0, one can deduce that

$$\lim_{t \to \infty} (Kx(t) - w(t)) = \lim_{t \to \infty} [Kx(t) - (Ez(t) + My(t))] = \lim_{t \to \infty} [K - (ET + MC)] x(t)$$
$$= [K - (ET + MC)] \lim_{t \to \infty} x(t)$$

To guarantee (6.61), it must hold that $K - (ET + MC) = 0$. $\square$

How to determine the minimum order of a functional state observer is a very difficult problem. If $K \in \mathbb{R}^{1 \times n}$, the observer has dimension $r = \nu - 1$ where ν is the observability index of (A, C). Generally, when the value of q is relatively large, $\nu - 1$ is much smaller than $n - q$, which implies the order of a functional state observer is less than that of a reduced-order observer.

6.4.4 Dynamical Systems with State Observer Feedback

The existence of a state observer makes it convenient in engineering applications. However, the feedback with a state observer is different from the feedback directly from states of practical systems. It should be concerned with the influence after introducing the state feedback with a state observer into a dynamical system.

Consider a completely controllable and observable linear time-invariant system (6.43). Without loss of generality, construct the reduced-order state observer (6.58) and (6.59) for this system. Furthermore, introduce a state feedback control law

$$u(t) = K\hat{x}(t) + v(t) \tag{6.62}$$

From (6.43), (6.58), (6.59) and (6.62), one can deduce the augmented system

$$\begin{cases} \begin{bmatrix} \dot{x}(t) \\ \dot{z}(t) \end{bmatrix} = \begin{bmatrix} A + BKQ_1C & BKQ_2 \\ GC + HKQ_1C & F + HKQ_2 \end{bmatrix} \begin{bmatrix} x(t) \\ z(t) \end{bmatrix} + \begin{bmatrix} B \\ H \end{bmatrix} v(t) \\ y(t) = \begin{bmatrix} C & 0 \end{bmatrix} \begin{bmatrix} x(t) \\ z(t) \end{bmatrix} \end{cases} \tag{6.63}$$

The configuration of this system is shown in Fig.6.8.

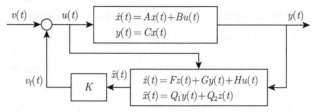

Fig.6.8 A state feedback system with a state-observer

Introduce a transformation matrix

$$\bar{T} = \begin{bmatrix} I & 0 \\ -T & I \end{bmatrix}$$

and it can be calculated that

$$\bar{T}^{-1} = \begin{bmatrix} I & 0 \\ T & I \end{bmatrix}$$

where T is given in (6.59).

Perform the equivalence transformation with

$$\begin{bmatrix} \bar{x}(t) \\ \bar{z}(t) \end{bmatrix} = \bar{T} \begin{bmatrix} x(t) \\ z(t) \end{bmatrix}$$

the system (6.63) is transformed into

$$\begin{bmatrix} \dot{\bar{x}}(t) \\ \dot{\bar{z}}(t) \end{bmatrix} = \begin{bmatrix} A+BK & BKQ_2 \\ 0 & F \end{bmatrix} \begin{bmatrix} \bar{x}(t) \\ \bar{z}(t) \end{bmatrix} + \begin{bmatrix} B \\ 0 \end{bmatrix} v(t), \quad y(t) = \begin{bmatrix} C & 0 \end{bmatrix} \begin{bmatrix} \bar{x}(t) \\ \bar{z}(t) \end{bmatrix} \quad (6.64)$$

It can be seen that the characteristic polynomial of the closed-loop system (6.64) is the product of the ones of $A+BK$ and F. Therefore, the introduction of an observer does not affect the eigenvalues of $A+BK$ designed by the state feedback. Meanwhile, the state feedback has no influence on the eigenvalues of the observer. One can conclude that the design of the control law (6.62) can be carried out independently of the design of the observer (6.58) and (6.59). This is referred to as the *separation property*. When designing other observers presented in this section, this conclusion is still valid.

The introduction of an observer will never change the closed-loop transfer function matrix of a system. In fact, when adopting the direct state feedback $u(t) = Kx(t) + v(t)$, the transfer function matrix of the closed-loop system is

$$G_K(s) = C(sI - A - BK)^{-1}B$$

It is known an equivalent transformation does not change the transfer function matrix. Therefore, the transfer function of the system (6.63) is

$$\hat{G}_K(s) = \begin{bmatrix} C & 0 \end{bmatrix} \begin{bmatrix} sI - A - BK & -BKQ_2 \\ 0 & sI - F \end{bmatrix}^{-1} \begin{bmatrix} B \\ 0 \end{bmatrix} = C(sI - A - BK)^{-1}B = G_K(s)$$

6.4 State Observer

Generally, the robustness of a state feedback system with an observer is worse than that of the direct state feedback. In practice, the negative real parts of the eigenvalues of an observer are often assigned $2 \sim 3$ times larger than those of $A + BK$. It is worth mentioning that if the eigenvalues of an observer are too large, the state observer will show a differential effect, which will introduce unexpected high-frequency interference.

Example 6.19 Consider a linear time-invariant system

$$\dot{x}(t) = \begin{bmatrix} 0 & 1 & 0 & 0 \\ 0 & 0 & -2 & 0 \\ 0 & 0 & -3 & 1 \\ 0 & 0 & 4 & 0 \end{bmatrix} x(t) + \begin{bmatrix} 0 & 0 \\ 0 & 1 \\ 0 & 1 \\ -1 & 0 \end{bmatrix} u(t),\ y(t) = \begin{bmatrix} 1 & 0 & 0 & 0 \\ 0 & 1 & 0 & 0 \end{bmatrix} x(t)$$

(1) Design $K\hat{x}(t) + v(t)$ such that $\lambda_1^* = -1$, $\lambda_{2,3}^* = -1$ and $\lambda_4^* = -2$.

Apparently, the desired characteristic polynomial is

$$\alpha_c(s) = (s+1)(s+1-j)(s+1+j)(s+2) = s^4 + 5s^3 + 10s^2 + 10s + 4$$

Consider the feedback gain is

$$K = \begin{bmatrix} k_1 & k_2 & k_3 & k_4 \\ k_5 & k_6 & k_7 & k_8 \end{bmatrix}$$

and then one can obtain that

$$A + BK = \begin{bmatrix} 0 & 1 & 0 & 0 \\ k_5 & k_6 & k_7 - 2 & k_8 \\ k_5 & k_6 & k_7 - 3 & k_8 + 1 \\ -k_1 & -k_2 & -k_3 + 4 & -k_4 \end{bmatrix} = \begin{bmatrix} 0 & 1 & 0 & 0 \\ 0 & 0 & 1 & 0 \\ 0 & 0 & 0 & 1 \\ -4 & -10 & -10 & -5 \end{bmatrix}$$

It can be further computed that

$$K = \begin{bmatrix} 4 & 10 & 14 & 5 \\ 0 & 0 & 3 & 0 \end{bmatrix}$$

(2) Design a reduced-order observer.

It is noted that the system is with the form of (6.50) and rank $C = 2$. In this case, one can obtain that

$$A_{11} = \begin{bmatrix} 0 & 1 \\ 0 & 0 \end{bmatrix},\ A_{12} = \begin{bmatrix} 0 & 0 \\ -2 & 0 \end{bmatrix},\ A_{21} = \begin{bmatrix} 0 & 0 \\ 0 & 0 \end{bmatrix},\ A_{22} = \begin{bmatrix} -3 & 1 \\ 4 & 0 \end{bmatrix}$$

$$B_1 = \begin{bmatrix} 0 & 0 \\ 0 & 1 \end{bmatrix},\ B_2 = \begin{bmatrix} 0 & 1 \\ -1 & 0 \end{bmatrix},\ C_1 = \begin{bmatrix} 1 & 0 \\ 0 & 1 \end{bmatrix},\ C_2 = \begin{bmatrix} 0 & 0 \\ 0 & 0 \end{bmatrix}$$

Consider the observer gain is

$$G_2 = \begin{bmatrix} g_1 & g_2 \\ g_3 & g_4 \end{bmatrix}$$

and then it holds that

$$A_{22} - G_2 A_{12} = \begin{bmatrix} -3+2g_2 & 1 \\ 4+2g_4 & 0 \end{bmatrix}$$

Assign eigenvalues of the observer at -5 and -6, one can select

$$G_2 = \begin{bmatrix} 0 & -4 \\ 0 & -17 \end{bmatrix}$$

It can be computed that

$$A_{22} - G_2 A_{12} = \begin{bmatrix} -11 & 1 \\ -30 & 0 \end{bmatrix}$$

Furthermore, one can obtain that

$$(A_{22} - G_2 A_{12})G_2 + (A_{21} - G_2 A_{11}) = (A_{22} - G_2 A_{12})G_2 = \begin{bmatrix} 0 & 27 \\ 0 & 120 \end{bmatrix}$$

and

$$B_2 - G_2 B_1 = \begin{bmatrix} 0 & 1 \\ -1 & 0 \end{bmatrix} - \begin{bmatrix} 0 & -4 \\ 0 & -17 \end{bmatrix}\begin{bmatrix} 0 & 0 \\ 0 & 1 \end{bmatrix} = \begin{bmatrix} 0 & 5 \\ -1 & 17 \end{bmatrix}$$

Therefore, the desired reduced-order observer is

$$\dot{z}(t) = \begin{bmatrix} -11 & 1 \\ -30 & 0 \end{bmatrix} z(t) + \begin{bmatrix} 0 & 27 \\ 0 & 120 \end{bmatrix} y(t) + \begin{bmatrix} 0 & 5 \\ -1 & 17 \end{bmatrix} u(t)$$

$$\hat{x}(t) = \begin{bmatrix} y(t) \\ z(t) + G_2 y(t) \end{bmatrix} = \begin{bmatrix} I & 0 \\ G_2 & I \end{bmatrix} \begin{bmatrix} y(t) \\ z(t) \end{bmatrix} = \begin{bmatrix} 1 & 0 & 0 & 0 \\ 0 & 1 & 0 & 0 \\ 0 & -4 & 1 & 0 \\ 0 & -17 & 0 & 1 \end{bmatrix} \begin{bmatrix} y(t) \\ z(t) \end{bmatrix}$$

The expected control law with $\hat{x}(t)$ is

$$u(t) = \begin{bmatrix} 4 & 10 & 14 & 5 \\ 0 & 0 & 3 & 0 \end{bmatrix} \hat{x}(t) + v(t)$$

As a matter of fact, a state feedback system with an observer can be viewed as an output-feedback system which has series and parallel compensators. For example, consider a state feedback system with a reduced-order state observer displayed in Fig.6.8. Regard the observer as a linear time-invariant system where $y(t)$ and $u(t)$ are inputs while $v_f(t)$ is the output. The system displayed in Fig.6.9 is equivalent to the one displayed in Fig.6.9, where $G_1(s)$ denotes the transfer function matrix from $u(t)$ to $v_f(t)$, and $G_2(s)$ is the one from $y(t)$ to $v_f(t)$. Based on $v_f(t) = K\hat{x}(t)$, there exists

$$G_1(s) = KQ_2(sI - F)^{-1}H$$

and
$$G_2(s) = KQ_1 + KQ_2(sI - F)^{-1}G$$

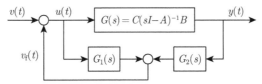

Fig.6.9 An equivalent system of the one shown in Fig.6.8

Simplify the block diagram shown in Fig.6.9, one can obtain the equivalent system shown in Fig.6.10, which can be further simplified as the system shown in Fig.6.11. It is an output feedback system with a series compensator $G_b(s)$ and an output feedback parallel compensator $G_f(s)$, where
$$G_f(s) = G_2(s) = KQ_1 + KQ_2(sI - F)^{-1}G$$
and
$$G_b(s) = (I - G_1(s))^{-1}$$

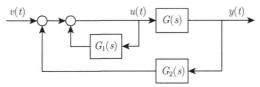

Fig.6.10 An equivalent system of the one shown in Fig.6.9

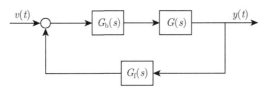

Fig.6.11 An equivalent system of the one shown in Fig.6.10

It can be further verified that the transfer function matrix of the closed-loop system with an observer does not change, i.e.,
$$\hat{G}_K(s) = C(sI - A - BK)^{-1}B$$

Assignments

6.1 Using PBH criterion to prove Theorem 6.1 and Theorem 6.5.

6.2 Consider the

$$\dot{x}(t) = \begin{bmatrix} 2 & 1 & & \\ & 2 & & \\ & & -1 & \\ & & & -1 \end{bmatrix} x(t) + \begin{bmatrix} 0 \\ 1 \\ 1 \\ 1 \end{bmatrix} u(t)$$

Discuss whether a state feedback matrix can be determined to assign eigenvalues at $(-2,-2,-2,-2)$, $(-2,-2,-1,-1)$ or $(-2,-2,-2,-1)$.

6.3 Suppose that the transfer function of a system is

$$g(s) = \frac{(s-1)(s-2)}{(s+1)(s-2)(s+3)}$$

Discuss whether a state feedback can be adopted to change the transfer function to

$$g_f(s) = \frac{(s-1)}{(s+2)(s+3)}$$

6.4 Consider the system

$$\dot{x}(t) = \begin{bmatrix} 1 & 1 & 0 \\ 0 & 1 & 0 \\ 0 & 0 & -2 \end{bmatrix} x(t) + \begin{bmatrix} 0 & 0 \\ 1 & 0 \\ 0 & -1 \end{bmatrix} u(t)$$

Determine the state feedback gain such that the poles of the closed-loop system are placed at $\lambda_1 = -1$ and $\lambda_{2,3} = -2 \pm j$.

6.5 Consider the system

$$\dot{x}(t) = \begin{bmatrix} 0 & 1 & 0 & 0 \\ 0 & 0 & 1 & 0 \\ -3 & 1 & 2 & 3 \\ 2 & 1 & 0 & 0 \end{bmatrix} x(t) + \begin{bmatrix} 0 & 0 \\ 0 & 0 \\ 1 & 1 \\ 0 & 1 \end{bmatrix} u(t)$$

Determine the state feedback gain such that the poles of the closed-loop system are placed at $\lambda_1 = -3$, $\lambda_{2,3} = -1 \pm j$ and $\lambda_4 = -2$.

6.6 Judge whether the following systems can be stabilized using state feedback, and whether the systems are detectable.

(1) $\dot{x}(t) = \begin{bmatrix} 1 & 3 \\ 2 & 1 \end{bmatrix} x(t) + \begin{bmatrix} 0 \\ 1 \end{bmatrix} u(t)$, $y(t) = \begin{bmatrix} 0 & 1 \end{bmatrix} x(t)$.

(2) $\dot{x}(t) = \begin{bmatrix} 1 & 0 & 0 \\ 0 & -2 & 1 \\ 0 & 0 & -2 \end{bmatrix} x(t) + \begin{bmatrix} 1 & 0 \\ 0 & 1 \\ 0 & 0 \end{bmatrix} u(t)$, $y(t) = \begin{bmatrix} 1 & 0 & 0 \end{bmatrix} x(t)$.

6.7 Judge whether the following system can be stabilized with output feedback,

$$\dot{x}(t) = \begin{bmatrix} 4 & 0 & 0 \\ 0 & -2 & 1 \\ 0 & 0 & -2 \end{bmatrix} x(t) + \begin{bmatrix} 1 & 0 \\ 0 & 1 \\ 0 & 0 \end{bmatrix} u(t), \ y(t) = \begin{bmatrix} 1 & 0 \\ 1 & 1 \\ 2 & 4 \end{bmatrix} x(t)$$

6.8 Judge whether the following system can be decoupled with state feedback, and whether the following system can be decoupled with output feedback.

$$\dot{x}(t) = \begin{bmatrix} 3 & 1 & 0 \\ 0 & 0 & -1 \\ 0 & 1 & -1 \end{bmatrix} x(t) + \begin{bmatrix} 0 & 0 \\ 1 & 0 \\ 0 & 1 \end{bmatrix} u(t), \ y(t) = \begin{bmatrix} 2 & -1 & 1 \\ 0 & 2 & 1 \end{bmatrix} x(t)$$

6.9 Consider the following dynamical system,

$$\dot{x}(t) = \begin{bmatrix} -1 & 0 & 0 \\ 0 & -2 & -4 \\ 1 & 0 & 0 \end{bmatrix} x(t) + \begin{bmatrix} 1 & 0 \\ 0 & 1 \\ 0 & 1 \end{bmatrix} u(t), \ y(t) = \begin{bmatrix} 1 & 0 & 0 \\ -1 & 0 & 1 \end{bmatrix} x(t)$$

(1) Find out the state feedback matrix which dynamically decouples the system.
(2) Find out the state feedback matrix which statically decouples the system.
(3) Discuss whether the poles of the system can be placed at -1, -2 and -3 while decoupling.

6.10 Consider the following dynamical system,

$$\dot{x}(t) = \begin{bmatrix} 0 & 0 & 1 \\ 0 & 0 & 0 \\ 0 & 1 & 0 \end{bmatrix} x(t) + \begin{bmatrix} 1 & 0 \\ 0 & 1 \\ 0 & 0 \end{bmatrix} u(t), \ y(t) = \begin{bmatrix} 1 & 0 & 1 \\ 0 & 1 & 0 \end{bmatrix} x(t)$$

Design the state feedback control law to place the poles at -1, -2 and -3.

6.11 Consider the following dynamical system,

$$\dot{x}(t) = \begin{bmatrix} -1 & -2 & -2 \\ 0 & -1 & 1 \\ 1 & 0 & -1 \end{bmatrix} x(t) + \begin{bmatrix} 2 \\ 0 \\ 1 \end{bmatrix} u(t), \ y(t) = \begin{bmatrix} 1 & 1 & 0 \end{bmatrix} x(t)$$

Design the full-order state observer and the reduced-order state observer, and place all poles of the observer at -3.

6.12 Consider the system describing the aircraft dynamics. Thereinto, its longitudinal motion can be expressed by

$$\dot{x}(t) = \begin{bmatrix} -0.0188 & 11.5959 & 0 & -32.2 \\ -0.0007 & -0.5357 & 1 & 0 \\ 0.000048 & -0.4944 & -0.4935 & 0 \\ 0 & 0 & 1 & 0 \end{bmatrix} x(t) + \begin{bmatrix} 0 \\ 0 \\ -0.5632 \\ 0 \end{bmatrix} u(t)$$

(1) For the state space representation of the longitudinal motion of the fighter AFTI-16, design a linear state feedback control law so that the eigenvalues of closed-loop system are at $-1.25 \pm j2.2651$ and $-0.01 \pm j0.095$.

(2) Let $y(t) = Cx(t)$ with $C = [\ 0\ \ 0\ \ 1\ \ 0\ ]$. Design a full-order state observer with eigenvalues at 0, -0.421, -0.587 and -1.

(3) Let the system be compensated via the state feedback control law $u(t) = K\hat{x}(t) + v(t)$, where $\hat{x}(t)$ is the output of the state estimator. Derive the state-space representation and the transfer function between $y(t)$ and $r(t)$ of the compensated system. Is the system fully controllable from $r(t)$? Explain it.